OPTIMIZATION IN MEDICINE AND BIOLOGY

ENGINEERING AND MANAGEMENT INNOVATION SERIES

Hamid R. Parsaei and Ali K. Kamrani, Series Advisors
University of Houston, Houston, TX

Facility Logistics: Approaches and Solutions to Next Generation Challenges
Maher Lahmar
ISBN: 0-8493-8518-0

Simulation of Industrial Systems: Discrete Event Simulation Using Excel/VBA
David Elizandro and Hamdy Taha
ISBN: 1-4200-6744-3

Additional Titles in RESOURCE MANAGEMENT SERIES

Rightsizing Inventory
by Joseph L. Aiello
ISBN: 0-8493-8515-6

Integral Logistics Management: Operations and Supply Chain Management in Comprehensive Value-Added Networks, Third Edition
by Paul Schönsleben
ISBN: 1-4200-5194-6

Supply Chain Cost Control Using Activity-Based Management
Sameer Kumar and Matthew Zander
ISBN: 0-8493-8215-7

Financial Models and Tools for Managing Lean Manufacturing
Sameer Kumar and David Meade
ISBN: 0-8493-9185-7

RFID in the Supply Chain
Judith M. Myerson
ISBN: 0-8493-3018-1

Handbook of Supply Chain Management, Second Edition
by James B. Ayers
ISBN: 0-8493-3160-9

The Portal to Lean Production: Principles & Practices for Doing More With Less
by John Nicholas and Avi Soni
ISBN: 0-8493-5031-X

Supply Market Intelligence: A Managerial Handbook for Building Sourcing Strategies
by Robert B. Handfield
ISBN: 0-8493-2789-X

The Small Manufacturer's Toolkit: A Guide to Selecting the Techniques and Systems to Help You Win
by Steve Novak
ISBN: 0-8493-2883-7

Velocity Management in Logistics and Distribution: Lessons from the Military to Secure the Speed of Business
by Joseph L. Walden
ISBN: 0-8493-2859-4

Supply Chain for Liquids: Out of the Box Approaches to Liquid Logistics
by Wally Klatch
ISBN: 0-8493-2853-5

Supply Chain Architecture: A Blueprint for Networking the Flow of Material, Information, and Cash
by William T. Walker
ISBN: 1-57444-357-7

ERP: Tools, Techniques, and Applications for Integrating the Supply Chain
by Carol A. Ptak with Eli Schragenheim
ISBN: 1-57444-358-5

Introduction to e-Supply Chain Management: Engaging Technology to Build Market-Winning Business Partnerships
by David C. Ross
ISBN: 1-57444-324-0

Supply Chain Networks and Business Process Orientation
by Kevin P. McCormack and William C. Johnson with William T. Walker
ISBN: 1-57444-327-5

Collaborative Manufacturing: Using Real-Time Information to Support the Supply Chain
by Michael McClellan
ISBN: 1-57444-341-0

The Supply Chain Manager's Problem-Solver: Maximizing the Value of Collaboration and Technology
by Charles C. Poirier
ISBN: 1-57444-335-6

Lean Performance ERP Project Management: Implementing the Virtual Lean Enterprise, Second Edition
by Brian J. Carroll
ISBN: 0-8493-0532-2

Integrated Learning for ERP Success: A Learning Requirements Planning Approach
by Karl M. Kapp, with William F. Latham and Hester N. Ford-Latham
ISBN: 1-57444-296-1

Basics of Supply Chain Management
by Lawrence D. Fredendall and Ed Hill
ISBN: 1-57444-120-5

Lean Manufacturing: Tools, Techniques, and How to Use Them
by William M. Feld
ISBN: 1-57444-297-X

Back to Basics: Your Guide to Manufacturing Excellence
by Steven A. Melnyk and R.T. Chris Christensen
ISBN: 1-57444-279-1

Enterprise Resource Planning and Beyond: Integrating Your Entire Organization
by Gary A. Langenwalter
ISBN: 1-57444-260-0
ISBN: 0-8493-8515-6

OPTIMIZATION IN MEDICINE AND BIOLOGY

Edited by Gino J. Lim and Eva K. Lee

Auerbach Publications
Taylor & Francis Group
New York London

CRC Press is an imprint of the
Taylor & Francis Group, an **informa** business

Auerbach Publications
Taylor & Francis Group
6000 Broken Sound Parkway NW, Suite 300
Boca Raton, FL 33487-2742

Printed in the United States of America on acid-free paper
10 9 8 7 6 5 4 3 2 1

International Standard Book Number-13: 978-0-8493-0563-4 (Hardcover)

Library of Congress Cataloging-in-Publication Data

Optimization in medicine and biology / edited by Gino J. Lim and Eva K. Lee.
p. ; cm. -- (Engineering management innovation series ; 3)
Includes bibliographical references and index.
ISBN-13: 978-0-8493-0563-4 (hardcover : alk. paper)
ISBN-10: 0-8493-0563-2 (hardcover : alk. paper)
1. Medicine--Mathematical models. 2. Biology--Mathematical models. 3. Mathematical optimization. I. Lim, Gino J. II. Lee, Eva K. III. Title. IV. Series.
[DNLM: 1. Mathematics. 2. Medicine. 3. Computational Biology--methods. 4. Decision Making, Computer-Assisted. 5. Patient Care Planning. QT 35 O62 2008]

R853.M3O68 2008
610--dc22 2007026356

Visit the Taylor & Francis Web site at
http://www.taylorandfrancis.com

and the Auerbach Web site at
http://www.auerbach-publications.com

Contents

Preface ix

Acknowledgments xiii

Editors xv

Contributors xvii

About the Contributors xxi

Part I: Medicine

1 Classification and Disease Prediction via Mathematical Programming 3

EVA K. LEE AND TSUNG-LIN WU

2 Using Influence Diagrams in Cost-Effectiveness Analysis for Medical Decisions 61

RAM S. DURISETI

3 Non-Bayesian Classification to Obtain High Quality Clinical Decisions 95

RAM S. DURISETI

4 Optimizing Pediatric Vaccine Formularies 117

SHANE N. HALL, SHELDON H. JACOBSON, AND EDWARD C. SEWELL

5 Optimal Spending on HIV Prevention and Treatment: A Framework for Evaluating Cost-Effectiveness with Example Application to the India AIDS Initiative 147

MARGARET L. BRANDEAU, ELISA F. LONG, DAVID W. HUTTON, AND DOUGLAS K. OWENS

6 Optimization over Graphs for Kidney Paired Donation 177

SOMMER E. GENTRY

7 **Introduction to Radiation Therapy Planning Optimization** 197

GINO J. LIM

8 **Beam Orientation Optimization Methods in Intensity Modulated Radiation Therapy Treatment Planning** 223

DIONNE M. ALEMAN, H. EDWIN ROMEIJN, AND JAMES F. DEMPSEY

9 **Multileaf Collimator Shape Matrix Decomposition** 253

THOMAS KALINOWSKI

10 **Optimal Planning for Radiation Therapy** 287

MARK LANGER, RONALD RARDIN, AND ALI TUNCEL

Part II: Biology

11 **Introduction to Systems Biology for Mathematical Programmers** 311

EIVIND ALMAAS, ALLEN HOLDER, AND KEVIN LIVINGSTONE

12 **Algorithms for Genomics Analysis** 355

EVA K. LEE AND KAPIL GUPTA

13 **Computational Methods for Probe Design and Selection** 395

CLÁUDIO N. MENESES, PANOS M. PARDALOS, AND MICHELLE A. RAGLE

14 **Implementation of Logical Analysis of Data for Oligo Probe Selection** 415

IN-YONG JANG, KWANGSOO KIM, AND HONG SEO RYOO

15 **New Dihedral Angle Measure for Protein Secondary Prediction** 439

MOON K. KIM, YUNHO JANG, AND J. MACGREGOR SMITH

16 **Optimization of Tumor Virotherapy with Recombinant Measles Viruses** 467

ŽELJKO BAJZER, THOMAS CARR, DAVID DINGLI, AND KREŠIMIR JOSIĆ

17 **Combating Microbial Resistance to Antimicrobial Agents through Dosing Regimen Optimization** 493

MICHAEL NIKOLAOU AND VINCENT H. TAM

Appendix: Tutorial Guide to Mixed-Integer Programming Models and Solution Techniques **521**
J. COLE SMITH AND Z. CANER TAŞKIN

Index **549**

Preface

Optimization has long been a cornerstone for the advancement of various industrial, government, military, and healthcare applications, and it plays an increasingly important role in modern medicine and biological investigations. Many medical and biological problems can be formulated into mathematical models and can be analyzed using sophisticated optimization and computational techniques. Optimization is a rewarding field that offers challenging problems for applied mathematicians, basic scientists, computer scientists, engineers, medical professionals, and physicists.

This book covers a collection of emerging and exciting optimization applications and advances to the fields of medicine and biology, ranging from disease prediction and control, cancer therapeutics, to DNA sequencing, protein structure analysis, and drug design and development.

The book is divided into two parts and has an appendix. Part I covers various topics in medicine, and Part II focuses on biology. Each part is organized in such a way that a basic overview to the field is given followed by specific detailed applications. All chapters are written by select leading experts in their research areas, and cover basic models, theories, and computational approaches to the most advanced applications. The practice of optimization largely depends on good modeling techniques as well as efficient and robust algorithms. Therefore, some of the chapters are dedicated to modeling techniques, whereas other chapters address computational algorithms for solving these very complex models, or a combination of both. Because of the wide range of levels of technical diversity, a typical chapter begins with a simple discussion of the topic at hand to familiarize the reader. Further, appropriate figures are included to help the readers better understand the subject.

The first three chapters in Part I focus on optimization techniques for disease prediction and medical decision-making processes. Here, mathematical programming classification models and support vector machine techniques are introduced, and influence diagrams are described. Specifically, Chapter 1 provides an overview of advances in mathematical programming for classification. It summarizes linear, nonlinear, integer programming, and support vector machine models; and model characteristics and computational challenges. The authors then provide first-hand successful applications of their classification models to medicine, including prediction of diseases including skin diseases, cancer, and cardiovascular diseases; automated drug delivery; and image recognition. Chapter 2 describes medical decision making, quality metrics, and influence diagrams in clinical cost-effectiveness analysis. The author describes clinical models for venous thromboembolic disease and analyzes strategies for cost-effectiveness in diagnosing pulmonary embolism. Chapter 3 discusses a novel application of support vector machines in patients presenting to an emergency department with chest pain.

The next three chapters discuss applications of optimization to medical delivery. Chapter 4 addresses techniques for optimizing pediatric vaccine formularies. The authors present two discrete optimization models that illuminate alternatives and choices by selecting a vaccine formulary that minimizes the cost of fully immunizing a child and that limits the amount of extra immunization (i.e., extra doses of vaccine) for any given childhood immunization schedule. The authors analyze the computational complexity of the models, and present several optimization algorithms—both exact and heuristic—for solving these models. In Chapter 5, a model-based framework for evaluating the cost-effectiveness of HIV prevention and treatment programs is discussed. A framework of this type can help identify the most cost-effective programs in a portfolio of HIV interventions, thus guiding the allocation (and possible reallocation) of scarce funds. Chapter 6 describes a graph-theoretical approach for optimizing kidney paired donation. The author introduces both the medical background and mathematical tools used to find a societal optimal allocation of organs for this type of transplant.

The last four chapters in Part I, Chapters 7, 8, 9, and 10, are dedicated to recent advances in radiation therapy cancer treatment planning models and solution algorithms. Chapter 7 introduces the background in radiation therapy, including three-dimensional conventional conformal radiation therapy (3DCRT), intensity-modulated radiation therapy (IMRT), tomotherapy, and proton therapy. Chapter 8 focuses on techniques for beam angle selection in IMRT, whereas Chapter 9 provides matrix decomposition techniques for leaf-sequencing in IMRT treatment delivery. Chapter 10 offers an overview of various issues within treatment planning models, including dose calculation, clinical objectives, and considerations related to constraints; the choice of optimization approaches; the determination of leaf segments for delivery; and 4D and fractionated planning.

Part II focuses on optimization in biology. Chapter 11 begins with an introduction to systems biology for mathematical programmers. The chapter covers various areas including gene regulatory networks, protein interaction networks, and metabolic networks. Chapter 12 touches upon popular algorithms for genomic analysis, in particular for phylogenetic analysis and multiple sequence alignment. The authors then introduce a graph-theoretical approach that generalizes the complexity of various classes of sequencing problems in computer science. The model allows the study of a wide group of genomic analysis problems, including phylogenetic analysis, multiple sequence alignment, DNA sequencing, and sequence comparison problems.

The next two chapters are related to probes in biological experiments. Specifically, Chapter 13 examines computational issues related to probe design and selection. Properties of probes such as the melting temperature, specificity, existence of secondary structures, and probe length are discussed in detail. The authors highlight various algorithms and software packages that have been developed to aid in the design and selection of probes in practice. Chapter 14 focuses on a logical analysis framework for selecting short oligo probes for genotyping applications.

The last three chapters deal with various classes of biological models. Chapter 15 introduces a new dihedral angle measure for protein secondary prediction. Specifically, the Steiner tree is used for the structure analysis of amino acids. Chapter 16 provides an optimization approach for tumor virotherapy with recombinant measles viruses. The last chapter describes the design of effective dosing regimens that suppress the emergence and proliferation of resistant microbial populations. The authors provide a comprehensive presentation of their recent theoretical and experimental work on a mathematical modeling framework that can be used to optimize the design of such dosing regimens.

As described in various chapters throughout the book, integer programming (IP) is often used for modeling within various medical and biological applications. A short tutorial chapter on integer programming is included in the appendix.

Although optimization in medicine and biology is too broad to be addressed entirely within one edited book, the chapters in this book highlight some of the most recent advances in optimization techniques that are relevant to solving complex problems arising from medical and biological research. We hope that this book will facilitate strong collaborative relationships among optimization researchers and medical and biological professionals.

This book will benefit medical professionals and researchers who desire to understand and explore basic optimization concepts and their potential use in clinical and laboratory research, as well as their potential to accelerate medical and biological advances. It also serves as an invaluable reference for optimization researchers to prepare and start research projects in medicine and biology. Further, it can be used as a textbook in courses such as optimization in medicine and biology for senior undergraduate and graduate students in engineering, operations research, mathematics, computer science, and biology. Finally, it is an excellent reference/text book for graduate students who choose to pursue research, leading to a thesis, in this area.

Gino J. Lim
Eva K. Lee

Acknowledgments

We would like to thank the authors who contributed to this book; without their commitment this book would not have been possible. We would also like to thank all medical researchers who work tirelessly around the clock to improve and save human lives. Their passion and dedication inspired us to work on this book.

We are grateful to the reviewers of various chapters for their comments in improving the content of the book. Our special thanks go to Ray O'Connell, a senior editor from Auerbach Publishing, for his patience and guidance, and offering us the opportunity to complete this book. Josh Reese and Araby Abdel-Rahman, students at the University of Houston, deserve recognition for their logistic assistance.

Editors

Gino J. Lim, PhD is an assistant professor of industrial engineering at the University of Houston. His research interests are in the fields of optimization models and computational algorithms, operations research applications in health systems, and emergency planning and logistics. He has a keen interest in developing optimization techniques for solving large-scale decision-making problems in medicine, biology, healthcare delivery, and transportation networks. He received the Pierskalla Best Paper award for his work on Gamma Knife radiotherapy optimization for patients with brain cancer. He is the founding director of the Systems Optimization and Computing Laboratory (SOCL) at the University of Houston. His current research projects include developing solution techniques for the p-median problem, radiation treatment planning optimization, Markov decision process approaches for transportation problems, emergency evacuation planning and management, and process control. His research projects have been funded by the Department of Homeland Security, Texas Department of Transportation, the University of Houston, the University of Texas MD Anderson Cancer Center, and Varian Medical Systems, Inc. Dr Lim received both his MS and PhD in industrial engineering from the University of Wisconsin–Madison.

Eva K. Lee, PhD is an associate professor in the H. Milton Stewart School of Industrial and Systems Engineering at Georgia Institute of Technology, where she is the director of the Center for Operations Research in Medicine and HealthCare. Dr Lee works in the area of large-scale optimization and computing, focusing on applications to medicine, biology, healthcare, biodefense, and logistics. In medicine and biology, her work involves health risk prediction, disease diagnosis and early intervention, optimal treatment design and drug delivery, and outcome prediction. In healthcare and biodefense, she focuses on efficient and cost-effective healthcare delivery and operations logistics within healthcare systems and processes, and on emergency response and pandemic preparedness, planning, and dispensing for large regional populations. Dr Lee received her MS and PhD in computational and applied mathematics from Rice University.

Contributors

Dionne M. Aleman
Department of Mechanical and Industrial Engineering
University of Toronto
Toronto, Ontario, Canada

Eivind Almaas
Microbial Systems Biology, Biosciences, and Biotechnology Division
Lawrence Livermore National Laboratory
Livermore, California

Željko Bajzer
Department of Biochemistry and Molecular Biology
Mayo Clinic College of Medicine
Rochester, Minnesota

Margaret L. Brandeau
Department of Management Science and Engineering
Stanford University
Stanford, California

Thomas Carr
Department of Mathematics
Southern Methodist University
Dallas, Texas

James F. Dempsey
Department of Radiation Oncology
University of Florida
Gainesville, Florida

David Dingli
Division of Hematology
Mayo Clinic College of Medicine
Rochester, Minnesota

Ram S. Duriseti
Department of Veterans Affairs Palo Alto Health Care System
Palo Alto, California
and
Department of Management Science and Engineering and Division of Emergency Medicine
Department of Surgery
Stanford University
Stanford, California

Sommer E. Gentry
Department of Mathematics
United States Naval Academy
Annapolis, Maryland
and
Department of Surgery
Johns Hopkins University School of Medicine
Baltimore, Maryland

Kapil Gupta
Center for Operations Research in Medicine and HealthCare
School of Industrial and Systems Engineering
Georgia Institute of Technology
Atlanta, Georgia

Shane N. Hall
Department of Operational Sciences
Air Force Institute of Technology
Wright Patterson Air Force Base, Ohio

Allen Holder
Department of Mathematics
Trinity University
San Antonio, Texas

David W. Hutton
Department of Management Science and Engineering
Stanford University
Stanford, California

Sheldon H. Jacobson
Simulation and Optimization Laboratory
Department of Computer Science
University of Illinois at Urbana–Champaign
Urbana, Illinois

In-Yong Jang
Division of Information
Management Engineering
Korea University
Seoul, South Korea

Yunho Jang
Department of Mechanical and Industrial Engineering
University of Massachusetts
Amherst, Massachusetts

Krešimir Josić
Department of Mathematics
University of Houston
Houston, Texas

Thomas Kalinowski
Department of Mathematics
University of Rostock
Rostock, Germany

Kwangsoo Kim
Division of Information Management Engineering
Korea University
Seoul, South Korea

Moon K. Kim
Department of Mechanical and Industrial Engineering
University of Massachusetts
Amherst, Massachusetts

Mark Langer
Department of Radiation Oncology
Indiana University
Indianapolis, Indiana

Eva K. Lee
Center for Operations Research in Medicine and HealthCare
School of Industrial and Systems Engineering
Georgia Institute of Technology
Atlanta, Georgia

Gino J. Lim
Department of Industrial Engineering
University of Houston
Houston, Texas

Kevin Livingstone
Department of Biology
Trinity University
San Antonio, Texas

Elisa F. Long
Department of Management Science and Engineering
Stanford University
Stanford, California

Cláudio N. Meneses
Department of Transportation Engineering
University of Sao Paulo
Sao Paulo, Brazil

Michael Nikolaou
Department of Chemical and Biomolecular Engineering
University of Houston
Houston, Texas

Douglas K. Owens
VA Palo Alto Health Care System
Palo Alto, California
and
Center for Primary Care and Outcomes Research
Department of Medicine and Department of Health Research and Policy
Stanford University Medical School
Stanford, California

Panos M. Pardalos
Department of Industrial and Systems Engineering
University of Florida
Gainesville, Florida

Michelle A. Ragle
Department of Industrial and Systems Engineering
University of Florida
Gainesville, Florida

Ronald Rardin
Department of Industrial Engineering
University of Arkansas
Fayetteville, Arkansas

H. Edwin Romeijn
Department of Industrial and Systems Engineering
University of Florida
Gainesville, Florida

Hong Seo Ryoo
Division of Information Management Engineering
Korea University
Seoul, South Korea

Edward C. Sewell
Department of Mathematics and Statistics
Southern Illinois University
Edwardsville, Illinois

J. Cole Smith
Department of Industrial and Systems Engineering
University of Florida
Gainesville, Florida

J. MacGregor Smith
Department of Mechanical and Industrial Engineering
University of Massachusetts
Amherst, Massachusetts

Vincent H. Tam
Department of Clinical Sciences and Administration
University of Houston College of Pharmacy
Houston, Texas

Z. Caner Taşkın
Department of Industrial and Systems Engineering
University of Florida
Gainesville, Florida

Ali Tuncel
School of Industrial Engineering
Purdue University
West Lafayette, Indiana

Tsung-Lin Wu
Center for Operations Research in Medicine and HealthCare
School of Industrial and Systems Engineering
Georgia Institute of Technology
Atlanta, Georgia

About the Contributors

Dionne M. Aleman, PhD is an assistant professor in the Department of Mechanical and Industrial Engineering at the University of Toronto in Toronto, Ontario, Canada. Her research topic focuses on medical applications of operations research techniques, and she hopes to better educate operations researchers on important issues in medical science, where optimization can be employed. She received her MS and PhD in industrial and systems engineering from the University of Florida in Gainesville, Florida.

Eivind Almaas earned a PhD in physics in 2002 from the Ohio State University under the supervision of Professor David Stroud. The focus of his research is the application of complex network theory and statistical physics approaches to biology, with particular emphasis on microbial systems such as whole-cell level metabolism, gene regulation, and protein interactions. He holds a research scientist position and directs a team in the biosciences and biotechnology division at the Lawrence Livermore National Laboratory.

Željko Bajzer, PhD is professor of biophysics and codirector of biomathematics resource at Mayo Clinic College of Medicine. He started his academic career as a theoretical physicist, doing research in nuclear physics. Gradually, he became interested in the mathematical modeling of biological systems and worked in this field for more than 30 years. He published numerous scientific papers, many of which are related to tumor growth kinetics, which since 1983 is one of his most prominent interests. He has also done research in applications of mathematics to analysis of fluorescence decay in macromolecules, to enzyme kinetics, radioactive tracer kinetics, and to modeling of myocardial circulation. He is involved in teaching courses at Mayo Graduate School.

Margaret L. Brandeau, PhD is professor of management science and engineering and of medicine at Stanford University. Her work focuses on the application of mathematical and economic tools to problems in the area of health policy. Her recent work has focused on HIV prevention and treatment, bioterrorism preparedness, and control of infectious diseases. She was awarded the Presidential Young Investigator Award from the National Science Foundation and the Pierskalla Prize from the Institute for Operations Research and Management Science (INFORMS) for research excellence in healthcare management science.

Thomas Carr, PhD is an associate professor of mathematics at Southern Methodist University. His research interests are dynamical systems theory, bifurcation theory, and asymptotic and perturbation methods applied to laser instabilities, epidemiology, and coupled oscillators.

James F. Dempsey, PhD is an associate professor in the Department of Radiation Oncology and an affiliate member of the Department of Nuclear and Radiological engineering at the University of Florida. His current research interests are in the area of real-time image-guided radiation therapy and optimized radiation therapy. He received his PhD in nuclear chemistry from Washington University in St. Louis. He is the chief science officer and founder of ViewRay Incorporated and inventor of the company's real-time magnetic resonance imaging (MRI)-guided radiation therapy technology.

David Dingli, MD, PhD, FRCP, FACP, FIBiol is a physician-scientist at Mayo Clinic College of Medicine. He is board certified in hematology and internal medicine, and has a specific interest in hematopoietic tumors and bone marrow transplantation. During his graduate training, he studied the use of replication-competent viruses as novel cancer therapeutic agents. He has published several articles on virus engineering and the use of trackable, replication-competent viruses for cancer therapy. Tumor virotherapy and mathematical modeling continue to be his major focus in the laboratory. His interest in mathematical modeling of biological systems has taken him to the Santa Fe Institute, the University of Tennessee, and the Program for Evolutionary Dynamics at Harvard University, where he was a visiting scholar for two years.

Ram S. Duriseti, MD, PhD is a clinical assistant professor and medical informatics director at Stanford University Hospitals and Clinics. He completed his doctoral work at Stanford University in management science and engineering with a concentration in decision and risk analysis with an emphasis on computational modeling of medical decisions in 2007. He currently serves as one of the medical informatics directors for the Stanford Hospital Clinical Information System initiative. He regularly consults for companies seeking to develop intelligent healthcare-related applications. He is board certified in Emergency Medicine and practices actively in both academic and community settings.

Sommer E. Gentry, PhD is an assistant professor of mathematics at the U.S. Naval Academy and is also affiliated with the division of transplantation at the Johns Hopkins University School of Medicine. She holds degrees from Stanford University and Massachusetts Institute of Technology, and received the prestigious Department of Energy Computational Science Graduate Fellowship. Dr Gentry's work applies operations research to designing better healthcare systems, with a focus on transplantation and organ allocation.

She serves as an adviser to both the United States and Canada in their efforts to create national paired donation registries.

Kapil Gupta is a PhD student in School of Industrial and Systems Engineering at Georgia Institute of Technology. His research expertise lies in the application of mathematical programming to develop efficient and robust algorithms for large scale decision problems. He has specifically concentrated on sequence analysis problems arising in Computational Genomics and Evolutionary Biology. He holds a BS in chemical engineering from Indian Institute of Technology, Madras and an MS in operations research from Georgia Institute of Technology.

Shane N. Hall is a commissioned officer in the United States Air Force (USAF), and is an assistant professor of operations research at the Air Force Institute of Technology, Wright—Patterson Air Force Base, Ohio. He attended Brigham Young University in Provo, Utah, and received a BS in mathematics and his officer commission in April 1997. In March 2000, he received his MS in operations research from the Air Force Institute of Technology, Wright—Patterson Air Force Base, Ohio. He also has several years of experience as an analyst for the USAF. In October 2006, he received his PhD in industrial engineering from the University of Illinois at Urbana—Champaign where his dissertation work was on the design and analysis of pediatric vaccine formularies using optimization and heuristics. His research interests include discrete optimization, heuristics, and algorithm design with healthcare and military applications.

Allen Holder earned a PhD in applied mathematics in 1998 from the University of Colorado at Denver under the tutelage of Dr Harvey Greenberg. His primary interests are in optimization and its applications in medicine, biology, and economics. He holds joint positions in the departments of radiation oncology and radiological sciences at the University of Texas Health Science Center at San Antonio. He won the 2000 William Pierskalla Award for his work in optimizing radiotherapy treatments and has a strong background in undergraduate research. He further won a 2003 Moving Spirit Award from the Institute for Operations Research and Management Science.

David W. Hutton is a doctoral student in management science and engineering at Stanford University, with a concentration in policy and strategy. His research interests include mathematical modeling of complexity and uncertainty surrounding decisions regarding infectious disease prevention and control. Previously, he worked in the enterprise software industry, assisting call center managers in optimizing their workforce with improved processes and technology. He has received a graduate fellowship from Stanford University.

Sheldon H. Jacobson is a professor, Willett faculty scholar, and director of the Simulation and Optimization Laboratory at the University of Illinois at Urbana—Champaign. He has a BSc and MSc (both in mathematics) from McGill University, and an MS and PhD (both in operations research and industrial engineering) from Cornell University.

Since 1996, he has been applying operation research methodologies to address healthcare problems associated with pediatric immunization and vaccination economics, pediatric vaccine pricing, and pediatric vaccine stockpile economics. He has received numerous awards for his research, including a best paper award in IIE Transactions Focused Issue on Operations Engineering and a John Simon Guggenheim Memorial Foundation Fellowship. His healthcare research has been published in a wide spectrum of operations research and medical journals, including *Health Care Management Science*, *Journal of the Operational Research Society*, *Pediatric Infectious Disease Journal*, and *Vaccine*, among others. He has briefed the Advisory Committee on Immunization Practice (ACIP), the committee that provides guidance to the Secretary of the Department of Health and Human Services, on issues related to immunization policy in the United States.

He has also worked to transition his research into a publicly available Web site, www.vaccineselection.com, which has been widely used by both government and private sector organizations. He has received research funding from several government agencies and industrial partners, including the National Science Foundation and the Air Force Office of Scientific Research.

In-Yong Jang received his MS in industrial engineering from Korea University in Seoul, Korea, in February 2006 and has since worked in the solution development team of Samsung SDS Co., Ltd., in South Korea. His primary research interest lies in the area of Boolean logic and combinatorial optimization-based supervised machine learning methods to decision analysis.

Yunho Jang is a PhD student at the University of Massachusetts at Amherst. His current research interests are focused on computational structural biology and multi-scale polymer composites based on robot kinematics and elastic network representation.

Krešimir Josić received his PhD in mathematics from the Pennsylvania State University (USA) in 1999. He was a visiting assistant professor and member of the Center for BioDynamics at Boston University until 2002. He is currently an associate professor in the Department of Mathematics at the

University of Houston and is a member of the Gulf Coast Consortium for Theoretical and Computational Neuroscience. Besides working on models of virotherapy, with members of the symmetry and dynamics group, he studies the effects of architecture on network dynamics and examines the impact of dendritic morphology on information processing with experimental colleagues.

Thomas Kalinowski, PhD is a researcher and lecturer in mathematical optimization and combinatorics at the University of Rostock. He has published some articles on the optimization of intensity modulated radiation therapy (IMRT) and on optimization problems in VLSI design. He was awarded the Joachim-Jungius Award of the University of Rostock for his PhD thesis on IMRT using multileaf collimators. He has also participated in several combinatorics conferences and workshops and presented results in extremal set theory and hypergraph theory.

Kwangsoo Kim is a PhD student of the Graduate School of Information Management and Security at Korea University in Seoul, Korea. He received his MS in industrial engineering from Korea University in February 2006 and is currently working on advancing theory and algorithms of global optimization for practical decision-making problems in bioinformatics, supervised machine learning, and pattern classification.

Moon K. Kim, PhD is an assistant professor of the Department of Mechanical and Industrial Engineering, University of Massachusetts at Amherst. He received his BS and MS in mechanical engineering from Seoul National University, Korea, in 1997 and 1999, respectively, and his MSE and PhD from the Johns Hopkins University in 2002 and 2004, respectively. His current research interests are focused on computational structural biology based on robot kinematics including protein dynamics, protein folding prediction, and drug design for conformational diseases.

Mark Langer, MD is a professor of clinical radiation oncology at Indiana University School of Medicine with an adjunct appointment at Purdue University, and serves as the residency program director. His research is centered on problems in radiation oncology optimization and planning. Dr Langer's publications have focused on the application of mathematical programming techniques to radiation oncology treatment plan and delivery scheme design. He has also investigated problems in combinatorial geometry in the definition of unstable targets, and the application of group theory to the handling of fluence perturbation symmetries in intensity modulated radiotherapy. He has served as co-organizer of the joint NSF/NIH workshop on operations research and radiotherapy, and served as chairman of the operations research and radiation therapy collaborative working group.

Kevin Livingstone, PhD is an assistant professor in the biology department at Trinity University in San Antonio, Texas. His educational background is in plant genetics and evolution and statistics. He is now working to increase mathematical applications in the biological sciences at Trinity University in both teaching and research.

Elisa F. Long is a doctoral student in management science and engineering at Stanford University, with a concentration in decision analysis and risk analysis. Her research interests include mathematical modeling of infectious diseases and evaluating the cost-effectiveness of health interventions in resource-limited settings. She previously worked as a summer consultant at Strategic Decisions Group, where she assisted in developing asset valuation models for a biotechnology firm. She also received the Seth Bonder Scholarship for Applied Operations Research in Health Services, sponsored by the Institute for Operations Research and Management Sciences (INFORMS).

Cláudio N. Meneses is a researcher at the University of Sao Paulo, Brazil. He has expertise in computational biology and is currently involved in developing software for scheduling systems.

Michael Nikolaou, PhD is currently a professor in the Department of Chemical and Biomolecular Engineering at the University of Houston. He received a diploma from the National Technical University, Athens, Greece, in 1984, and a PhD from the University of California, Los Angeles, in 1989, both in chemical engineering. He started his academic career at Texas A&M University in 1989, and held a visiting scientist position at Massachusetts Institute of Technology (MIT) in 1995 before moving to the University of Houston in 1997. Dr Nikolaou's research interests are in computer-aided systems engineering, with emphasis on process modeling, identification, monitoring, and control, and their applications to a diverse gamut of industries, including chemicals, petroleum, food, microelectronics, and biomedicine. His research has been extramurally funded by government as well as industry, where he has consulted widely. Dr Nikolaou has also published extensively. A dedicated researcher and teacher, his two dozen former graduate students now hold successful positions in industry and academia.

Douglas K. Owens, MD, MS is a senior investigator at the Veterans Affairs Palo Alto Health Care System, and a professor of medicine and of health research and policy at Stanford University. He directs the program on clinical decision making and guideline development at the Center for Primary Care and Outcomes Research at Stanford University, and also directs the Stanford University-UCSF Evidence Based Practice Center. His research focuses on technology assessment, methods for clinical decision making, cost-effectiveness analysis, and guideline development.

Panos M. Pardalos is a distinguished professor of industrial and systems engineering at the University of Florida. He is also an affiliated faculty member of the computer science department, the Hellenic Studies Center, and the Biomedical Engineering Program. He is also the codirector of the Center for Applied Optimization.

Dr Pardalos obtained a PhD from the University of Minnesota in computer and information sciences. He has held visiting appointments at Princeton University, DIMACS Center, Institute of Mathematics and Applications, Fields Institute, AT&T Labs Research, Trier University, Linkoping Institute of Technology, and universities in Greece.

He has received numerous awards including University of Florida Research Foundation Professor, UF Doctoral Dissertation Advisor/Mentoring Award, Foreign Member of the Royal Academy of Doctors (Spain), Foreign Member Lithuanian Academy of Sciences, Foreign Member of the Ukrainian Academy of Sciences, Foreign Member of the Petrovskaya Academy of Sciences and Arts (Russia), and Honorary Member of the Mongolian Academy of Sciences.

Dr Pardalos received the degree of honorary doctor from Lobachevski University (Russia), he is a fellow of AAAS, a fellow of INFORMS, and in 2001 he was awarded the Greek National Award and a gold medal for operations research.

Dr Pardalos is a world leading expert in global and combinatorial optimization. He is the editor-in-chief of the *Journal of Global Optimization*, *Journal of Optimization Letters*, and *Computational Management Science*. In addition, he is the managing editor of several book series, and a member of the editorial board of several international journals. He is the author of eight books and the editor of several books. He has written numerous articles and developed several well-known software packages. His research is supported by the National Science Foundation and other government organizations. His recent research interests include network design problems, optimization in telecommunications, e-commerce, data mining, biomedical applications, and massive computing.

Dr Pardalos has been an invited lecturer at many universities and research institutes around the world. He has also organized several international conferences.

Michelle A. Ragle, PhD received her PhD in industrial and systems engineering at the University of Florida. She has been teaching mathematics at Okaloosa-Walton College in Northwest Florida for nine years. Before teaching, she worked as a software engineer in both private industry and government contracting. Her research interests include bioinformatics and combinatorial optimization.

Ronald Rardin is John and Mary Lib White Systems Integration Chair and distinguished professor of industrial engineering at the University of Arkansas, Fayetteville. He also heads the university's new Center on Innovation in Healthcare Logistics, targeting innovations in supply chain and material flow aspects of healthcare operations in collaboration with Wal-Mart, Blue Cross Blue Shield, and other partners. Professor Rardin recently retired as professor emeritus of industrial engineering at Purdue University after directing the Purdue Energy Modeling Research Groups and playing a leading role in its Regenstrief Center for Healthcare Engineering. He also served a rotation from 2000–2003 as program director for operations research and service enterprise engineering at the National Science Foundation including founding the program to foster research in service industries, especially healthcare. Dr Rardin obtained his BA and MPA from the University of Kansas, and after working in city government, consulting, and distribution for five years, he obtained a PhD at Georgia Institute of Technology. His current teaching and research interests center on large-scale optimization modeling and algorithms, especially their applications in healthcare delivery. He is an award-winning teacher of those topics, and coauthor of numerous research papers and two comprehensive textbooks: a graduate textbook *Discrete Optimization*, published in 1988, and a comprehensive undergraduate textbook *Optimization in Operations Research* in 1998.

H. Edwin Romeijn is a professor in the Department of Industrial and Systems Engineering and an affiliate member of the Department of Radiation Oncology at the University of Florida. His current research interests are in the area of optimization theory and applications in supply chain management and medicine. He received his PhD in operations research and his MS in econometrics from Erasmus University Rotterdam in the Netherlands. He has published over 50 articles in peer-reviewed journals and is on the editorial boards of the *Journal of Global Optimization* and the *International Journal of Inventory Research*.

Hong Seo Ryoo, PhD is an associate professor of the Graduate School of Information Management and Security and of the division of information management engineering at Korea University in Seoul, Korea. He received his PhD in industrial engineering from the University of Illinois at Urbana-Champaign in May 1999 and has held tenure-track assistant professor positions in the Department of Mathematics at Towson University in Baltimore, Maryland, and the Department of Mechanical and Industrial

Engineering at the University of Illinois at Chicago, Illinois, before joining the faculty of Korea University in September 2003. With expertise in mathematical programming and global optimization, he is currently interested in developing optimization theory, models, and algorithms for supervised machine learning, bioinformatics, and portfolio optimization, broadly speaking.

Edward C. Sewell is professor of mathematics and statistics at Southern Illinois University at Edwardsville. He holds a BS in applied mathematics from the University of Missouri at Rolla, an MS in mathematics from St. Louis University, and a PhD in operations research from Cornell University. His current research interests are combinatorial optimization and health applications. His papers have appeared in various journals such as *Mathematical Programming*, *Discrete Mathematics*, *Naval Research Logistics*, *Journal of Operations Management*, and *Journal of Combinatorial Theory, Series B*.

J. Cole Smith, PhD is an associate professor of industrial and systems engineering at the University of Florida. His research interests lie in optimization, with a focus on exact algorithms for challenging discrete optimization problems. His theoretical interests lie in polyhedral analysis of mixed-integer programming problems and large-scale integer programming techniques. His research has been applied in medical, transportation, logistics, telecommunications, military, and security domains. Dr Smith's research has been funded by the Defense Advanced Research Projects Agency, the Air Force Office of Scientific Research, and the Office of Naval Research, with the latter grant awarded under the competitive Young Investigator Program.

J. MacGregor Smith, PhD is a professor of mechanical and industrial engineering at the University of Massachusetts, Amherst. He conducts research on topological network design, stochastic network design and analysis, and facility layout and location problems. In particular, he is doing research on Steiner minimal trees in 3D, applications of Steiner trees to minimum energy configurations (MECs), and protein modeling. Also, he is working on state-dependent queuing network analysis and finite buffer queuing network models, and quadratic assignment and set packing problems. Applications include the design and layout of manufacturing plants, healthcare facilities, and many other production and service-oriented systems. Professor Smith and his graduate students have carried out many and varied projects for manufacturing and service industries in and around Massachusetts. One of the unique modeling tools developed in his research is concerned with dynamic traffic flow models using queuing theory and queuing networks, as is shown on his Web page: http://www.ecs.umass.edu/mie/faculty/smith/. Dr Smith has carried out sponsored research for the National Science Foundation, the Digital Equipment Corporation, IBM, and the U.S. Army Natick Research and Development Laboratories. He is a member of the Institute for Operations Research and Management Sciences (INFORMS), Association for Computing Machinery (ACM), Society for Industrial and Applied Mathematics (SIAM), The Mathematical Programming Society, and Sigma Xi.

Vincent H. Tam, PharmD is an assistant professor at the University of Houston College of Pharmacy in Houston, Texas. He is board certified in pharmacotherapy with added qualifications in infectious diseases. He has over 40 peer-reviewed publications in antimicrobial pharmacokinetics/pharmacodynamics and infectious disease therapeutics. He is on the editorial boards of *Antimicrobial Agents and Chemotherapy* and *Diagnostic Microbiology and Infectious Diseases*. He has also served as a journal reviewer for the *American Journal of Health-System Pharmacy*, *Annals of Pharmacotherapy*, *Clinical Infectious Diseases*, and *Pharmacotherapy*.

Z. Caner Taşkın is a PhD student at the Department of Industrial and Systems Engineering, University of Florida. He received his BS and MS in industrial engineering from Boğazici University, İstanbul. Before starting his doctorate study, he worked for ICRON Technologies as a product consultant, where he took a role in advanced planning and scheduling (APS) projects for customers in several industries including steel, automotive, electronics, and glass manufacturing industries. His research interests include integer programming, combinatorial optimization, and medical applications of operations research.

Ali Tuncel is a PhD candidate in the Department of Industrial Engineering at Purdue University with a primary research interest in optimization of radiation therapy planning and delivery. Before starting his PhD program, he was an internal consultant at United Airlines for over four years, working on various applications development projects in the areas of supply chain management, crew scheduling, market forecasting, and aircraft scheduling. He holds a BS in industrial engineering from Dokuz Eylul University and an MS in industrial engineering from Purdue University.

Tsung-Lin Wu is a doctoral student in industrial and systems engineering at Georgia Institute of Technology. His research interests include optimization and medical applications.

MEDICINE I

Chapter 1

Classification and Disease Prediction via Mathematical Programming

Eva K. Lee and Tsung-Lin Wu

CONTENTS

1.1 Introduction 5
  1.1.1 Pattern Recognition, Discriminant Analysis, and Statistical Pattern Classification 5
  1.1.2 Supervised Learning, Training, and Cross-Validation 6
  1.1.3 Bayesian Inference and Classification 7
  1.1.4 Discriminant Functions 8
1.2 Mathematical Programming Approaches 10
  1.2.1 Linear Programming Classification Models 11
    1.2.1.1 Two-Group Classification 11
    1.2.1.2 Multigroup Classification 17
  1.2.2 Mixed Integer Programming Classification Models 20
    1.2.2.1 Two-Group Classification 20
    1.2.2.2 Multigroup Classification 23
  1.2.3 Nonlinear Programming Classification Models 25
  1.2.4 Support Vector Machine 29
1.3 MIP-Based Multigroup Classification Models and Applications to Medicine and Biology 31
  1.3.1 Discrete Support Vector Machine Predictive Models 32

1.3.1.1 Modeling of Reserved-Judgment Region for General Groups 32
1.3.1.2 Mixed Integer Programming Formulations 34
1.3.1.3 Model Variations 36
1.3.1.4 Validation of Model and Computational Effort 40
1.3.2 Classification Results on Real-World Biological and Medical Applications 40
1.3.2.1 Determining the Type of Erythemato-Squamous Disease 41
1.3.2.2 Predicting Presence/Absence of Heart Disease 42
1.3.2.3 Predicting Aberrant CpG Island Methylation in Human Cancer 42
1.3.2.4 Discriminant Analysis of Cell Motility and Morphology Data in Human Lung Carcinoma 44
1.3.2.5 Ultrasonic-Assisted Cell Disruption for Drug Delivery 45
1.3.2.6 Identification of Tumor Shape and Volume in Treatment of Sarcoma 45
1.3.2.7 Discriminant Analysis of Biomarkers for Prediction of Early Atherosclerosis 46
1.3.2.8 Fingerprinting Native and Angiogenic Microvascular Networks through Pattern Recognition and Discriminant Analysis of Functional Perfusion Data 47
1.3.2.9 Prediction of Protein Localization Sites 48
1.3.2.10 Pattern Recognition in Satellite Images for Determining Types of Soil 49
1.3.3 Further Advances 50
1.4 Progress and Challenges 50
1.5 Other Methods 50
1.6 Summary and Conclusion 52
Acknowledgment 54
References 54

Abstract In this chapter, we present classification models based on mathematical programming approaches. We first provide an overview on various mathematical programming approaches, including linear programming, mixed integer programming, nonlinear programming, and support vector machines. Next, we present our effort of novel optimization-based classification models that are general purpose and suitable for developing predictive rules for large heterogeneous biological and medical datasets. Our predictive model simultaneously incorporates (1) the ability to classify any number of distinct groups; (2) the ability to incorporate heterogeneous types of attributes as input; (3) a high-dimensional data transformation that eliminates noise and errors in biological data; (4) the ability to incorporate constraints to limit the rate of misclassification, and a reserved-judgment region that provides a safeguard against overtraining (which tends to lead to high misclassification rates from the resulting predictive rule); and

(5) successive multistage classification capability to handle data points placed in the reserved-judgment region. To illustrate the power and flexibility of the classification model and solution engine, and its multigroup prediction capability, application of the predictive model to a broad class of biological and medical problems is described. Applications include the differential diagnosis of the type of erythemato-squamous diseases; predicting presence/absence of heart disease; genomic analysis and prediction of aberrant CpG island meythlation in human cancer; discriminant analysis of motility and morphology data in human lung carcinoma; prediction of ultrasonic cell disruption for drug delivery; identification of tumor shape and volume in treatment of sarcoma; multistage discriminant analysis of biomarkers for prediction of early atherosclerois; fingerprinting of native and angiogenic microvascular networks for early diagnosis of diabetes, aging, macular degeneracy, and tumor metastasis; prediction of protein localization sites; and pattern recognition of satellite images in classification of soil types. In all these applications, the predictive model yields correct classification rates ranging from 80 to 100 percent. This provides motivation for pursuing its use as a medical diagnostic, monitoring, and decision-making tool.

1.1 Introduction

Classification is a fundamental machine learning task whereby rules are developed for the allocation of independent observations to groups. Classic examples of applications include medical diagnosis—the allocation of patients to disease classes based on symptoms and lab tests, and credit screening—the acceptance or rejection of credit applications based on applicant data. Data is collected concerning observations with known group membership. This training data is used to develop rules for the classification of future observations with unknown group membership.

In this section, we briefly describe some terminologies related to classification, and provide a brief organization of the materials written in this chapter.

1.1.1 Pattern Recognition, Discriminant Analysis, and Statistical Pattern Classification

Cognitive science is the science of learning, knowing, and reasoning. Pattern recognition is a broad field within cognitive science that is concerned with the process of recognizing, identifying, and categorizing input information. These areas intersect with computer science, particularly in the closely related areas of artificial intelligence, machine learning, and statistical pattern recognition. Artificial intelligence is associated with constructing machines and systems that reflect human abilities in cognition. Machine learning refers to how these machines and systems replicate the learning process, which is often achieved by seeking and discovering patterns in data, or statistical pattern recognition.

Discriminant analysis is the process of discriminating between categories or populations. Associated with discriminant analysis as a statistical tool are the tasks of determining the features that best discriminate between populations, and the process of classifying new objects based on these features. The former is often called feature selection and the latter is referred to as statistical pattern classification. This work will be largely concerned with the development of a viable statistical pattern classifier.

As with many computationally intensive tasks, recent advances in computing power have led to a sharp increase in the interest and application of discriminant analysis techniques. The reader is referred to Duda et al. [25] for an introduction to various techniques for pattern classification, and to Zopounidis et al. [121] for examples of applications of pattern classification.

1.1.2 Supervised Learning, Training, and Cross-Validation

An entity or observation is essentially a data point as commonly understood in statistics. In the framework of statistical pattern classification, an entity is a set of quantitative measurements (or qualitative measurements expressed quantitatively) of attributes for a particular object. As an example, in medical diagnosis an entity could be the various blood chemistry levels of a patient. With each entity is associated one or more groups (or populations, classes, categories) to which it belongs. Continuing with the medical diagnosis example, the groups could be the various classes of heart disease. Statistical classification seeks to determine rules for associating entities with the groups to which they belong. Ideally, these associations align with the associations that human reasoning would produce based on information gathered on objects and their apparent categories.

Supervised learning is the process of developing classification rules based on entities for which the classification is already known. Note that the process implies that the populations are already well defined. Unsupervised learning is the process of discovering patterns from unlabeled entities and thereby discovering and describing the underlying populations. Models derived using supervised learning can be used for both functions of discriminant analysis—feature selection and classification. The model that we consider is a method for supervised learning, so we assume that populations are previously defined.

The set of entities with known classification that is used to develop classification rules is the training set. The training set may be partitioned so that some entities are withheld during the model-development process, also known as the training of the model. The withheld entities form a test set that is used to determine the validity of the model, a process known as cross-validation. Entities from the test set are subjected to the rules of classification to measure the performance of the rules on entities with unknown group membership.

Validation of classification models is often performed using m-fold cross-validation where the data with known classification is partitioned into m *folds* (subsets) of approximately equal size. The classification model is trained m times, with the mth fold withheld during each run for testing. The performance of the model is evaluated by the classification accuracy on the m test folds, and can be represented using a classification matrix or confusion matrix.

The classification matrix is a square matrix with the number of rows and columns equal to the number of groups. The ijth entry of the classification matrix contains the number or proportion of test entities from group i that were classified by the model as belonging to group j. Therefore, the number or proportion of correctly classified entities are contained in the diagonal elements of the classification matrix, and the number or proportion of misclassified entities are in the off-diagonal entries.

1.1.3 Bayesian Inference and Classification

The popularity of Bayesian inference has risen drastically over the past several decades, perhaps in part due to its suitability for statistical learning. The reader is referred to O'Hagan's volume [92] for a thorough treatment of Bayesian inference. Bayesian inference is usually contrasted against classical inference, though in practice they often imply the same methodology.

The Bayesian method relies on a subjective view of probability, as opposed to the frequentist view upon which classical inference is based [92]. A subjective probability describes a degree of belief in a proposition held by the investigator based on some information. A frequency probability describes the likelihood of an event given an infinite number of trials.

In Bayesian statistics, inferences are based on the posterior distribution. The posterior distribution is the product of the prior probability and the likelihood function. The prior probability distribution represents the initial degree of belief in a proposition, often before empirical data is considered. The likelihood function describes the likelihood that the behavior is exhibited, given that the proposition is true. The posterior distribution describes the likelihood that the proposition is true, given the observed behavior.

Suppose we have a proposition or random variable θ about which we would like to make inferences, and data x. Application of Bayes' theorem gives

$$dF(\theta|x) = \frac{dF(\theta)dF(x|\theta)}{dF(x)}$$

Here, F denotes the (cumulative) distribution function. For ease of conceptualization, assume that F is differentiable, then $dF = f$, and the above equality can be rewritten as

$$f(\theta|x) = \frac{f(\theta)f(x|\theta)}{f(x)}.$$

For classification, a prior probability function $\pi(g)$ describes the likelihood that an entity is allocated to group g regardless of its exhibited feature values x. A group density function $f(x|g)$ describes the likelihood that an entity exhibits certain measurable attribute values, given that it belongs to population g. The posterior distribution for a group $P(g|x)$ is given by the product of the prior probability and group density function, normalized over the groups to obtain a unit probability over all groups. The observation x is allocated to group h if $h = \arg\max_{g \in \mathcal{G}} P(g|x) = \arg\max_{g \in \mathcal{G}} \frac{\pi(g)f(x|g)}{\sum_{j \in \mathcal{G}} \pi(j)f(x|j)}$, where $\mathcal{G}$ denotes the set of groups.

1.1.4 Discriminant Functions

Most classification methods can be described in terms of discriminant functions. A discriminant function takes as input an observation and returns information about the classification of the observation. For data from a set of groups $\mathcal{G}$, an observation x is assigned to group h if $h = \arg\max_{g \in \mathcal{G}} l_g(x)$ where the functions l_g are the discriminant functions. Classification methods restrict the form of the discriminant functions, and training data is used to determine the values of parameters that define the functions.

The optimal classifier in the Bayesian framework can be described in terms of discriminant functions. Let $\pi_g = \pi(g)$ be the prior probability that an observation is allocated to group g and let $f_g(x) = f(x|g)$ be the likelihood that data x is drawn from population g. If we wish to minimize the probability of misclassification given x, then the optimal allocation for an entity is to the group $h = \arg\max_{g \in \mathcal{G}} P(g|x) = \arg\max_{g \in \mathcal{G}} \frac{\pi_g f_g(x)}{\sum_{j \in \mathcal{G}} \pi_j f_j(x)}$. Under the Bayesian framework,

$$P(g|x) = \frac{\pi_g f(x|g)}{f(x)} = \frac{\pi_g f(x|g)}{\sum_{j \in \mathcal{G}} \pi_j f(x|j)}$$

The discriminant functions can be $l_g(x) = P(g|x)$ for $g \in \mathcal{G}$. The same classification rule is given by $l_g(x) = \pi_g f(x|g)$ and $l_g(x) = \log f(x|g) + \log \pi_g$. The problem then becomes finding the form of the prior functions and likelihood functions that match the data.

If the data is multivariate normal with equal covariance matrices ($f(x|g) \sim N(\mu_g, \Sigma)$), then a linear discriminant function is optimal:

$$
\begin{aligned}
l_g(x) &= \log f(x|g) + \log \pi_g \\
&= -1/2(x-\mu_g)^T \Sigma^{-1}(x-\mu_g) - 1/2 \log \left|\sum_g\right| - d/2 \log 2\pi + \log \pi_g \\
&= w_g^T x + w_{g0}
\end{aligned}
$$

where d is the number of attributes, $w_g = \Sigma^{-1}\mu_g$, and $w_{g0} = -1/2\mu_g^T \Sigma^{-1}\mu_g + \log \pi_g + x^T \Sigma^{-1} x - d/2 \log 2\pi$. Note that the last two terms of w_{g0} are constant for all g and need not be calculated. When there are two groups ($\mathcal{G} = \{1, 2\}$) and the priors are equal ($\pi_1 = \pi_2$), the discriminant rule is equivalent to Fisher's linear discriminant rule [30]. Fisher's rule can also be derived, as it was by Fisher, by choosing w so that $\frac{(w^T\mu_1 - w^T\mu_2)^2}{w^T \Sigma w}$ is maximized.

These linear and quadratic discriminant functions are often applied to datasets that are not multivariate normal or continuous (see Ref. [98], pp. 234–235) by using approximations for the means and covariances. Regardless, these models are parametric in that they incorporate assumptions about the distribution of the data. Fisher's linear discriminant is nonparametric because no assumptions are made about the underlying distribution of the data. Thus, for a special case, a parametric and nonparametric model coincide to produce the same discriminant rule. The linear discriminant function derived above is also called the Homoscedastic Model, and the quadratic discriminant function is called the Heteroscedastic Model. The exact form of discriminant functions in the Bayesian framework can be derived for other distributions [25].

Some classification methods are essentially the methods for finding coefficients for linear discriminant functions. In other words, they seek coefficients w_g and constants w_{g0} such that $l_g(x) = w_g x + w_{g0}$, $g \in \mathcal{G}$, is an optimal set of discriminant functions. The criteria for optimality is different for different methods. Linear discriminant functions project the data onto a linear subspace and then discriminate between entities in that subspace. For example, Fisher's linear discriminant projects two-group data on an optimal line, and discriminates on that line. A good linear subspace may not exist for data with overlapping distributions between groups and therefore the data will not be classified accurately using these methods. The hyperplanes defined by the discriminant functions form boundaries between the group regions. A large portion of the literature concerning the use of mathematical programming models for classification describe methods for finding coefficients of linear discriminant functions [121].

Other classification methods seek to determine parameters to establish quadratic discriminant functions. The general form of a quadratic discriminant function is $l_g(x) = x^T W_g x + w_g^T x + w_{g0}$. The boundaries defining the group regions can assume any hyperquadric form, as can the Bayes decision rules for arbitrary multivariate normal distributions [25].

In this chapter, we survey the development and advances of classification models via the mathematical programming techniques, and summarize our experience in classification models applied to prediction in biological and medical applications. The rest of this chapter is organized as follows. Section 1.2 first provides a detailed overview of the development and advances of mathematical programming-based classification models, including linear programming (LP), mixed integer programming (MIP), nonlinear programming, and support vector machine (SVM) approaches. In Section 1.3, we describe our effort in developing optimization-based multigroup multistage discriminant analysis predictive models for classification. The use of the predictive models on various biological and medical problems are presented. Section 1.4 provides several tables to summarize the progress of mathematical programming-based classification models and their characteristics. This is followed by a brief description of other classification methods in Section 1.5, and summary and concluding remarks in Section 1.6.

1.2 Mathematical Programming Approaches

Mathematical programming methods for statistical pattern classification emerged in the 1960s, gained popularity in the 1980s, and have grown drastically since. Most of the mathematical programming approaches are nonparametric, which has been cited as an advantage when analyzing contaminated datasets over methods that require assumptions about the distribution of the data [107]. Most of the literature about mathematical programming methods are concerned with either using mathematical programming to determine the coefficients of linear discriminant functions or with SVMs.

The following notation will be used. The subscripts i, j, and k are used for the observation, attribute, and group, respectively. Let x_{ij} be the value of attribute j of observation i. Let m be the number of attributes, K be the number of groups, G_k represent the set of data from group k, M be a big positive number, and ϵ be a small positive number. The abbreviation urs is used in reference to a variable to denote "unrestricted in sign."

1.2.1 Linear Programming Classification Models

The use of linear programs to determine the coefficients of linear discriminant functions has been widely studied [31,46,50,74]. The methods determine the coefficients for different objectives, including minimizing the sum of the distances to the separating hyperplane, minimizing the maximum distance of an observation to the hyperplane, and minimizing other measures of badness of fit or maximizing measures of goodness of fit.

1.2.1.1 Two-Group Classification

One of the earliest LP classification models was proposed by Mangasarian [74] to construct a hyperplane to separate two groups of data. Separation by a nonlinear surface using LP was also proposed when the surface parameters appear linearly. Two sets of points may be inseparable by one hyperplane or surface through a single-step LP approach, but they can be strictly separated by more planes or surfaces via a multistep LP approach (Mangasarian [75]). In Ref. [75], real problems with up to 117 data points, 10 attributes, and 3 groups were solved. The three-group separation was achieved by separating group 1 from groups 2 and 3, and then group 2 from group 3.

Studies of LP Models for the discriminant problem in the early 1980s were carried out by Hand [47], Freed and Glover [31,32], and Bajgier and Hill [5]. Three LP Models for the two-group classification problem, including minimizing the sum of deviations (MSD), minimizing the maximum deviation (MMD), and minimizing the sum of interior distances (MSID) were proposed. Freed and Glover [33] provided computational studies of these models where the test conditions involved normal and non-normal populations.

MSD (Minimize the sum of deviations)

$$
\begin{aligned}
\text{Min} \quad & \sum_i d_i \\
\text{s.t.} \quad & w_0 + \sum_j x_{ij} w_j - d_i \le 0 \quad \forall i \in G_1 \\
& w_0 + \sum_j x_{ij} w_j + d_i \ge 0 \quad \forall i \in G_2 \\
& w_j \text{ urs} \quad \forall j \\
& d_i \ge 0 \quad \forall i
\end{aligned}
$$

MMD (Minimize the maximum deviation)

$$\text{Min} \quad d$$

$$\text{s.t.} \quad w_0 + \sum_j x_{ij} w_j - d \le 0 \quad \forall i \in G_1$$

$$w_0 + \sum_j x_{ij} w_j + d \ge 0 \quad \forall i \in G_2$$

$$w_j \text{ urs} \quad \forall j$$

$$d \ge 0$$

MSID (Minimize the sum of interior distances)

$$\text{Min} \quad pd - \sum_i e_i$$

$$\text{s.t.} \quad w_0 + \sum_j x_{ij} w_j - d + e_i \le 0 \quad \forall i \in G_1$$

$$w_0 + \sum_j x_{ij} w_j + d - e_i \ge 0 \quad \forall i \in G_2$$

$$w_j \text{ urs} \quad \forall j$$

$$d \ge 0$$

$$e_i \ge 0 \quad \forall i$$

where p is a weight constant.

The objective function of the MSD Model is the L_1-norm distance while the objective function of MMD is the L_∞-norm distance. They are special cases of L_p-norm classification [50,108].

In some models, the constant term of the hyperplane is a fixed number instead of a decision variable. The model MSD^0 shown below is an example where the cut-off score b replaces w_0 in the formulation. The same replacement could be used in other formulations.

MSD^0 (Minimize the sum of deviations with constant cutoff score)

$$\begin{aligned} \text{Min} \quad & \sum_i d_i \\ \text{s.t.} \quad & \sum_j x_{ij} w_j - d_i \leq b \quad \forall i \in G_1 \\ & \sum_j x_{ij} w_j + d_i \geq b \quad \forall i \in G_2 \\ & w_j \text{ urs} \quad \forall j \\ & d_i \geq 0 \quad \forall i \end{aligned}$$

A gap can be introduced between the two regions determined by the separating hyperplane to prevent degenerate solutions. Take MSD as an example, the separation constraints become

$$w_0 + \sum_j x_{ij} w_j - d_i \leq -\epsilon \quad \forall i \in G_1$$

$$w_0 + \sum_j x_{ij} w_j + d_i \geq \epsilon \quad \forall i \in G_2$$

The small number ϵ can be normalized to 1.

Besides introducing a gap, another normalization approach is to include constraints such as $\sum_{j=0}^{m} w_j = 1$ or $\sum_{j=1}^{m} w_j = 1$ in the LP Models to avoid unbounded or trivial solutions.

Specifically, Glover et al. [45] gave the Hybrid Model, as follows:

Hybrid Model

$$\begin{aligned} \text{Min} \quad & pd + \sum_i p_i d_i - qe - \sum_i q_i e_i \\ \text{s.t.} \quad & w_0 + \sum_j x_{ij} w_j - d - d_i + e + e_i = 0 \quad \forall i \in G_1 \\ & w_0 + \sum_j x_{ij} w_j + d + d_i - e - e_i = 0 \quad \forall i \in G_2 \end{aligned}$$

$$w_j \text{ urs } \quad \forall j$$

$$d, e \geq 0$$

$$d_i, e_i \geq 0 \quad \forall i$$

where p, p_i, q, and q_i are the cost for different deviations. Including different combinations of deviation terms in the objective function then leads to variant models.

Joachimsthaler and Stam [50] review and summarize LP formulations applied to two-group classification problems in discriminant analysis, including MSD, MMD, MSID, MIP Models, and the Hybrid Model. They summarize the performance of the LP methods together with the traditional classification methods such as Fisher's linear discriminant function (LDF) [30], Smith's quadratic discriminant function (QDF) [106], and a logistic discriminant method. In their review, MSD sometimes but not uniformly improves classification accuracy, compared with traditional methods. On the other hand, MMD is found to be inferior to MSD. Erenguc and Koehler [27] present a unified survey of LP Models and their experimental results, in which the LP Models include several versions of MSD, MMD, MSID, and Hybrid Models. Rubin [99] provides experimental results of comparing these LP Models with Fisher's LDF and Smith's QDF. He concludes that QDF performs best when the data follows normal distributions and that QDF could be the benchmark when seeking situations for advantageous LP methods. In summary, the above review papers [27,50,99] describe previous work on LP classification models and their comparison with traditional methods. However, it is difficult to make definitive statements about conditions under which one LP Model is superior to others, as stated in Ref. [107].

Stam and Ungar [110] introduce a software package RAGNU, a utility program in conjunction with the LINDO optimization software, for solving two-group classification problems using LP-based methods. LP formulations such as MSD, MMD, MSID, Hybrid Models, and their variants are contained in the package.

There are some difficulties in LP-based formulations, in that some models could result in unbounded, trivial, or unacceptable solutions [34,87], but possible remedies are proposed. Koehler [51–53] and Xiao [114,115] characterize the conditions of unacceptable solutions in two-group LP discriminant Models, including MSD, MMD, MISD, the Hybrid Model, and their variants. Glover [44] proposes the normalization constraint, $\sum_{j=1}^{m}(-|G_2|\sum_{i\in G_1} x_{ij} + |G_1|\sum_{i\in G_2} x_{ij})w_j = 1$, which is more effective and reliable. Rubin [100] examines the separation failure for two-group models and suggests to apply the models twice, reversing the group designations the second time. Xiao and Feng [116] propose a regularization method to avoid multiple solutions in LP discriminant analysis by adding the term $\epsilon \sum_{j=1}^{m} w_j^2$ in the objective functions.

Bennett and Mangasarian [9] propose the following model that minimizes the average of the deviations, which is called robust linear programming (RLP).

RLP (Robust linear programming)

$$\text{Min} \quad \frac{1}{|G_1|}\sum_{i\in G_1} d_i + \frac{1}{|G_2|}\sum_{i\in G_2} d_i$$

$$\text{s.t.} \quad w_0 + \sum_j x_{ij}w_j - d_i \le -1 \quad \forall i \in G_1$$

$$w_0 + \sum_j x_{ij}w_j + d_i \ge 1 \quad \forall i \in G_2$$

$$w_j \text{ urs} \quad \forall j$$

$$d_i \ge 0 \quad \forall i$$

It is shown that this model gives the null solution $w_1 = \cdots = w_m = 0$ if, and only if, $\frac{1}{|G_1|}\sum_{i\in G_1} x_{ij} = \frac{1}{|G_2|}\sum_{i\in G_2} x_{ij}$ for all j, in which case the solution $w_1 = \cdots = w_m = 0$ is guaranteed to be not unique. Data of different diseases is tested by the proposed classification methods, as in most of Mangasarian's papers.

Mangasarian et al. [86] describe two applications of LP Models in the field of breast cancer research, one in diagnosis and the other in prognosis. The first application is to discriminate benign from malignant breast lumps, while the second application is to predict when breast cancer is likely to recur. Both of them work successfully in clinical practice. The RLP Model [9] together with the multisurface method tree algorithm (MSMT) [8] is used in the diagnostic system.

Duarte Silva and Stam [104] include the second-order (i.e., quadratic and cross-product) terms of the attribute values in the LP-based models such as MSD and Hybrid Models and compare them with linear Models, Fisher's LDF, and Smith's QDF. The results of the simulation experiments show that the methods that include second-order terms perform much better than first-order methods, given that the data substantially violates the multivariate normality assumption. Wanarat and Pavur [113] investigate the effect of the inclusion of the second-order terms in the MSD, MIP, and Hybrid Models when sample size is small to moderate. However, the simulation study shows that second-order terms may not always improve the performance of a first-order LP Model even with data configurations that are more appropriately classified by Smith's QDF. Another result of the simulation study is that inclusion of the cross-product terms may hurt the model's accuracy, whereas omission of these terms causes the model to be not invariant with respect to a nonsingular transformation of the data.

Pavur [94] studies the effect of the position of the contaminated normal data in the two-group classification problem. The methods for comparison in their study include MSD, MM (described in the MIP part), Fisher's LDF, Smith's QDF, and nearest-neighbor models. The nontraditional methods such as LP Models have the potential for outperforming the standard parametric procedures when non-normality is present, but this study shows that no one model is consistently superior in all cases.

Asparoukhov and Stam [3] propose LP and MIP Models to solve the two-group classification problem where the attributes are binary. In this case, the training data can be partitioned into multinomial cells, allowing for a substantial reduction in the number of variables and constraints. The proposed models not only have the usual geometric interpretation but also possess a strong probabilistic foundation. Let s be the index of the cells, n_{1s}, n_{2s} be the number of data points in cell s from groups 1 and 2, respectively, and $(b_{s1}, \ldots, b_{sm})$ be the binary digits representing cell s. The model shown below is the LP Model of minimizing the sum of deviations for two-group classification with binary attributes.

Cell convensional MSD

$$\text{Min} \quad \sum_{s:\, n_{1s}+n_{2s}>0} (n_{1s}d_{1s} + n_{2s}d_{2s})$$

$$\text{s.t.} \quad w_0 + \sum_j b_{sj}w_j - d_{1s} \leq 0 \quad \forall s : n_{1s} > 0$$

$$w_0 + \sum_j b_{sj}w_j + d_{2s} > 0 \quad \forall s : n_{2s} > 0$$

$$w_j \text{ urs} \quad \forall j$$

$$d_{1s}, d_{2s} \geq 0 \quad \forall s$$

Binary attributes are usually found in medical diagnoses data. In this study, three real datasets about disease discrimination are tested: developing postoperative pulmonary embolism or not, having dissecting aneurysm or other diseases, and suffering from posttraumatic epilepsy or not. In these datasets, the MIP Model for binary attributes (BMIP), which will be described later, performs better than other LP Models or traditional methods.

1.2.1.2 Multigroup Classification

Freed and Glover [32] extend the LP classification models from two-group to multigroup problems. One formulation that uses a single discriminant function is given below:

$$\text{Min} \quad \sum_{k=1}^{K-1} c_k \alpha_k$$

$$\text{s.t.} \quad \sum_j x_{ij} w_j \leq U_k \quad \forall i \in G_k \; \forall k$$

$$\sum_j x_{ij} w_j \geq L_k \quad \forall i \in G_k \; \forall k$$

$$U_k + \epsilon \leq L_{k+1} + \alpha_k \quad \forall k = 1, \ldots, K-1$$

$$w_j \text{ urs} \quad \forall j$$

$$U_k, L_k \text{ urs} \quad \forall k$$

$$\alpha_k \text{ urs} \quad \forall k = 1, \ldots, K-1$$

where the number ϵ could be normalized to be 1, and c_k is the misclassification cost. However, single function classification is not as flexible and general as multiple function classification. Another extension from the two-group case to multigroup in Ref. [32] is to solve two-group LP Models for all pairs of groups and determine classification rules based on these solutions. However, in some cases, the group assignment is not clear and the resulting classification scheme may be suboptimal [107].

For the multigroup discrimination problem, Bennett and Mangasarian [10] define the piecewise-linear separability of data from K groups as the following: The data from K groups is piecewise-linear separable if, and only if, there exist $(w_0^k, w_1^k, \ldots, w_m^k) \in R^{m+1}$, $k = 1, \ldots, K$, such that $w_0^h + \sum_j x_{ij} w_j^h \geq w_0^k + \sum_j x_{ij} w_j^k + 1$, $\forall i \in G_h$ $\forall h, k \neq h$. The following LP will generate a piecewise-linear separation for the K groups if one exists, otherwise it will generate an error-minimizing separation.

$$\text{Min} \quad \sum_h \sum_{k \neq h} \frac{1}{|G_h|} \sum_{i \in G_h} d_i^{hk}$$

$$\text{s.t.} \quad d_i^{hk} \geq -\left(w_0^h + \sum_j x_{ij} w_j^h\right) + \left(w_0^k + \sum_j x_{ij} w_j^k\right) + 1$$

$$\forall i \in G_h \quad \forall h, k \neq h$$

$$w_j^k \text{ urs} \quad \forall j, k$$

$$d_i^{hk} \geq 0 \quad \forall i \in G_h \; \forall h, k \neq h$$

The method is tested in three datasets. It performs pretty well in two of the datasets that are totally (or almost totally) piecewise-linear separable. The classification result is not good in the third dataset, which is inherently more difficult. However, by combining the MSMT [8], the performance improves.

Gochet et al. [46] introduce an LP Model for the general multigroup classification problem. The method separates the data with several hyperplanes by sequentially solving LPs. The vectors w^k, $k = 1, \ldots, K$, are estimated for the classification decision rule. The rule is to classify an observation i into group s where $s = \arg\max_k \{w_0^k + \sum_j x_{ij} w_j^k\}$.

Suppose, observation i is from group h. Denote the goodness of fit for observation i with respect to group k as

$$G_{hk}^i(w^h, w^k) = \left[\left(w_0^h + \sum_j x_{ij} w_j^h\right) - \left(w_0^k + \sum_j x_{ij} w_j^k\right)\right]^+$$

where $[a]^+ = \max\{0, a\}$.

Likewise, denote the badness of fit for observation i with respect to group k as

$$B_{hk}^i(w^h, w^k) = \left[\left(w_0^h + \sum_j x_{ij} w_h^h\right) - \left(w_0^k + \sum_j x_{ij} w_j^k\right)\right]^-$$

where $[a]^- = -\min\{0, a\}$.

The total goodness of fit and total badness of fit are then defined as

$$G(w) = G(w^1, \ldots, w^K) = \sum_h \sum_{k \neq h} \sum_{i \in G_h} G^i_{hk}(w^h, w^k)$$

$$B(w) = B(w^1, \ldots, w^K) = \sum_h \sum_{k \neq h} \sum_{i \in G_h} B^i_{hk}(w^h, w^k)$$

The LP is to minimize the total badness of fit, subject to a normalization equation, in which $q > 0$

$$\begin{aligned} &\text{Min} \quad B(w) \\ &\text{s.t.} \quad G(w) - B(w) = q \\ &\qquad w \text{ urs} \end{aligned}$$

Expanding $G(w)$ and $B(w)$ and substituting $G^i_{hk}(w^h, w^k)$ and $B^i_{hk}(w^h, w^k)$ by γ^i_{hk} and β^i_{hk}, respectively, the LP becomes

$$\text{Min} \quad \sum_h \sum_{k \neq h} \sum_{i \in G_h} \beta^i_{hk}$$

$$\text{s.t.} \quad \left(w_0^h + \sum_j x_{ij} w_j^h\right) - \left(w_0^k + \sum_j x_{ij} w_j^k\right) = \gamma^i_{hk} - \beta^i_{hk}$$

$$\forall i \in G_h \ \forall h, k \neq h$$

$$\sum_h \sum_{k \neq h} \sum_{i \in G_h} (\gamma^i_{hk} - \beta^i_{hk}) = q$$

$$w_j^k \text{ urs} \quad \forall j, k$$

$$\gamma^i_{hk}, \beta^i_{hk} \geq 0 \quad \forall i \in G_h \ \forall h, k \neq h$$

The classification results for two real datasets show that this model can compete with Fisher's LDF and the nonparametric k-nearest-neighbor method.

The LP-based models for classification problems highlighted above are all nonparametric models. In Section 1.3, we describe LP-based and MIP-based classification models that utilize a parametric multigroup discriminant analysis approach [39,40,60,63]. These latter models have been employed successfully in various multigroup disease diagnosis and biological/medical prediction problems [16,28,29,56,57,59,60,64,65].

1.2.2 Mixed Integer Programming Classification Models

Although LP offers a polynomial-time computational guarantee, MIP allows more flexibility in (among other things) modeling misclassified observations or misclassification costs.

1.2.2.1 Two-Group Classification

In the two-group classification problem, binary variables can be used in the formulation to track and minimize the exact number of misclassifications. Such an objective function is also considered as the L_0-norm criterion [107].

MM (Minimizing the number of misclassifications)

$$\text{Min} \quad \sum_i z_i$$

$$\text{s.t.} \quad w_0 + \sum_j x_{ij} w_j \leq M z_i \quad \forall i \in G_1$$

$$w_0 + \sum_j x_{ij} w_j \geq -M z_i \quad \forall i \in G_2$$

$$w_j \text{ urs} \quad \forall j$$

$$z_i \in \{0, 1\} \quad \forall i$$

The vector w is required to be a nonzero vector to prevent the trivial solution.

In the MIP formulation, the objective function could include the deviation terms, such as those in the Hybrid Models, as well as the number of misclassifications [5]; or it could represent expected cost of misclassification [1,6,101,105]. In particular, there are some variant versions of the basic model.

Stam and Joachimsthaler [109] study the classification performance of MM and compare it with MSD, Fisher's LDF, and Smith's QDF. In some cases, the MM Model performs better, but in some cases it does not. MIP formulations are in the review studies of Joachimsthaler and Stam [50] and

Erenguc and Koehler [27], and contained in the software developed by Stam and Ungar [110]. Computational experiments show that the MIP Model performs better when the group overlap is higher [50,109], although it is still not easy to reach general conclusions [107].

Because the MIP Model is NP-hard, exact algorithms and heuristics are proposed to solve it efficiently. Koehler and Erenguc [54] develop a procedure to solve MM in which the condition of nonzero w is replaced by the requirement of at least one violation of the constraints $w_0 + \sum_j x_{ij}w_j \leq 0$ for $i \in G_1$ or $w_0 + \sum_j x_{ij}w_j \geq 0$ for $i \in G_2$. Banks and Abad [6] solve the MIP of minimizing the expected cost of misclassification by an LP-based algorithm. Abad and Banks [1] develop three heuristic procedures to the problem of minimizing the expected cost of misclassification. They also include the interaction terms of the attributes in the data and apply the heuristics [7]. Duarte Silva and Stam [105] introduce the divide and conquer algorithm for the classification problem of minimizing the misclassification cost by solving MIP and LP subproblems. Rubin [101] solves the same problem by using a decomposition approach, and tests this procedure on some datasets, including two breast cancer datasets. Yanev and Balev [119] propose exact and heuristic algorithms for solving MM, which are based on some specific properties of the vertices of a polyhedral set neatly connected with the model.

For the two-group classification problem where the attributes are binary, Asparoukhov and Stam [3] propose LP and MIP Models that partition the data into multinomial cells and result in fewer number of variables and constraints. Let s be the index of the cells, n_{1s}, n_{2s} be the number of data points in cell s from groups 1 and 2, respectively, and $(b_{s1}, \ldots, b_{sm})$ be the binary digits representing cell s. Below is the MIP Model for binary attributes (BMIP), which performs best in three real datasets in Ref. [3].

BMIP

$$\text{Min} \quad \sum_{s:\, n_{1s}+n_{2s}>0} \{|n_{1s} - n_{2s}|z_s + \min(n_{1s}, n_{2s})\}$$

$$\text{s.t.} \quad w_0 + \sum_j b_{sj}w_j \leq Mz_s \quad \forall s : n_{1s} \geq n_{2s};\ n_{1s} > 0$$

$$w_0 + \sum_j b_{sj}w_j > -Mz_s \quad \forall s : n_{1s} < n_{2s}$$

$$w_j \text{ urs} \quad \forall j$$

$$z_s \in \{0, 1\} \quad \forall s : n_{1s} + n_{2s} > 0$$

Pavur et al. [96] include different secondary goals in the model MM and compare their misclassification rates. A new secondary goal is proposed, which maximizes the difference between the means of the discriminant scores of the two groups. In this model, the term $-\delta$ is added to the minimization objective function as a secondary goal with a constant multiplier while the constraint $\sum_j \bar{x}_j^{(2)} w_j - \sum_j \bar{x}_j^{(1)} w_j \geq \delta$ is included, where $\bar{x}_j^{(k)} = \frac{1}{|G_k|} \sum_{i \in G_k} x_{ij} \ \forall j$, for $k = 1, 2$. The results of simulation study show that an MIP Model with the proposed secondary goal has better performance than other studied models.

Glen [42] proposes IP techniques for normalization in the two-group discriminant analysis models. One technique is to add the constraint $\sum_{j=1}^{m} |w_j| = 1$. In the proposed model, w_j for $j = 1, \ldots, m$ is represented by $w_j = w_j^+ - w_j^-$, where $w_j^+, w_j^- \geq 0$, and binary variables δ_j and γ_j are defined such that $\delta_j = 1 \Leftrightarrow w_j^+ \geq \epsilon$ and $\gamma_j = 1 \Leftrightarrow w_j^- \geq \epsilon$. The IP normalization technique is applied to MSD and MMD, and the MSD version is presented below.

MSD—with IP normalization

$$\text{Min} \quad \sum_i d_i$$

$$\text{s.t.} \quad w_0 + \sum_{j=1}^{m} x_{ij}(w_j^+ - w_j^-) - d_i \leq 0 \quad \forall i \in G_1$$

$$w_0 + \sum_{j=1}^{m} x_{ij}(w_j^+ - w_j^-) + d_i \geq 0 \quad \forall i \in G_2$$

$$\sum_{j=1}^{m} (w_j^+ + w_j^-) = 1$$

$$w_j^+ - \epsilon\delta_j \geq 0 \quad \forall j = 1, \ldots, m$$

$$w_j^+ - \delta_j \leq 0 \quad \forall j = 1, \ldots, m$$

$$w_j^- - \epsilon\gamma_j \geq 0 \quad \forall j = 1, \ldots, m$$

$$w_j^- - \gamma_j \leq 0 \quad \forall j = 1, \ldots, m$$

$$\delta_j + \gamma_j \leq 1 \quad \forall j = 1, \ldots, m$$

$$w_0 \text{ urs}$$

$$w_j^+, w_j^- \geq 0 \quad \forall j = 1, \ldots, m$$

$$d_i \geq 0 \quad \forall i$$

$$\delta_j, \gamma_j \in \{0, 1\} \quad \forall j = 1, \ldots, m$$

The variable coefficients of the discriminant function generated by the models are invariant under origin shifts. The proposed models are validated using two datasets from Refs. [45,87]. The models are also extended for attribute selection by adding the constraint $\sum_{j=1}^{m}(\delta_j + \gamma_j) = p$, which allows only a constant number, p, of attributes to be used for classification.

Glen [43] develops MIP Models that determine the thresholds for forming dichotomous variables as well as the discriminant function coefficients, w_j. For each continuous attribute to be formed as a dichotomous attribute, the model finds the threshold among possible thresholds while determining the separating hyperplane and optimizing the objective function such as MSD or minimizing the number of misclassifications. Computational results of a real dataset and some simulated datasets show that the MSD Model with dichotomous categorical variable formation can improve classification performance. The reason for the potential of this technique is that the generated linear discriminant function is a nonlinear function of the original variables.

1.2.2.2 Multigroup Classification

Gehrlein [41] proposes MIP formulations of minimizing the total number of misclassifications in the multigroup classification problem. He gives both a single function classification scheme and a multiple function classification scheme, as follows:

GSFC (General single function classification—minimizing the number of misclassifications)

$$\text{Min} \quad \sum_i z_i$$

$$\text{s.t.} \quad w_0 + \sum_j x_{ij} w_j - M z_i \leq U_k \quad \forall i \in G_k$$

$$w_0 + \sum_j x_{ij} w_j + M z_i \geq L_k \quad \forall i \in G_k$$

$$U_k - L_k \geq \delta' \quad \forall k$$

$$\left.\begin{aligned} L_g - U_k + My_{gk} &\geq \delta \\ L_k - U_g + My_{kg} &\geq \delta \\ y_{gk} + y_{kg} &= 1 \end{aligned}\right\} \quad \forall g, k,\ g \neq k$$

$$w_j \text{ urs} \quad \forall j$$

$$U_k, L_k \text{ urs} \quad \forall k$$

$$z_i \in \{0, 1\} \quad \forall i$$

$$y_{gk} \in \{0, 1\} \quad \forall g, k,\ g \neq k$$

where

U_k, L_k denote the upper and lower endpoints of the interval assigned to group k

$y_{gk} = 1$ if the interval associated with group g precedes that with group k

$y_{gk} = 0$ otherwise.

The constant δ' is the minimum width of an interval of a group and the constant δ is the minimum gap between adjacent intervals.

GMFC (General multiple function classification—minimizing the number of misclassifications)

$$\text{Min} \quad \sum_i z_i$$

$$\text{s.t.} \quad w_0^h + \sum_j x_{ij} w_j^h - w_0^k - \sum_j x_{ij} w_j^k + Mz_i \geq \epsilon \quad \forall i \in G_h,\ \forall h, k \neq h$$

$$w_j^k \text{ urs} \quad \forall j, k$$

$$z_i \in \{0, 1\} \quad \forall i$$

Both models work successfully on the iris dataset provided by Fisher [30].

Pavur [93] solves the multigroup classification problem by sequentially solving GSFC in one dimension each time. Linear discriminant functions are generated by successively solving GSFC with the added constraints that all linear discriminants are uncorrelated to each other for the total dataset. This procedure could be repeated for the number of dimensions that is believed to be enough. According to simulation results, this procedure substantially improves the GSFC Model and sometimes outperforms GMFC, Fisher's LDF, or Smith's QDF.

To solve the three-group classification problem more efficiently, Loucopoulos and Pavur [71] make a slight modification on GSFC and propose the model MIP3G, which also minimizes the number of misclassifications. Compared to GSFC, MIP3G is also a single-function classification model, but it reduces the possible group orderings from six to three in the formulation and thus becomes more efficient. Loucopoulos and Pavur [72] report the results of a simulation experiment on the performance of GMFC, MIG3G, Fisher's LDF, and Smith's QDF for three-group classification problem with small training samples. Second-order terms are also considered in the experiment. Simulation results show that GMFC and MIP3G can outperform the parametric procedures in some non-normal datasets and that the inclusion of second-order terms can improve the performance of MIP3G in some datasets. Pavur and Loucopoulos [95] investigate the effect of the gap size in the MIP3G Model for the three-group classification problem. A simulation study illustrates that for fairly separable data, or data with small sample sizes, a nonzero-gap model can improve the performance. A possible reason for this result is that the zero-gap model may be over-fitting the data.

Gallagher et al. [39,40] Lee et al. [63], and Lee [59,60] propose MIP Models, both heuristic and exact, as a computational approach to solving the constrained discriminant method described by Anderson [2]. These models are described in detail in Section 1.3.

1.2.3 Nonlinear Programming Classification Models

Nonlinear programming approaches are natural extensions for some of the LP-based models. Thus far, nonlinear programming approaches have been developed for two-group classification.

Stam and Joachimsthaler [108] propose a class of nonlinear programming methods to solve the two-group classification problem under the L_p-norm objective criterion. This is an extension of MSD and MMD, for which the objectives are the L_1-norm and L_∞-norm, respectively.

Minimize the general L_p-norm distance

$$\text{Min} \quad \left(\sum_i d_i^p \right)^{1/p}$$

$$\text{s.t.} \quad \sum_j x_{ij} w_j - d_i \le b \quad \forall i \in G_1$$

$$\sum_j x_{ij} w_j + d_i \ge b \quad \forall i \in G_2$$

$$w_j \text{ urs} \quad \forall j$$

$$d_i \geq 0 \quad \forall i$$

The simulation results show that, in addition to the L_1-norm and L_∞-norm, it is worth the effort to compute other L_p-norm objectives. Restricting the analysis to $1 \leq p \leq 3$, plus $p = \infty$, is recommended. This method is reviewed by Joachimsthaler and Stam [50] and Erenguc and Koehler [27].

Mangasarian et al. [85] propose a nonconvex model for the two-group classification problem:

$$\text{Min} \quad d^1 + d^2$$

$$\text{s.t.} \quad \sum_j x_{ij} w_j - d^1 \leq 0 \quad \forall i \in G_1$$

$$\sum_j x_{ij} w_j + d^2 \geq 0 \quad \forall i \in G_2$$

$$\max_{j=1,\ldots,m} |w_j| = 1$$

$$w_j \text{ urs} \quad \forall j$$

$$d^1, d^2 \text{ urs}$$

This model can be solved in polynomial-time by solving $2m$ linear programs, which generate a sequence of parallel planes, resulting in a piecewise-linear nonconvex discriminant function. The model works successfully in clinical practice for the diagnosis of breast cancer.

Further, Mangasarian [76] also formulates the problem of minimizing the number of misclassifications as a linear program with equilibrium constraints (LPEC) instead of the MIP Model MM described previously.

MM–LPEC (Minimizing the number of misclassifications – linear program with equilibrium constraints)

$$\text{Min} \quad \sum_{i \in G_1 \cup G_2} z_i$$

$$\text{s.t.} \quad w_0 + \sum_j x_{ij} w_j - d_i \leq -1 \quad \forall i \in G_1$$

$$z_i(w_0 + \sum_j x_{ij}w_j - d_i + 1) = 0 \quad \forall i \in G_1$$

$$w_0 + \sum_j x_{ij}w_j + d_i \geq 1 \quad \forall i \in G_2$$

$$z_i\left(w_0 + \sum_j x_{ij}w_j + d_i - 1\right) = 0 \quad \forall i \in G_2$$

$$d_i(1 - z_i) = 0 \quad \forall i \in G_1 \cup G_2$$

$$0 \leq z_i \leq 1 \quad \forall i \in G_1 \cup G_2$$

$$d_i \geq 0 \quad \forall i \in G_1 \cup G_2$$

$$w_j \text{ urs} \quad \forall j$$

The general LPEC can be converted to an exact penalty problem with a quadratic objective and linear constraints. A stepless Frank–Wolfe-type algorithm is proposed for the penalty problem, terminating at a stationary point or a global solution. This method is called the parametric misclassification minimization (PMM) procedure, and numerical testing is included in Ref. [77].

To illustrate the next model, we first define the step function $s : R \to \{0, 1\}$ as

$$s(u) = \begin{cases} 1 & \text{if } u > 0 \\ 0 & \text{if } u \leq 0. \end{cases}$$

The problem of minimizing the number of misclassifications is equivalent to

$$\text{Min} \quad \sum_{i \in G_1 \cup G_2} s(d_i)$$

$$\text{s.t.} \quad w_0 + \sum_j x_{ij}w_j - d_i \leq -1 \quad \forall i \in G_1$$

$$w_0 + \sum_j x_{ij}w_j + d_i \geq 1 \quad \forall i \in G_2$$

$$d_i \geq 0 \quad \forall i \in G_1 \cup G_2$$

$$w_j \text{ urs} \quad \forall j$$

Mangasarian [77] proposes a simple concave approximation of the step function for nonnegative variables: $t(u, \alpha) = 1 - e^{-\alpha u}$, where $\alpha > 0, u \geq 0$. Let $\alpha > 0$ and approximate $s(d_i)$ by $t(d_i, \alpha)$. The problem then reduces to minimizing a smooth concave function bounded below on a nonempty polyhedron, which has a minimum at a vertex of the feasible region. A finite successive linearization algorithm (SLA) is proposed, terminating at a stationary point or a global solution. Numerical tests of SLA are done and compared with the PMM procedure described above. The results show that the much simpler SLA obtains a separation that is almost as good as PMM in considerably less computing time.

Chen and Mangasarian [21] propose an algorithm on a defined hybrid misclassification minimization problem, which is more computationally tractable than the NP-hard misclassification minimization problem. The basic idea of the hybrid approach is to obtain iteratively w_0 and $(w_1, \ldots, w_m)$ of the separating hyperplane: (1) for a fixed w_0, solve RLP [9] to determine $(w_1, \ldots, w_m)$; and (2) for this $(w_1, \ldots, w_m)$, solve the one-dimensional misclassification minimization problem to determine w_0. Comparison of the hybrid method is made with the RLP method and the PMM procedure. The hybrid method performs better in the testing sets of the 10-fold cross-validation and is much faster than PMM.

Mangasarian [78] proposes the model of minimizing the sum of arbitrary-norm distances of misclassified points to the separating hyperplane. For a general norm $||\cdot||$ on R^m, the dual norm $||\cdot||'$ on R^m is defined as $||x||' = \max_{||y||=1} x^T y$. Define $[a]^+ = \max\{0, a\}$ and let $w = (w_1, \ldots, w_m)$. The formulation can then be written as

$$\text{Min} \quad \sum_{i \in G_1} \Big[w_0 + \sum_j x_{ij} w_j \Big]^+ + \sum_{i \in G_2} \Big[-w_0 - \sum_j x_{ij} w_j \Big]^+$$

$$\text{s.t.} \quad ||w||' = 1$$

$$w_0, w \text{ urs}$$

The problem is to minimize a convex function on a unit sphere. A related decision problem to this minimization problem is shown to be NP-complete, except for $p = 1$. For a general p-norm, the minimization problem can be transformed via an exact penalty formulation to minimizing the sum of a convex function and a bilinear function on a convex set.

1.2.4 Support Vector Machine

An SVM is a type of mathematical programming approach [112]. It has been widely studied, and has become popular in many application fields in recent years. The introductory description of SVMs given here is summarized from the tutorial by Burges [20]. To maintain consistency with SVM studies in published literature, the notation used below is slightly different than the notation used to describe the mathematical programming methods in earlier sections.

In the two-group separable case, the objective function is to maximize the margin of a separating hyperplane, $2/||w||$, which is equivalent to minimizing $||w||^2$.

$$\begin{aligned} \text{Min} \quad & w^T w \\ \text{s.t.} \quad & x_i^T w + b \geq +1 \quad \text{for } y_i = +1 \\ & x_i^T w + b \leq -1 \quad \text{for } y_i = -1 \\ & w, b \text{ urs} \end{aligned}$$

where

$x_i \in R^m$ represents the values of attributes of observation i
$y_i \in \{-1, 1\}$ represents the group of observation i.

This problem can be solved by solving its Wolfe dual problem.

$$\begin{aligned} \text{Max} \quad & \sum_i \alpha_i - \frac{1}{2} \sum_{i,j} \alpha_i \alpha_j y_i y_j x_i^T x_j \\ \text{s.t.} \quad & \sum_i \alpha_i y_i = 0 \\ & \alpha_i \geq 0 \quad \forall i \end{aligned}$$

Here, α_i is the Lagrange multiplier for the training point i, and the points with $\alpha_i > 0$ are called the support vectors (analogous to the support of a hyperplane, and thus the introduction of the name support vector). The primal solution w is given by $w = \sum_i \alpha_i y_i x_i$. b can be computed by solving $y_i(w^T x_i + b) - 1 = 0$ for any i with $\alpha_i > 0$.

For the non-separable case, slack variables ξ_i are introduced to handle the errors. Let C be the penalty for the errors. The problem becomes

$$\text{Min} \quad \frac{1}{2}w^T w + C\left(\sum_i \xi_i\right)^k$$

$$\text{s.t.} \quad x_i^T w + b \geq +1 - \xi_i \quad \text{for } y_i = +1$$

$$x_i^T w + b \leq -1 + \xi_i \quad \text{for } y_i = -1$$

$$w, b \text{ urs}$$

$$\xi_i \geq 0 \quad \forall i$$

When k is chosen to be 1, neither the ξ_i's nor their Lagrange multipliers appear in the Wolfe dual problem.

$$\text{Max} \quad \sum_i \alpha_i - \frac{1}{2}\sum_{i,j} \alpha_i \alpha_j y_i y_j x_i^T x_j$$

$$\text{s.t.} \quad \sum_i \alpha_i y_i = 0$$

$$0 \leq \alpha_i \leq C \quad \forall i$$

The data points can be separated nonlinearly by mapping the data into some higher dimensional space and applying linear SVM to the mapped data. Instead of knowing explicitly the mapping Φ, SVM needs only the dot products of two transformed data points $\Phi(x_i)\cdot\Phi(x_j)$. The kernel function K is introduced such that $K(x_i, x_j) = \Phi(x_i)\cdot\Phi(x_j)$. Replacing $x_i^T x_j$ by $K(x_i, x_j)$ in the above problem, the separation becomes nonlinear while the problem to be solved remains a quadratic program. In testing a new data point x after training, the sign of the function $f(x)$ is computed to determine the group of x:

$$f(x) = \sum_{i=1}^{N_S} \alpha_i y_i \Phi(s_i)\cdot\Phi(x) + b = \sum_{i=1}^{N_S} \alpha_i y_i K(s_i, x) + b$$

where

s_i's are the support vectors

N_s is the number of support vectors. Again the explicit form of $\Phi(x)$ is avoided.

Mangasarian provides a general mathematical programming framework for SVM, called generalized support vector machine (GSVM) [79,83]. Special cases can be derived from GSVM, including the standard SVM.

Many SVM-type methods have been developed by Mangasarian and other authors to solve huge-sized classification problems more efficiently. These methods include successive overrelaxation for SVM [82], proximal SVM [36,38], smooth SVM [68], reduced SVM [67], Lagrangian SVM [84], incremental SVMs [37], and other methods [13,81]. Mangasarian summarizes some of the developments in Ref. [80]. Examples of applications of SVM include breast cancer studies [69,70] and genome research [73].

Hsu and Lin [49] compare different methods for multigroup classification using SVMs. Three methods studied are based on several binary classifiers: one-against-one, one-against-all, and directed acyclic graph (DAG) SVM. The other two methods studied are altogether methods with decomposition implementation. The experiment results show that the one-against-one and DAG methods are more suitable for practical use than the other methods. Lee et al. [66] propose a generic approach to multigroup problems with some theoretical properties, and the proposed method is well applied to microarray data for cancer classification and satellite radiance profiles for cloud classification.

Gallagher et al. [39,40] and Lee et al. [63] offer the first discrete SVM for multigroup classification with reserved judgement. The approach has been successfully applied to a diverse variety of biological and medical applications (see Section 1.3).

1.3 MIP-Based Multigroup Classification Models and Applications to Medicine and Biology

Commonly used methods for classification, such as linear discriminant functions, decision trees, mathematical programming approaches, SVMs, and artificial neural networks (ANN), can be viewed as attempts at approximating a Bayes optimal rule for classification; that is, a rule that maximizes (minimizes) the total probability of correct classification (misclassification). Even if a Bayes optimal rule is known, intergroup misclassification rates may be higher than desired. For example, in a population that is mostly healthy, a Bayes optimal rule for medical diagnosis might misdiagnose sick patients as healthy to maximize total probability of correct diagnosis. As a remedy, a constrained discriminant rule that limits the misclassification rate is appealing.

Assuming that the group density functions and prior probabilities are known, Anderson [2] showed that an optimal rule for the problem of maximizing the probability of correct classification subject to constraints on the misclassification probabilities must be of a specific form when discriminating among multiple groups with a simplified model. The formulae in Anderson's result depend on a set of parameters satisfying a complex relationship between the density functions, the prior probabilities, and the bounds on the misclassification probabilities. Establishing a viable mathematical model to describe Anderson's result, and finding values for these parameters that yield an optimal rule are challenging tasks. The first computational models utilizing Anderson's formulae were proposed in Refs. [39,40].

1.3.1 Discrete Support Vector Machine Predictive Models

As part of the work carried out at Georgia Institute of Technology's Center for Operations Research in Medicine, we have developed a general-purpose discriminant analysis modeling framework and computational engine that are applicable to a wide variety of applications, including biological, biomedical, and logistics problems. Utilizing the technology of large-scale discrete optimization and SVMs, we have developed novel classification models that simultaneously include the following features: (1) the ability to classify any number of distinct groups; (2) the ability to incorporate heterogeneous types of attributes as input; (3) a high-dimensional data transformation that eliminates noise and errors in biological data; (4) constraints to limit the rate of misclassification, and a reserved-judgment region that provides a safeguard against overtraining (which tends to lead to high misclassification rates from the resulting predictive rule); and (5) successive multistage classification capability to handle data points placed in the reserved-judgment region. Studies involving tumor volume identification, ultrasonic cell disruption in drug delivery, lung tumor cell motility analysis, CpG island aberrant methylation in human cancer, predicting early atherosclerosis using biomarkers, and fingerprinting native and angiogenic microvascular networks using functional perfusion data indicate that our approach is adaptable and can produce effective and reliable predictive rules for various biomedical and biobehavioral phenomena [16,28,29,56,57,59,60,64,65].

On the basis of the description in Refs. [39,40,59,60,63], we summarize below some of the classification models we have developed.

1.3.1.1 Modeling of Reserved-Judgment Region for General Groups

When the population densities and prior probabilities are known, the constrained rules with a reject option (reserved judgment), based on Anderson's results, call for finding a partition $\{R_0, \ldots, R_G\}$ of $\mathbb{R}^k$ that maximizes the probability of correct allocation subject to constraints on the misclassification probabilities; i.e.,

$$\text{Max} \quad \sum_{g=1}^{G} \pi_g \int_{\mathrm{R_g}} f_g(w)\,\mathrm{d}w \tag{1.1}$$

$$\text{s.t.} \quad \int_{\mathrm{R_g}} f_h(w)\mathrm{d}w \leq \alpha_{hg}, \quad h, g = 1, \ldots, G, h \neq g \tag{1.2}$$

where
f_h, $h \in \{1, \ldots, G\}$, are the group conditional density functions
π_g denotes the prior probability that a randomly selected entity is from group g, $g \in \{1, \ldots, G\}$
α_{hg}, $h \neq g$, are constants between zero and one.

Under quite general assumptions, it was shown that there exist unique (up to a set of measure zero) nonnegative constants λ_{ih}, i, $h \in \{1, \ldots, G\}$, $i \neq h$, such that the optimal rule is given by

$$R_g = \{x \in \mathbb{R}^k : L_g(x) = \max_{h \in \{0,1,\ldots,G\}} L_h(x)\}, \quad g = 0, \ldots, G \tag{1.3}$$

where

$$L_0(x) = 0 \tag{1.4}$$

$$L_h(x) = \pi_h f_h(x) - \sum_{i=1, i \neq h}^{G} \lambda_{ih} f_i(x), \quad h = 1, \ldots, G \tag{1.5}$$

For $G = 2$, the optimal solution can be modeled rather straightforward. However, finding optimal λ_{ih}'s for the general case, $G \geq 3$, is a difficult problem, with the difficulty increasing as G increases. Our model offers an avenue for modeling and finding the optimal solution in the general case. It is the first such model to be computationally viable [39,40].

Before proceeding, we note that R_g can be written as $R_g = \{x \in \mathbb{R}^k : L_g(x) \geq L_h(x)$ for all $h \in \{0, \ldots, G\}$. So, because $L_g(x) \geq L_h(x)$ if, and only if, $(1/\sum_{t=1}^{G} f_t(x))L_g(x) \geq (1/\sum_{t=1}^{G} f_t(x))L_h(x)$, the functions L_h, $h = 1, \ldots G$, can be redefined as

$$L_h(x) = \pi_h p_h(x) - \sum_{i=1, i \neq h}^{G} \lambda_{ih} p_i(x), \quad h = 1, \ldots, G \tag{1.6}$$

where $p_i(x) = f_i(x)/\sum_{t=1}^{G} f_t(x)$. We assume that L_h is defined as in Equation 1.6 in our model.

1.3.1.2 Mixed Integer Programming Formulations

Assume that we are given a training sample of N entities whose group classifications are known; say n_g entities are in group g, where $\sum_{g=1}^{G} n_g = N$. Let the k dimensional vectors x^{gj}, $g = 1, \ldots, G$, $j = 1, \ldots, n_g$, contain the measurements on k available characteristics of the entities. Our procedure for deriving a discriminant rule proceeds in two stages. The first stage is to use the training sample to compute estimates, $\hat{f}_h$, either parametrically or nonparametrically, of the density functions f_h (see Ref. [89]) and estimates, $\hat{\pi}_h$, of the prior probabilities π_h, $h = 1, \ldots, G$. The second stage is to determine the optimal λ_{ih}'s given these estimates. This stage requires being able to estimate the probabilities of correct classification and misclassification for any candidate set of λ_{ih}'s. One could, in theory, substitute the estimated densities and prior probabilities into Equation 1.5, and directly use the resulting regions R_g in the integral expressions given in Equations 1.1 and 1.2. This would involve, even in simple cases such as normally distributed groups, the numerical evaluation of k-dimensional integrals at each step of a search for the optimal λ_{ih}'s. Therefore, we have designed an alternative approach. After substituting the $\hat{f}_h$'s and $\hat{\pi}_h$'s into Equation 1.5, we simply calculate the proportion of training sample points that fall in each of the regions $R_1, \ldots, R_G$. The MIP Models discussed below attempt to maximize the proportion of training sample points correctly classified although satisfying constraints on the proportions of training sample points misclassified. This approach has two advantages. First, it avoids having to evaluate the potentially difficult integrals in Equations 1.1 and 1.2. Second, it is nonparametric in controlling the training sample misclassification probabilities. That is, even if the densities are poorly estimated (by assuming, for example, normal densities for non-normal data), the constraints are still satisfied for the training sample. Better estimates of the densities may allow a higher correct classification rate to be achieved, but the constraints will be satisfied even if poor estimates are used. Unlike most SVM Models that minimize the sum of errors, our objective is driven by the number of correct classifications, and will not be biased by the distance of the entities from the supporting hyperplane.

A word of caution is in order. In traditional unconstrained discriminant analysis, the true probability of correct classification of a given discriminant rule tends to be smaller than the rate of correct classification for the training sample from which it was derived. One would expect to observe such an effect for the method described herein as well. In addition, one would expect to observe an analogous effect with regard to constraints on misclassification probabilities—the true probabilities are likely to be greater than any limits imposed on the proportions of training sample misclassifications. Hence, the α_{hg} parameters should be carefully chosen for the application in hand.

Our first model is a nonlinear 0/1 MIP Model with the nonlinearity appearing in the constraints. Model 1 maximizes the number of correct classifications of the given N training entities. Similarly, the constraints on the misclassification probabilities are modeled by ensuring that the number of group g training entities in region R_h is less than or equal to a prespecified percentage, $\alpha_{hg}(0 < \alpha_{hg} < 1)$, of the total number, n_g, of group g entities, $h,\ g \in \{1, \dots, G\},\ h \neq g$.

For notational convenience, let $\mathbf{G} = \{1, \dots, G\}$ and $\mathbf{N}_g = \{1, \dots, n_g\}$, for $g \in \mathbf{G}$. Also, analogous to the definition of p_i, define $\hat{p}_i$ by $\hat{p}_i = \hat{f}_i(x)/\sum_{t=1}^{G}\hat{f}_t(x)$. In our model, we use binary indicator variables to denote the group classification of entities. Mathematically, let u_{hgj} be a binary variable indicating whether or not x^{gj} lies in region R_h; i.e., whether or not the jth entity from group g is allocated to group h. Then Model 1 can be written as follows:

DAMIP

$$\text{Max} \sum_{g \in G} \sum_{j \in N_g} u_{ggj}$$

s.t.

$$L_{hgj} = \hat{\pi}_h \hat{p}_h(x^{gj}) - \sum_{i \in G \backslash h} \lambda_{ih} \hat{p}_i(x^{gj}), \quad h, g \in \mathbf{G},\ j \in \mathbf{N}_g \tag{1.7}$$

$$y_{gj} = \max\{0, L_{hgj} : h = 1, \dots, G\}, \quad g \in \mathbf{G},\ j \in \mathbf{N}_g \tag{1.8}$$

$$y_{gj} - L_{ggj} \leq M(1 - u_{ggj}), \quad g \in \mathbf{G},\ j \in \mathbf{N}_g \tag{1.9}$$

$$y_{gj} - L_{hgj} \geq \varepsilon(1 - u_{hgj}), \quad h, g \in \mathbf{G},\ j \in \mathbf{N}_g,\ h \neq g \tag{1.10}$$

$$\sum_{j \in N_g} u_{hgj} \leq \lfloor \alpha_{hg} n_g \rfloor, \quad h, g \in \mathbf{G},\ h \neq g \tag{1.11}$$

$$-\infty < L_{hgj} < \infty, \quad y_{gj} \geq 0, \quad \lambda_{ih} \geq 0, \quad u_{hgj} \in \{0, 1\}$$

Constraint given in Equation 1.7 defines the variable L_{hgj} as the value of the function L_h evaluated at x^{gj}. Therefore, the continuous variable y_{gj}, defined in constraint given by Equation 1.8, represents $\max\{L_h(x^{gj}) : h = 0, \dots, G\}$; and consequently, x^{gj} lies in region R_h if, and only if, $y_{gj} = L_{hgj}$. The binary variable u_{hgj} is used to indicate whether or not x^{gj} lies in region R_h; i.e., whether or not the jth entity from group g is allocated to group h. In particular, constraint given by Equation 1.9, together with the objective, force u_{ggj} to be 1 if, and only if, the jth entity from group g is correctly allocated to group g; and constraints given by Equations 1.10 and 1.11 ensure that at most $\lfloor \alpha_{hg} n_g \rfloor$ (i.e., the greatest integer less than or equal to $\alpha_{hg} n_g$) group g entities are allocated to group h, $h \neq g$. One caveat regarding the indicator variables u_{hgj} is that although the condition $u_{hgj} = 0$, $h \neq g$, implies (by Equation 1.10) that $x^{gj} \notin R_h$, the converse need not hold. As a consequence, the number of misclassifications may be overcounted. However, in our preliminary numerical study we found that the actual amount of overcounting is minimal. One could force the converse (thus, $u_{hgj} = 1$ if, and only if, $x^{gj} \in R_h$) by adding constraints $y_{gj} - L_{hgj} \leq M(1 - u_{hgj})$, for example. Finally, we note that the parameters M and ε are extraneous to the discriminant analysis problem itself, but are needed in the model to control the indicator variables u_{hgj}. The intention is for M and ε to be large and small positive constants, respectively.

1.3.1.3 Model Variations

We explore different variations in the model to grasp the quality of the solution and the associated computational effort.

A first variation involves transforming Model 1 to an equivalent linear mixed integer model. In particular, Model 2 replaces the N constraints defined in Equation 1.8 with the following system of $3GN + 2N$ constraints:

$$y_{gj} \geq L_{hgj}, \quad h, g \in \mathbf{G},\ j \in \mathbf{N}_g \tag{1.12}$$

$$\tilde{y}_{hgj} - L_{hgj} \leq M(1 - v_{hgj}), \quad h, g \in \mathbf{G},\ j \in \mathbf{N}_g \tag{1.13}$$

$$\tilde{y}_{hgj} \leq \hat{\pi}_h \hat{p}_h(x^{gj}) v_{hgj}, \quad h, g \in \mathbf{G},\ j \in \mathbf{N}_g \tag{1.14}$$

$$\sum_{h \in G} v_{hgj} \leq 1, \quad g \in \mathbf{G},\ j \in \mathbf{N}_g \tag{1.15}$$

$$\sum_{h \in G} \tilde{y}_{hgj} = y_{gj}, \quad g \in \mathbf{G},\ j \in \mathbf{N}_g \tag{1.16}$$

where $\tilde{y}_{hgj} \geq 0$ and $v_{hgj} \in \{0, 1\}$, $h, g \in \mathbf{G}, j \in \mathbf{N}_g$. These constraints, together with the non-negativity of y_{gj} force $y_{gj} = \max\{0, L_{hgj} : h = 1, \ldots, G\}$.

The second variation involves transforming Model 1 to a heuristic linear MIP Model. This is done by replacing the nonlinear constraint (Equation 1.8) with $y_{gj} \geq L_{hgj}$, $h, g \in \mathbf{G}, j \in \mathbf{N}_g$, and including penalty terms in the objective function. In particular, Model 3 has the objective

$$\text{Max} \sum_{g \in G} \sum_{j \in N_g} \beta u_{ggj} - \sum_{g \in G} \sum_{j \in N_g} \gamma y_{gj}$$

where β and γ are positive constants. This model is heuristic in that there is nothing to force $y_{gj} = \max\{0, L_{hgj} : h = 1, \ldots, G\}$. However, because in addition to trying to force as many u_{ggj}'s to one as possible, the objective in Model 3 also tries to make the y_{gj}'s as small as possible, and the optimizer tends to drive y_{gj} toward $\max\{0, L_{hgj} : h = 1, \ldots, G\}$. We remark that β and γ could be stratified by group (i.e., introduce possibly distinct $\beta_g, \gamma_g, g \in \mathbf{G}$) to model the relative importance of certain groups to be correctly classified.

A reasonable modification to Models 1, 2, and 3 involves relaxing the constraints specified by Equation 1.11. Rather than placing restrictions on the number of type g training entities classified into group h, for all h, $g \in \mathbf{G}, h \neq g$, one could simply place an upper bound on the total number of misclassified training entities. In this case, the $G(G-1)$ constraints specified by Equation 1.11 would be replaced by the single constraint

$$\sum_{g \in G} \sum_{h \in G \setminus \{g\}} \sum_{j \in N_g} u_{hgj} \leq \lfloor \alpha N \rfloor \tag{1.17}$$

where α is a constant between 0 and 1. We will refer to Models 1, 2, and 3, modified in this way, as Models 1T, 2T, and 3T, respectively. Of course, other modifications are also possible. For instance, one could place restrictions on the total number of type g points misclassified for each $g \in \mathbf{G}$. Thus, in place of the constraints specified in Equation 1.17, one would include the constraints $\sum_{h \in G \setminus \{g\}} \sum_{j \in N_g} u_{hgj} \leq \lfloor \alpha_g N \rfloor$, $g \in \mathbf{G}$, where $0 < \alpha_g < 1$.

We also explore a heuristic linear model of Model 1. In particular, consider the linear program (DALP):

$$\text{Max} \quad \sum_{g \in G} \sum_{j \in N_g} (c_1 w_{gj} + c_2 y_{gj}) \tag{1.18}$$

s.t.

$$L_{hgj} = \pi_h \hat{p}_h(x^{gj}) - \sum_{i \in G \setminus h} \lambda_{ih} \hat{p}_i(x^{gj}), \quad h, g \in \mathbf{G},\ j \in \mathbf{N}_g \tag{1.19}$$

$$L_{ggj} - L_{hgj} + w_{gj} \geq 0, \quad h, g \in \mathbf{G},\ h \neq g,\ j \in \mathbf{N}_g \tag{1.20}$$

$$L_{ggj} + w_{gj} \geq 0, \quad g \in \mathbf{G},\ j \in \mathbf{N}_g, \tag{1.21}$$

$$-L_{hgj} + y_{gj} \geq 0, \quad h, g \in \mathbf{G},\ j \in \mathbf{N}_g \tag{1.22}$$

$$-\infty < L_{hgj} < \infty,\ w_{gj},\ y_{gj},\ \lambda_{ih} \geq 0$$

Constraint given by Equation 1.19 defines the variable L_{hgj} as the value of the function L_h evaluated at x^{gj}. As the optimization solver searches through the set of feasible solutions, the λ_{ih} variables will vary, causing the L_{hgj} variables to assume different values. Constraints given by Equation 1.20 through 1.22 link the objective-function variables with the L_{hgj} variables in such a way that correct classification of training entities, and allocation of training entities into the reserved-judgment region are captured by the objective-function variables. In particular, if the optimization solver drives w_{gj} to zero for some g,j pair, then constraints given by Equations 1.20 and 1.21 imply that $L_{ggj} = \max\{0, L_{hgj} : h \in \mathbf{G}\}$. Hence, the jth entity from group g is correctly classified. If, on the other hand, the optimal solution yields $y_{gj} = 0$ for some g,j pair, then constraint given by Equation 1.22 implies that $\max\{0, L_{hgj} : h \in \mathbf{G}\} = 0$. Thus, the jth entity from group g is placed in the reserved-judgment region. (Of course, it is possible for both w_{gj} and y_{gj} to be zero. One should decide before solving the linear program how to interpret the classification in such cases.) If both w_{gj} and y_{gj} are positive, the jth entity from group g is misclassified.

The optimal solution yields a set of λ_{ih}'s that best allocates the training entities (i.e., best in terms of minimizing the penalty objective function). The optimal λ_{ih}'s can then be used to define the functions L_h, $h \in G$, which in turn can be used to classify a new entity with feature vector $x \in \mathbb{R}^k$ by simply computing the index at which $\max\{L_h(x) : h \in \{0, 1, \ldots, G\}\}$ is achieved.

Note that Model DALP places no a priori bound on the number of misclassified training entities. However, because the objective is to minimize a weighted combination of the variables w_{gj} and y_{gj}, the optimizer will attempt to drive these variables to zero. Thus, the optimizer is, in essence, attempting either to correctly classify training entities ($w_{gj} = 0$), or to place them in the reserved-judgment region ($y_{gj} = 0$). By varying the weights c_1 and c_2, one has a means of controlling the optimizer's emphasis for correctly classifying training entities versus placing them in the reserved-judgment region. If $c_2/c_1 < 1$, the optimizer will tend to place a greater emphasis on driving the w_{gj} variables to zero than driving the y_{gj} variables to zero (conversely, if $c_2/c_1 > 1$). Hence, when $c_2/c_1 < 1$, one should expect to get relatively more entities correctly classified, fewer placed in the reserved-judgment region, and more misclassified, than when $c_2/c_1 > 1$. An extreme case is when $c_2 = 0$. In this case, there is no emphasis on driving y_{gj} to zero (the reserved-judgment region is thus ignored), and the full emphasis of the optimizer is to drive w_{gj} to zero.

Table 1.1 Model Size

Model	Type	Constraints	Total Variables	0/1 Variables
1	Nonlinear MIP	$2GN + N + G(G-1)$	$2GN + N + G(G-1)$	GN
2	Linear MIP	$5GN + 2N + G(G-1)$	$4GN + N + G(G-1)$	$2GN$
3	Linear MIP	$3GN + G(G-1)$	$2GN + N + G(G-1)$	GN
1T	Nonlinear MIP	$2GN + N + 1$	$2GN + N + G(G-1)$	GN
2T	Linear MIP	$5GN + 2N + 1$	$4GN + N + G(G-1)$	$2GN$
3T	Linear MIP	$3GN + 1$	$2GN + N + G(G-1)$	GN
DALP	Linear program	$3GN$	$NG + N + G(G-1)$	0

Table 1.1 summarizes the number of constraints, the total number of variables, and the number of 0/1 variables in each of the discrete SVM Models, and in the heuristic LP Model (DALP). Clearly, even for moderately sized discriminant analysis problems, the MIP instances are relatively large. Also, note that Model 2 is larger than Model 3, both in terms of the number of constraints and the number of variables. However, it is important to bear in mind that the difficulty of solving an MIP problem cannot, in general, be predicted solely by its size; problem structure has a direct and substantial bearing on the effort required to find optimal solutions. The LP relaxation of these MIP Models pose computational challenges as commercial LP solvers return (optimal) LP solutions that are infeasible, due to the equality constraints, and the use of big M and small ε in the formulation.

It is interesting to note that the set of feasible solutions for Model 2 is tighter than that for Model 3. In particular, if F_i denotes the set of feasible solutions of Model i, then

$$F_1 = \{(L, \lambda, u, y) : \text{ there exists } \tilde{y}, v \text{ such that } (L, \lambda, u, y, \tilde{y}, v) \in F_2\} \subsetneq F_3 \quad (1.23)$$

The novelties of the classification models developed herein include (1) they are suitable for discriminant analysis given any number of groups; (2) they accept heterogeneous types of attributes as input; (3) they use a parametric approach to reduce high-dimensional attribute spaces; and (4) they allow constraints on the number of misclassifications, and utilize a reserved judgment to facilitate the reduction of misclassifications. The latter point opens the possibility of performing multistage analysis.

Clearly, the advantage of an LP Model over an MIP Model is that the associated problem instances are computationally much easier to solve. However, the most important criterion in judging a method for obtaining discriminant rules is how the rules perform in correctly classifying new unseen entities. Once the rule is developed, applying it to a new entity to determine its group is trivial. Extensive computational experiments have been performed to gauge the qualities of solutions of different models [17,18,40,59,60,63].

1.3.1.4 Validation of Model and Computational Effort

We performed 10-fold cross-validation, and designed simulation and comparison studies on our models. Results reported in [40,63] demonstrate that our approach works well when applied to both simulated data and datasets from the machine learning database repository [91]. In particular, our methods compare favorably and at times superior to other mathematical programming methods, including the general single function classification (GSFC) model by Gehrlein [41], and the LP Model by Gochet et al. [46], as well as Fisher's LDF, artificial neural networks, quadratic discriminant analysis, tree classification, and other SVMs, on real biological and medical data.

1.3.2 Classification Results on Real-World Biological and Medical Applications

The main objective in discriminant analysis is to derive rules that can be used to classify entities into groups. Computationally, the challenge lies in the effort expended to develop such a rule. Once the rule is developed, applying it to a new entity to determine its group is trivial. Feasible solutions obtained from our classification models correspond to predictive rules. Empirical results [40,63] indicate that the resulting classification model instances are computationally very challenging, and even intractable by competitive commercial MIP solvers. However, the resulting predictive rules prove to be very promising, offering correct classification rates on new unknown data ranging from 80 to 100 percent on various types of biological/medical problems. Our results indicate that the general-purpose classification framework that we have designed has the potential to be a very powerful predictive method for clinical settings.

The choice of MIP as the underlying modeling and optimization technology for our SVM classification model is guided by the desire to simultaneously incorporate a variety of important and desirable properties of predictive models within a general framework. MIP itself allows for incorporation of continuous and discrete variables, and linear and nonlinear constraints, providing a flexible and powerful modeling environment.

Our mathematical modeling and computational algorithm design shows great promise as the resulting predictive rules are able to produce higher rates of correct classification on new biological data (with unknown group status) compared to existing classification methods. This is partly due to the transformation of raw data via the set of constraints given in Equation 1.7. Although most mathematical programming approaches directly determine the hyperplanes of separation using raw data, our approach transforms the raw data via a probabilistic model, before the determination of the supporting hyperplanes. Further, the separation is driven by maximizing the sum of binary variables (representing correct classification or not of entities), instead of maximizing the margins between groups, or minimizing a sum of errors (representing distances of entities from hyperplanes), as in other SVMs. The combination of these two strategies offers better classification capability. Noise in the transformed data is not as profound as in raw data. And the magnitudes of the errors do not skew the determination of the separating hyperplanes, as all entities have equal importance when correct classification is being counted.

To highlight the broad applicability of our approach, below, we briefly summarize the application of our predictive models and solution algorithms to ten different biological problems. Each of the projects was carried out in close partnership with experimental biologists or clinicians. Applications to finance and other industry applications are described elsewhere [18,40,63].

1.3.2.1 Determining the Type of Erythemato-Squamous Disease

The differential diagnosis of erythemato-squamous diseases is an important problem in dermatology [60]. They all share the clinical features of erythema and scaling, with very little differences. The six groups are psoriasis, seboreic dermatitis, lichen planus, pityriasis rosea, cronic dermatitis, and pityriasis rubra pilaris. Usually, a biopsy is necessary for the diagnosis but unfortunately these diseases share many histopathological features as well. Another difficulty for the differential diagnosis is that a disease may show the features of another disease at the beginning stage and may have the characteristic features at the following stages [91].

The six groups consist of 366 subjects (112,61,72,49,52,20, respectively) with 34 clinical attributes. Patients were first evaluated clinically with 12 features. Later, skin samples were taken for the evaluation of 22 histopathological features. The values of the histopathological features are determined by an analysis of the samples under a microscope. The 34 attributes include (1) clinical attributes: erythema, scaling, definite borders, itching, koebner phenomenon, polygonal papules, follicular papules, oral mucosal involvement, knee and elbow involvement, scalp involvement, family history, and age; and (2) histopathological attributes: melanin incontinence, eosinophils in the infiltrate, PNL infiltrate, fibrosis of the papillary dermis, exocytosis, acanthosis, hyperkeratosis, parakeratosis, clubbing of the rete ridges, elongation of the rete ridges, thinning of the suprapapillary epidermis, spongiform pustule, munro microabscess, focal hypergranulosis, disappearance of the granular layer, vacuolization and damage of basal layer, spongiosis, sawtooth appearance of retes, follicular horn plug, perifollicular parakeratosis, inflammatory infiltrate, monoluclear and band-like infiltrate.

Our multigroup classification model selected 27 discriminatory attributes, and successfully classified the patients into six groups, each with an unbiased correct classification of greater than 93 percent (with 100 percent correct rate for groups 1,3,5,6) with an average overall accuracy of 98 percent. Using 250 subjects to develop the rule, and testing the remaining 116 patients, we obtain a prediction accuracy of 91 percent.

1.3.2.2 Predicting Presence/Absence of Heart Disease

The four databases concerning heart disease diagnosis were collected by Dr. Janosi of Hungarian Institute of Cardiology, Budapest; Dr. Steinbrunn of University Hospital, Zurich; Dr. Pfisterer of University Hospital, Basel, Switzerland; and Dr. Detrano of V.A. Medical Center, Long Beach and Cleveland Clinic Foundation [60]. Each database contains the same 76 attributes. The goal field refers to the presence of heart disease in the patient. The classification attempts to distinguish presence (values 1,2,3,4, involving a total of 509 subjects) from absence (value 0, involving 411 subjects) [91]. The attributes include demographics, physio-cardiovascular conditions, traditional risk factors, family history, personal lifestyle, and cardiovascular exercise measurements. This dataset has posed some challenges to past analysis via various classification approaches, resulting in less than 80 percent correct classification. Applying our classification models without reserved judgment, we obtain 79 and 85 precent correct classification for each group, respectively. To gauze the usefulness of multistage analysis, we apply two-stage classification. In the first stage, 14 attributes were selected as discriminatory. 135 Group absence subjects were placed into the reserved-judgment region, with 85 percent of the remaining were classified as Group absence correctly; while 286 Group presence subjects were placed into the reserved-judgment region, and 91 percent of the remaining classified correctly into the Group presence. In the second stage, 11 attributes were selected with 100 and 229 classified into Group absence and presence, respectively. Combining the two stages, we obtained a correct classification of 82 and 85 percent, respectively, for diagnosis of absence or presence of heart disease. Figure 1.1 illustrates the two-stage classification.

1.3.2.3 Predicting Aberrant CpG Island Methylation in Human Cancer

More details of this subsection are found in Refs. [28,29]. Epigenetic silencing associated with aberrant methylation of promoter region CpG islands is one mechanism leading to loss of tumor suppressor function

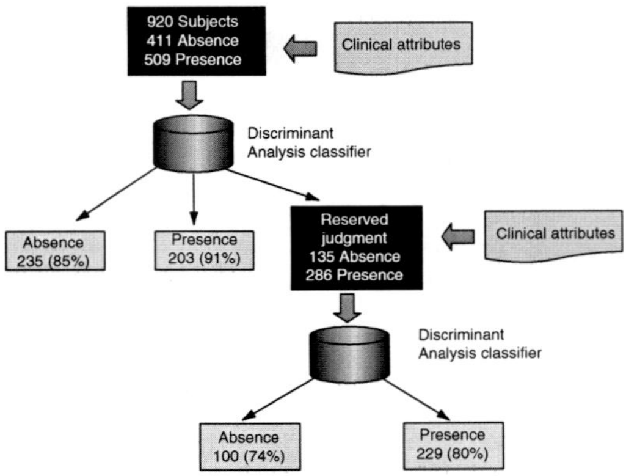

Figure 1.1 A tree diagram for two-stage classification and prediction of heart disease.

in human cancer. Profiling of CpG island methylation indicates that some genes are more frequently methylated than others, and that each tumor type is associated with a unique set of methylated genes. However, little is known about why certain genes succumb to this aberrant event. To address this question, we used Restriction Landmark Genome Scanning (RLGS) to analyze the susceptibility of 1749 unselected CpG islands to de novo methylation driven by overexpression of DNMT1. We found that although the overall incidence of CpG island methylation was increased in cells overexpressing DNMT1, not all loci were equally affected. The majority of CpG islands (69.9 percent) were resistant to de novo methylation, regardless of DNMT1 overexpression. In contrast, we identified a subset of methylation-prone CpG islands (3.8 percent) that were consistently hypermethylated in multiple DNMT1 overexpressing clones. Methylation-prone and methylation-resistant CpG islands were not significantly different with respect to size, C+G content, CpG frequency, chromosomal location, or gene- or promoter-association. To discriminate methylation-prone from methylation-resistant CpG islands, we developed a novel DNA pattern recognition model and algorithm [61], and coupled our predictive model described herein with the patterns found. We were able to derive a classification function based on the frequency of seven novel sequence patterns that was capable of discriminating methylation-prone from methylation-resistant CpG islands with 90 percent correctness upon cross-validation, and 85 percent accuracy when tested against blind CpG islands unknown to us on the methylation status. The data indicates that CpG islands differ in their intrinsic susceptibility to de novo methylation, and suggests that the propensity for a CpG island to become aberrantly methylated can be predicted based on its sequence context.

The significance of this research is twofold. First, the identification of sequence patterns/attributes that distinguish methylation-prone CpG islands will lead to a better understanding of the basic mechanisms underlying aberrant CpG island methylation. Because genes that are silenced by methylation are otherwise structurally sound, the potential for reactivating these genes by blocking or reversing the methylation process represents an exciting new molecular target for chemotherapeutic intervention. A better understanding of the factors that contribute to aberrant methylation, including the identification of sequence elements that may act to target aberrant methylation, will be an important step in achieving this long-term goal. Also, the classification of the more than 29,000 known (but as yet unclassified) CpG islands in human chromosomes will provide an important resource for the identification of novel gene targets for further study as potential molecular markers that could impact on both cancer prevention and treatment. Extensive RLGS fingerprint information (and thus potential training sets of methylated CpG islands) already exists for a number of human tumor types, including breast, brain, lung, leukemias, hepatocellular carcinomas, and PNET [23,24,35,102]. Thus, the methods and tools developed are directly applicable to CpG island methylation data derived from human tumors. Moreover, new microarray-based techniques capable of profiling more than 7000 CpG islands have been developed and applied to human breast cancers [15,117,118]. We are uniquely poised to take advantage of the tumor CpG island methylation profile information that will likely be generated using these techniques over the next several years. Thus, our general-predictive modeling framework has the potential to lead to improved diagnosis and prognosis and treatment planning for cancer patients.

1.3.2.4 Discriminant Analysis of Cell Motility and Morphology Data in Human Lung Carcinoma

For more details on this subsection refer to Ref. [16]. This study focuses on the differential effects of extracellular matrix proteins on the motility and morphology of human lung epidermoid carcinoma cells. The behavior of carcinoma cells is contrasted with that of normal L-132 cells, resulting in a method for the prediction of metastatic potential. Data collected from time-lapsed videomicroscopy was used to simultaneously produce quantitative measures of motility and morphology. The data was subsequently analyzed using our discriminant analysis model and algorithm to discover relationships between motility, morphology, and substratum. Our discriminant analysis tools enabled the consideration of many more cell attributes than is customary in cell motility studies. The observations correlate with behaviors seen in vivo and suggest specific roles for the extracellular matrix proteins and their integrin receptors in metastasis. Cell translocation in vitro has been associated with malignancy, as has an elongated phenotype [120] and a rounded phenotype [97]. Our study suggests that extracellular matrix proteins contribute in different ways to the malignancy of cancer cells, and that multiple malignant phenotypes exist.

1.3.2.5 Ultrasonic-Assisted Cell Disruption for Drug Delivery

Reference [57] discusses this subsection in detail. Although biological effects of ultrasound must be avoided for safe diagnostic applications, ultrasound's ability to disrupt cell membranes has attracted interest as a method to facilitate drug and gene delivery. This preliminary study seeks to develop rules for predicting the degree of cell membrane disruption based on specified ultrasound parameters and measured acoustic signals. Too much ultrasound destroys cells, whereas cell membranes will not open up for absorption of macromolecules when too little ultrasound is applied. The key is to increase cell permeability to allow absorption of macromolecules, and to apply ultrasound transiently to disrupt viable cells so as to enable exogenous material to enter without cell damage. Thus our task is to uncover a predictive rule of ultrasound-mediated disruption of red blood cells using acoustic spectrums and measurements of cell permeability recorded in experiments.

Our predictive model and solver for generating prediction rules are applied to data obtained from a sequence of experiments on bovine red blood cells. For each experiment, the attributes consist of four ultrasound parameters, acoustic measurements at 400 frequencies, and a measure of cell membrane disruption. To avoid overtraining, various feature combinations of the 404 predictor variables are selected when developing the classification rule. The results indicate that the variable combination consisting of ultrasound exposure time and acoustic signals measured at the driving frequency and its higher harmonics yields the best rule, and our method compares favorably with classification tree and other ad hoc approaches, with correct classification rate of 80 percent upon cross-validation and 85 percent when classifying new unknown entities. Our methods used for deriving the prediction rules are broadly applicable, and could be used to develop prediction rules in other scenarios involving different cell types or tissues. These rules and the methods used to derive them could be used for real-time feedback about ultrasound's biological effects. For example, it could assist clinicians during a drug delivery process, or could be imported into an implantable device inside the body for automatic drug delivery and monitoring.

1.3.2.6 Identification of Tumor Shape and Volume in Treatment of Sarcoma

Reference [56] discusses this subsection in detail. This project involves the determination of tumor shape for adjuvant brachytherapy treatment of sarcoma, based on catheter images taken after surgery. In this application, the entities are overlapping consecutive triplets of catheter markings, each of which is used for determining the shape of the tumor contour. The triplets are to be classified into one of the two groups: Group 1 = triplets for which the middle catheter marking should be bypassed, and Group 2 = triplets for which the middle marking should not be bypassed. To develop and validate a classification rule, we used clinical data collected from 15 soft tissue sarcoma (STS) patients. Cumulatively, this comprised 620 triplets of catheter markings. By careful (and tedious) clinical analysis of the geometry of these triplets, 65 were determined to belong to Group 1, the bypass group, and 555 were determined to belong to Group 2, the do-not-bypass group.

A set of measurements associated with each triplet is then determined. The choice of what attributes to measure to best distinguish triplets as belonging to Group 1 or Group 2 is nontrivial. The attributes involved distance between each pair of markings, angles, and curvature formed by the three triplet markings. On the basis of the selected attributes, our predictive model was used to develop a classification rule. The resulting rule provides 98 percent correct classification on cross-validation, and was capable of correctly determining/predicting 95 percent of the shape of the tumor on new patients' data. We remark that the current clinical procedure requires manual outline based on markers in films of the tumor volume. This study was the first to use automatic construction of tumor shape for sarcoma adjuvant brachytherapy [56,62].

1.3.2.7 Discriminant Analysis of Biomarkers for Prediction of Early Atherosclerosis

More details on this subsection are found in Ref. [65]. Oxidative stress is an important etiologic factor in the pathogenesis of vascular disease. Oxidative stress results from an imbalance between injurious oxidant and protective antioxidant events in which the former predominate [88,103]. This results in the modification of proteins and DNA, alteration in gene expression, promotion of inflammation, and deterioration in endothelial function in the vessel wall, all processes that ultimately trigger or exacerbate the atherosclerotic process [22,111]. It was hypothesized that novel biomarkers of oxidative stress would predict early atherosclerosis in a relatively healthy nonsmoking population who are free from cardiovascular disease. One hundred and twenty-seven healthy nonsmokers, without known clinical atherosclerosis had carotid intima media thickness (IMT) measured using ultrasound. Plasma oxidative stress was estimated by measuring plasma lipid hydroperoxides using the determination of reactive oxygen metabolites (d-ROMs) test. Clinical measurements include traditional risk factors including age, sex, low density lipoprotein (LDL), high density lipoprotein

(HDL), triglycerides, cholesterol, body mass index (BMI), hypertension, diabetes mellitus, smoking history, family history of CAD, Framingham risk score, and Hs-CRP.

For this prediction, the patients are first clustered into two groups: (Group 1: IMT >= 0.68, Group 2: IMT < 0.68). On the basis of this separator, 30 patients belong to Group 1 and 97 belong to Group 2. Through each iteration, the classification method trains and learns from the input training set and returns the most discriminatory patterns among the 14 clinical measurements; ultimately resulting in the development of a prediction rule based on observed values of these discriminatory patterns among the patient data. Using all 127 patients as a training set, the predictive model identified age, sex, BMI, HDLc, Fhx CAD < 60, Hs-CRP, and d-ROM as discriminatory attributes that together provide unbiased correct classification of 90 percent and 93 percent for Group 1 (IMT >= 0.68) and Group 2 patients (IMT < 0.68), respectively. To further test the power of the classification method for correctly predicting the IMT status on new/unseen patients, we randomly selected a smaller patient training set of size 90. The predictive rule from this training set yields 80 percent and 89 percent correct rates for predicting the remaining 37 patients into Group 1 and Group 2, respectively. The importance of d-ROM as a discriminatory predictor for IMT status was confirmed during the machine learning process; this biomarker was selected in every iteration as the machine learned and trained to develop a predictive rule to correctly classify patients in the training set. We also performed predictive analysis using Framingham Risk Score and d-ROM; in this case the unbiased correct classification rates (for the 127 individuals) for Groups 1 and 2 are 77 percent and 84 percent, respectively. This is the first study to illustrate that this measure of oxidative stress can be effectively used along with traditional risk factors to generate a predictive rule that can potentially serve as an inexpensive clinical diagnostic tool for prediction of early atherosclerosis.

1.3.2.8 Fingerprinting Native and Angiogenic Microvascular Networks through Pattern Recognition and Discriminant Analysis of Functional Perfusion Data

Reference [64] discusses this subsection in detail. The cardiovascular system provides oxygen and nutrients to the entire body. Pathological conditions that impair normal microvascular perfusion can result in tissue ischemia, with potentially serious clinical effects. Conversely, development of new vascular structures fuels the progression of cancer, macular degeneration, and atherosclerosis. Fluorescence-microangiography offers superb imaging of the functional perfusion of new and existent microvasculature, but quantitative analysis of the complex capillary patterns is challenging. We developed an automated pattern-recognition algorithm to systematically analyze the microvascular networks, and then apply our classification model herein to generate a predictive rule. The pattern-recognition algorithm identifies the complex vascular branching patterns, and the predictive rule demonstrates, respectively, 100 percent and 91 percent correct classification on perturbed (diseased) and normal-tissue perfusion. We confirmed that transplantation of normal bone marrow to mice in which genetic deficiency resulted in impaired angiogenesis eliminated predicted differences and restored normal-tissue perfusion patterns (with 100 percent correctness). The pattern recognition and classification method offers an elegant solution for the automated fingerprinting of microvascular networks that could contribute to better understanding of angiogenic mechanisms and be utilized to diagnose and monitor microvascular deficiencies. Such information would be valuable for early detection and monitoring of functional abnormalities before they produce obvious and lasting effects, which may include improper perfusion of tissue or support of tumor development.

The algorithm can be used to discriminate between the angiogenic response in a native healthy specimen compared to groups with impairment due to age or chemical or other genetic deficiency. Similarly, it can be applied to analyze angiogenic responses as a result of various treatments. This will serve two important goals. First, the identification of discriminatory patterns/attributes that distinguish angiogenesis status will lead to a better understanding of the basic mechanisms underlying this process. Because therapeutic control of angiogenesis could influence physiological and pathological processes such as wound and tissue repairing, cancer progression and metastasis, or macular degeneration, the ability to understand it under different conditions will offer new insight in developing novel therapeutic interventions, monitoring and treatment, especially in aging and heart disease. Thus, our study and the results form the foundation of a valuable diagnostic tool for changes in the functionality of the microvasculature and for discovery of drugs that alter the angiogenic response. The methods can be applied to tumor diagnosis, monitoring, and prognosis. In particular, it will be possible to derive microangiographic fingerprints to acquire specific microvascular patterns associated with early stages of tumor development. Such angioprinting could become an extremely helpful early diagnostic modality, especially for easily accessible tumors such as skin cancer.

1.3.2.9 Prediction of Protein Localization Sites

The protein localization database consists of eight groups with a total of 336 instances (143, 77, 52, 35, 20, 5, 2, 2, respectively) with seven attributes [91]. The eight groups are eight localization sites of protein, including cp (cytoplasm), im (inner membrane without signal sequence), pp (perisplasm), imU (inner membrane, uncleavable signal sequence), om (outer membrane), omL (outer membrane lipoprotein), imL (inner membrane lipoprotein), and imS (inner membrane, cleavable signal sequence). However, the last four groups are taken out from our classification experiment as the population sizes are too small to ensure significance.

The seven attributes include mcg (McGeoch's method for signal sequence recognition), gvh (von Heijne's method for signal sequence recognition), lip (von Heijne's Signal Peptidase II consensus sequence score), chg (presence of charge on N-terminus of predicted lipoproteins), aac (score of discriminant analysis of the amino acid content of outer membrane and periplasmic proteins), alm1 (score of the ALOM membrane spanning region prediction program), and alm2 (score of ALOM program after excluding putative cleavable signal regions from the sequence).

In the classification we use four groups, 307 instances, with seven attributes. Our classification model selected the discriminatory patterns mcg, gvh, alm1, and alm2 to form the predictive rule with unbiased correct classification rates of 89 precent, compared to the results of 81 percent by other classification models [48].

1.3.2.10 Pattern Recognition in Satellite Images for Determining Types of Soil

The satellite database consists of the multispectral values of pixels in 3 × 3 neighborhoods in a satellite image, and the classification associated with the central pixel in each neighborhood. The aim is to predict this classification, given the multispectral values. In the sample database, the class of a pixel is coded as a number. There are six groups with 4435 samples in the training dataset and 2000 samples in testing dataset; and each sample entity has 36 attributes describing the spectral bands of the image [91].

The original Landsat Multi-Spectral Scanner (MSS) image data for this database was generated from data purchased from NASA by the Australian Centre for Remote Sensing. The Landsat satellite data is one of the many sources of information available for a scene. The interpretation of a scene by integrating spatial data of diverse types and resolutions including multispectral and radar data, maps indicating topography, land use, etc., is expected to assume significant importance with the onset of an era characterized by integrative approaches to remote sensing (for example, NASA's Earth Observing System commencing this decade).

One frame of Landsat MSS imagery consists of four digital images of the same scene in different spectral bands. Two of these are in the visible region (corresponding approximately to green and red regions of the visible spectrum) and two are in the (near) infrared. Each pixel is an 8-bit binary word, with 0 corresponding to black and 255 to white. The spatial resolution of a pixel is about 80m × 80m. Each image contains 2340 × 3380 such pixels.

The database is a (tiny) subarea of a scene, consisting of 82 × 100 pixels. Each line of data corresponds to a 3 × 3 square neighborhood of pixels completely contained within the 82 × 100 subarea. Each line contains the pixel values in the four spectral bands (converted to ASCII) of each of the 9 pixels in the 3 × 3 neighborhood and a number indicating the classification label of the central pixel. The number is a code for the following six groups: red soil, cotton crop, gray soil, damp gray soil, soil with vegetation stubble, and very damp gray soil. Running our classification model, 17 discriminatory attributes were selected to form the classification rule, producing an unbiased prediction with 85 percent accuracy.

1.3.3 Further Advances

Brooks and Lee (2007) [18,19] devised other variations of the basic DAMIP Model. They also showed that DAMIP is strongly universally consistent (in some sense) with very good rates of convergence from Vapnik and Chervonenkis theory. A polynomial-time algorithm for discriminating between two populations with the DAMIP Model was developed, and DAMIP was shown to be NP-complete for a general number of groups. The proof demonstrating NP-completeness employs results used in generating edges of the conflict graph [4,11,12,55]. Exploiting the necessary and sufficient conditions that identify edges in the conflict graph is the central contribution to the improvement in solution performance over industry-standard software. The conflict graph is the basis for various valid inequalities, a branching scheme, and for conditions under which integer variables are fixed for all solutions. Additional solution methods are identified, which include a heuristic for finding solutions at nodes in the branch-and-bound tree, upper bounds for model parameters, and necessary conditions for edges in the conflict hypergraph [26,58]. Further, we have concluded that DAMIP is a computationally feasible, consistent, stable, robust, and accurate classifier.

1.4 Progress and Challenges

We summarize below (Table 1.2) the mathematical programming techniques used in classification problems as reviewed in this chapter.

As noted by current research effort, multigroup classification remains NP-complete and much work is needed to design effective models as well as to derive novel and efficient computational algorithms to solve these multigroup instances.

1.5 Other Methods

Although most classification methods can be described in terms of discriminant functions, some methods are not trained in the paradigm of

Table 1.2 Progress in Mathematical Programming-Based Classification Models

Mathematical Programming Methods	References
Linear Programming	
Two-group classification	
Separate data by hyperplanes	[74,75]
Minimizing the sum of deviations (MSD), minimizing the maximum deviation (MMD), and minimizing the sum of interior distances (MSID)	[5,31,32,33,47,99]
Hybrid Model	[45,99]
Review	[27,50,107]
Software	[110]
Issues about normalization	[34,44,51–53,87,100,114–116]
Robust linear programming (RLP)	[9,86]
Inclusion of second-order terms	[104,113]
Effect of the position of outliers	[94]
Binary attributes	[3]
Multigroup classification	
Single function classification	[32]
Multiple function classification	[10,46]
Classification with reserved-judgment region using LP	[39,40,60,63]
Mixed Integer Programming	
Two-group classification	
Minimizing the number of misclassifications	[1,5,6,7,54,101,105,109,119]
Review	[27,50,107]
Software	[110]
Secondary goals	[96]
Binary attributes	[3]
Normalization and attribute selection	[42]
Dichotomous categorical variable formation	[43]
Multigroup classification	
Three-group classification	[71,72,95]
Multigroup classification	[39,40,41,59,60,93]
Multigroup classification with reserved-judgment region	[18,39,40,59,60]
Nonlinear Programming	
Two-group classification	
L_p-norm criterion	[108]
Review	[27,50,107]
Piecewise-linear nonconvex discriminant function	[85]
Minimizing the number of misclassifications	[21,76,77]
Minimizing the sum of arbitrary-norm distances	[78]
Support Vector Machine	
Introduction and tutorial	[20,112]
Generalized SVM	[79,83]
Methods for huge-size problems	[13,36–38,67,68,80,81, 82,84]
Multigroup SVM	[18,38–40,49,59,60,63,66]

determining coefficients or parameters for functions of a predefined form. These methods include classification and regression trees (CART), nearest-neighbor methods, and neural networks.

Classification and regression trees [14] are nonparametric approaches to prediction. Classification trees seek to develop classification rules based on successive binary partitions of observations based on attribute values. Regression trees also employ rules consisting of binary partitions, but are used to predict continuous responses.

The rules generated by classification trees are easily viewable by plotting them in a tree-like structure from which the name arises. A test entity may be classified using rules in a tree plot by first comparing the entity's data with the root node of the tree. If the root node condition is satisfied by the data for a particular entity, the left branch is followed to another node; otherwise, the right branch is followed to another node. The data from the observation is compared to conditions at subsequent nodes until a leaf node is reached.

Nearest-neighbor methods begin by establishing a set of labeled prototype observations. The nearest-neighbor classification rule assigns test entities to groups according to the group membership of the nearest prototype. Different measures of distance may be used. The k-nearest-neighbor rule assigns entities to groups according to the group membership of the k-nearest prototypes.

Neural networks are classification models that can also be interpreted in terms of discriminant functions, though they are used in a way that does not require finding an analytic form for the functions [25]. Neural networks are trained by considering one observation at a time, modifying the classification procedure slightly with each iteration.

1.6 Summary and Conclusion

In this chapter, we presented an overview of mathematical programming-based classification models and analyzed their development and advances in recent years. Many mathematical programming methods are geared toward two-group analysis only, and performance is often compared to Fisher's linear discriminant or Smith's quadratic discriminant. It has been noted that these methods can be used for multiple group analysis by finding $G(G-1)/2$ discriminants for each pair of groups (one-against-one) or by finding G discriminants for each group versus the remaining data (one-against-all), but these approaches can lead to ambiguous classification rules [25].

Mathematical programming methods developed for multiple group analysis are described [10,32,39,40,41,46,63,93]. Multiple group formulations for

SVMs have been proposed and tested [18,36,40,49,59,60,66], but are still considered computationally intensive [49]. The one-against-one and one-against-all methods with SVMs have been successfully applied [49,90].

We also discussed a class of multigroup general-purpose predictive models that we have developed based on the technology of large-scale optimization and SVMs [18,17,39,40,59,60,63]. Our models seek to maximize the correct classification rate while constraining the number of misclassifications in each group. The models incorporate the following features: (1) the ability to classify any number of distinct groups; (2) allow incorporation of heterogeneous types of attributes as input; (3) a high-dimensional data transformation that eliminates noise and errors in biological data; (4) constraining the misclassification in each group and a reserved-judgment region that provides a safeguard against overtraining (which tends to lead to high misclassification rates from the resulting predictive rule); and (5) successive multistage classification capability to handle data points placed in the reserved-judgment region. The performance and predictive power of the classification models is validated through a broad class of biological and medical applications.

Classification models are critical to medical advances as they can be used in genomic, cell, molecular, and system level analyses to assist in early prediction, diagnosis, and detection of disease, as well as for intervention and monitoring. As shown in the CpG island study for human cancer, such prediction and diagnosis opens up novel therapeutic sites for early intervention. The ultrasound application illustrates its application to a novel drug delivery mechanism, assisting clinicians during a drug delivery process, or in devising implantable devices into the body for automated drug delivery and monitoring. The lung cancer cell motility offers an understanding of how cancer cells behave under different protein media, thus assisting in the identification of potential gene therapy and target treatment. Prediction of the shape of a cancer tumor bed provides a personalized treatment design, replacing manual estimates by sophisticated computer predictive models. Prediction of early atherosclerosis through inexpensive biomarker measurements and traditional risk factors can serve as a potential clinical diagnostic tool for routine physical and health maintenance, alerting doctors and patients to the need for early intervention to prevent serious vascular disease. Fingerprinting of microvascular networks opens up the possibility for early diagnosis of perturbed systems in the body that may trigger disease (e.g., genetic deficiency, diabetes, aging, obesity, macular degeneracy, and tumor formation), identify target sites for treatment, and monitoring prognosis and success of treatment. Determining the type of erythemato-squamous disease and the presence/absence of heart disease helps clinicians to correctly diagnose and effectively treat patients.

Thus, classification models serve as a basis for predictive medicine where the desire is to diagnose early and provide personalized target intervention. This has the potential to reduce healthcare costs, improve success of treatment, and improve quality-of-life of patients.

Acknowledgment

This research was partially supported by the National Science Foundation.

References

[1] P.L. Abad and W.J. Banks. New LP based heuristics for the classification problem. *European Journal of Operational Research*, 67:88–100, 1993.

[2] J.A. Anderson. Constrained discrimination between *k* populations. *Journal of the Royal Statistical Society. Series B (Methodological)*, 31(1):123–139, 1969.

[3] O.K. Asparoukhov and A. Stam. Mathematical programming formulations for two-group classification with binary variables. *Annals of Operations Research*, 74:89–112, 1997.

[4] A. Atamturk. Conflict graphs and flow models for mixed-integer linear optimization problems. PhD thesis, School of Industrial and Systems Engineering, Georgia Institute of Technology, Atlanta, Georgia, 1998.

[5] S.M. Bajgier and A.V. Hill. An experimental comparison of statistical and linear programming approaches to the discriminant problem. *Decision Sciences*, 13:604–618, 1982.

[6] W.J. Banks and P.L. Abad. An efficient optimal solution algorithm for the classification problem. *Decision Sciences*, 22:1008–1023, 1991.

[7] W.J. Banks and P.L. Abad. On the performance of linear programming heuristics applied on a quadratic transformation in the classification problem. *European Journal of Operational Research*, 74:23–28, 1994.

[8] K.P. Bennett. Decision tree construction via linear programming. In M. Evans, editor, Proceedings of the 4th Midwest Artificial Intelligence and Cognitive Science Society Conference, pp. 97–101, 1992.

[9] K.P. Bennett and O.L. Mangasarian. Robust linear programming discrimination of two linearly inseparable sets. *Optimization Methods and Software*, 1:23–34, 1992.

[10] K.P. Bennett and O.L. Mangasarian. Multicategory discrimination via linear programming. *Optimization Methods and Software*, 3:27–39, 1994.

[11] R.E. Bixby and E.K. Lee. Solving a truck dispatching scheduling problem using branch-and-cut. *Operations Research*, 46:355–367, 1998.

[12] R. Borndörfer. Aspects of set packing, partitioning and covering. PhD thesis, Technischen Universität Berlin, Berlin, Germany, 1997.

[13] P.S. Bradley and O.L. Mangasarian. Massive data discrimination via linear support vector machines. *Optimization Methods and Software*, 13(1):1–10, 2000.

[14] L. Breiman, J.H. Friedman, R.A. Olshen, and C.J. Stone. *Classification and Regression Trees*. Wadsworth & Brooks/Cole Advanced Books & Software, Pacific Grove, California, 1984.

[15] G.J. Brock, T.H. Huang, C.M. Chen, and K.J. Johnson. A novel technique for the identification of CpG islands exhibiting altered methylation patterns (ICEAMP). *Nucleic Acids Research*, 29:e123, 2001.

[16] J.P. Brooks, A. Wright, C. Zhu, and E.K. Lee. Discriminant analysis of motility and morphology data from human lung carcinoma cells placed on purified extracellular matrix proteins. *Annals of Biomedical Engineering*, Submitted 2007.

[17] J.P. Brooks and E.K. Lee. Mixed integer programming constrained discrimination model for credit screening. Proceedings of the 2007 Spring Simulation Multiconference, Business and Industry Symposium, Norfolk, VA, March 2007. ACM Digital Library, pp. 1–6.

[18] J.P. Brooks and E.K. Lee. Solving a mixed-integer programming formulation of a multi-category constrained discrimination model. Proceedings of the 2006 INFORMS Workshop on Artificial Intelligence and Data Mining, Pittsburgh, PA, Nov. 2006.

[19] J.P. Brooks and E.K. Lee. Analysis of the consistency of a mixed integer programming-based multi-category constrained discriminant model. Submitted, 2007.

[20] C.J.C. Burges. A tutorial on support vector machines for pattern recognition. *Data Mining and Knowledge Discovery*, 2:121–167, 1998.

[21] C. Chen and O.L. Mangasarian. Hybrid misclassification minimization. *Advances in Computational Mathematics*, 5:127–136, 1996.

[22] M. Chevion, E. Berenshtein, and E.R. Stadtman. Human studies related to Protein oxidation: Protein carbonyl content as a marker of damage. *Free Radical Research*, 33(Suppl):S99–S108, 2000.

[23] J.F. Costello, M.C. Fruhwald, D.J. Smiraglia, L.J. Rush, G.P. Robertson, X. Gao, F.A. Wright, J.D. Feramisco, P. Peltomaki, J.C. Lang, D.E. Schuller, L. Yu, C.D. Bloomfield, M.A. Caligiuri, A. Yates, R. Nishikawa, H.H. Su, N.J. Petrelli, X. Zhang, M.S. O'Dorisio, W.A. Held, W.K. Cavenee, and C. Plass. Aberrant CpG-island methylation has non-random and tumour-type-specific patterns. *Nature Genetics*, 24:132–138, 2000.

[24] J.F. Costello, C. Plass, and W.K. Cavenee. Aberrant methylation of genes in low-grade astrocytomas. *Brain Tumor Pathology*, 17:49–56, 2000.

[25] R.O. Duda, P.E. Hart, and D.G. Stork. *Pattern Classification*. Wiley, New York, 2001.

[26] T. Easton, K. Hooker, and E.K. Lee. Facets of the independent set plytope. *Mathematical Programming, Series B*, 98:177–199, 2003.

[27] S.S. Erenguc and G.J. Koehler. Survey of mathematical programming models and experimental results for linear discriminant analysis. *Managerial and Decision Economics*, 11:215–225, 1990.

[28] F.A. Feltus, E.K. Lee, J.F. Costello, C. Plass, and P.M. Vertino. Predicting aberrant CpG island methylation. *Proceedings of the National Academy of Sciences*, 100:12253–12258, 2003.

[29] F.A. Feltus, E.K. Lee, J.F. Costello, C. Plass, and P.M. Vertino. DNA signatures associated with CpG island methylation states. *Genomics*, 87:572–579, 2006.

[30] R.A. Fisher. The use of multiple measurements in taxonomic problems. *Annals of Eugenics*, 7:179–188, 1936.

[31] N. Freed and F. Glover. A linear programming approach to the discriminant problem. *Decision Sciences*, 12:68–74, 1981.

[32] N. Freed and F. Glover. Simple but powerful goal programming models for discriminant problems. *European Journal of Operational Research*, 7:44–60, 1981.

[33] N. Freed and F. Glover. Evaluating alternative linear programming models to solve the two-group discriminant problem. *Decision Sciences*, 17:151–162, 1986.

[34] N. Freed and F. Glover. Resolving certain difficulties and improving the classification power of LP discriminant analysis formulations. *Decision Sciences*, 17:589–595, 1986.

[35] M.C. Fruhwald, M.S. O'Dorisio, L.J. Rush, J.L. Reiter, D.J. Smiraglia, G. Wenger, J.F. Costello, P.S. White, R. Krahe, G.M. Brodeur, and C. Plass. Gene amplification in NETs/medulloblastomas: Mapping of a novel amplified gene within the MYCN amplicon. *Journal of Medical Genetics*, 37:501–509, 2000.

[36] G.M. Fung and O.L. Mangasarian. Proximal support vector machine classifiers. In Proceedings KDD-2001, San Francisco, August 26–29 2001.

[37] G.M. Fung and O.L. Mangasarian. Incremental support vector machine classification. In R. Grossman, H. Mannila, and R. Motwani, editors, Proceedings of the Second SIAM International Conference on Data Mining, pp. 247–260, SIAM, Philadelphia, 2002.

[38] G.M. Fung and O.L. Mangasarian. Multicategory proximal support vector machine classifiers. *Machine Learning*, 59:77–97, 2005.

[39] R.J. Gallagher, E.K. Lee, and D.A. Patterson. An optimization model for constrained discriminant analysis and numerical experiments with iris, thyroid, and heart disease datasets. In Proceedings of the 1996 American Medical Informatics Association, October 1996.

[40] R.J. Gallagher, E.K. Lee, and D.A. Patterson. Constrained discriminant analysis via 0/1 mixed integer programming. *Annals of Operations Research*, 74:65–88, 1997.

[41] W.V. Gehrlein. General mathematical programming formulations for the statistical classification problem. *Operations Research Letters*, 5(6):299–304, 1986.

[42] J.J. Glen. Integer programming methods for normalisation and variable selection in mathematical programming discriminant analysis models. *Journal of the Operational Research Society*, 50:1043–1053, 1999.

[43] J.J. Glen. Dichotomous categorical variable formation in mathematical programming discriminant analysis models. *Naval Research Logistics*, 51:575–596, 2004.

[44] F. Glover. Improved linear programming models for discriminant analysis. *Decision Sciences*, 21:771–785, 1990.

[45] F. Glover, S. Keene, and B. Duea. A new class of models for the discriminant problem. *Decision Sciences*, 19:269–280, 1988.

[46] W. Gochet, A. Stam, V. Srinivasan, and S. Chen. Multigroup discriminant analysis using linear programming. *Operations Research*, 45(2):213–225, 1997.

[47] D.J. Hand. *Discrimination and Classification*. John Wiley, New York, 1981.

[48] P. Horton and K. Nakai. A probablistic classification system for predicting the cellular localization sites of proteins. In Proceedings of the Fourth International Conference on Intelligent Systems for Molecular Biology, pp. 109–115, St. Louis, Missouri, 1996.

[49] C.-W. Hsu and C.-J. Lin. A comparison of methods for multiclass support vector machines. *IEEE Transactions on Neural Networks*, 13(2):415–425, 2002.

[50] E.A. Joachimsthaler and A. Stam. Mathematical programming approaches for the classification problem in two-group discriminant analysis. *Multivariate Behavioral Research*, 25(4):427–454, 1990.

[51] G.J. Koehler. Characterization of unacceptable solutions in LP discriminant analysis. *Decision Sciences*, 20:239–257, 1989.

[52] G.J. Koehler. Unacceptable solutions and the hybrid discriminant model. *Decision Sciences*, 20:844–848, 1989.

[53] G.J. Koehler. A response to Xiao's "necessary and sufficient conditions of unacceptable solutions in LP discriminant analysls": Something is amiss. *Decision Sciences*, 25:331–333, 1994.

[54] G.J. Koehler and S.S. Erenguc. Minimizing misclassifications in linear discriminant analysis. *Decision Sciences*, 21:63–85, 1990.

[55] E.K. Lee. Solving a truck dispatching scheduling problem using branch-and-cut. PhD thesis, Computational and Applied Mathematics, Rice University, Houston, Texas, 1993.

[56] E.K. Lee, A.Y.C. Fung, J.P. Brooks, and M. Zaider. Automated planning volume definition in soft-tissue sarcoma adjuvant brachytherapy. *Biology in Physics and Medicine*, 47:1891–1910, 2002.

[57] E.K. Lee, R.J. Gallagher, A.M. Campbell, and M.R. Prausnitz. Prediction of ultrasound-mediated disruption of cell membranes using machine learning techniques and statistical analysis of acoustic spectra. *IEEE Transactions on Biomedical Engineering*, 51:1–9, 2004.

[58] E.K. Lee and S. Maheshwary. Conflict hypergraphs in integer programming. Technical Report, Georgia Institute of Technology, 2006. submitted.

[59] E.K. Lee. Discriminant analysis and predictive models in medicine. In S.J. Deng, editor, *Interdisciplinary Research in Management Science, Finance, and HealthCare*. Peking University Press, 2006. To appear.

[60] E.K. Lee. Large-scale optimization-based classification models in medicine and biology. *Annals of Biomedical Engineering, Systems Biology and Bioinformatics*, 35(6):1095–1109, 2007.

[61] E.K. Lee, T. Easton, and K. Gupta. Novel evolutionary models and applications to sequence alignment problems. *Annals of Operations Research, Operations Research in Medicine—Computing and Optimization in Medicine and Life Sciences*, 148:167–187, 2006.

[62] E.K. Lee, A.Y.C. Fung, and M. Zaider. Automated planning volume contouring in soft-tissue sarcoma adjuvant brachytherapy treatment. *International Journal of Radiation, Oncology, Biology and Physics*, 51:391, 2001.

[63] E.K. Lee, R.J. Gallagher, and D.A. Patterson. A linear programming approach to discriminant analysis with a reserved-judgment region. *INFORMS Journal on Computing*, 15(1):23–41, 2003.

[64] E.K. Lee, S. Jagannathan, C. Johnson, and Z.S. Galis. Fingerprinting native and angiogenic microvascular networks through pattern recognition and discriminant analysis of functional perfusion data. Submitted, 2006.

[65] E.K. Lee, T.L. Wu, S. Ashfaq, D.P. Jones, S.D. Rhodes, W.S. Weintrau, C.H. Hopper, V. Vaccarino, D.G. Harrison, and A.A. Quyyumi. Prediction of early atherosclerosis in healthy adults via novel markers of oxidative stress and d-ROMs. Working paper, 2007.

[66] Y. Lee, Y. Lin, and G. Wahba. Multicategory support vector machines: Theory and application to the classification of microarray data and satellite radiance data. *Journal of the American Statistical Association*, 99:67–81, 2004.

[67] Y.-J. Lee and O.L. Mangasarian. RSVM: Reduced support vector machines. In Proceedings of the SIAM International Conference on Data Mining, Chicago, April 5–7, 2001.

[68] Y.-J. Lee and O.L. Mangasarian. SSVM: A smooth support vector machine for classification. *Computational Optimization and Applications*, 20(1):5–22, 2001.

[69] Y.-J. Lee, O.L. Mangasarian, and W.H. Wolberg. Breast cancer survival and chemotherapy: A support vector machine analysis. In *DIMACS Series in Discrete Mathematical and Theoretical Computer Science*, volume 55, American Mathematical Society, 2000, pp. 1–10.

[70] Y.-J. Lee, O.L. Mangasarian, and W.H. Wolberg. Survival-time classification of breast cancer patients. *Computational Optimization and Applications*, 25:151–166, 2003.

[71] C. Loucopoulos and R. Pavur. Computational characteristics of a new mathematical programming model for the three-group discriminant problem. *Computers and Operations Research*, 24(2):179–191, 1997.

[72] C. Loucopoulos and R. Pavur. Experimental evaluation of the classificatory performance of mathematical programming approaches to the three-group discriminant problem: The case of small samples. *Annals of Operations Research*, 74:191–209, 1997.

[73] P.P. Luedi, A.J. Hartemink, and R.L. Jirtle. Genome-wide prediction of imprinted murine genes. *Genome Research*, 15:875–884, 2005.

[74] O.L. Mangasarian. Linear and nonlinear separation of patterns by linear programming. *Operations Research*, 13:444–452, 1965.

[75] O.L. Mangasarian. Multi-surface method of pattern separation. *IEEE Transactions on Information Theory*, 14(6):801–807, 1968.

[76] O.L. Mangasarian. Misclassification minimization. *Journal of Global Optimization*, 5:309–323, 1994.

[77] O.L. Mangasarian. Machine learning via polyhedral concave minimization. In H. Fischer, B. Riedmueller, and S. Schaeffler, editors, *Applied Mathematics and Parallel Computing—Festschrift for Klaus Ritter*, Physica-Verlag, Germany, 1996, pp. 175–188.

[78] O.L. Mangasarian. Arbitrary-norm separating plane. *Operations Research Letters*, 24:15–23, 1999.

[79] O.L. Mangasarian. Generalized support vector machines. In A.J. Smola, P. Bartlett, B. Schökopf, and D. Schuurmans, editors, *Advances in Large Margin Classifiers*, MIT Press, Cambridge, Massachusetts, 2000, pp. 135–146.

[80] O.L. Mangasarian. Data mining via support vector machines. In E.W. Sachs and R. Tichatschke, editors, *System Modeling and Optimization XX*, Kluwer Academic Publishers, Boston, 2003, pp. 91–112.

[81] O.L. Mangasarian. Support vector machine classification via parameterless robust linear programming. *Optimization Methods and Software*, 20:115–125, 2005.

[82] O.L. Mangasarian and D.R. Musicant. Successive overrelaxation for support vector machines. *IEEE Transactions on Neural Networks*, 10:1032–1037, 1999.

[83] O.L. Mangasarian and D.R. Musicant. Data discrimination via nonlinear generalized support vector machines. In M.C. Ferris, O.L. Mangasarian, and J.-S. Pang, editors, *Complementarity: Applications, Algorithms and Extensions*, Kluwer Academic Publishers, Boston, Massachusetts, 2001, pp. 233–251.

[84] O.L. Mangasarian and D.R. Musicant. Lagrangian support vector machines. *Journal of Machine Learning Research*, 1:161–177, 2001.

[85] O.L. Mangasarian, R. Setiono, and W.H. Wolberg. Pattern recognition via linear programming: Theory and application to medical diagnosis. In T.F. Coleman and Y. Li, editors, *Large-Scale Numerical Optimization*, SIAM, Philadelphia, Pennsylvania, 1990, pp. 22–31.

[86] O.L. Mangasarian, W.N. Street, and W.H. Wolberg. Breast cancer diagnosis and prognosis via linear programming. *Operations Research*, 43(4):570–577, 1995.

[87] E.P. Markowski and C.A. Markowski. Some difficulties and improvements in applying linear programming formulations to the discriminant problem. *Decision Sciences*, 16:237–247, 1985.

[88] J.M. McCord. The evolution of free radicals and oxidative stress. *The American Journal of Medicine*, 108:652–659, 2000.

[89] G.J. McLachlan. *Discriminant Analysis and Statistical Pattern Recognition*. Wiley, New York, 1992.

[90] K.-R. Müller, S. Mika, G. Rätsch, K. Tsuda, and B. Schölkopf. An introduction to kernel-based learning algorithms. *IEEE Transactions on Neural Networks*, 12(2):181–201, March 2001.

[91] P.M. Murphy and D.W. Aha. UCI Repository of machine learning databases [http:/www.ics.uci.edu/-mlearn/MLRepository.html], Department of Information and Computer Science, University of California, Irvine, California.

[92] A. O'Hagan. *Kendall's Advanced Theory of Statistics: Bayesian Inference*, volume 2B. Halsted Press, New York, 1994.

[93] R. Pavur. Dimensionality representation of linear discriminant function space for the multiple-group problem: An MIP approach. *Annals of Operations Research*, 74:37–50, 1997.

[94] R. Pavur. A comparative study of the effect of the position of outliers on classical and nontraditional approaches to the two-group classification problem. *European Journal of Operational Research*, 136:603–615, 2002.

[95] R. Pavur and C. Loucopoulos. Evaluating the effect of gap size in a single function mathematical programming model for the three-group classification problem. *Journal of the Operational Research Society*, 52:896–904, 2001.

[96] R. Pavur, P. Wanarat, and C. Loucopoulos. Examination of the classificatory performance of MIP models with secondary goals for the two-group discriminant problem. *Annals of Operations Research*, 74:173–189, 1997.

[97] A. Raz and A. Ben-Zéev. Cell contact and architecture of malignant cells and their relationship to metastasis. *Cancer and Metastasis Reviews*, 6:3–21, 1987.

[98] A.C. Rencher. *Multivariate Statistical Inference and Application*. Wiley, New York, 1998.

[99] P.A. Rubin. A comparison of linear programming and parametric approaches to the two-group discriminant problem. *Decision Sciences*, 21:373–386, 1990.

[100] P.A. Rubin. Separation failure in linear programming discriminant models. *Decision Sciences*, 22:519–535, 1991.

[101] P.A. Rubin. Solving mixed integer classification problems by decomposition. *Annals of Operations Research*, 74:51–64, 1997.

[102] L.J. Rush, Z. Dai, D.J. Smiraglia, X. Gao, F.A. Wright, M. Fruhwald, J.F. Costello, W.A. Held, L. Yu, R. Krahe, J.E. Kolitz, C.D. Bloomfield, M.A. Caligiuri, and C. Plass. Novel methylation targets in de novo acute myeloid leukemia with prevalence of chromosome 11 loci. *Blood*, 97:3226–3233, 2001.

[103] H. Sies. Oxidative stress: Introductory comments. In H. Sies, editor, *Oxidative Stress*, Academic Press, London, pp. 1–8, 1985.

[104] A.P. Duarte Silva and A. Stam. Second order mathematical programming formulations for discriminant analysis. *European Journal of Operational Research*, 72:4–22, 1994.

[105] A.P. Duarte Silva and A. Stam. A mixed integer programming algorithm for minimizing the training sample misclassification cost in two-group classification. *Annals of Operations Research*, 74:129–157, 1997.

[106] C.A.B. Smith. Some examples of discrimination. *Annals of Eugenics*, 13:272–282, 1947.

[107] A. Stam. Nontraditional approaches to statistical classification: Some perspectives on l_p-norm methods. *Annals of Operations Research*, 74:1–36, 1997.

[108] A. Stam and E.A. Joachimsthaler. Solving the classification problem in discriminant analysis via linear and nonlinear programming methods. *Decision Sciences*, 20:285–293, 1989.

[109] A. Stam and E.A. Joachimsthaler. A comparison of a robust mixed-integer approach to existing methods for establishing classification rules for the discriminant problem. *European Journal of Operational Research*, 46:113–122, 1990.

[110] A. Stam and D.R. Ungar. RAGNU: A microcomputer package for two-group mathematical programming-based nonparametric classification. *European Journal of Operational Research*, 86:374–388, 1995.

[111] S. Tahara, M. Matsuo, and T. Kaneko. Age-related changes in oxidative damage to lipids and DNA in rat skin. *Mechanisms of Ageing and Development*, 122:415–426, 2001.

[112] V. Vapnik. *The Nature of Statistical Learning Theory*. Springer-Verlag, New York, 1995.

[113] P. Wanarat and R. Pavur. Examining the effect of second-order terms in mathematical programming approaches to the classification problem. *European Journal of Operational Research*, 93:582–601, 1996.

[114] B. Xiao. Necessary and sufficient conditions of unacceptable solutions in LP discriminant analysis. *Decision Sciences*, 24:699–712, 1993.

[115] B. Xiao. Decision power and solutions of LP discriminant models: Rejoinder. *Decision Sciences*, 25:335–336, 1994.

[116] B. Xiao and Y. Feng. Alternative discriminant vectors in LP models and a regularization method. *Annals of Operations Research*, 74:113–127, 1997.

[117] P.S. Yan, C.M. Chen, H. Shi, F. Rahmatpanah, S.H. Wei, C.W. Caldwell, and T.H. Huang. Dissecting complex epigenetic alterations in breast cancer using CpG island microarrays. *Cancer Research*, 61:8375–8380, 2001.

[118] P.S. Yan, M.R. Perry, D.E. Laux, A.L. Asare, C.W. Caldwell, and T.H. Huang. CpG island arrays: An application toward deciphering epigenetic signatures of breast cancer. *Clinical Cancer Research*, 6:1432–1438, 2000.

[119] N. Yanev and S. Balev. A combinatorial approach to the classification problem. *European Journal of Operational Research*, 115:339–350, 1999.

[120] A. Zimmermann and H.U. Keller. Locomotion of tumor cells as an element of invasion and metastasis. *Biomedicine and Pharmacotherapy*, 41:337–344, 1987.

[121] C. Zopounidis and M. Doumpos. Multicriteria classification and sorting methods: A literature review. *European Journal of Operational Research*, 138: 229–246, 2002.

Chapter 2

Using Influence Diagrams in Cost-Effectiveness Analysis for Medical Decisions

Ram S. Duriseti

CONTENTS

2.1 Introduction 62
2.2 Formalized Decision Making and Quality Metrics 62
2.3 Influence Diagrams in Clinical Cost-Effectiveness Analysis 63
2.4 Model Scope and the Structure of Influence Diagrams: Implications for Clinical Decision Problems 65
2.4.1 Introduction 65
2.4.2 Structural Characteristics of Clinical IDs 67
2.4.3 Computational Complexity in Influence Diagrams 68
2.4.4 Reducing the Computational Burden and Model Scope 69
2.4.5 Illustrative Example of Active Path Analysis and Model Scope 73
2.4.6 Determining Problem Scope in Medical Decision Problems 78
2.5 Introduction to Venous Thromboembolic Disease and the D-Dimer Assay 79
2.5.1 Clinical Background and Epidemiology 79
2.5.2 Diagnosis and Treatment of Venous Thromboembolism 80
2.5.3 ELISA D-Dimer Test 81

2.6 Cost-Effectiveness of Strategies for Diagnosing Pulmonary Embolism: The Recommendation Provided Depends on Model Scope 82
2.6.1 Overview of the Limited Model and Results 82
2.6.2 Extended D-Dimer Model Results in a Change in Policy 86
2.7 Summary 89
References 89

2.1 Introduction

Recent decades have seen tremendous advances in our ability to diagnose complex diseases, intervene in various diseases at stages and to degrees previously impossible, and sustain life in the face of crippling pathology. In many cases, our ability to detect abnormalities has advanced beyond our understanding of the significance of what we find. Although longevity and quality of life have increased, significant resources are often expended to enhance human life. As a consequence, decision making regarding diagnostic tools and interventions has become increasingly complicated. Within the art of medicine, there is an ever growing need for formalized normative decision making.

2.2 Formalized Decision Making and Quality Metrics

We begin with the premise that formalized decision making with normative recommendations is preferred to ad hoc decision making in the face of limited resources. Formalized decision making has several advantages. It provides a reproducible process, which is amenable to critical review. It provides a transparent view into the assumptions and contributing variables for a decision. The clear description of assumptions creates a framework, whereby advances in the state of information can be incorporated into the decision-making process to assess the impact on the policy recommendation. Formalized decision making can identify the consequences of alternative decisions, thus helping decision makers make an informed, high-quality decision.

All normative decision making requires a value metric. In medical decision problems, maximizing health state is a crucial metric. We assume that we can attach a measure to the quality of an individual's health state. In the same way that people are assumed to have preference probabilities over different outcomes in classical normative decision making, we assume that defined health states exist and that patients can quantify preferences for these health states relative to a state of perfect health. When we normalize this health state's presence for one year, we refer to it as a quality-adjusted life year (QALY). A QALY therefore represents the relative value of living for one year in the current state of health. The QALY value of living for one year in a perfect state of health is 1.0 and 0 for death. This paradigm has gained wide acceptance.[1]

The formalized valuation of different health states is an ongoing process.[2,3] Valuation of the quality of health states can be flawed due to an individual's cognitive biases,[4] and by failures to use established methodological guidelines.[4,5] However, independent quality estimates of health states by patients in different studies tend to correlate well with each other. Furthermore, the normative results of many cost-effectiveness studies tend to be insensitive to minor changes in quality estimates.[6]

Health outcomes cannot be the sole metric of concern in medical decision-making problems. With health care expenditures exceeding 15 percent of GDP in the United States and growing at an annual rate of roughly 7–8 percent, optimizing health resource allocation is imperative.[7–9]

For a patient with a diagnosed disease, costs can be driven down by limiting resource utilization to treatments tailored to the disease process. Such tailored therapy can maximize efficacy while avoiding unnecessary costs. However, most patients do not present with a clearly defined disease process. Most patients present with a set of symptoms, which can be attributed to a large number of diseases. Consequently, additional costs are incurred in diagnosis and evaluation of patients.

During the initial medical evaluation of a patient complaint, various uncertainties interact to create a complex probabilistic framework. Consider the evaluation of a patient presenting with chest pain. Chest pain can be caused by physical disease of any number of organ systems stretching from the throat down to the mid-abdomen, or it can be associated with a number of psychiatric conditions.[10,11] The clinician must differentiate between a dizzying assortment of possibilities with disparate outcomes in a logical and an efficient fashion. Although clinicians can develop tremendous clarity and informed expert judgment, the quality of this judgment can be highly variable. Ad hoc decision making in such complex environments can be unduly dominated by anecdotal evidence, hunches, and other cognitive biases. Thus, complex medical diagnostic problems are best modeled in a formalized fashion, where the assumptions and data can be critiqued and modified as new knowledge is discovered.

2.3 Influence Diagrams in Clinical Cost-Effectiveness Analysis

Because clinical uncertainty plays so heavily into unnecessary or costly testing, cost-effectiveness analyses of clinical problems must include the uncertainty seen by the clinician. Bayesian networks provide an intuitive and natural means of representing the various uncertainties and the relationships between them in a clinical decision problem.

Bayesian networks are directed acyclic graphs that represent probabilistic relationships between uncertainties. Influence diagrams (IDs) represent such probabilistic networks with the addition of utility measures for different outcomes represented in the decision problem. Broadly speaking, every node can have a set of ancestor nodes, descendant nodes, parent nodes, and nodes which are children.

The nodes in an ID can be uncertainties, functional nodes, decision nodes, or utility nodes. The uncertainty nodes can be observed or unobserved. Observed uncertainties represent random variables, whose state has been revealed. A functional node has a fixed functional relationship to its parents.

In an ID, the utility node can be used to represent any utility value for the set of outcomes considered. In a medical decision-making problem, utilities are usually costs in present value dollars and health outcomes (usually QALYs) discounted to the present. Costs are usually computed as incremental costs; they represent the additional cost incurred by a particular policy beyond the costs inherent to all policies.[12] For example, an electrocardiogram (ECG) may be considered the test that all patients receive and, therefore, the base policy. All other policies are assessed for the associated incremental change in the utility measure(s) relative to the base policy.

We keep costs and valuation of health outcomes separate. Some authors convert between QALYs and costs using a set cost per QALY value.[12,13] Although this might be appropriate for some analyses, it is not a recommended part of medical cost-effectiveness analysis.[12] There are two main reasons for this. First, there is no universally agreed upon conversion constant. Second, when utility measures are subsumed into costs, insights regarding the relative merits of alternative policies may be lost. This can happen in several ways. Suppose that a policy is defined as the set of choices representing a choice at every decision in the diagram for certain background state of information. When a policy Π^* is the lowest cost policy with the best outcomes among all policies considered, we have found a dominant policy. More frequently, however, we find that alternative policies occur on an efficient frontier over the dimensions of cost and health outcomes (Figure 2.1). Policies on this efficient frontier can have a lower cost than other policies with the same health outcome or they result in a better health outcome than other policies with the same cost.

When comparing policies on the efficient frontier, we can analyze the trade-offs between policies. Consider two policies located along the efficient frontier: Π_i and Π_j. Suppose that Π_i costs less than Π_j whereas Π_j results in higher QALYs. When the cost and QALY measures are separated, we can determine an incremental cost-effectiveness ratio (ICER) associated

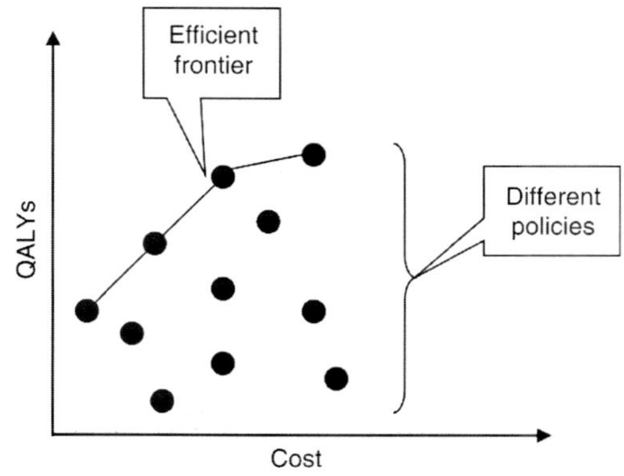

Figure 2.1 Efficient frontier of policies in (cost, QALY) space.

with switching from Π_i to Π_j. This is the cost per QALY gained to move from policy Π_i to Π_j:

ICER is given by the following expression:

$$\text{ICER} = \frac{(\text{Cost of policy } j - \text{Cost of policy } i)}{(\text{QALYs achieved by policy } j - \text{QALYs achieved by policy } i)}$$
$$= \frac{[\text{Cost}(\Pi_j) - \text{Cost}(\Pi_i)]}{[\text{QALY}(\Pi_j) - \text{QALY}(\Pi_i)]}$$

The ICER allows one to represent the economic trade-off (cost per incremental health outcome achieved) between any two policies under comparison.

Whether one uses a cost/QALY conversion factor or not, when one policy dominates all others (the case of Π^* above and shown in Figure 2.2), a dominant policy will always be the best policy. However, even in this case, when one converts the QALYs to costs, one can lose the powerful insight that no matter what the cost/QALY conversion factor is, Π^* will remain the best policy.

2.4 Model Scope and the Structure of Influence Diagrams: Implications for Clinical Decision Problems

2.4.1 Introduction

Frequently, model complexity contributes to difficulties in the acceptance and adoption of recommendations from cost-effectiveness analyses.

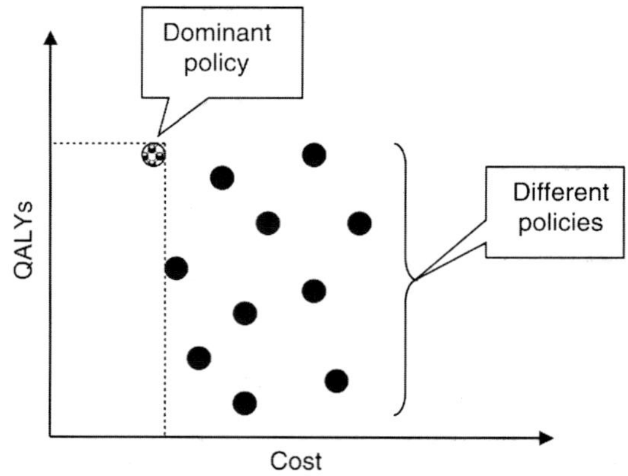

Figure 2.2 Extended dominance of other policies by Π*.

The analyst must strike a balance between model realism (enough detail to credibly address the clinical question posed) and model tractability. IDs can be structured to mitigate problems with computational complexity. However, structural modifications to an ID can affect the applicability of the model to the decision problem under consideration. In this section we discuss how the inherent conflict between model complexity and clinical applicability can be mitigated. We discuss a structural condition that uses the ID to identify at the outset the variables (and the relationships between them) that must be included to model the clinical question under consideration.

The standard nomenclature and icons for nodes in IDs are shown in Figure 2.3. In contrast to an unobserved uncertainty, an observed uncertainty is a random variable whose value has been set to a particular value in its distribution. At deterministic node represents a functional relationship where the functional arguments are from various nodes in the graph. A decision node represents a choice between alternatives chosen by the decision maker. A utility node defines the value measures for a particular policy in the model.

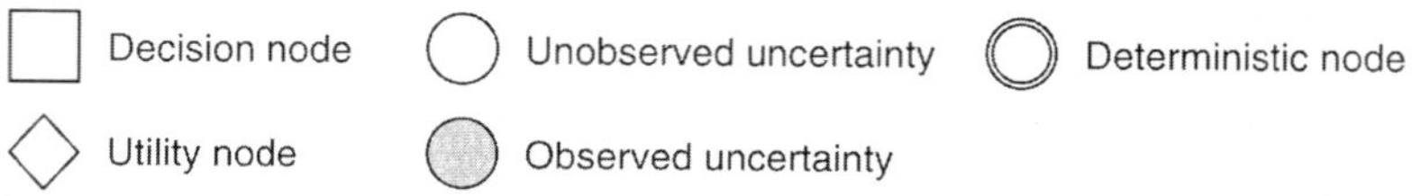

Figure 2.3 Legend for influence diagram (ID) icons.

2.4.2 Structural Characteristics of Clinical IDs

One of the most difficult aspects of cost-effectiveness analysis is deciding which variables to include in a decision problem and which to exclude. In clinical decision problems, large numbers of uncertainties are the norm rather than the exception. Disease processes with vastly different outcomes present with similar clinical symptoms. For example, in patients presenting with chest pain, a complaint of generalized weakness can be because of any number of etiologies including, among others, progressive frailty in the elderly, infectious causes, and strokes.[14,15] Therefore, in a graphical representation of this notion, for a given set of symptoms S, there are many edges from different disease processes $\{D_1, \ldots, D_n\}$ to S (Figure 2.4).

Moreover, a particular disease can cause a wide variety of symptoms. For example, cardiac ischemia (insufficient arterial blood flow to the heart to meet metabolic needs) can present with symptoms as wide ranging and nondescript as weakness, shortness of breath, nausea, lightheadedness, etc., in addition to chest pain. Therefore, for a given disease process D, many different symptoms $\{S_1, \ldots, S_n\}$ have edges from the node D (Figure 2.5). Furthermore, the probability of having one disease affects the probability of having another disease (e.g., see the horizontal edge between D_i and D_j in Figure 2.5).

Owing to all of these factors, IDs for clinical decision problems are highly connected graphs.

A large set of highly connected variables introduce problems of three types. First, as the number of variables under consideration increases, the computational complexity of the solution to the ID increases. Second, as model complexity increases, even efficient graphical representations of the decision problem can be difficult to interpret and validate. In a highly connected graph, the various dependencies and independencies implied by the relationships between variables can be difficult to follow. Third, joint distributions over a large number of variables are rarely available from the literature and can be difficult to elicit from experts.

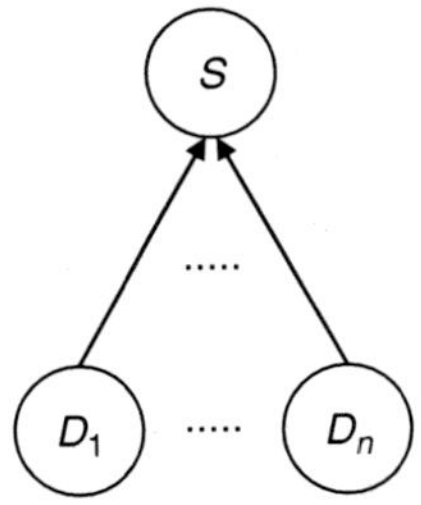

Figure 2.4 Different diseases have edges to the same set of symptoms.

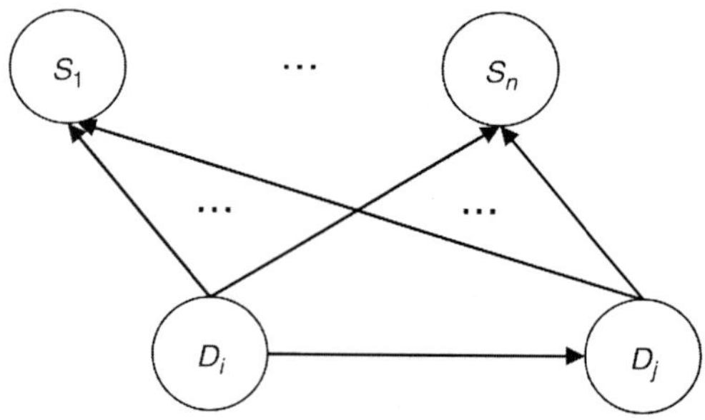

Figure 2.5 Different sets of symptoms have edges to the same disease and diseases are related.

In the remainder of this section, we discuss the reasons why computational complexity increases rapidly with the number of nodes and paths in an ID. We discuss methods, which can simplify a problem, and explain how some simplification methods can unduly compromise the accuracy of the cost-effectiveness analysis. Finally, we present structural conditions, derived from an earlier work on Bayesian networks and IDs[16–19] that can be used to determine when simplification by disregarding certain variables in the ID will not compromise the accuracy of the computational results for the decision problem under consideration.

2.4.3 Computational Complexity in Influence Diagrams

For a discussion of computational complexity in IDs, we must formalize the notion of trails and paths in the graph. Every node can have a set of ancestor nodes, descendant nodes, parent nodes, and nodes which are children. As in the traditional genealogic sense, all parents are also ancestors and all children are also descendants. In a directed acyclic graph, we can define a path in the graph as a directed sequence of edges and nodes traversed when navigating between any two nodes in the graph. A path is a directed trail. We define parents and children of a node using the concept of paths:

Parents of a node: The parents of a node X in a directed acyclic graph are those nodes from which node X has directed paths of length one.

Children of a node: The children of a node X in a directed acyclic graph are those nodes to which node X has directed paths of length one.

In an ID, the more connections there are between variables, the larger the size of the resulting conditional probability tables (CPTs) and the greater the computational requirements. In Figure 2.6, suppose that W, X, Y, and Z are discrete random variables, and that each variable has three potential states. In Figure 2.6, node X has two parents (W and Y) whereas the node

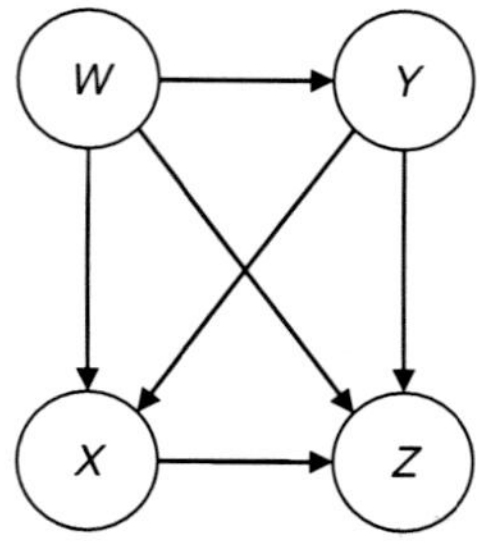

Figure 2.6 Example of exponential expansion of conditional probability table (CPT) size.

Z has three parents (W, X, and Y). The size of the CPTs of X and Z will therefore be quite different.

The CPT of X is for $P\{X|W, Y\}$ and it can be represented as a 9×3 table, whereas the CPT of Z is for $P\{Z|W, X, Y\}$ which must be represented by a 27×3 table. The size of the CPT for a given node increases exponentially with the number of parents of the node. Thus, even clinical IDs with a modest number of nodes can have significant computational complexity because of their highly connected nature.

Decisions add another layer of complexity. When an ID includes multiple decisions, a particular policy is defined by a particular combination of choices for the decisions in the diagram along with observations pertinent to those decision choices. Observations, in this context, are uncertainties whose state is known before a decision is made. For example, suppose a particular ID has a decision about ordering a lab test (with a choice to order the test or not to order the test) and a decision about ordering a radiographic test (with a choice between one of R different imaging strategies). Assuming that no observations are made before any of the decisions, a policy will be defined by a combination of a decision about ordering the lab test and a decision about the imaging strategy. Therefore, this ID has $2 \times R$ possible policies (i.e., decision choices). Complexity increases exponentially in both the number of choices for and the number of decisions in the ID.

Concerns about computational tractability in clinical decision problems often lead to specific efforts to minimize the number of uncertainties or decisions.

2.4.4 Reducing the Computational Burden and Model Scope

We define the scope of a clinical decision model to be the range of clinical situations that are represented by a decision problem. Scope is related to the context of the problem represented by an ID, but is not completely defined by the context alone. The context of the decision problem describes the background state of information (which might not be explicitly represented in the model). For example, the context of patients presenting to the emergency department (ED) with chest pain differs markedly from the context of patients presenting to a primary care physician's office for chest pain.

Model scope is much more than the background state of information or a catalogue of the decisions and uncertainties in the problem. Model scope depends not only on the uncertainties, observations, and decisions considered in the ID but also on the relationships between the variables in the model. Thus, modifications to the relationships between elements of an ID can directly affect the model scope. Conversely, model scope can be viewed as the clinical question postulated by the analyst that, in turn, informs the structure of the ID. This bidirectional feedback between the structure of the ID and the decision question answered by the model is central to understanding how attempts to control complexity affect the model scope and vice versa.

Although reduction in complexity may be required to render a problem computationally tractable, this should ideally occur without altering the scope. One obvious means by which to simplify a problem is to remove irrelevant variables. For an irrelevant variable, no distribution in the graph is affected by the probability distribution over the variable. It is rare, however, that a thoughtful construction of a decision problem would include irrelevant nodes in an ID. More commonly, nodes in an ID and the relationship between the nodes represent important factors that affect the problem under consideration. Consequently, attempts to reduce the complexity of a network either by node removal or altering the relationship between nodes can result in a model scope that does not accurately reflect the problem originally under consideration. Subsequent cost-effectiveness analysis using such a model can generate policy recommendations that are not globally preferred in the scope of the true clinical problem at hand.

Other techniques that can be used to reduce the number of uncertainties and decisions include construction of causal graphs and cutset conditioning (the so-called case-based reasoning).[20,21]

With causal reasoning, one reasons from cause to effect so as to generate probability distributions over an effect given a cause. With a causal construction, one is sometimes able to remove relationships between nodes that might otherwise complicate the graph. For example, in creating an ID to model health outcomes of patients with chest pain, we might start with a model, which has an edge between aortic dissection (AD) and acute myocardial infarction (AMI) because patients with myocardial infarction (MI) are potentially at increased risk of AD and vice versa. Suppose we can identify the fundamental cause for the disease process of AD and a similar fundamental cause for MI. In such a case, we can remove the relationship in the ID between MI and AD, thereby simplifying the CPTs of MI and AD and reducing the computational burden. Unfortunately, available data often does not accommodate a causal construction.

With case-based reasoning, the ID is solved for a particular value of an uncertainty or decision node. The ID is then solved for each possible value that the uncertainty or decision can attain. For example, in an ID representing chest pain and its causes, one would analyze the ID only for cases where the chest pain is due to cardiac ischemia and then individually for each other potential cause for the chest pain in the model.

Even with such techniques, most clinical decision problems remain complicated. Fortunately, we can leverage structural characteristics of IDs to quickly and accurately determine the appropriate variable set for the clinical problem under consideration and therefore determine the model scope for the clinical problem. These structural characteristics involve an understanding of active trails and conditional independence.

Definition of active trail: An active trail in graph G exists between two nodes X and Y in G if knowing something about X affects the probability distribution over Y. In effect, when there is an active trail, there exists a probability distribution in G where X and Y are dependent: $P_G\{X\} \neq P_G\{X|Y\}$. An active trail can go through multiple nodes in the graph G as long as each node appears only once in the description of the path.

Definition of conditional independence: Given nodes X, Y, and Z in graph G, X is conditionally independent of Z if there is a distribution in G, where X and Z are independent if the value of Y is known: $P\{X|Y,Z\} = P\{X|Y\}$.

This definition of an active trail derives from the principle of d-separation.[19] If no active trails exist between nodes X and Y in graph G, nodes X and Y are said to be d-separated in the graph G. A more formal discussion of d-separation is beyond the scope of this chapter.[19] Active paths are directed active trails.

We can manipulate the active paths for a variable as a means to simplify a decision problem. This alters the structural relationships between nodes in the graph, and leads to a more manageable requisite set of nodes (the nodes about which we require knowledge to solve the decision problem).[16] When done properly, such manipulation keeps the scope of the model aligned with the clinical problem under consideration.

If we want to remove a variable from consideration, without affecting the usefulness and credibility of the model, we need to evaluate and, if possible, eliminate the active paths to the utility node from that variable.

With the removal of this d-separated node, the result is a simpler model that is still potentially appropriate for the clinical question under consideration. If active paths from the candidate variable to the utility node remain despite careful consideration, removal of the variable will affect the policy recommendation and, depending on the degree and nature of the impact, adversely impact the applicability of the resulting policy recommendation.

To specify the process for evaluating active trails for a node, we first define a model ID G'. Once the ID is defined, we provide a formal statement of the conditions by which a node can be potentially removed, and then present an illustrative example.

Construction of ID G'

Consider an ID G. We define a set uncertainties $X = \{X_1 \cup X_2 \cup \cdots \cup X_n\}$ which are not in G. The uncertainty x_{ij} is the jth uncertainty in the set of uncertainties X_i. Define G' where nodes(G') = nodes(G) $\cup X$ and there is a nonempty set of trails between X and G. Within G' is a set of utility nodes $U = \{u_1, \ldots, u_n\}$, a set of decision nodes $D = \{d_1, \ldots, d_n\}$ and a set of uncertainties $Y = \{Y_1 \cup Y_2 \cup \cdots \cup Y_n\}$ all of which are in the ID G'. Recall that $X \subseteq$ nodes(G'). The uncertainties X and Y are random variables describing some element in the state of the world. The decisions D are actions that the decision maker can take. Therefore, each $u_k \in U$ has a utility value defined by the set of actions and uncertainties relevant to u_k: Utility value of $u_k = f_U(x_k, d_k)$. Consider some $X_j \subseteq X$ where the nodes in X_j may or may not be observed. Define the set $Y_O \subseteq Y$, where the states of all nodes in Y_O are observed before any $d_i \in D$. Note that $\{X \cap Y\} = \emptyset$.

Suppose we want to eliminate the influence of uncertainty x_{ij} on the computational results of the ID G', the uncertainty x_{ij} can be removed from consideration if conditional independence properties along all trails to any utility node from x_{ij} render the joint probability distribution of the variables along the trail independent of x_{ij}.

Structural conditions to exclude an uncertainty's impact on utility in G'

We can remove an uncertainty x_{ij} from G' without affecting the maximum expected utility (MEU) of the ID G if any of the following conditions is met

- Condition A: paths from x_{ij} to any $u_i \in U$ all go through some $y_{ij} \in Y_O$ (where Y_O is the set of uncertainties in Y, whose state is set and observed).
- Condition B: paths from x_{ij} to any $u_i \in U$ which go through some $d_{ij} \in D_j$, also go through a $x_{im} \in X$, which is observed before decision d_{ij} is made (i.e., the x_{im} is a parent node of any such d_{ij} and x_{im}'s value is known).
- Condition C: for any x_{ik} and y_j on the trail τ_j between x_{ij} and any $u_j \in U$, where y_j is a parent of x_{ij} and a parent of x_{ik} on the trail

τ_j then all such y_j are observed (i.e., $y_j \in Y_O$); this condition must hold for all trails τ_j, which run between x_{ij} and any $u_j \in U$.

Formal proofs for these conditions are beyond the scope of this chapter, but the theory that forms the basis of the conditions is attributable to several authors and is predicated on the notion of d-separation (and, therefore, conditional independence or context specific independence).[16–18] Shachter discussed an efficient computational algorithm (of order O[n], where n is the number of nodes in the network) to identify the existence of any active path between two nodes.[16] Using this algorithm to identify the active paths for a particular node, one can quickly identify the required variables for a particular construction of the ID. One can then alter the relationship between variables in the diagram to inactivate any active paths between the candidate variable for exclusion and utility nodes in the ID. If the conditions are met, we can remove the variable without affecting the relevance of the computational result for the clinical scope of the problem at hand.

2.4.5 Illustrative Example of Active Path Analysis and Model Scope

To illustrate an interaction between the structure of an ID and the model's scope, we consider a simplified version of the clinical problem of evaluating and diagnosing patients who present with symptoms that might be caused by an AD.

The aorta, the major artery from the heart, supplies oxygenated blood to the entire body. An AD is a tear in the aortic tissue that is often rapidly fatal (the actor John Ritter recently died from an AD). If the patient survives to make it to the ED, the mortality is very high if the condition is left untreated. Diagnosis of AD requires aortography for definitive diagnosis. Currently, the most common method of aortography is a computed tomographic (CT) scan of the aorta with intravenous contrast (a radio opaque dye that is injected into the blood stream for radiographic imaging studies). CTs are expensive, potentially time consuming in a busy clinical setting, and not without risk. Treatment for AD is also expensive, and incurs significant morbidity and mortality. Consequently, neither the test (CT) nor the treatment should be ordered without careful consideration.

Further complicating the problem, the presenting symptoms for AD are not unique to this condition. The symptoms and signs, which we denote by S, can be caused by many different disease processes; we denote these other diseases (OD) by OD = $\{OD_1, \ldots, OD_n\}$. The set OD includes conditions, such as muscle strains, pleurisy, pneumonia, pulmonary mass, anxiety, cardiac ischemia, pulmonary embolism (PE), gastro-esophageal reflux (heartburn), and biliary colic (gallstone-related pain), among others. In this example, OD and S are groups of nodes, but one can consider AD to

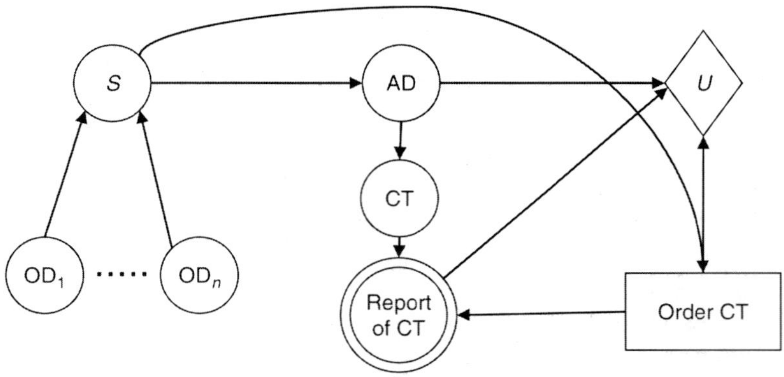

Figure 2.7 Influence diagram (ID) for the simplified aortic dissection (AD) network.
***Note*: AD = aortic dissection, S = symptoms, OD_i = other disease i, which can cause the symptoms S, CT = CT aortography, order CT = decision to order the CT aortography, and U = utility.**

be a group of nodes as well without any loss of generality. The specific question we pose is what is the value of CT for diagnosing AD in patients who present with signs and symptoms in S? Figure 2.7 shows the ID that represents this problem.

We assume the symptoms S can also be caused by OD. The CPT for S will have a different row for each combination of OD_i which lead to S. For each of the different possible values of S (i.e., for each possible set of symptoms) we have a probability of AD. The node AD represents the presence or absence of AD. The CPT of the node AD encodes the probability that the patient has AD for a given set of symptoms S. Both AD and a decision to order a CT affect the costs and health outcomes for the patient; these influences are represented by the arrows to the utility node U from AD and Report of CT. Notice that the decision order CT only affects whether one obtains a Report of CT. The outcome of the CT depends on its sensitivity and specificity in the presence or absence of an AD. The decision about whether to order a CT does not change the performance characteristics of the imaging study in the setting of an AD.

In Figure 2.7, the arrow from S to the decision order CT implies that information about all of the patient's symptoms is available to the decision maker before making a decision to order the CT. In reality, the clinician looks for symptoms that might indicate an AD; symptoms that do not affect the risk of an AD are not relevant to this problem. Moreover, some symptoms are not observable by the clinician

Suppose that a set of clinical criteria have been developed to risk stratify patients who may have an AD. Clinical criteria are simply a set of observed

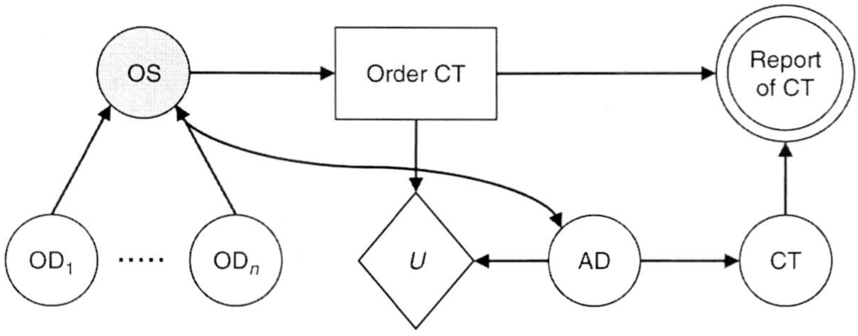

Figure 2.8 Aortic dissection (AD) influence diagram (ID) with observed clinical criteria.
***Note*: AD = aortic dissection, OS = observed symptoms for clinical decision rule, OD_i = other disease *i* which can cause the symptoms *S*, CT = CT aortography, order CT = decision to order the CT aortography, and *U* = utility.**

symptoms (OS) where OS $\subseteq S$. We assume that symptoms in $\{S \backslash \text{OS}\}$ do not have an appreciable impact on the risk of AD. Figure 2.8 shows an ID, which includes this new representation. Because OS are observed, they are represented in Figure 2.8 as a shaded observed uncertainty node. Because these symptoms are observed before the decision to order a CT scan is made, the diagram shows an arrow that goes from OS to the decision order CT. Such paths from uncertainties to decision nodes are often referred to as informational edges.[22]

In Figure 2.7, there is an active path from OD $\rightarrow U$ through S and AD: the distribution for the uncertainty AD is affected by the distribution for the uncertainty OD such that

$$P\{D=d\} = \sum_{s}\sum_{\text{od}} P\{D=d\,|S=s\}P\{S=s\,|\text{OD}=\text{od}\}P\{\text{OD}=\text{od}\}$$

In this notation, $P\{\text{AD} = \text{ad}\}$ is the probability that the random variable AD equals ad. In Figure 2.8, there is no active path between OD and U. The distribution for the uncertainty AD is not affected by the distribution for the uncertainty OD. Because the state of OS is observed:

$$P\{\text{OS}=\text{os}\,|\text{OD}=\text{od}\} = P\{\text{OS}=\text{os}\}$$

Therefore,

$$P\{\text{AD}=\text{ad}\} = \sum_{\text{os}}\sum_{\text{od}} P\{\text{AD}=\text{ad}|\text{OS}=\text{os}\}\; P\{\text{OS}=\text{os}|\text{OD}=\text{od}\}$$

$$= \sum_{\text{os}} P\{\text{AD}=\text{ad}|\text{OS}=\text{os}\}\; P\{\text{OS}=\text{os}\}$$

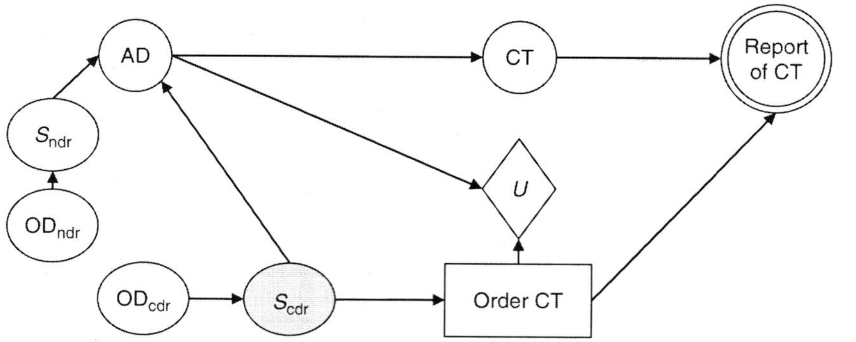

Figure 2.9 Aortic dissection (AD) influence diagram (ID) with symptoms outside of the clinical decision rule.
***Note*: AD = aortic dissection, OS = observed symptoms for clinical decision rule, OD_{cdr} = other diseases which cause the symptoms in S_{cdr}, OD_{ndr} = other diseases which cause the symptoms in S_{ndr}, CT = CT aortography, order CT = decision to order the CT aortography, and U = utility.**

Assuming that no variables other than these shown are relevant to the clinical problem, observing the clinical criteria (OS) allows the analysis to exclude consideration of the impact of other variables in OD.

Now consider the case where the OS do not include the full range of symptoms that affect the probability that the patient has an AD. Suppose that symptoms associated with AD can be grouped into two sets, those that are observable and therefore used as part of the clinical decision rule (S_{cdr}) and those not used as part of the decision rule (S_{ndr}). We assume $S = \{S_{cdr} \cup S_{ndr}\}$ and $\{S_{cdr} \cap S_{ndr}\} = \emptyset$. Likewise, suppose that the other diseases can be grouped into two sets: those that produce symptoms in the clinical decision rule (OD_{cdr}) and those that produce symptoms not in the decision rule (OD_{ndr}). In this simplified example, we make a strong assumption that OD_{cdr} and OD_{ndr} are mutually exclusive and collective exhaustive subsets of OD. As discussed earlier, such a strong assumption is not generally valid in most clinical decision problems. The new decision problem is depicted in Figure 2.9.

In Figure 2.9, with observation of S_{cdr}, the clinical decision problem can exclude consideration of diseases in OD_{cdr}, but not OD_{ndr} with observation of S_{cdr}. If we want to exclude both OD_{cdr} and OD_{ndr}, we would have to observe the symptoms in S_{ndr} as well as those in S_{cdr}.

So far, we have not considered the possibility that other diseases can be detected by a CT ordered for AD. Such detection can affect the utility measures. For example, an esophageal perforation (EP), which is a hole in the wall of the esophagus, can cause symptoms very similar to those caused by AD. Although not as dramatic as AD, EPs can have a significant and measurable impact on health outcomes and costs.

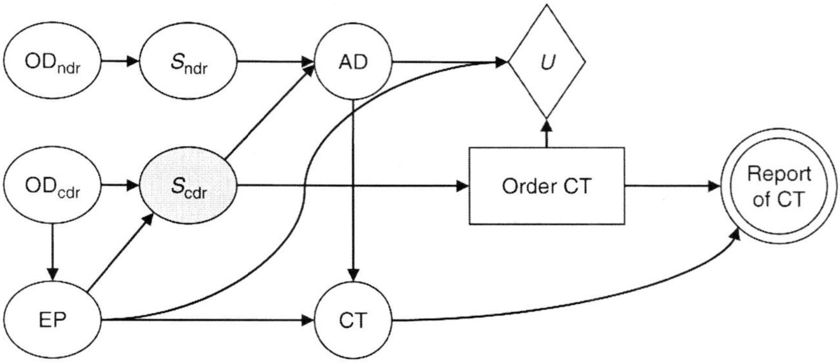

Figure 2.10 Aortic dissection (AD) influence diagram (ID) with the possibility that CT may detect an esophageal perforation (EP).
***Note*: AD = aortic dissection, EP = esophageal perforation, OS = observed symptoms for clinical decision rule, OD_{cdr} = other diseases which cause the symptoms in S_{cdr}, OD_{ndr} = other diseases which cause the symptoms in S_{ndr}, CT = CT aortography, order CT = decision to order the CT aortography, and *U* = utility.**

Suppose that EP is the only disease in OD that could be detected by a CT ordered for AD. Figure 2.10 shows an ID that allows for the possibility that CT may detect an EP. It is possible that some diseases in OD_{cdr} affect the probability that the patient has an EP; this influence is represented by the arrow between OD_{cdr} and PE. In this formulation of the decision problem, $OD = \{OD_{cdr} \cup OD_{ndr} \cup EP\}$ and $\{OD_{cdr} \cap OD_{ndr} \cap EP\} = \emptyset$. The effect of EP on the utility is represented by an arrow from the node EP to the utility node.

The question posed by this construction is more general that the previous question about the value of CT in diagnosing AD. Now we are asking: What is the value of CT in patients who present with signs and symptoms suggestive of AD? With this reformulation, we are no longer just concerned about the value of a CT for AD, but rather about the value of CT for modeled disease processes that lead to signs and symptoms suggestive of AD. We are not asking: What are the costs and health outcomes in patients who present with signs and symptoms suggestive of AD for whom a CT is considered? For this latter question, we would have to take into account the effect on utility of not just EP, but of all other diseases in OD.

In Figure 2.10, the path from EP to the utility node and to the CT result renders the variable for EP relevant to the computation of the ID. Because OD_{cdr} affects the distribution for the presence of EP, when analyzing this ID, one cannot disregard the diseases in OD_{cdr}. This is in contrast to the model depicted in Figure 2.9, where one can disregard the diseases in

OD_{cdr}. By explicitly considering EP in the utility model, we create an active path for both EP and OD_{cdr} to the utility node. The inclusion of EP in the analysis makes relevant certain variables that were previously irrelevant to the analysis. All of the CPTs involved in determining the interrelationship between OD_{cdr}, previously of no concern in Figure 2.9, are now required to solve the model in Figure 2.10. Although this increase in complexity may be regrettable, it is unavoidable if the model is meant to provide an accurate representation of the decision problem postulated.

2.4.6 Determining Problem Scope in Medical Decision Problems

Armed with the above conditions and the application of a path-finding algorithm, a visual appraisal of the relationship between certain variables and the utility node/nodes can quickly determine if a variable of interest can be excluded from consideration. In the simple example of Section 2.4.5, visual appraisal is sufficient. In more complex IDs, a formalized path-finding algorithm is critical.[16]

If inclusion of a variable and inspection for the conditions in the node removal theorem reveal an active path from the newly included variable and at least one utility node then one has two potential options: one can either include the variable, or render inactive all paths to the utility node from the newly included variable. This latter process can be subtle, as depicted in the model for Figure 2.10. There, the introduction of the EP node created an active trail for OD_{cdr} that had previously been excluded from consideration upon observing S_{cdr}.

As shown, inactivating a path can be as simple as observing the state of some variable: in our example of Figure 2.9, observation of the symptoms S_{cdr} would inactivate any active trails from OD_{cdr} to the utility node. This amounts to setting the value of some set of nodes to a particular value. Once these observations have been made, the model is applicable only to instances of the decision problem where the variables of interest represented by the observed nodes take on the observed values. In effect, redefining the decision problem's requisite set[16,17] of variables redefines the model scope.

Altering the model scope based upon observation of certain variables does not necessarily nullify the applicability of the resulting model. If the variables are truly observable by the decision maker in the clinical situation, the model retains its clinical applicability. The following necessary condition extends our previous necessary conditions for excluding an uncertainty beyond the notion of d-separation alone.

Conditions to exclude an uncertainty's impact on utility and model scope in G'

Uncertainty x_{ij} will not impact the computational result of the ID and will not alter the applicability of the model when

- Uncertainty x_{ij} is d-separable from the utility node.
- If d-separation of x_{ij} involves observation of a descendant d_{ij} of x_{ij}, d_{ij} is an observable state of nature by the decision maker.

In Figure 2.10, by restricting the utility analysis to a subset of the nodes—in this case, to AD and EP rather than AD and OD—we are, in effect, disregarding the impact of other diseases in OD that could affect health or cost outcomes. If the effect on utility of the diseases in OD is represented, the OD_{cdr} disease nodes would have active paths of length one to the utility node that does not go through the S_{cdr} node. In such a case, even the observation of S_{cdr} would not render the active paths, inactive from diseases in OD_{cdr} to the utility node. The absence of an active path directly to the utility node means that excluding OD_{cdr} from the value analysis could adversely affect the applicability of the model's result for certain decision questions. This is not the case for the decision question of Figure 2.9. Disregarding OD_{cdr} is only truly a violation of the conditions we establish, if we are using the results to address a question other than specific question about the utility of the CT for diagnosing AD and other diseases amenable to diagnosis with a CT which present with signs and symptoms suggestive of AD.

In subsequent sections, we construct a clinical model for a common and controversial clinical decision problem. We show how failure to apply the techniques detailed above results in a cost-effectiveness model that provides a policy recommendation that is dramatically different from the policy recommendation accruing from a model with a properly circumscribed scope.

2.5 Introduction to Venous Thromboembolic Disease and the D-Dimer Assay

In this section, we provide a brief introduction to venous thromboembolic (VTE) disease to lay the foundation for the clinical model.

2.5.1 Clinical Background and Epidemiology

VTE disease consists mainly of two pathological processes: PE and DVT (deep venous thrombosis). A DVT is a blood clot that has formed in a non-superficial vein in an extremity (e.g., leg or arm). The majority of

DVTs form in the veins of the pelvis and lower extremity, but roughly 2–4 percent form in the upper extremity.[23–26] DVTs can remain in the vein of formation, causing local complications, such as pain, swelling, and even compromise of arterial blood supply (ischemia). DVTs can also dislodge and migrate.[27,28] When they migrate, they almost always lodge in the lungs as PEs. VTE occurs with a higher incidence than average as part of several disease processes, and can also occur as a consequence of inactivity, surgery, or trauma.[29]

PEs can be fatal, symptomatic and non-fatal, or asymptomatic.[30–32] Studies suggest that roughly 10 percent of patients with PEs die within the first hour of the embolic event. Some experts have hypothesized that up to 50 percent of patients with PE die within the first hour, but this contention is not well substantiated by available evidence.[33–35] The incidence of PE increases with age: incidence is 30 per 100,000 people per year in the <35 year old age group, increasing to 500 per 100,000 people per year in the >70 year old age group. In the general American population, the rate is roughly 110 per 100,000 (for a total of 330,000 PEs per year in the United States).[35–37] National statistics suggest that up to 50 percent of PEs go undiagnosed.

VTE, and thus PE, is associated with many different risk factors.[38–41] Certain chronic medical conditions, such as end-stage liver disease and congestive heart failure with associated right heart failure, predispose patients to VTE.[42,43] Other patients are susceptible because of genetic variation in various procoagulant and anticoagulant proteins in their system: a condition referred to as thrombophilia. Certain genetic variants of these proteins upset the homeostatic mechanisms for clot formation and clot breakdown, thereby predisposing patients to VTE.[44] Cancer, for reasons that are not entirely clear, predisposes patients to VTE as well. Certain types of cancer are much more associated with VTE risk than others. The cancer-related relative risk for VTE is accrued independent of age-related increases in risk.[35,45–47]

2.5.2 Diagnosis and Treatment of Venous Thromboembolism

Pulmonary embolisms (PEs) have a variable presentation and require resource-intensive diagnostic modalities. The mortality rate of 22–35 percent in known untreated cases[13,32] drops to roughly 10 percent with anticoagulation.[32] Some authors question the validity of these high mortality rates in the population of patients presenting to the ED. Calder et al. estimate the mortality and recurrence rates of untreated PE in ambulatory patients presenting to the ED to be 5 percent.[48] Although the vast majority of PEs arise from DVTs, these DVTs are not uniformly detected in patients with PE.[49,50] Consequently, simple compression ultrasound (CUS) testing of the lower extremities is not considered a reliable stand-alone test for the diagnostic exclusion of PE.[51]

Direct imaging studies for PE that are readily available in most urban EDs are computed tomographic angiogram (CTA) and the ventilation/perfusion (V/Q) scan. VQ scan results are categorized under the prospective investigation of pulmonary embolism diagnosis (PIOPED) classification system as very low (near normal), low, intermediate, and high. Choosing the VQ result that triggers therapy is an active clinical decision.[52–54]

The treatment for VTE is systemic anticoagulation. In the acute phase, this usually involves some form of heparin (delivered either intravenously or by injection). In the subacute and chronic phase, oral anticoagulation is the mainstay of therapy in most cases except where coincident medical conditions (i.e., pregnancy), allergy, or genetic variation mandate ongoing use of heparin. The appropriate duration of therapy is controversial. Current recommendations call for a minimum of three–six months of therapy, with lifetime therapy for recurrent cases.[55–60]

2.5.3 ELISA D-Dimer Test

Several highly sensitive serum markers for the breakdown products of blood clots exist to aid in the evaluation of patients with suspected VTE. One of the most widely used quantitative assays is the VIDAS enzyme-linked immunosorbent assay (ELISA) D-dimer test.[12,61]

Consensus panels have provided recommendations as to which clinical scenarios are appropriate for use of such a test. Such recommendations are predicated on opinions of an acceptable posterior probability of disease, rather than health outcome value measures.[62] From a clinical standpoint, a probability helps establish a measure of certainty, but does not explicitly identify the decision with the most value: such a consideration must include a notion of costs as well as health outcomes.

Some authors have analyzed the cost-effectiveness of CTA for the diagnosis of PE, without considering the use of D-dimer.[63,64] One study analyzed the cost-effectiveness of PE diagnostic strategies including D-dimer,[13] but did not consider the impact of different cutoffs for either the VQ scan or the D-dimer result, nor the use of a CTP/CTV alone with D-dimer.

On the basis of the performance of two cutoff values (350 and 500 ng/mL) for the VIDAS D-dimer assay,[61,65–67] we fitted the assay's behavior to a gamma distribution over populations with and without VTE (Figure 2.11). The estimated probability mass functions (PMFs) for D-dimer levels in patients with and without VTE are plotted in Figure 2.11.

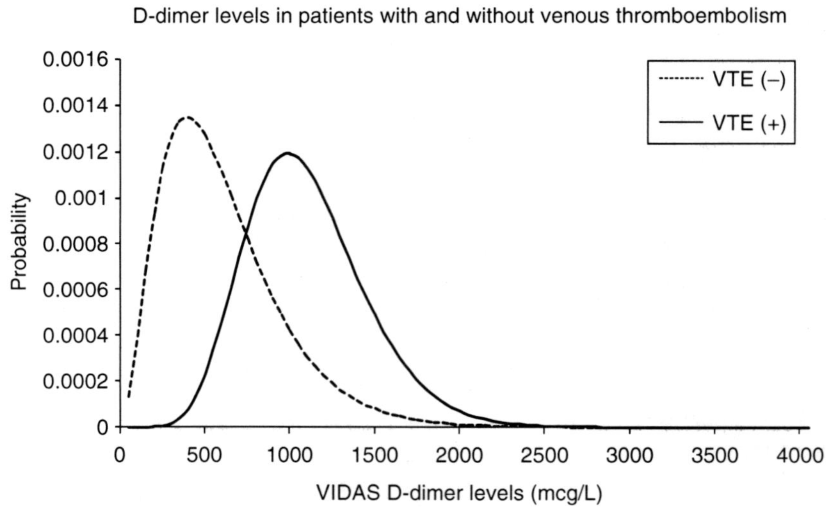

Figure 2.11 Estimated gamma distributions for D-dimer levels in patients with and without venous thromboembolic (VTE).

The distributions allowed us to estimate the sensitivity and specificity of three other cutoff values for the VIDAS D-dimer assay (Table 2.1). The gamma distribution is a probability distribution for nonnegative values. It has two critical parameters, whose degrees of freedom can be fully specified with the associated sensitivity and specificity of two cutoff values. Figure 2.12 plots the receiver operating characteristic (ROC) curve of the quantitative D-dimer assay for VTE.

2.6 Cost-Effectiveness of Strategies for Diagnosing Pulmonary Embolism: The Recommendation Provided Depends on Model Scope

2.6.1 Overview of the Limited Model and Results

We applied a sequential decision model to evaluate the cost-effectiveness of different testing and imaging strategies for patients in different PE risk categories. Clinical prediction rules, developed by Wells and colleagues,[68] help clinicians stratify patients into risk categories based on a scoring rule. Wells' scoring rule has been prospectively validated by several authors and has been found to be at least as effective as other available clinical

Table 2.1 Conditional Probabilities for VIDAS ELISA D-Dimer Assay[a]

D-Dimer Level	Pr{VTE + |D-Dimer}	Pr{VTE − |D-Dimer}	Sensitivity for Pulmonary Embolism (PE)	Specificity for Pulmonary Embolism (PE)
Cutoff I (200 mcg/L)	0.00005	0.99995	0.99998	0.08312
Cutoff II (350 mcg/L)	0.00689	0.99311	0.99820	0.25954
Cutoff III (500 mcg/L)	0.03944	0.96056	0.98119	0.45825
Cutoff IV (650 mcg/L)	0.11168	0.88832	0.92071	0.63065
Cutoff V (800 mcg/L)	0.20817	0.79183	0.79997	0.76085

[a] These probabilities represent the percent of patients with or without venous thromboembolic (VTE) who will have a D-dimer value less than or equal to the listed value (thus, it is the cumulative distribution of the conditional probability). The published sensitivity for cutoff II is 99.5 percent with a specificity of 27 percent.[61,65–67] The published sensitivity for cutoff III is 98 percent with a specificity of 47 percent.[61,65–67]

prediction rules.[68–72] By allowing for different cutoff values for the VIDAS ELISA D-dimer assay and VQ scans, actively framing clinical assessment into the model, and explicitly representing the causal relationship between DVT and PE, we extend previous work to more accurately reflect the clinical decision and associated outcome measures in a cost-effectiveness framework.

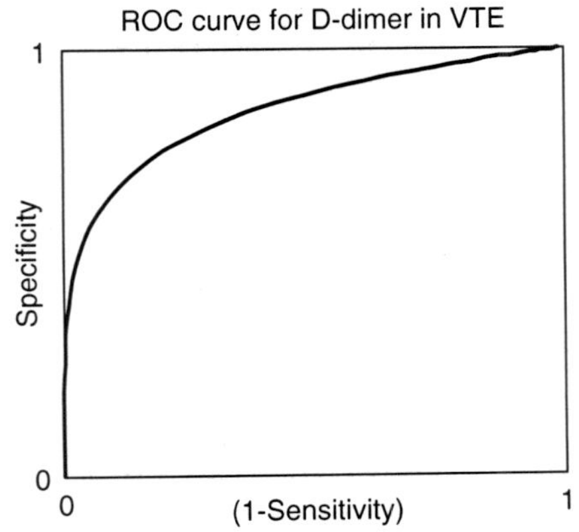

Figure 2.12 Receiver operator characteristic (ROC) curve of D-dimer for venous thromboembolic (VTE) disease.

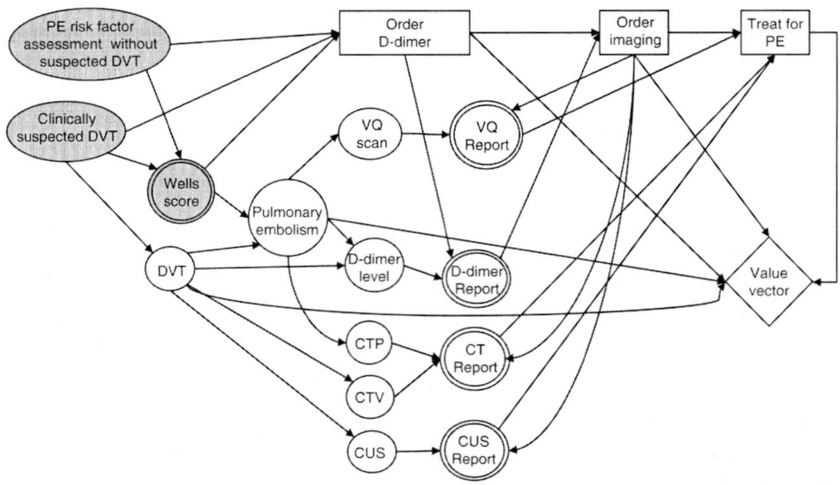

Figure 2.13 Directed acyclic graph of the influence diagram (ID).
***Note*: The diagram depicts the complexity of this decision-making problem. The structure and the implied order of probability assessment are partially dictated by the conditional probabilities available in the literature.**

We assumed the following sequence of events. A patient with a suspected PE is given an initial clinical evaluation (history, exam, vital signs, 12-lead ECG, and chest radiograph) and a Wells score. The decision to order a test is made. If a D-dimer is ordered, the clinician decides on the cutoff value to trigger an imaging test (if any). If a VQ scan is involved, the cutoff value of the VQ scan is chosen for when anticoagulation and hospitalization for PE is initiated. Similar to many tests, neither the D-dimer nor the VQ scan delivers a binary indication (normal or abnormal); our sequential model incorporates that notion. Treatment is delivered based on the results of the imaging test.

We modeled the sequential decision process based on the ID depicted in Figure 2.13. The ID implies an assessment order on the relevant probabilities and explicit conditional independence assumptions. The ID includes the relevant observations, decisions, and the associated value of a diagnostic policy. The ID is constructed in the causal direction for intuitive reasons and to simplify assessment[20]. The probability of PE is conditioned on the Wells pretest risk category (low, moderate, or high). The test indication is rendered conditionally independent of the specific population under consideration, once the probabilities of PE and DVT are established.[73] In this way, differences in the prevalence of PE in the Wells study population and the prevalence of PE in populations used for various studies on the accuracy of a particular imaging test can be used in the same probabilistic network.[16,68,74,75] Using the terminology of Section 2.4, the study population of the Wells study is d-separated from the prevalence of PE given a particular Wells risk category.

We assumed that a CTA may include a pulmonary portion (CTP) and a deep venous portion (CTV). The decisions concerning which cutoffs to choose for a D-dimer and for a VQ scan are subsumed in the decision to order the test. The node labeled as value vector includes an incremental cost component and a QALY component.

Tests are assumed to be conditionally independent. Whenever we use a D-dimer with an imaging test, we perform the imaging test only if the D-dimer test is positive. Within a compound imaging strategy, where CUS is considered along with a VQ or a CTP, we assumed that the VQ scan or the CTP is performed only if the CUS is negative.

The details of the parameter estimates and the methodology for the computational model have been published and are beyond the scope of the current discussion.[76] Figure 2.14 shows an example of a policy plot in (cost, QALY) space resulting from this model for the base case on a single patient in a high Wells pretest category.

For the limited model, for patients with suspected PE, the use of CTP/CTV without D-dimer is a robust strategy for diagnosis of PE under

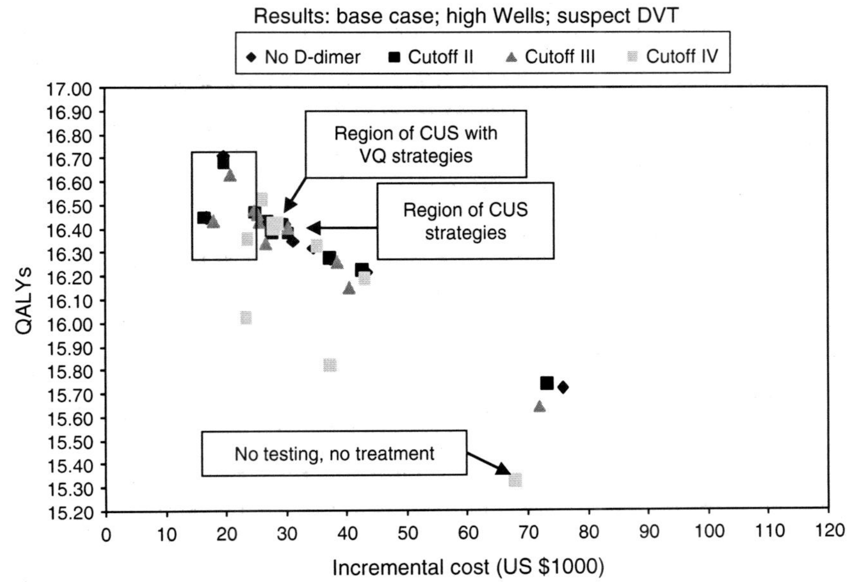

Figure 2.14 Results under base case assumptions for high Wells category patients with suspected DVT.

a wide range of assumptions. Although it is almost never a dominant strategy, it is frequently preferred over a wide range of assumptions and, when it is not, CTA remains part of a dominant mixed strategy with CUS and D-dimer use. When CTA is not available, D-dimer (cutoff II or III) with CUS/VQ is the dominant strategy for high and moderate Wells category patients, although D-dimer cutoff IV with CUS is the dominant strategy for low Wells category patients. The analysis was potentially limited by the need to estimate the sensitivity and specificity of various D-dimer cutoff values with the gamma distribution and potentially insufficient breadth and depth of the sensitivity analyses performed.

2.6.2 Extended D-Dimer Model Results in a Change in Policy

Although the results from the limited model are robust to sensitivity analysis on a wide range of parameters, we suspected that they may depend critically on the breadth of the clinical situation assumed: diagnosis of PE in a patient suspected of having PE. In a typical acute care setting, PE is usually one of several competing diagnoses in patients presenting with chest pain or shortness of breath. When considering the use of a D-dimer in this larger patient population, the ease and relatively low cost of the D-dimer assay might well make it cost effective. The scope of this latter question is much different than that posed in the limited model.

We developed an extended model to assess the value of a quantitative D-dimer in diagnosing PE among patients presenting to an ED with chest pain or shortness of breath (Figure 2.15). This model has two major additions to the previous model. Both of these additions represent aggregation points for input from a subgraph of the extended model. These aggregation points are a clear other cause for the signs and symptoms and other causes for an elevated D-dimer.

Some patients who present with symptoms attributable to VTE have a seemingly clear explanation for their symptoms outside of VTE. If these potential non-VTE explanatory causes for the presenting symptoms are not subsumed in the pretest clinical decision rule for PE (Wells criteria) then there is an active trail from these explanatory causes to the utility node. Thus, the probability that the patient has a PE varies with the alternative disease process under consideration. A recent study demonstrated that the probability of PE not only differs if there is an alternative diagnosis but also varies with the nature of the alternative diagnosis under consideration (i.e., it varies depending upon the clinician's suspicion of anxiety, an upper respiratory infection, a muscle strain, emphysema, etc.).[77,78] From a Bayesian standpoint, observing another likely disease process in the patient

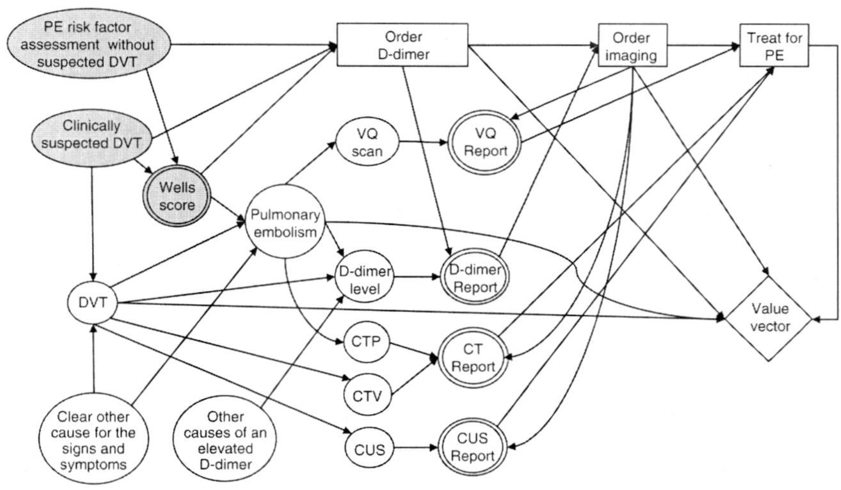

Figure 2.15 Extended D-dimer model (abbreviated depiction).
***Note*: The nodes, clear other cause for the signs and symptoms and other causes of an elevated D-dimer are aggregation nodes that represent the joint influence of sets of nodes. They are depicted here as single nodes for diagrammatic reasons only.**

that could be the cause of the chest pain or shortness of breath reduces the probability that PE is a candidate process.

Other disease processes such as acute coronary syndrome (ACS) and AD can affect D-dimer results.[79–86] ACS encompasses the spectrum of diseases caused by acutely restricted delivery of oxygenated blood to the heart muscle. The extreme version of this is a heart attack or AMI. Because a CTA dedicated for the discovery of PE is executed to optimize the sensitivity and specificity for PE, such a CTA rarely detects an AD. Therefore, the sensitivity for AD of a PE-directed CTA is markedly lower than the sensitivity of an AD-directed CTA. The degree of crossover detection increases marginally with more modern imaging equipment.[87,88] The performance characteristics for CT that we use in our model are predicated on current widely available and utilized CT technology.

On the basis of initial results from this extended model, using techniques for encoding the limited model, it appears that the D-dimer test is always valuable in the base case with different cutoffs for different levels of pretest risk of PE. Presumably, this is due to alterations in the probability of PE and a positive D-dimer result which previously had not influenced the computation. Had we considered the various potential inputs that were amenable to modeling from the outset, the nodes in the extended model for clear other cause for the signs and symptoms and other causes for an elevated D-dimer would have been noted to have active paths to the utility

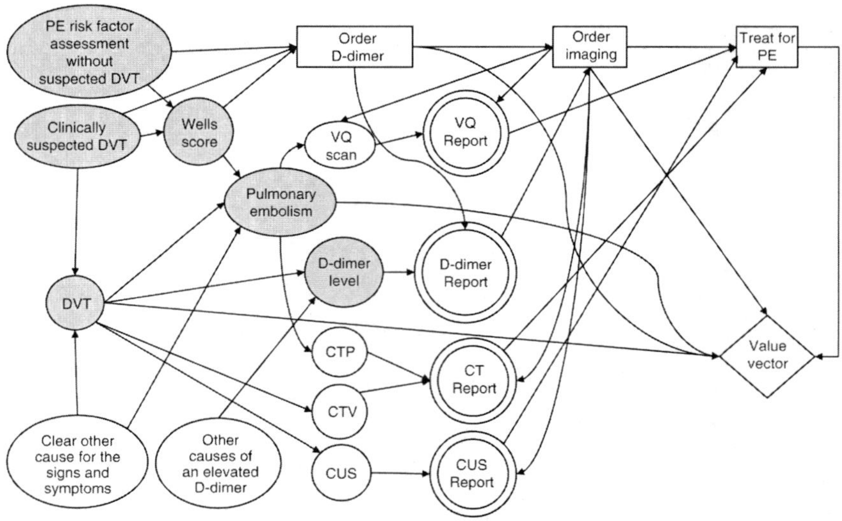

Figure 2.16 Requisite observations on the extended model to exclude the new inputs to the sub-graph representing the limited model.
***Note*: If we observe the shaded nodes in this diagram, we can render other potential causes of the patient's signs and symptoms and other causes of an elevated D-dimer irrelevant to the computational results of the model.**

node. In other words, we would have seen that knowing something about this set of nodes tells us something about the value of the D-dimer test in the evaluation of patients for PE. Our analysis would have indicated that removing these nodes would compromise the computational results of the model.

We can make a set of observations on uncertainties in the extended model that would block any active paths from the additional nodes in the limited model (Figure 2.16).

If we observe the presence or absence of DVT, PE, and the D-dimer level, we create a situation in which the clinician needs to know nothing about clear other cause for the signs and symptoms and other causes for an elevated D-dimer. The formulation in Figure 2.16 has two major problems. First, the clinician has no clairvoyance on whether the patient has PE or DVT, and what the D-dimer level will be. If the clinician knows the patient has a PE or DVT, providing a lifesaving and life-prolonging treatment is in order; and if the patient has no DVT and no PE, providing treatment is unnecessary and wasteful (and possibly harmful). Second, even if the requisite uncertainties could be observed, the resulting model does not answer the fundamental question under consideration: is use of a quantitative D-dimer assay cost effective in the evaluation of patients presenting to the ED with signs and symptoms that might be suggestive of PE?

Although the second point is related to the first point, they are actually different notions: one has to do with the feasibility of certain observations and the other has to do with the applicability and acceptability of a model with certain observations.

2.7 Summary

We have shown how an ID can provide an intuitive graphical framework for understanding cost-effectiveness problems while also providing a computational framework for their solution. Because computation complexity increases rapidly with the number of modeled variables and decisions, a balance must be struck between clinical relevance (model complexity) and minimizing computational burden. We have shown how IDs can be used to formally test conditions allowing for the inclusion and exclusion of certain variables in attempts to simplify very complex models without impacting the validity of the analysis. We created the notion of scope in this context. There is interplay between the structural characteristics of the ID and the scope of the problem, the analyst is seeking to answer.

The techniques outlined in this chapter are meant to manage this interplay to ensure that the policy recommendations from the cost-effectiveness analysis are applied to the appropriate decision question. The principles we have presented could be applied to any probabilistic decision problem and are especially helpful when precision with respect to the question critically determines the utility of the resulting analysis.

References

1. Neumann, P. J., Greenberg, D., Olchanski, N. V., Stone, P. W., and Rosen, A. B. Growth and quality of the cost-utility literature, 1976–2001. *Value in Health*, **8**, 3–9, 2005.
2. Tengs, T.O. and Wallace, A. One thousand health-related quality-of-life estimates. *Medical Care*, **38**, 583–637, 2000.
3. Revicki, D. A. National health preference data: A useful resource for health services research. *Medical Decision Making*, **26**, 310–312, 2006.
4. Kostopoulou, O. The transient nature of utilities and health preferences. *Medical Decision Making*, **26**, 304–306, 2006.
5. Neumann, P. J., Zinner, D. E., and Wright, J. C. Are methods for estimating QALYs in cost-effectiveness analyses improving? *Medical Decision Making*, **17**, 402–408, 1997.
6. Chapman, R. H., Marc, B., Milton, C. W., Jane, C. W., Sue, G., and Peter, J. N. When does quality-adjusting life-years matter in cost-effectiveness analysis? *Health Economics*, **13**, 429–436, 2004.
7. Ginsburg, P. B., Strunk, B. C., Banker, M. I., and Cookson, J. P. Tracking health care costs: Spending growth remains stable at high rate in 2005. *Data Bulletin No. 33 (Center for Studies in Health Systems and Change)*, 33, 1–2, 2006.

8. Anderson, G. F. Controlling U.S. health spending: Opportunities for academic health centers. *Academic Medicine*, **81**, 807–811, 2006.
9. Cutler, D. M., Rosen, A. B., and Vijan S. The value of medical spending in the United States, 1960–2000. *The New England Journal of Medicine*, **355**, 920–927, 2006.
10. Knockaert, D. C., Buntinx, F., Stoens, N., Bruyninckx, R., and Delooz, H. Chest pain in the emergency department: The broad spectrum of causes. *European Journal of Emergency Medicine*, **9**, 25–30, 2002.
11. Weissman, I. A., Dickinson, C. Z., Dworkin, H. J., O'Neill, W. W., and Juni, J. E. Evaluation of chest pain in the emergency department. *Annals of Internal Medicine*, **121**, 976–978, 1994.
12. Gold, M. R., Siegel, J. E., Russel, L. B., and Weinstein, M. C. *Cost-Effectiveness in Health and Medicine*, Oxford University Press, New York, 1996.
13. Perrier, A., Nendaz, M., Sarasin, F., Howarth, N., and Bounameaux, H. Cost-effectiveness analysis of diagnostic strategies for suspected pulmonary embolism including helical computed tomography. *American Journal of Respiratory and Critical Care Medicine*, **167**, 39–44, 2003.
14. Thompson, D., Eitel, D., Fernandes, C. M. B., Pines, J. M., Amsterdam, J., and Davidson, S. J. Coded chief complaints—Automated analysis of free-text complaints. *Academic Emergency Medicine*, **13**, 774–782, 2006.
15. National Hospital Ambulatory Medical Care (N. A. M. C.) Survey: 2001 emergency department summary. *Advanced Data*, 1–29, 2003.
16. Shachter, R. D. Bayes-Ball: The rational pastime (for determining irrelevance and requisite information in belief networks and influence diagrams), in G. F. Cooper, S. Moral (eds.), *Uncertainty in Artificial Intelligence: Proceedings of the Fourteenth Conference*, pp. 480–487, Morgan Kaufmann, San Francisco, CA, 1998.
17. Shachter, R. D. Efficient value of infomation computation, in: K. B. Laskey, H. Prade (eds.), *Uncertainty in Artificial Intelligence: Proceedings of the Fifteenth Conference*, pp. 594–601, Morgan Kaufmann, San Mateo, CA, 1999.
18. Pearl, J., Geiger, D., and Verma, T. In *Influence Diagrams, Belief Nets, and Decision Analysis*, Oliver, R. M. and Smith, J. Q. (Eds.), John Wiley & Sons, 1990, pp. 67–87.
19. Geiger, D., Verma, T., and Pearl, J. d-Separation: From theorems to algorithms, in *Fifth Workshop on Uncertainty in Artificial Intelligence*, pp. 118–125, University of Windsor, Ontario, 1989.
20. Heckerman, D. and Shachter, R. Decision-theoretic foundations for causal reasoning. *Journal of Artificial Intelligence Research*, **3**, 405–430, 1995.
21. Darwiche, A. Recursive conditioning. *Artificial Intelligence*, 5–41, 2001.
22. Howard, R. In *Influence Diagrams, Belief Nets and Decision Analysis*, Oliver, R. M. and Smith, J. Q. (Eds.), John Wiley & Sons, 1990, pp. 3–23.
23. Hingorani, A., Ascher, E., Marks, N., Schutzer, R., Mutyala, M., Yorkovich, W., and Jacob, T. Morbidity and mortality associated with brachial vein thrombosis. *Annals of Vascular Surgery*, **20**, 297–300, 2006.
24. Hingorani, A., Ascher, E., Markevich, N., Schutzer, R. W., Kallakuri, S., Mutyala, M., Nahata, S., Yorkovich, W., and Jacob, T. Prospective evaluation of combined upper and lower extremity DVT. *Vascular and Endovascular Surgery*, **40**, 131–134, 2006.

25. Joffe, H. V., Kucher, N., Tapson, V. F., and Goldhaber, S. Z. Upper-extremity deep vein thrombosis: A prospective registry of 592 patients. *Circulation*, **110**, 1605–1611, 2004.
26. Malhotra, S. and Punia, V. P. S. Upper extremity deep vein thrombosis. *Journal of the Association of Physicians of India*, **52**, 237–241, 2004.
27. Markel, A. Origin and natural history of deep vein thrombosis of the legs. *Seminars in Vascular Medicine*, **5**, 65–74, 2005.
28. Prandoni, P., Villalta, S., Bagatella, P., Rossi, L., Marchiori, A., Piccioli, A., Bernardi, E., Girolami, B., Simioni, P., and Girolami, A. The clinical course of deep-vein thrombosis. Prospective long-term follow-up of 528 symptomatic patients. *Haematologica*, **82**, 423–428, 1997.
29. Cushman, M., Tsai, A. W., White, R. H., Heckbert, S. R., Rosamond, W. D., Enright, P., and Folsom, A. R. Deep vein thrombosis and pulmonary embolism in two cohorts: The longitudinal investigation of thromboembolism etiology. *The American Journal of Medicine*, **117**, 19–25, 2004.
30. Kearon, C. Natural history of venous thromboembolism. *Circulation*, **107**, I22–I30, 2003.
31. Bulger, C. M., Jacobs, C., and Patel, N. H. Epidemiology of acute deep vein thrombosis. *Techniques in Vascular and Interventional Radiology*, **7**, 50–54, 2004.
32. Dalen, J. and Alpert, J. Natural history of pulmonary embolism. *Progressive Cardiovascular Diseases*, **17**, 259–270, 1975.
33. Stein, P. D., Kayali, F., and Olson, R. E. Estimated case fatality rate of pulmonary embolism, 1979 to 1998. *The American Journal of Cardiology*, **93**, 1197–1199, 2004.
34. Stein, P. D., Kayali, F., and Olson, R. E. Regional differences in rates of diagnosis and mortality of pulmonary thromboembolism. *The American Journal of Cardiology*, **93**, 1194–1197, 2004.
35. White, R. H. The epidemiology of venous thromboembolism. *Circulation*, **107**, I4–I8, 2003.
36. Anderson, F. A population-based perspective of the hospital incidence and case-fatality rates of deep vein thrombosis and pulmonary embolism. The Worcester DVT Study. *Archives of Internal Medicine*, **151**, 933–938, 1991.
37. Silverstein, M. D., Heit, J. A., Mohr, D. N., Petterson, T. M., O'Fallon, W. M., and Melton, L. J. Trends in the incidence of deep vein thrombosis and pulmonary embolism: A 25-year population-based study. *Archives of Internal Medicine*, **158**, 585–593, 1998.
38. Michota, F. Venous thromboembolism: Epidemiology, characteristics, and consequences. *Clinical Cornerstone*, **7**, 8–15, 2005.
39. Heit, J. A. Venous thromboembolism: Disease burden, outcomes and risk factors. *Journal of Thrombosis and Haemostasis*, **3**, 1611–1617, 2005.
40. Kroegel, C. and Reisseg, A. Principle mechanisms underlying venous thromboembolism: Epidemiology, risk factors, pathophysiology and pathogenesis. *Respiration*, **70**, 7–30, 2003.
41. Ageno, W., Squizzato, A., Garcia, D., and Imberti, D. Epidemiology and risk factors of venous thromboembolism. *Seminars in Thrombosis and Hemostasis*, **32**, 651–658, 2006.

42. Beemath, A., Stein, P., Skaf, E., Al Sibae, M., and Alesh, I. Risk of venous thromboembolism in patients hospitalized with heart failure. *The American Journal of Cardiology*, **98**, 793–795, 2006.
43. Northup, P., McMahon, M. M., Ruhl, A. P., Altschuler, S. E., Volk-Bednarz, A., Caldwell, S. H., and Berg, C. L. Coagulopathy does not fully protect hospitalized cirrhosis patients from peripheral venous thromboembolism. *American Journal of Gastroenterology*, **101**, 1524–1528, 2006.
44. Simioni, P., Tormene, D., Spiezia, L., Tognin, G., Rossetto, V., Radu, C., and Prandoni, P. Inherited thrombophilia and venous thromboembolism. *Seminars in Thrombosis and Hemostasis*, **32**, 700–708 2006.
45. Kearon, C. Natural history of venous thromboembolism. *Circulation*, **107**, 122–130, 2003.
46. Monreal, M., Munoz, F. J., Rosa, V., Romero, C., Roman, P., Di Micco, P., and Prandoni, P., Upper extremity DVT in oncological patients: Analysis of risk factors. Data from the RIETE registry, *Experimental Oncology*, **28**, 245–247, 2006.
47. Joung, S. and Bridget, R. Venous thromboembolism in cancer patients in Christchurch, 1995–1999. *The New Zealand Medical Journal*, **115**, 257–260, 2002.
48. Calder, K. K. H., Henderson, M., and Sean, O. The mortality of untreated pulmonary embolism in emergency department patients. *Annals of Emergency Medicine*, **45**, 302–310, 2005.
49. Barrellier, M., Lezin, B., Landy, S., and Le Hello, C. Prevalence of duplex ultrasonography detectable venous thrombosis in patients with suspected or acute pulmonary embolism. *Journal of Malaysian Vascular Surgery*, **26**, 23–30, 2001.
50. Arcelus, J. I., Carpini, J. A., Monreal, M., Suárez, C., and González-Fajardo, J. A. The management and outcome of acute venous thromboembolism: A prospective registry including 4011 patients. *Journal of Vascular Surgery*, **38**, 916–922, 2003.
51. Gaitini, D. Current approaches and controversial issues in the diagnosis of deep vein thrombosis via duplex Doppler ultrasound. *Journal of Clinical Ultrasound*, **34**, 289–297, 2006.
52. PIOPED. Ventilation-perfusion scintigraphy in the prospective investigation of pulmonary embolism diagnosis (PIOPED) study. *Journal of Nuclear Medicine*, **34**, 1109–1126, 1993.
53. PIOPED. Comprehensive analysis of the results of the prospective investigation of pulmonary embolism diagnosis (PIOPED) study. *Journal of Nuclear Medicine*, **36**, 2380–2387, 1995.
54. PIOPED. Value of the ventilation/perfusion scan in acute pulmonary embolism. Results of the prospective investigation of pulmonary embolism diagnosis (PIOPED). The PIOPED investigators. *JAMA*, **263**, 2753–2759, 1990.
55. Huisman, M. V. and Henri, B. Treating patients with venous thromboembolism: Initial strategies and long-term secondary prevention. *Seminars in Vascular Medicine*, **5**, 276–284, 2005.
56. Cosmi, B. and Gualtiero, P. Extended treatment for venous thromboembolism: How long is long enough? *Current Hematology Reports*, **3**, 375–381, 2004.
57. Cosmi, B. and Gualtiero, P. Oral anticoagulant therapy in venous thromboembolism. *Seminars in Vascular Medicine*, **3**, 303–314, 2003.
58. Keeling, D. Duration of anticoagulation: Decision making based on absolute risk. *Blood Reviews*, **20**, 173–178, 2006.
59. Pinede, L., Ninet, J., and Duhaut, P. For the Investigators of the Durée Optimale du Traitement Antivitamines K (DOTAVK) study. Comparison of 3 and

6 months of oral anticoagulant therapy after a first episode of proximal deep vein thrombosis or pulmonary embolism and comparison of 6 and 12 weeks of therapy after isolated calf deep vein thrombosis. *Circulation*, **103**, 2453–2460, 2001.

60. Schulman, S., Lindmarker, P., Holmström, M., Lärfars, G., Carlsson, A., Nicol, P., Svensson, E., Ljungberg, B., Viering, S., Nordlander, S., Leijd, B., Jahed, K., Hjorth, M., Linder, O., and Beckman, M. Post-thrombotic syndrome, recurrence, and death 10 years after the first episode of venous thromboembolism treated with warfarin for 6 weeks or 6 months. *Journal of Thrombosis and Haemostasis*, **4**, 734–742, 2006.
61. de Moerloose, P., Bounameaux, H., Perrier, A., and Reber, G. Performances of the VIDAS D-dimer new assay for the exclusion of venous thromboembolism. *Thrombosis and Haemostasis*, **85**, 185–186, 2001.
62. Agency for Healthcare Research and Quality (AHRQ). Diagnosis and treatment of deep venous thrombosis and pulmonary embolism. Summary, Evidence Report/Technology Assessment: Number 68. AHRQ Publication Number 03-E012, January 2003, Rockville, MD.
63. van Erkel, A., van Rossum, A., Bloem, J., Kievit, J., and Pattynama, P. Spiral CT angiography for suspected pulmonary embolism: A cost-effectiveness analysis. *Radiology*, **201**, 209–236, 1996.
64. Paterson, D. and Schwartzman, K. Strategies incorporating spiral CT for the diagnosis of acute pulmonary embolism: A cost-effectiveness analysis. *Chest*, **199**, 1791–1800, 2001.
65. de Monye, W., Sanson, B., Buller, H., Pattynama, P., and Huisman, M. The performance of two rapid quantitative D-dimer assays in 287 patients with clinically suspected pulmonary embolism. *Thrombosis Research*, **107**, 283–286, 2002.
66. Duet, M., Benelhadj, S., Kedra, W., Vilain, D., Ajzenberg, C., Elkharrat, D., Drouet, L., Soria, C., and Mundler, O. A new quantitative D-dimer assay appropriate in emergency: Reliability of the assay for pulmonary embolism exclusion diagnosis. *Thrombosis Research*, **91**, 1–5, 1998.
67. bioMerieux. Vidas D-Dimer New (DD2). *Package Insert* 2002.
68. Wells, P. S., Ginsberg, J. S., Anderson, D. R., Kearon, C., Gent, M., Turpie, A. G., Bormanis, J., Weitz, J., Chamberlain, M., Bowie, D., Barnes, D., and Hirsh, J. Use of a clinical model for safe management of patients with suspected pulmonary embolism. *Annals of Internal Medicine*, **129**, 997–1005, 1998.
69. Kruip, M., Slob, M., Schijen, J., van der Heul, C., and Buller, H. Use of a clinical decision rule in combination with D-dimer concentration in diagnostic workup of patients with suspected pulmonary embolism: A prospective management study. *Archives of International Medicine*, **162**, 1631–1635, 2002.
70. Moores, L., Collen, J., Woods, K., and Shorr, A. Practical utility of clinical prediction rules for suspected acute pulmonary embolism in a large academic institution. *Thrombosis Research*, **113**, 1–6, 2004.
71. Cornuz, J., Ghali, W. A., Hayoz, D., Stoianov, R., Depairon, M., and Yersin, B. Clinical prediction of deep venous thrombosis using two risk assessment methods in combination with rapid quantitative D-dimer testing. *American Journal of Medicine*, **112**, 198–203, 2002.
72. Anand, S., Wells, P. S., Hunt, D., Brill-Edwards, P., Cook, D., and Ginsberg, J. S. Does this patient have deep vein thrombosis? *JAMA*, **279**, 1094–1099, 1998.
73. Pearl, J. Fusion, propogation, and structuring in belief networks. *Artificial Intelligence*, **29**, 241–288, 1986.

74. Wells, P., Anderson, D. R., Rodger, M., Ginsberg, J. S., Kearon, C., Gent, M., Turpie, A. G., Bormanis, J., Weitz, J., Chamberlani, M., Bowie, D., Barnes, D., and Hirsh, J. Derivation of a simple clinical model to categorize patients probability of pulmonary embolism: Increasing the models utility with the SimpliRED D-dimer. *Thrombosis Haemostasis*, **83**, 416–420, 2000.
75. Wolf, S. J., Mc Cubbin, T. R., Feldhaus, K. M., Faragher, J. P., and Adcock, D. M. Prospective validation of Wells Criteria in the evaluation of patients with suspected pulmonary embolism. *Annals of Emergency Medicine*, **44**, 503–510, 2004.
76. Duriseti, R., Shachter, R., and Brandeau, M. Value of quantitative D-dimer assays in identifying pulmonary embolism: Implications from a sequential decision model. *Academic Emergency Medicine*, **13**, 755–766, 2006.
77. Kabrhel, C., Mc Afee, A. T., and Goldhaber, S. Z. The probability of pulmonary embolism is a function of the diagnoses considered most likely before testing. *Academic Emergency Medicine*, **13**, 471–474, 2006.
78. Kohn, M. A., Kwan, E., Gupta, M., and Tabas, J. A. Prevalence of acute myocardial infarction and other serious diagnoses in patients presenting to an urban emergency department with chest pain. *Journal of Emergency Medicine*, **29**, 383–390, 2005.
79. Ohlmann, P., Faure, A., Morel, O., Petit, H., Kabbaj, H., Meyer, N., Cheneau, E., Jesel, L., Epailly, E., Desprez, D., Grunebaum, L., Schneider, F., Roul, G., Mazzucotteli, J., Eisenmann, B., and Bareiss, P. Diagnostic and prognostic value of circulating D-dimers in patients with acute aortic dissection. *Critical Care Medicine*, **34**, 1358–1364, 2006.
80. Shilon, Y., Bar-Gil, S. A., Rudensky, B., Yinnon, A. M., Margalit, M., Sulkes, J., and Shitrit, D. A rapid quantitative D-dimer assay at admission correlates with the severity of community acquired pneumonia. *Blood Coagulation & Fibrinolysis*, **14**, 745–748, 2003.
81. Querol-Ribelles, J. M., Tenias, J. M., Grau, E., Querol-Borras, J. M., Climent, J. L., Gomez, E., and Martinez, I. Plasma D-dimer levels correlate with outcomes in patients with community-acquired pneumonia. *Chest*, **126**, 1087–1092, 2004.
82. Weber, T., Hogler, S., Auer, J., Berent, R., Lassnig, E., Kvas, E., and Eber, B. D-dimer in acute aortic dissection. *Chest*, **123**, 1375–1378, 2003.
83. Shitrit, D., Bar-Gil, S. A., Rudensky, B., Sulkes, J., and Tzviony, D. Determinants of ELISA D-dimer sensitivity for unstable angina pectoris as defined by coronary catheterization. *American Journal of Hematology*, **76**, 121–125, 2004.
84. Akutsu, K., Sato, N., Yamamoto, T., Morita, N., Takagi, H., Fujita, N., Tanaka, K., and Takano, T. A rapid bedside D-dimer assay (cardiac D-dimer) for screening of clinically suspected acute aortic dissection. *Circulation Journal*, **69**, 397–403, 2005.
85. Sbarouni, K., Georgiadou, K., Marathias, K., Geroulanos, K., and Kremastinos, K. D-dimer and BNP levels in acute aortic dissection. *International Journal of Cardiology*, January 2007.
86. Kiernan, T. J. Aortic dissection and elevated D-dimers—an important clinical link. *International Journal of Cardiology*, **114**, E77–E78, 2007.
87. Johnson, T. R. C., Konstantin, N., Wintersperger, B. J., Knez, A., Boekstegers, P., Reiser, M. F., and Becker, C. R. ECG-gated 64-MDCT angiography in the differential diagnosis of acute chest pain. *American Journal of Roentgenology*, **188**, 76–82, 2007.
88. Chiles, C. C. and Jeffrey, J. Vascular diseases of the thorax: Evaluation with multidetector CT. *The Radiologic Clinics of North America*, **43**, 543–569, 2005.

Chapter 3

Non-Bayesian Classification to Obtain High Quality Clinical Decisions

Ram S. Duriseti

CONTENTS

3.1 Introduction 96
3.2 Introduction to SVMs 97
3.3 Asymmetric Cost SVMs for High Risk Clinical Decisions 99
3.4 Clinical Background 101
3.5 Methods 102
3.5.1 Descriptive Statistics of Data Set 102
3.5.2 Data Labeling 102
3.5.3 Preprocessing and Incomplete Data 103
3.5.4 Scaling of Data 103
3.5.5 Asymmetric Cost Regularization 103
3.5.6 Choosing a Kernel 104
3.5.7 Setting Parameter Values 105
3.5.8 Principal Components Analysis 106
3.5.9 Assessing the Accuracy of the Decision Function 106
3.6 Results 107
3.6.1 Optimal Penalties (C_1^* and C_2^*) 107
3.6.2 Performance 107
3.6.3 Feature Selection 107

3.7 Discussion 108
3.8 Summary 109
Appendix: Argument for a Biased Classifier 109
References 111

3.1 Introduction

The clinician's goal is to make a high quality decision. From the clinician's standpoint, a quality decision involves not only accuracy but also a degree of certainty about the accuracy of the decision: the clinician would like to make an accurate decision, and do so with a high degree of confidence. In this chapter, we discuss a novel application of a statistical learning technique called a support vector machine (SVM) to the domain of patients presenting to an emergency department (ED) with chest pain.

We have discussed elsewhere in this text how creation of utility assessments along with measures of uncertainty represented in the Bayesian Framework of an influence diagram (ID) is well suited to the clinical decision-making task. However, the same features of IDs that make them useful for clinical decision making also make them difficult to implement.

Bayesian modeling, as used for IDs, has gained wide acceptance for clinical applications. However, data for high quality Bayesian Models are not always readily available. Moreover, the highly structured nature of Bayesian networks requires significant knowledge of the relationships between variables.[1–3]

Some would argue that modeling the expert's knowledge, in the best case, only duplicates what the expert already knows intuitively and, in the worst case, encodes the cognitive biases of the expert. In the domains of assessing patients for pulmonary embolism (PE) and the diagnosis of acute coronary syndrome (ACS) among patients presenting to the ED with chest pain or related symptoms, the experience of the practitioner has not been shown to correlate with improved earlier probability assessments on the risk of PE or ACS.[4–11]

One can use structured data to validate the underlying probabilistic assumptions in a Bayesian network. In this manner, the assumptions of experts are vetted and either validated or dismissed. Many data-driven methodologies exist for learning the structure of networks and the nature of the relationships between variables in the networks.[12] However, data-driven methods to validate Bayesian Models usually suffer from a paucity of data to verify the independence assumptions of the network and the probability assignments to different conditional and joint distributions.

Consequently, although Bayesian Models have an advantage over SVMs and other machine learning techniques in that the relationships between variables under consideration are explicitly represented (unlike the black box of statistical learning techniques such as SVMs),[1] the need to explicitly represent the relationships between variables in a Bayesian Model can also be a weakness. SVMs avoid some of these pitfalls.

3.2 Introduction to SVMs

An SVM is a non-Bayesian method that takes a labeled data set (a label is assigned to each of the n categories of instances), features of a data set (e.g., patient variables, such as height, weight, ethnicity, etc.), ascertains the relative importance of those features to accurately label data instances using a portion of a data set (the training data), tests the resulting decision model on some fraction of the remaining data, and then classifies any new instances of data into any one of the n categories. If data instances can fall into only two categories (i.e., disease is present or absent), an SVM maps a new data instance into one of the two categories. This mapping can be correct or incorrect. In this way, for a binary classification problem, an SVM will create true positives, false positives, true negatives, and false negatives. The degree to which an SVM can properly label all instances of data in a particular domain refers to the separability of the data. In a perfectly separable data set (Figure 3.1a), each instance can be classified correctly with no errors. Some data sets are not perfectly separable in the chosen feature space (Figure 3.1b). When there are t features, the partition is a $(t - 1)$ dimensional object (if the partition is linear, it is a $(t - 1)$ dimensional hyperplane).

The data instances in the data set of Figure 3.1a are perfectly separable; in Figure 3.1b, they are not. For simplicity, a feature space of only two features is shown in this example. This data set is linearly separable because a simple line partitions the data set correctly. A nonlinear partition could also be used. Sometimes increasing the number of features or changing

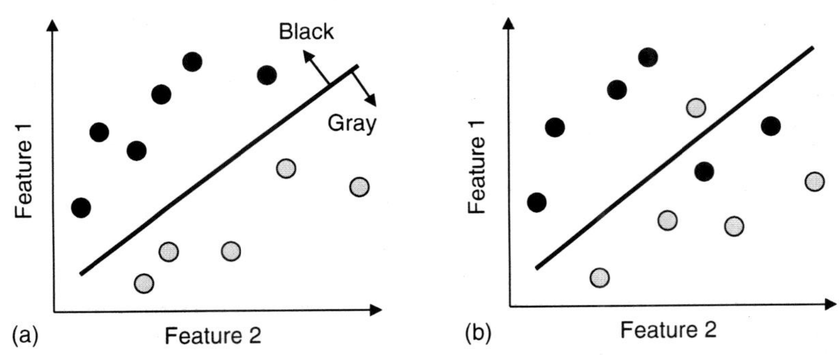

Figure 3.1 Separable versus inseparable data sets.

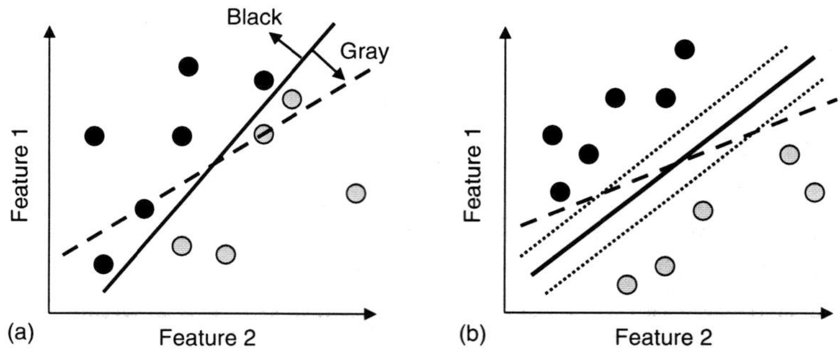

Figure 3.2 Classification with an error margin.

the geometry of the partition can help to make an inseparable data set separable.

The classification of data instances by an SVM is designed not only to just properly classify instances but also to classify each instance with a high degree of certainty. To achieve this, an SVM allows for an error margin. The algorithm tries to optimize classification while also trying to maximize distance from the error margins. If there is no margin for error, when data instances lie close to the classifier line (the classifier can be any n-dimensional shape, usually represented as a kernel function), small changes in the orientation of the classifier can result in changes to the classification of data instances. In Figure 3.2a, the data is perfectly separable using the solid line, but a marginal change in the placement of the classifier (the dashed line in Figure 3.2a) misclassifies a number of instances in the data set. Figure 3.2b shows a classifier with error margins. The weights applied to the features are such that the average distance from the error lines is maximized. Small changes in the positioning of the partition will not result in misclassification. In this way, a measure of confidence can be attached to the classification. In the formulation of the optimization problem for feature weights, C is the penalty incurred for a misclassification.

When the algorithm tunes the feature weights in the setting of error margins, minor perturbations in the classifier position will not result in classification errors.

Using the following notations,

m = the number of events in the data set
α_i = the Lagrangian for the ith training example
C = the penalty
w = vector of feature weights
b = intercept for the decision function
$x^{(i)}$ = the ith event in the data set

$y^{(i)}$ = the data label for the ith training example where $y^{(i)} \in \{-1, +1\}$

$\left(x^{(i)}, y^{(i)}\right)$ = the data vector and the label for the ith element of the data set

we can represent the primal optimization problem as

$$\operatorname*{Min}_{w,b} = \frac{1}{2} \|w\|_2$$

$$\text{s.t.:} \quad y^{(i)}\left(w^T x^{(i)} + b\right) \geq 1 \quad \text{for } i = 1, \ldots, m$$

$$0 \leq \alpha_i \leq C$$

The primal optimization problem has a dual. We derive it from the Lagrangian of the primal. The form of the dual is

$$\operatorname*{Max}_{\alpha} W(\alpha) = \sum_{i=1}^{m} \alpha_i - \frac{1}{2} \sum_{i=1}^{m} \sum_{j=1}^{m} y^{(i)} y^{(j)} \alpha_i \alpha_j \left(x^{(i)}\right)^T x^{(j)}$$

$$\text{s.t.:} \quad \sum_{i=1}^{m} \alpha_i y^{(i)} = 0$$

$$0 \leq \alpha_i \leq C$$

SVMs are non-Bayesian: they do not depend upon beliefs about the state of the world. The inclusion of a feature reflects a belief about the potential relevance of the feature, but places no assumption on the importance of the feature. In fact, the weight assigned by an SVM for a feature may be zero, suggesting that the feature is not helpful in the classification problem under consideration.

3.3 Asymmetric Cost SVMs for High Risk Clinical Decisions

In this chapter, we present a method that uses SVMs with nonhomogenous penalties for incorrect classifications in inseparable data sets. This method is intended to maximize the sensitivity of the classification rather than both the sensitivity and the specificity. For many high risk clinical decisions, data sets are not perfectly separable. In addition, the consequences of false positive classifications are not the same as the consequences of false negative classifications: the costs of mistakenly sending a patient with ACS home, in terms of money and health outcomes, are not comparable to the costs of admitting a patient unnecessarily because of a mistaken suspicion of

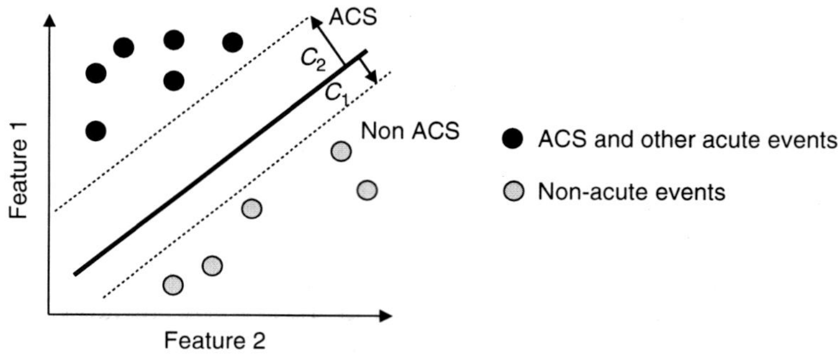

Figure 3.3 The asymmetric cost SVM.

ACS (Appendix).[13,14] This concept is portrayed in Figure 3.3 for the case of ACS.

We applied the asymmetric cost SVM to a data set of patients who presented to an urban academic ED with signs and symptoms (e.g., chest pain and shortness of breath) that could be attributable to ACS.

Non-acute events include not only non-ACS events, but also other events such as AD or PE which benefit from admission.

With asymmetric penalties, the dual takes the form

$$\operatorname*{Max}_{\alpha} W(\alpha) = \sum_{i=1}^{m} \alpha_i - \frac{1}{2}\sum_{i=1}^{m}\sum_{j=1}^{m} y^{(i)} y^{(j)} \alpha_i \alpha_j \left(x^{(i)}\right)^T x^{(j)}$$

$$\text{s.t.:} \quad \sum_{i=1}^{m} \alpha_i y^{(i)} = 0$$

$$0 \le \alpha_i \le C_1 \quad \text{for} \quad y^{(i)} = +1$$

$$0 \le \alpha_i \le C_2 \quad \text{for} \quad y^{(i)} = -1$$

Unlike traditional SVM formulations, this formulation uses asymmetric penalties. We assume $C_2 > C_1 > 0$.

In the next section, we apply the asymmetric cost SVM to a clinical data set that is traditionally inseparable and where the consequences of incorrect decisions are sizable: patients with chest pain or other symptoms referable to ACS. It is traditionally one of the most difficult decisions in clinical medicine.

3.4 Clinical Background

Chest pain is one of the most common presenting complaints of patients to EDs nationally.[15–17] As with PEs, the range of diseases that could cause chest pain varies widely in both severity and organ system involved (from psychosomatic disease to acid reflux to potentially lethal cardiovascular pathology), complicating the clinical assessment.[7,18] National data suggests that only 11 percent of patients who present to EDs with the complaint of chest pain are later found to have ACS or another diagnosis requiring admission.[16] Moreover, of admitted patients, only 15–20 percent are thought to have chest pain related to acute ischemic cardiovascular pathology.[7,16,18–22] Unnecessary admissions incur tremendous costs to the system without identifiable benefit.[23] In 1997, more than three million patients were admitted to U.S. hospitals with chest pain. Conservative estimates of the total annual costs to the system for those not found to have an ischemic etiology for the pain are well over $3 billion.[24] However, injudicious discharge of these patients home can result in major patient morbidity and mortality.[22,25] Untreated myocardial infarctions have a six month mortality of 25 percent, a figure very similar to estimates of mortality for untreated PEs.[5,25–30] Several studies estimate the rate of discharge of patients with chest pain due to ACS to be 4–4.4 percent.[31,32]

Different computational techniques have been used to differentiate patients with chest pain.[33,34] Goldman et al. imply that no one (not even the healthy male in his mid-twenties) is safe for discharge home based upon an ad hoc threshold for posterior probability of disease (usually less than 1 percent).[33] The 1 percent rule has not been established with any sort of methodological rigor beyond expert opinion.[7,35]

Despite tremendous clinical advances, the rate of missed myocardial ischemia has remained roughly 4 percent since 1996.[33,35] Neural networks, Bayesian methods, and computer-designed decision rules have had variable efficacy in improving on the sensitivity of experienced clinicians.[9,35–43] These methodologies have improved specificity to levels up to ~88 percent (from a base of ~30 percent), but often at considerable cost in sensitivity, decreasing it to 80 percent from 88 percent. Often, practitioners do not accept such methodologies. Hollander et al. found that out of 432 patients enrolled in a study, feedback to physicians derived with a neural network affected decisions in only two cases.[35]

Expert clinical opinion has major limitations as well. Ting et al. found that each year of postgraduate clinical experience resulted in a 1.4 increased odds of admitting a patient with suspected ischemic chest pain, but with no increase in the detection of myocardial.[11] Dreiseitl et al. analyzed four standard statistical computing techniques to identify which case features, of a possible 43 features in a data set, are most predictive of ischemic causes for chest pain. Different analytic methods selected different features. Eleven features were selected as important by most algorithms. Interestingly, a consulting cardiologist identified only three of the eleven selected features, and five of the nine features identified by the cardiologist as important were not selected by any of the learning algorithms.[41]

In this chapter, we demonstrate how one would apply an SVM to the classification of patients presenting to the ED with a complaint potentially stemming from ischemic heart disease. A review of the published literature found no published work on the use of SVMs for this clinical problem.[44]

Clinical evidence suggests that in a large enough data set, perfect classification is not possible. Thus, success in classification should be measured by a reduction in false positives without untoward effects on sensitivity. Such a method of classification can yield greater economic and health benefits than existing clinical methods: cases found by other methods, including expert judgment, will not be missed whereas fewer patients who do not need treatment will receive the unnecessary intervention.

3.5 Methods

We modified a data set obtained from the Department of Emergency Medicine of the Hospital of the University of Pennsylvania. The data set contained data on 4356 patients seen in the ED from July 1999 through December 2002 who complained of chest pain or symptoms potentially referable to ACS.

3.5.1 Descriptive Statistics of Data Set

Of the 4356, 20 percent (873) were diagnosed with ischemic heart disease or a related disease process benefiting from admission. Of these 873, less than 1 percent had an acute process such as AD or PE as the putative nonischemic cause for the patient's discomfort. Demographics are presented in Table 3.1.

3.5.2 Data Labeling

We labeled data based upon a final WHO diagnosis. We considered two possible labels (thus, a creating a binary classification): (1) patients with pathology that requires admission (admit), (2) patients who did not require admission (discharge). Patients who require admission are ones who either have a final diagnosis of acute myocardial infarction (AMI), unstable angina (USA) (where AMI and USA are points on the spectrum of ACS), AD

Table 3.1 Demographic Descriptive Statistics of the Data Set

Demographic Parameter	Percentage of Data Set (Percent)
Black	70
White	26
Asian	2
Hispanic	1
Other	1
Female	59

(a total of two patients), PE (a total of ten patients), or experience a major complication within 30 days of being seen in the ED.

3.5.3 Preprocessing and Incomplete Data

In the data set, each patient did not necessarily have a value for each of the 88 features. We replaced incomplete data with a maximum likelihood estimate of the feature value, estimated over patients with the same label. We divided data categories such as blood pressure and heart rate that demonstrate non-monotonic behavior into n variables with binary labels. The 88 features we identified are shown in Table 3.2.

3.5.4 Scaling of Data

To avoid domination of other parameters by one parameter in the computation of matrix inner products, some feature values had to be scaled. Several authors have commented on this, but this is not a formal requirement of an SVM application.[45] In the data set we used, this is potentially an issue. The data matrix is sparse with mostly binary indicator variables, but with several variables that had routine values exceeding 5000. Therefore, we scaled all data values linearly on the [0,1] interval by dividing by the maximum value for each parameter.

3.5.5 Asymmetric Cost Regularization

Using the SVM methodology with asymmetric penalties (Figure 3.3), C_2 is the penalty incurred when a patient with an admit label is misclassified (a data example justifying admission is labeled for discharge—a false negative) and C_1 is the penalty incurred when a patient with a discharge label is misclassified (a data example justifying discharge is labeled for admission—a false positive).

Table 3.2 Features Identified after Preprocessing of Data Set

Variable	Variable Domain	Variable	Variable Domain
Age	Positive integers	Radiation to arms	Two binary variables
Number of visits	Positive integers	Radiation of pain	Five binary variables
High pulse	Binary variable	Cardiac exam findings	Three binary variables
Normal pulse	Binary variable	Associated symptoms	Six binary variables
Low pulse	Binary variable	Cocaine abuse	Binary variable
High systolic	Binary variable	Amphetamine abuse	Binary variable
Low systolic	Binary variable	Personal medical history	Eight binary variables
High diastolic	Binary variable	Family medical history	Binary variable
Low diastolic	Binary variable	ECG findings	Ten binary variables
Fever	Binary variable	Initial cardiac markers	Three real-valued variables
Ethnicity	Five binary variables	Urine cocaine detected	Binary variable
Female	Binary variable	Medication history	Eight binary variables
Chest pain duration	Positive integers	ACITIPI score	Real-valued variable
Time since onset	Positive integers	Goldman score	Positive integers
Location of chest pain	Eight binary variables	Chest pain constant	Binary variable
Involvement of arms	Two binary variables	Relationship to palpation	Binary variable
Involvement of jaw	Binary variable	Change with exertion	Binary variable
Involvement of legs	Binary variable	Chest pain main	Binary variable
Involvement of back	Binary variable	Qualityof pain	Binary variable

3.5.6 *Choosing a Kernel*

The kernel is the partition used to separate data instances in the feature space. In our case, the feature space has 88 dimensions. The data set is not completely separable with a linear kernel for any value of C_1 or C_2. Consequently, we considered nonlinear feature spaces including radial basis and polynomial kernels. We used two methods to select and construct kernels (K): we selected kernels to minimize alignment (A) to each other while maximizing alignment of the component kernels to the data ($Y^{(i)}$). For the second method of kernel selection, we iterated over different types of kernels in preprocessing step to choose the kernel, which maximized the objective function of the dual. The definition of alignment used is attributable to Cristianini et al.[46]

$$A\left(\mathrm{K}_i, YY^T\right) = \frac{\left\langle \mathrm{K}_i, YY^T \right\rangle}{\sqrt{\langle \mathrm{K}_i, \mathrm{K}_i \rangle \left\langle YY^T, YY^T \right\rangle}}$$

Linear kernel: XX^T

$$A\left(\mathrm{K}_i, \mathrm{K}_j\right) = \frac{\langle \mathrm{K}_i, \mathrm{K}_i \rangle}{\sqrt{\langle \mathrm{K}_i, \mathrm{K}_i \rangle \langle \mathrm{K}_i, \mathrm{K}_i \rangle}}$$

Radial basis kernel:

$$\exp - \left\{ \left[\gamma \left(x_i - x_j\right) \left(x_i - x_j\right) \right]^T \right\}$$

where K_i and K_i are kernel matrices

Polynomial kernel:

$$\mathrm{K}(X) = \left(XX^T + 1\right)^p XX^T$$

In performing this task, three kernels were considered (with the understanding that the linear kernel is a degenerate case of the radial basis kernel). The alignment method is well suited to selecting the best value for p of the polynomial kernel.

3.5.7 Setting Parameter Values

The values of C_1, the linear function $C_2 = f(C_1)$, and the value gamma (γ) for the radial basis kernel are not known a priori. Furthermore, of the three fundamental kernel forms considered, nothing clearly identified one kernel as better than another. Therefore, we solved the following parameter optimization problem with hold-out cross validation (HOCV) using 7.5 percent of the data set:

$$\underset{\theta}{\arg\max} \left\{ \begin{array}{l} \underset{C_1 f(C_1), \gamma, K(x)}{\arg\max} \left\{ \left(\mu(\text{sensitivity}) + \text{specificity}\right) \middle| \max\limits_{\alpha} W(\alpha) \right\} \\ \underset{C_1 f(C_1), \gamma, K(x)}{\arg\max} \left\{ (1 - \text{error}) \middle| \max\limits_{\alpha} W(\alpha) \right\} \end{array} \right\}$$

where

$$W(\alpha) = \sum_{i=1}^{m} \alpha_i - \frac{1}{2} \sum_{i=1}^{m} \sum_{j=1}^{m} y^{(i)} y^{(j)} \alpha_i \alpha_j \left(x^{(i)}\right)^T x^{(j)}$$

$$\text{s.t.:} \quad \sum_{i=1}^{m} \alpha_i y^{(i)} = 0$$

$$0 \leq \alpha_i \leq C_1 \quad \text{for } y^{(i)} = +1$$

$$0 \leq \alpha_i \leq C_2 \quad \text{for } y^{(i)} = -1$$

$$\mu \in \{\Re : \mu \in [1, 1.3]\}$$

The variable μ above formalizes the notion that the model is intended to emphasize sensitivity over specificity in the non-separable case. No mixed kernels were considered in the course of this optimization. To solve the parameter optimization problem, we performed a binary search over the value space. We used varying proportions of the remaining data set (the 92.5 percent of data not used in the parameter optimization) to form the training set used to optimize the feature weights in the dual optimization problem.

3.5.8 Principal Components Analysis

We analyzed the data with and without dimensional reduction using principal components analysis (PCA). Owing to CPU time limitations, we did not optimize parameters in the reduced feature spaces.

3.5.9 Assessing the Accuracy of the Decision Function

The output from the optimization posed above is a vector of weights for each of the 88 features identified in the data set. We applied the vector of weights in the decision function $h(x^{(i)})$ to data instances that were not used for training. The output of the decision function is either +1 or −1:

$$h\left(x^{(i)}\right) = g\left(w^T x^{(i)} + b\right) \in \{1, -1\} \quad \text{where } g(z) = \begin{cases} 1 \text{ if } z \geq 0 \\ -1 \text{ otherwise} \end{cases}$$

For a data instance and its label: $\left(x^{(i)}, y^{(i)}\right)$

$$\text{if}\left[h\left(x^{(i)}\right)y^{(i)}\right] = \begin{cases} 1 \Rightarrow \text{correct classification} \\ -1 \Rightarrow \text{incorrect classification} \end{cases}$$

Whenever we encounter a new patient, one can create a feature vector for the patient encounter, apply the decision function with the feature weight vector obtained from the optimization problem, and classify the new patient as either an admit or a discharge. By applying the decision function on the portion of the data set that was not used to train the algorithm, we can estimate the sensitivity and specificity of the classification rule given the feature set used.

3.6 Results

3.6.1 *Optimal Penalties* (C_1^* *and* C_2^*)

From the parameter optimization step, we estimated that $C_1^* = 2.0$. The functional relationship between C_2^* and C_1^* is $C_2^* = f^*(C_1^*) = 32 \times C_1^*$. As a validation method for the value of C_1^* selected, we constructed a receiver operator curve (ROC) over different training sample sizes for a set value of C_1 and $C_2 = f(C_1)$. The largest area under the curve is achieved for values approximately in the range of $C_1 = 2$, $C_2 = 32 \times C_1$. The match is not exact, but the proximity of the estimates validates the optimization results.

3.6.2 *Performance*

The best test set performance achieved 93.5 percent sensitivity and a 43.5 percent specificity after training with 30 percent of the data set. The overall error rate (misclassification of either type—error = false positives + false negatives) was 27.5 percent. The sensitivity and specificity of the feature weights on the training set were 94.6 percent and 46.4 percent, respectively. Using an SVM with symmetric penalties on our data set, we achieved 85 percent sensitivity and 75 percent specificity with an overall error rate of 14 percent. There was no systematic pattern to the errors when one takes into account that the asymmetric cost algorithm is designed to bias towards a lower false negative rate.

3.6.3 *Feature Selection*

The true impact of feature selection could not be accurately assessed because parameters were not optimized before applying PCA to the data. Generally, feature selection performed best when the top ten or eleven features were selected.

3.7 Discussion

The data in the data set was highly non-separable. This is consistent with the observations of experienced clinicians and the published literature on the topic. In fact, this was the original motivation for an asymmetric cost SVM. If the data set were perfectly separable, we could simply optimize for error rate. Earlier attempts at risk stratification methods for patients presenting to an ED with complaints potentially referable to ACS or another acute process have also been unsuccessful in perfectly categorizing cases (either retrospectively or prospectively).[9,33,36,40,41,43,47–51] This general characteristic of non-separability, more than any other, may be what makes the problem of diagnosing concerning causes of chest pain in the ED one of the most difficult decision-making problems in medicine.

The full potential of the asymmetric cost SVM was likely not realized in this study because many parameters were not fully optimized before the training run. Despite this, the asymmetric cost SVM was able to classify patients into a category for discharge at a rate higher than current clinical practice. The associated drop in sensitivity may not (depending on the pretest assignment of disease probability) meet the standard of a posterior probability of disease, which is less than 1 percent. The asymmetric cost SVM, achieved a higher sensitivity than did a symmetric penalty SVM. Although this came at the cost of a lower specificity, the use of asymmetric penalties generated the desired behavior by preferentially tuning the feature weights to increase sensitivity. The higher error rate as sensitivity is increased with an associated drop in specificity occurs because true positives (presence of ACS) are much less frequent than true negatives in any data set that samples from a general population of patients with and without disease.

Our analysis has several limitations. When optimizing parameters, the training loop was terminated if it did not converge within 2000 runs. Allowing further iterations or requiring convergence might improve our results. One way to facilitate this is to incorporate an improvement to Platt's algorithm identified by Keerthi et al. that is particularly helpful for large values of C (in our case, C_1 and C_2).[52] Furthermore, more analysis of the impact of different types of kernels, including compound kernels, on the performance of the algorithm needs to be carried out.

It is entirely possible that the feature space over which data was collected was inadequate to facilitate accurate classification. There were, for example, some patient distinctions routinely used by some clinicians, which were not a part of the feature space. A consensus approach among experienced clinicians should be used to establish what types of data should be collected. Additionally, the demographics of this data set is not the representative of most EDs in the United States. Thus, for example, the feature weight estimates we derived may not apply to a suburban Bay Area ED.

3.8 Summary

The purpose of this chapter was to demonstrate the application of a novel variation on a non-Bayesian classification method that could be used in clinical decision problems. Many decision makers are uncomfortable with assigning utilities to outcomes and encoding probabilistic relationships between variables under consideration because they believe that their estimates are not accurate. Indeed, even for experienced decision makers in clinical medicine, such fears are often well founded.[4,6,8,10,11,53,54] In medical decision problems, these tasks are critical because errors of omission with regards to testing or treatment are usually penalized (in terms of future costs and health outcomes) much more heavily than errors of commission (unnecessary testing or treatment). The asymmetric penalty SVM presented here provides a methodology whereby a clinician can utilize a data-driven methodology to classify patients. This methodology precludes the need for consensus probabilities and assumptions about the relationships between parameters. The drawback of the methodology is that the relationships between various features are not transparent. This, however, does not make the optimized decision function any less useful.

Appendix

Argument for a Biased Classifier

The computations below by no means represent a proper cost-effectiveness analysis. It is intended only as an introduction to how and why one might

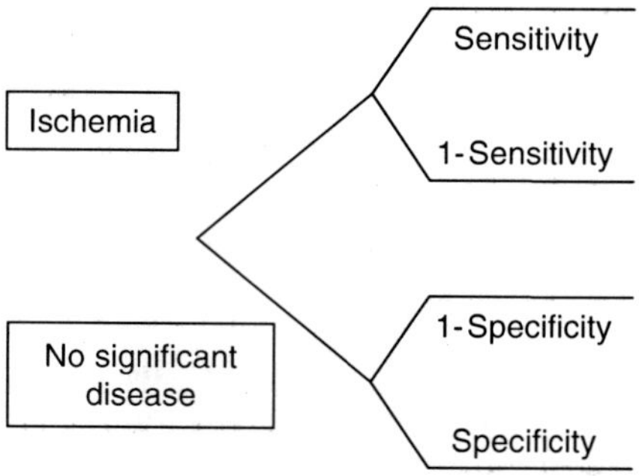

Figure A.3.1 A simple decision tree for the detection of ACS (myocardial ischemia).

want a biased classifier and why a simple notion of accuracy does not fully capture the difficulties inherent to the decision problem. Based upon and average age of presentation for patients with chest pain to the ED of 50 years old, an average remaining life span of 25 years, a 3 percent discount rate that is applicable to both costs incurred and QALYs, average settlement costs for a missed acute cardiac ischemic event of $250,000,[55,56] symmetric costs for immediate and delayed therapy (a conservative assumption as the costs are likely greater for delay of therapy with a missed diagnosis),[22,57] a 25 percent six month mortality for an unrecognized acute myocardial ischemic event, and a 10 percent six month mortality for a recognized acute myocardial event.

Assume risk neutrality for simplicity and that P (ischemia | chest pain, in ED) = 0.11. With the assumptions noted above, the expected utility analysis
shows that the biased classifier is preferred.
Biased classifier (asymmetric penalties) certain equivalent
$CE_{QALYs} = 17.52$ QALYs
$CE_{cost} = \$10,973$
Unbiased classifier (symmetric penalties) certain equivalent
$CE_{QALYs} = 17.43$ QALYs
$CE_{Cost} = \$15,905$

References

1. Lucas, P. Bayesian analysis, pattern analysis, and data mining in health care. *Current Opinion in Critical Care*, **10**, 399–403, 2004.
2. Luciani, D., Marchesi, M., and Bertolini, G. The role of Bayesian Networks in the diagnosis of pulmonary embolism. *Journal of Thrombosis and Haemostasis*, **1**, 698–707, 2003.
3. Sajda, P. Machine learning for detection and diagnosis of disease. *Annual Review of Biomedical Engineering*, **8**, 537–565, 2006.
4. Miller, C. D., Christopher, J. L., Sorabh, K., Abhinav, C., Charles, V. P., Brian, R. T., Judd, E. H., Brian, W. G., and James, W. H. Is the initial diagnostic impression of noncardiac chest pain adequate to exclude cardiac disease? *Annals of Emergency Medicine*, **44**, 565–574, 2004.
5. Stein, P. D., Kayali, F., and Olson, R. E. Regional differences in rates of diagnosis and mortality of pulmonary thromboembolism. *The American Journal of Cardiology*, **93**, 1194–1197, 2004.
6. Kabrhel, C., Carlos, C. A., and Goldhaber, S. Z. Clinical gestalt and the diagnosis of pulmonary embolism: Does experience matter? *Chest*, **127**, 1627–1630, 2005.
7. Goldman, L. and Kirtane, A. J. Triage of patients with acute chest pain and possible cardiac ischemia: The elusive search for diagnostic perfection. *Annals of Internal Medicine*, **139**, 987–995, 2003.
8. Goodacre S. L. T., Morris F., and Campbell S. How useful are clinical features in the diagnosis of acute, undifferentiated chest pain? *Academic Emergency Medicine*, **9**, 203–208, 2002.
9. Selker, H. P., Griffith, J. L., Patil, S., Long, W. J., and D'Agostino, R. B. A comparison of performance of mathematical predictive methods for medical diagnosis: Identifying acute cardiac ischemia among emergency department patients. *Journal of Investigative Medicine*, **43**, 468–476, 1995.
10. Tierney, W. M., Fitzgerald, J., McHenry, R., Roth, B. J., Psaty, B., Stump, D. L., and Anderson, F. K. Physicians' estimates of the probability of myocardial infarction in emergency room patients with chest pain. *Medical Decision Making*, **6**, 12–17, 1986.
11. Ting, H. H., Lee, T. H., Soukup, J. R., Cook, E. F., Tosteson, A. N., Brand, D. A., Rouan, G. W., and Goldman, L. Impact of physician experience on triage of emergency room patients with acute chest pain at three teaching hospitals. *The American Journal of Medicine*, **91**, 401–408, 1991.
12. Wang, X. H., Zheng, B., Good,W. F., King, J. L., and Chang, Y. H. Computer-assisted diagnosis of breast cancer using a data-driven Bayesian belief network. *International Journal of Medical Informatics*, **54**, 115–126, 1999.
13. Gold, M. R., Siegel, J. E., Russell, L.B., and Weinstein, M. C. *Cost-Effectiveness in Health and Medicine*, Oxford University Press, New York, 1996.
14. Tengs, T. and Wallace, A. One thousand health-related quality-of-life estimates. *Medical Care*, **38**, 583–637, 2000.
15. National Hospital Ambulatory Medical Care (N.A.M.C.) Survey: 2001 emergency department summary. *Advanced Data*, 1–29, 2003.

16. Burt, C. W. Summary statistics for acute cardiac ischemia and chest pain visits to United States EDs, 1995–1996. *The American Journal of Emergency Medicine*, **17**, 552–559, 1999.
17. McCaig, L. F. and Burt, C. W. National Hospital Ambulatory Medical Care (N.A.M.C.) Survey: 2002 emergency department summary. *Advanced Data*, **340**, 1–34, 2004.
18. Knockaert, D. C., Buntinx, F., Stoens, N., Bruyninckx, R., and Delooz, H. Chest pain in the emergency department: The broad spectrum of causes. *European Journal of Emergency Medicine*, **9**, 25–30, 2002.
19. Baroni P, Guida, G., and Zanella, M. Managing uncertainty in diagnosis of acute coronary ischemia. *Artificial Intelligence in Medicine*, **23**, 129–147, 2001.
20. Karlson, B. W., Herlitz, J., Pettersson, P., Ekvall, H. E., and Hjalmarson, A. Patients admitted to the emergency room with symptoms indicative of acute myocardial infarction. *Journal of Internal Medicine*, **230**, 251–258, 1991.
21. Koukkunen, H., Pyorala, K., and Halinen, M. O. Low-risk patients with chest pain and without evidence of myocardial infarction may be safely discharged from emergency department. *European Heart Journal*, **25**, 329–334, 2004.
22. Weissman I. A., O'Neill, W. W., and Juni, J. E. Evaluation of chest pain in the emergency department. *Annals of Internal Medicine*, **121**, 976–978, 1994.
23. Bayley, M. D., Schwartz, J. S., Shofer, F. S., Weiner, M., Sites, F. D., Traber, K. B., and Hollander, J. E. The financial burden of emergency department congestion and hospital crowding for chest pain patients awaiting admission. *Annals of Emergency Medicine*, **45**, 110–117, 2005.
24. Roberts, R. R., Zalenski, R. J., Mensah, E. K., Rydman, R. J., Ciavarella, G., Gussow, L., Das, K., Kampe, L. M., Dickover, B., McDermott, M. F., Hart, A., Straus, H. E., Murphy, D. G., and Rao, R. Costs of an emergency department based accelerated diagnostic protocol vs hospitalization in patients with chest pain: A randomized controlled trial. *JAMA*, **278**, 1701–1702, 1997.
25. Lee, T. H., Rouan, G. W., Weisberg, M. C., Brand, D. A., Acampora, D., Stasiulewicz, C., Walshon, J., Terranova, G., Gottlieb, L., and Goldstein-Wayne, B. Clinical characteristics and natural history of patients with acute myocardial infarction sent home from the emergency room. *The American Journal of Cardiology*, **60**, 219–224, 1987.
26. Stein, P. D., Kayali, F., and Olson, R. E. Estimated case fatality rate of pulmonary embolism, 1979 to 1998. *The American Journal of Cardiology*, **93**, 1197–1199, 2004.
27. White, R. H. The epidemiology of venous thromboembolism. *Circulation*, **107**, I4–I8, 2003.
28. Ageno, W., Squizzato, A., Garcia, D., and Imberti, D. Epidemiology and risk factors of venous thromboembolism. *Seminars in Thrombosis and Hemostasis*, **32**, 651–658, 2006.
29. Cushman, M., Lindmarker, P., Johnsson, H., Juhlin-Dennfelt, A., and Jorfeldt, L. Deep vein thrombosis and pulmonary embolism in two cohorts: The longitudinal investigation of thromboembolism etiology. *The American Journal of Medicine*, **117**, 19–25, 2004.

30. Ribeiro, A., Lindmarker, P., Johnsson, H., Juhlin-Dannfelt, A., and Jorfeldt, L. Pulmonary embolism: One-year follow-up with echocardiography doppler and five-year survival analysis. *Circulation*, **99**, 1325–1330, 1999.
31. Pope, J. H., Aufderheide, T. P., Ruthazer, R., Woolard, R. H., Feldman, J. A., Beshansky, J., Griffith, J. L., and Selker, H. P. Missed diagnoses of acute cardiac ischemia in the emergency department. *New England Journal of Medicine*, **342**, 1163–1170, 2000.
32. Herr, C. H. The diagnosis of acute myocardial infarction in the emergency department, Part I. *Journal of Emergency Medicine*, **10**, 455–461, 1992.
33. Goldman, L., Weinberg, M., Weisberg, M., Olshen, R., Cook, E. F., Sargent, R. K., Lamas, G. A., Dennis, C., Wilson, C., Deckelbaum, L., Fineberg, H., and Stiratelli, R. A computer-derived protocol to aid in the diagnosis of emergency room patients with acute chest pain. *New England Journal of Medicine*, **307**, 588–596, 1982.
34. Mitchell, A. M., Garvey, J. L., Chandra, A., Diercks, D., Pollack, C. V., and Kline, J. A. Prospective multicenter study of quantitative pretest probability assessment to exclude acute coronary syndrome for patients evaluated in emergency department chest pain units. *Annals of Emergency Medicine*, **47**, 447, 2006.
35. Hollander, J. E., Sease, K., Sparano, D. M., Sites, F. D., Shofer, F. S., and Baxt, W. G. Effects of neural network feedback to physicians on admit/discharge decision for emergency department patients with chest pain. *Annals of Emergency Medicine*, **44**, 199–205, 2004.
36. Tsien, C. L., Fraser, H. S. F., Long, W. J., and Kennedy, R. L. Using classification tree and logistic regression methods to diagnose myocardial infarction. *Medinfo*, **9**, 493–497, 1998.
37. Baxt, W. G. Use of an artificial neural network for the diagnosis of myocardial infarction. *Annals of Internal Medicine*, **115**, 843–848, 1991.
38. Baxt, W. G., Shofer, F. S., Sites, F. D., and Hollander, J. E. A neural network aid for the early diagnosis of cardiac ischemia in patients presenting to the emergency department with chest pain. *Annals of Emergency Medicine*, **40**, 575–583, 2002.
39. Baxt W. G., Shofer, F. S., Sites, F. D., and Hollander, J. E. A neural computational aid to the diagnosis of acute myocardial infarction. *Annals of Emergency Medicine*, **39**, 366–373, 2002.
40. Kennedy R. L., Harrison, R. F., Burton, A. M., Fraser, H. S., Hamer, W. G., MacArthur, D., McAllum, R., and Steedman, D. J. An artificial neural network system for diagnosis of acute myocardial infarction (AMI) in the accident and emergency department: Evaluation and comparison with serum myoglobin measurements. *Computer Methods and Programs in Biomedicine*, **52**, 93–103, 1997.
41. Dreiseitl S., Ohno-Machado, L., and Vinterbo, S. Evaluating variable selection methods for diagnosis of myocardial infarction. Proceedings of the AMIA Symposium, Washington D.C., pp. 246–250, 1999.
42. Ellenius, J. and Groth, T. Transferability of neural network-based decision support algorithms for early assessment of chest-pain patients. *International Journal of Medical Informatics*, **60**, 1–20, 2000.

43. Selker, H. P., Beshansky, J. R., Griffith, J. L., Aufderheide, T. P., Ballin, D. S., Bernard, S. A., Crespo, S. G., Feldman, J. A., Fish, S. S., Gibler, W. B., Kiez, D. A., McNutt, R. A., Moulton, A. W., Ornato, J. P., Podrid P. J., Pope, J. H., Salem, D. N., Sayre, M. R., and Woolard, R. H. Use of the acute cardiac ischemia time-insensitive predictive instrument (ACI-TIPI) to assist with triage of patients with chest pain or other symptoms suggestive of acute cardiac ischemia. A multicenter, controlled clinical trial. *Annals of Internal Medicine*, **129**, 845–855, 1998.
44. El-Naqa I., Yang, Y., Wernick, M. N., Galatsanos N. P., and Nishikawa, R. M. A support vector machine approach for detection of microcalcifications. *IEEE Transactions on Medical Imaging*, **21**, 1552–1563, 2002.
45. Chih-Wei H, C. -C. C., Chih-Jen L. (Taipei, Taiwan, 2002). A practical guide to support vector classification, online publication. Accessed on November, 2005.
46. Cristianini, N., Shawe-Taylor, J. and Kandola, J. *On Kernel Target Alignment*, in Proceedings of the Neural Information Processing Systems, NIPS'01, pp. 367–373. M.I.T. Press, Cambridge, MA.
47. Aase, O., Jonsbu, J., Liestol, K., Rollag, A., and Erikssen, J. Decision support by computer analysis of selected case history variables in the emergency room among patients with acute chest pain. *European Heart Journal*, **14**, 433–440, 1993.
48. Aase, O. Clinical experience with a decision support computer program using Bayes' theorem to diagnose chest pain patients. *Cardiology*, **92**, 128–134, 1999.
49. Baxt, W. G. and Skora, J. Prospective validation of artificial neural network trained to identify acute myocardial infarction. *Lancet*, **347**, 12–15, 1996.
50. Komorowski, J. and Øhrn, A. Modelling prognostic power of cardiac tests using rough sets. *Artificial Intelligence in Medicine*, **15**, 167–191, 1999.
51. Qamar, A., McPherson, C., Babb, J., Bernstein, L., Werdmann, M., Yasick, D., and Zarich, S. The Goldman algorithm revisited: Prospective evaluation of a computer-derived algorithm versus unaided physician judgment in suspected acute myocardial infarction. *The American Heart Journal*, **138**, 705–709, 1999.
52. Keerthi, S. S., Shevade, S. K., Bhattacharyya C., and Murthy K. R. K. Improvements to Platt's SMO algorithm for SVM classifier design. *Neural Computation*, **13**, 628–649, 2001.
53. Kostopoulou, O. The transient nature of utilities and health preferences. *Medical Decision Making*, **26**, 304–306, 2006.
54. Duseja, R. F. J. Missed acute cardiac ischemia in the ED: Limitations of diagnostic testing. *The American Journal of Emergency Medicine*, **22**, 219–225, 2004.
55. Rusnak, R. A., Stair, T. O., Hansen, K., and Fastow, J. S. Litigation against the emergency physician: Common features in cases of missed myocardial infarction. *Annals of Emergency Medicine*, **18**, 1029–1034, 1989.
56. Taylor, H. Most doctors report tear of malpractice liability has harmed their ability to provide quality care: caused them to order unnecessary tests, provide

unnecessary treatment and make unnecessary referrals. online publication. Accessed on June, 2004. in *The Harris Poll*, Harris Interactive Inc., 2002.

57. Centers for Medicare and Medicaid Services (CMS), HHS. Medicare program; payments for new medical services and new technologies under the acute care hospital inpatient prospective payment system. Final rule. *Federal Register*, **66**, 46901–46925, 2001.

Chapter 4

Optimizing Pediatric Vaccine Formularies

Shane N. Hall, Sheldon H. Jacobson, and Edward C. Sewell

CONTENTS

4.1 Motivation and Introduction 118
4.2 Models for Optimizing Pediatric Vaccine Formularies 122
4.2.1 Vaccine Formulary Selection with Limited Budget Problem 123
4.2.2 Vaccine Formulary Selection with Restricted Extraimmunization Problem 123
4.2.3 Integer Programming Model for Vaccine Formulary Selection with Limited Budget Problem 126
4.2.4 Integer Programming Model for Vaccine Formulary Selection with Restricted Extraimmunization Problem 127
4.3 Computational Complexity, Algorithms, and Heuristics 128
4.3.1 Dynamic Programming Algorithm for VFSLBP(O) and VFSREP(O) 132
4.3.2 MAX Rounding Heuristic for VFSLBP(O) and VFSREP(O) 134
4.3.3 Greedy Heuristic for VFSLBP(O) 136
4.3.4 Greedy Heuristic for VFSREP(O) 136
4.4 Computational Comparison of Algorithms and Heuristics 137
4.5 Conclusion 142
Acknowledgments 143
References 143

Abstract Vaccination against infectious disease has been hailed as one of the great public health achievements of the twentieth century. However, the United States Recommended Childhood Immunization Schedule is becoming increasingly complex, often requiring numerous separate injections to be administered during a single well-baby office visit. To address the issue of vaccine delivery complexity, vaccine manufacturers have developed combination vaccines that immunize against several diseases with a single injection. These combination vaccines are creating new challenges, such as how these vaccines should be administered to ensure that immunity is safely achieved. Furthermore, these vaccines are also creating a combinatorial explosion of alternatives and choices for public health policy-makers and administrators, pediatricians, and parents/guardians. This chapter presents two discrete optimization models that illuminate these alternatives and choices by selecting a vaccine formulary that minimizes the cost of fully immunizing a child and that limits the amount of extraimmunization (i.e., extra doses of vaccine) for any given childhood immunization schedule. The cost of vaccinating a child contributes to the underimmunization of children, and extraimmunization poses biological risks, amplifies philosophical concerns with vaccination, and creates an unnecessary economic burden on society. This chapter also discusses the computational complexity of these models, presents several optimization algorithms—both exact and heuristic—for solving these models, and provides a computational comparison of these algorithms using the 2006 Recommended Childhood Immunization Schedule as well as several randomly generated childhood immunization schedules that may be representative of future childhood immunization schedules.

4.1 Motivation and Introduction

The World Health Organization (WHO) states that immunization against infectious diseases is one factor that has had the greatest impact on world health [28]. Immunization spares millions of children each year from contracting potentially debilitating (and sometimes fatal) infectious diseases, thereby avoiding an enormous cost burden (both tangible and intangible) on the individual child, family, and society at large [11]. For example, one estimate is that pediatric immunizations prevent three million worldwide deaths in children each year [12]. Furthermore, pediatric immunizations prevent an enormous cost burden (both tangible and intangible) for individual children, families, and society at large. For example, the 2005 National Immunization Survey, administered by the Centers for Disease Control and Prevention (CDC), estimates a savings of US$27 in direct and indirect costs for every dollar spent on vaccinating against diphtheria, tetanus, and pertussis [10].

Each year, based on recommendations from the Advisory Committee on Immunization Practices (ACIP) and the American Academy of Family Physicians (AAFP), the National Immunization Program (NIP) publishes a

Recommended Childhood Immunization Schedule that outlines vaccination requirements for children through adolescence [4]. The Recommended Childhood Immunization Schedule (see Figure 4.1) outlines the vaccines required to protect a child against several (currently 13) infectious diseases that pose a risk to children living in the United States. This schedule includes the number of required doses of each vaccine and the recommended age for each dose (D1 = Dose 1, D2 = Dose 2, etc.). For example, polio requires four doses of vaccine, where the third dose (D3) may be administered at age six months, twelve months, fifteen months, or eighteen months.

Each vaccine dose is typically administered by injection during a scheduled well-baby checkup at a healthcare clinic. For example, an infant child should receive a dose of vaccine for hepatitis B, diphtheria, tetanus, pertussis, *Haemophilus influenzae* type b, polio, and pneumococcus at their two-month well-baby checkup, resulting in as many as five injections. Furthermore, a fifteen-month-old child, under extreme conditions, could receive as many as eight injections in a single clinic visit. These examples demonstrate that the Recommended Childhood Immunization Schedule is becoming overly crowded and complex. Moreover, this situation will only worsen in the future as new diseases emerge or new vaccines are

<table>
<tr><th rowspan="2">Disease</th><th colspan="10">Period (age of child)</th></tr>
<tr><th>1
(Birth)</th><th>2
(1 month)</th><th>3
(2 months)</th><th>4
(4 months)</th><th>5
(6 months)</th><th>6
(12 months)</th><th>7
(15 months)</th><th>8
(18 months)</th><th>9
(24 months)</th><th>10
(4–6 years)</th></tr>
<tr><td>Hepatitis B</td><td>D1</td><td colspan="2">D2</td><td></td><td colspan="4">D3</td><td></td><td></td></tr>
<tr><td>Diphtheria, tetanus, pertussis</td><td></td><td></td><td>D1</td><td>D2</td><td>D3</td><td></td><td colspan="2">D4</td><td></td><td>D5</td></tr>
<tr><td>Haemophilus influenzae type b</td><td></td><td></td><td>D1</td><td>D2</td><td>D3</td><td colspan="2">D4</td><td></td><td></td><td></td></tr>
<tr><td>Polio</td><td></td><td></td><td>D1</td><td>D2</td><td colspan="4">D3</td><td></td><td>D4</td></tr>
<tr><td>Measles, mumps, rubella</td><td></td><td></td><td></td><td></td><td></td><td colspan="2">D1</td><td></td><td></td><td>D2</td></tr>
<tr><td>Varicella</td><td></td><td></td><td></td><td></td><td></td><td colspan="3">D1</td><td></td><td></td></tr>
<tr><td>Pneumococcus</td><td></td><td></td><td>D1</td><td>D2</td><td>D3</td><td colspan="2">D4</td><td></td><td></td><td></td></tr>
<tr><td>Influenza</td><td></td><td></td><td></td><td></td><td colspan="6">D1 (yearly)</td></tr>
<tr><td>Hepatitis A</td><td></td><td></td><td></td><td></td><td></td><td>D1</td><td></td><td>D2</td><td></td><td></td></tr>
</table>

Figure 4.1 United States 2006 Recommended Childhood Immunization Schedule through age 6 (excluding recommendations for selected populations).

developed. For example, four periods and three diseases have been added to the Recommended Childhood Immunization Schedule since 1995, and there are currently several vaccine products being marketed and tested for use in children [8,10,19].

Weniger [33] discusses several options that address the issues of vaccine injection overcrowding and schedule complexity. The most feasible (and simplest) option is the development and use of combination vaccines—a vaccine that combines several antigens (a substance that stimulates the production of an antibody, i.e., toxins, bacteria, foreign blood cells, and the cells of transplanted organs) into a single injection. Some combination vaccines are already commonly used, such as the DTaP vaccine, which combines diphtheria and tetanus toxoids with acellular pertussis vaccine. The ideal combination vaccine would combine antigens for every disease in the Recommended Childhood Immunization Schedule into a single vaccine, which could be administered at birth. However, developing such a vaccine is highly unlikely based on financial and biological constraints. For example, live vaccines (vaccines that inject living antigens) can interfere with each other by competing for binding sites. Nonetheless, several pediatric combination vaccines are now coming to market, and several more are being developed and tested for licensing in the United States [13,19]. Combination vaccines will alleviate the issue of vaccine injection overcrowding and also offer economic opportunities by being more affordable per dose and reducing the shipping, handling, and storage costs of vaccines [13]. However, combination vaccines also pose their own unique challenges, such as which antigens should be combined and how should these vaccines be administered to ensure that immunity is safely achieved and remains economically reasonable. Moreover, combination vaccines offer pediatricians, public health policy-makers and administrators, and parents/guardians additional alternatives and choices on how to best immunize a child, and hence, these choices further amplify the schedule complexity. In fact, as the Recommended Childhood Immunization Schedule continually evolves, new combination vaccines will lead to a combinatorial explosion of alternatives and choices for such individuals, each with a different cost. Therefore, determining the set of vaccines that minimize the cost of immunizing a child will become more challenging, and, hence, creates a unique opportunity for optimization methods to be used to help make informed decisions.

Optimization and other operations research techniques have been used to address pediatric immunization problems. Most of this research to date addresses the economics of pediatric vaccine formulary design, combination vaccine pricing, and vaccine wastage [21,23]. Weniger et al. [34] report the results of a pilot study that uses operations research methods to assess the economic value of vaccine formularies—the set of vaccines inventoried by an immunization clinic or pediatrician. Specifically, the

Recommended Childhood Immunization Schedule for a subset of diseases (diphtheria, tetanus, pertussis, *Haemophilus influenzae* type b, and hepatitis B) and a reduced set of periods (1 month, 2 months, 4 months, 6 months, 12–18 months, and 60 months) were modeled as an integer program (IP). The objective of this IP was to assist decision makers in determining the vaccine formulary that minimized the cost to fully immunize a child against all five diseases. They describe how the model may be used to determine the best value to vaccine purchasers and briefly describe how operations research models might help determine the economic value of new vaccines being researched and developed. Jacobson et al. [24] present a more rigorous presentation of this pilot study and demonstrate how the model selects different vaccine formularies depending on the desired economic criteria. Sewell et al. [30] embed the IP from the pilot study into a bisection algorithm [2] to "reverse engineer" the maximum inclusion prices (the maximum price at which a vaccine remains part of the optimal vaccine formulary) of four combination vaccines not yet licensed in the United States. Sewell and Jacobson [29] present a rigorous description of this study, including the complete IP model. This study shows how operations research can provide beneficial economic analysis to the pharmaceutical companies that develop and manufacture vaccines (see Refs. [21,31] for additional applications of this bisection algorithm). Jacobson and Sewell [22] extended the bisection/IP algorithm approach by including it with Monte Carlo simulation, thereby determining a probability distribution for the price of the four combination vaccines.

This chapter demonstrates how discrete optimization models can be used to address the cost of immunizing a child and the issue of extraimmunization. Extraimmunization means that a child receives antigens for a given disease over the recommended quantity and timing sequence. Because combination vaccines reduce the number of required injections and may be more economical, pediatricians, public health policy-makers and administrators, and parents/guardians will likely choose combination vaccines over multiple single-antigen vaccines. However, using combination vaccines may inject children with antigens they have already received in the recommended quantity and timing sequence. For example, injecting a child with a DTaP-HBV-IPV (diphtheria, tetanus, pertussis, hepatitis B, and polio) combination vaccine at age of four months would provide extraimmunization for hepatitis B, because (according to Figure 4.1) no dose of vaccine is required at that age. Such extraimmunization poses biological risks and amplifies philosophical concerns. Biologically, extraimmunization of some antigens increases the risk of adverse side effects. Such is the case with diphtheria and tetanus vaccines [7]. Philosophically, many people challenge the safety and effectiveness of vaccinating children and particularly object to the use of combination vaccines, because they believe that injecting children with multiple antigens simultaneously overwhelm their immune system; extraimmunization due to combination vaccines only increases these fears [9,13]. This philosophical barrier to vaccination is an increasing concern for pediatricians and public health administrators. For example, in a recent national survey of pediatricians, 54 percent had encountered parents over a 12-month period who refused to vaccinate their child, citing safety concerns as the top reason for this refusal [14]. In another survey, 70 percent of pediatricians had encountered a parent in the 12 months preceding the survey who refused at least one immunization for their child [12]. In addition to these biological and philosophical concerns, the economic toll of extraimmunization is significant. For example, the annual societal cost burden of providing one extra dose of vaccine for each child born in the United States is over \$28 million, which assumes a birth rate of 11,100 births per day (see Ref. [26]) and a vaccine cost of \$7, both of which are conservative estimates, where the vaccine cost estimates are based on the federal contract purchase prices of the least expensive pediatric vaccine available (see Ref. [3]).

Specifically, this chapter presents two optimization models that minimize the cost of immunizing a child and that limit the amount of extraimmunization for any given childhood immunization schedule, and is organized as follows. Section 4.2 formally presents these two models (formulated as decision problems and as a discrete optimization problems) that determine the set of vaccines (i.e., a vaccine formulary) that should be used in a clinical environment to satisfy any given childhood immunization schedule at minimum cost or while restricting extraimmunization. Section 4.3 briefly discusses the computational complexity of the decision/discrete optimization problems, and presents a description of a dynamic programming algorithm and two heuristics for solving the discrete optimization problems. Section 4.4 presents a computational comparison of these algorithms and heuristics, while Section 4.5 presents a brief conclusion of the models and issues being reported and discussed.

4.2 Models for Optimizing Pediatric Vaccine Formularies

This section presents a model formulation for a decision problem and a discrete optimization problem used to design a vaccine formulary that addresses the cost of satisfying a given childhood immunization schedule [16]. A model formulation for a decision problem and a discrete optimization problem used to design a vaccine formulary that satisfies a given childhood immunization schedule while restricting extraimmunization is also presented [17]. Some simplifications and extensions of the discrete optimization problems are also described.

Given a childhood immunization schedule, the first decision problem, termed the Vaccine Formulary Selection with Limited Budget Problem (VFSLBP), asks whether it is possible to design a vaccine formulary within a specified budget, and the second decision problem, termed the Vaccine Formulary Selection with Restricted Extraimmunization Problem (VFSREP), asks whether it is possible to design a vaccine formulary that restricts extraimmunization for a specified set of diseases. These problems are now formally stated.

4.2.1 Vaccine Formulary Selection with Limited Budget Problem

Given

A set of periods, $T = \{1, 2, \ldots, \tau\}$,
A set of diseases, $\mathrm{D} = \{1, 2, \ldots, \delta\}$,
A set of vaccines $\mathrm{V} = \{1, 2, \ldots, \upsilon\}$, available to be administered to immunize against the δ diseases,
The number of doses of a vaccine that must be administered for immunization against the δ diseases, $n_1, n_2, \ldots, n_\delta$,
The cost of each vaccine, $c_1, c_2, \ldots, c_\upsilon$,
A budget B,
A set of binary parameters that indicate which vaccines immunize against which diseases; therefore, $I_{\mathrm{vd}} = 1$ if vaccine $\mathrm{v} \in \mathrm{V}$ immunizes against disease $\mathrm{d} \in \mathrm{D}$, 0 otherwise,
A set of binary parameters that indicate the set of periods in which a particular dose of a vaccine may be administered to immunize against a disease; therefore, $P_{\mathrm{d}jt} = 1$ if in time $t \in T$, a vaccine may be administered to satisfy the jth dose, $j = 1, 2, \ldots, n_{\mathrm{d}}$, requirement for disease $\mathrm{d} \in \mathrm{D}$, 0 otherwise,

Question: Does there exist a set of vaccines from V that can be administered over the periods in T such that these vaccines immunize against all the diseases in D, at a total cost no greater than B (i.e., do there exist values for the binary variables X_{tv}, $t \in T$, $\mathrm{v} \in \mathrm{V}$, where $X_{tv} = 1$ if vaccine $\mathrm{v} \in \mathrm{V}$ is administered in time $t \in T$, 0 otherwise, such that for all diseases $\mathrm{d} \in \mathrm{D}$, $\Sigma_{t\in T} \Sigma_{v\in V} P_{\mathrm{d}jt} X_{tv} I_{vd} \geq 1$ for dose $j = 1, 2, \ldots, n_{\mathrm{d}}$ and $\Sigma_{t\in T} \Sigma_{v\in \mathrm{V}} c_{\mathrm{v}} X_{t\mathrm{v}} \leq B$)?

4.2.2 Vaccine Formulary Selection with Restricted Extraimmunization Problem

Given

Sets $T = \{1, 2, \ldots, \tau\}$, $\mathrm{D} = \{1, 2, \ldots, \delta\}$, and $\mathrm{V} = \{1, 2, \ldots, \upsilon\}$, as defined in VFSLBP,

Parameters $\{n_{\text{d}} : \text{d} \in \text{D}\}$, $\{I_{\text{vd}} : \text{v} \in \text{V}, \text{d} \in \text{D}\}$, $\{P_{\text{d}jt} : \text{d} \in \text{D}, j = 1, 2, \ldots, n_{\text{d}}, t \in T\}$, and $\{m_{\text{d}t} : \text{d} \in \text{D}, t \in T\}$ as defined in VFSLBP,

A set of diseases where extraimmunization is permitted, $\text{D}_{\text{E}} \subseteq \text{D}$, with $|\text{D}_{\text{E}}| = \delta_{\text{E}}$,

A set of diseases where extraimmunization is not permitted, $\text{D}_{\text{NE}} = \text{D} \backslash \text{D}_{\text{E}}$, with $|\text{D}_{\text{NE}}| = \delta_{\text{NE}}$,

A set of binary parameters that indicate the set of periods in which no dose of a vaccine may be administered to immunize against a disease where extraimmunization is not permitted; therefore, $R_{\text{d}t} = 1$ if in time $t \in T$, no dose of a vaccine may be administered to immunize against disease $\text{d} \in \text{D}_{\text{NE}}$, 0 otherwise, (i.e., for any disease $\text{d} \in \text{D}_{\text{NE}}$ and time $t \in T$, $R_{\text{d}t} = 1$ if and only if $Q_{\text{d}t} = 0$).

Question: Does there exist a set of vaccines from V that can be administered over the periods in T such that these vaccines immunize against all the diseases in D while restricting extraimmunization (i.e., do there exist values for the binary decision variables $X_{t\text{v}}$, $t \in T$, $\text{v} \in \text{V}$, where $X_{t\text{v}} = 1$ if vaccine $\text{v} \in \text{V}$ is administered in time $t \in T$, 0 otherwise, such that for all diseases $\text{d} \in \text{D}_{\text{E}}$, $\Sigma_{t \in T} \Sigma_{\text{v} \in \text{V}} P_{\text{d}jt} X_{tv} I_{v\text{d}} \geq 1$ for dose $j = 1, 2, \ldots, n_{\text{d}}$, and for all diseases $\text{d} \in \text{D}_{\text{NE}}$, $\Sigma_{t \in T} \Sigma_{\text{v} \in \text{V}} R_{\text{d}t} X_{tv} I_{v\text{d}} = 0$ and $\Sigma_{t \in T} \Sigma_{\text{v} \in \text{V}} P_{\text{d}jt} X_{tv} I_{v\text{d}} = 1$ for dose $j = 1, 2, \ldots, n_{\text{d}}$)?

In the formulations of VFSLBP and VFSREP, the given sets and parameters correspond to a childhood immunization schedule together with budget and vaccine cost information (for VFSLBP) or the specified set of diseases for which extraimmunization is restricted (for VFSREP). Unless otherwise stated, the phrase "Childhood Immunization Schedule" refers to an arbitrary general immunization schedule, whereas the phrase "Recommended Childhood Immunization Schedule" refers to the published CDC immunization schedule (depicted in Figure 4.1). Define $T_{\text{d}j} = \{t \in T : P_{\text{d}jt} = 1\}$ to be the time window for disease $\text{d} \in \text{D}$ and dose $j = 1, 2, \ldots, n_{\text{d}}$, which is the set of periods when dose $j = 1, 2, \ldots, n_{\text{d}}$, may be administered for disease $\text{d} \in \text{D}$. Unless otherwise stated, assume that for all diseases $\text{d} \in \text{D}$ and doses $j = 1, 2, \ldots, n_{\text{d}}$, the periods in $T_{\text{d}j}$ are consecutive. Moreover, assume that the time windows for disease $\text{d} \in \text{D}$ are pairwise mutually exclusive (i.e., $T_{\text{d}i} \cap T_{\text{d}j} = \emptyset$ for all $i, j = 1, 2, \ldots, n_{\text{d}}$, $i \neq j$). This means that the set of periods when dose j may be administered for disease $\text{d} \in \text{D}$ does not overlap with the set of periods when dose i may be administered, for all doses $i \neq j$. Observe, for a given disease $\text{d} \in \text{D}$, that all of the time windows in the 2006 Recommended Childhood Immunization Schedule are pairwise mutually exclusive, though this was not always the case in past schedules. For example, hepatitis B did not have pairwise mutually exclusive time windows in the 2005 Recommended Childhood Immunization Schedule [5]. Hall et al. [16,17] relax the assumption of pairwise mutually exclusive time windows to generalize these models. Note that the above assumption implies that the doses for all diseases $d \in D$ are sequentially ordered, which means that for all doses $j, k = 1, 2, \ldots, n_d$, $j < k$, there exists a time $t' \in T$ such that $P_{djt'} = 1$ and $P_{dkt} = 0$ for all $t \leq t'$, $t \in T$. Define the valency, denoted by Val(v), as the number of antigens contained in vaccine $v \in V$, and hence, $\text{Val}(v) = \Sigma_{d \in D} I_{vd}$. Combination vaccines are often referred to as multivalent vaccines, or simply multivalents, because $\text{Val}(v) \geq 2$ when $v \in V$ is a combination vaccine. Furthermore, vaccine $v \in V$, where $\text{Val}(v) = 1, 2, 3, 4, 5$, or 6 is often referred to as a monovalent, bivalent, trivalent, tetravalent, pentavalent, or hexavalent vaccine, respectively. In practice, the dose parameters, n_d and m_{dt}, and schedule parameters, P_{djt}, depend on biological constraints and are determined by the recommendations of the ACIP and AAFP [6]. Note that schedule parameters P_{djt} specify the periods when vaccination is permitted (or useful) for disease $d \in D$, while the schedule parameters R_{dt} (in VFSREP) specify the periods when vaccination is restricted for disease $d \in D_{NE}$. For example, assuming disease d = hepatitis B $\in D_{NE}$, Figure 4.1 implies $P_{djt} = 1(0)$ for time $t = 1, 2, 3, 5, 6, 7, 8(4, 9, 10)$ for some dose $j = 1, 2, 3$, and $R_{dt} = 1(0)$ for time $t = 4, 9, 10(1, 2, 3, 5, 6, 7, 8)$.

In the formulation of VFSLBP, c_v is a general parameter that quantifies the economic cost of vaccine $v \in V$. For example, Weniger et al. [34] considered the actual vaccine purchase price, the cost of preparing the vaccine by medical staff, and the cost of administration (needle/syringe, needle-free injections, or oral) for a given vaccine $v \in V$. The question in VFSLBP asks if there exists a vaccine formulary administered over the periods in T that satisfies a given childhood immunization schedule and is within the given budget B (i.e., a variable assignment for the binary variables X_{tv}, for all periods $t \in T$ and vaccines $v \in V$, that satisfies the individual dose requirements $[\Sigma_{t \in T} \Sigma_{v \in V} P_{djt} X_{tv} I_{vd} \geq 1$ for dose $j = 1, 2, \ldots, n_d]$ for each disease $d \in D$, and the budget constraint $[\Sigma_{t \in T} \Sigma_{v \in V} c_v X_{tv} \leq B]$).

In the formulation of VFSREP, the set D_{NE} is the set of diseases where extraimmunization is restricted based on biological or philosophical constraints, and, hence, may change for each child, on a case-by-case basis. The question in VFSREP asks if there exists a vaccine formulary administered over the periods in T that satisfies the given childhood immunization schedule and restricts extraimmunization for the diseases in the set D_{NE} (i.e., a variable assignment for the binary decision variables X_{tv}, for all periods $t \in T$ and vaccines $v \in V$, that satisfies the per dose requirements $[\Sigma_{t \in T} \Sigma_{v \in V} P_{djt} X_{tv} I_{vd} \geq 1$ for dose $j = 1, 2, \ldots, n_d]$ for each disease $d \in D_E$, and does not exceed the dosage requirements $[\Sigma_{t \in T} \Sigma_{v \in V} P_{djt} X_{tv} I_{vd} = 1$ for dose $j = 1, 2, \ldots, n_d]$ or provide a dose in a period when no dose of a vaccine may be administered $[\Sigma_{t \in T} \Sigma_{v \in V} R_{dt} X_{tv} I_{vd} = 0]$ for each disease $d \in D_{NE}$).

Both decision problems VFSLBP and VFSREP can be addressed by solving a discrete optimization problem. More specifically, the following binary integer program can be used to answer VFSLBP.

4.2.3 Integer Programming Model for Vaccine Formulary Selection with Limited Budget Problem

The Integer Programming Model for VFSLBP is represented as VFSLBP(O).

$$\text{Min} \sum_{t \in T} \sum_{\text{v} \in \text{V}} c_{\text{v}} X_{t\text{v}}, \tag{4.1}$$

s.t.

$$\sum_{t \in T} \sum_{\text{v} \in \text{V}} P_{\text{d}jt} X_{t\text{v}} I_{\text{vd}} \geq 1 \quad \text{for all d} \in \text{D}, \; j = 1, 2, \ldots, n_{\text{d}}, \tag{4.2}$$

$$X_{t\text{v}} \in \{0, 1\} \quad \text{for all } t \in T, \; \text{v} \in \text{V}, \tag{4.3}$$

where sets T, D, and V, parameters $\{c_{\text{v}}\}$, $\{n_{\text{d}}\}$, $\{P_{\text{d}jt}\}$, and $\{I_{\text{vd}}\}$, and variables $\{X_{t\text{v}}\}$ are defined in VFSLBP.

The objective function given by Equation 4.1 minimizes the total cost of the vaccine formulary subject to the dose requirements for each disease d $\in$ D. Therefore, if the minimum total cost is less than or equal to the specified budget B, then the answer to VFSLBP is yes. Constraint given by Equation 4.2 ensures that for each disease d $\in$ D, at least one vaccine that provides immunization for disease d $\in$ D is administered in some period when dose $j = 1, 2, \ldots, n_{\text{d}}$ may be administered. Constraint given by Equation 4.3 is the binary requirement for the decision variables.

To describe the discrete optimization problem for VFSREP, several additional parameters and variables are needed. Let

$\rho_{\text{d}} \in^{+}$ be the weight of extraimmunization for disease d $\in$ D_{NE} for all time $t \in T$ such that
$P_{\text{d}jt} = 1$ for some dose $j = 1, 2, \ldots, n_{\text{d}}$ (i.e., in periods when vaccination is permitted)
$\gamma_{\text{d}} \in Q^{+}$ be the weight of extraimmunization for disease d $\in$ D_{NE} for all periods $t \in T$ such that $R_{\text{d}t} = 1$ (i.e., in periods when vaccination is restricted),
$Z^{P}_{\text{d}j} \in \text{Z}^{+} \cap \{0\}$ be the number of extra vaccine doses administered for disease d $\in$ D_{NE} in all periods $t \in T$ such that $P_{\text{d}jt} = 1$ (i.e., in periods when vaccination is permitted), and

$Z_{\mathrm{d}}^{R} \in \mathrm{Z}^{+} \mathrm{V}\{0\}$ be the number of extra vaccine doses administered for disease $\mathrm{d} \in \mathrm{D}_{\mathrm{NE}}$ in all periods $t \in T$ such that $R_{\mathrm{d}t} = 1$, (i.e., in periods when vaccination is restricted)

where Q^{+} and Z^{+} V{0} correspond to the set of all positive rational numbers and the set of all nonnegative integers, respectively. Therefore, the following integer program can be used to answer VFSREP.

4.2.4 Integer Programming Model for Vaccine Formulary Selection with Restricted Extraimmunization Problem

The Integer Programming Model for VFSREP is represented as VFSREP(O).

$$\operatorname{Min} \sum_{\mathrm{d} \in \mathrm{D}_{\mathrm{NE}}} \left[\rho_{\mathrm{d}} \left(\sum_{j=1}^{n_{\mathrm{d}}} Z_{\mathrm{d}j}^{P} \right) + \gamma_{\mathrm{d}} Z_{\mathrm{d}}^{R} \right], \tag{4.4}$$

s.t.

$$\sum_{t \in T} \sum_{\mathrm{v} \in \mathrm{V}} P_{\mathrm{d}jt} X_{t\mathrm{v}} I_{\mathrm{vd}} \geq 1 \quad \text{for all } \mathrm{d} \in \mathrm{D}_{\mathrm{E}}, \quad j = 1, 2, \ldots, n_{\mathrm{d}}, \tag{4.5}$$

$$\sum_{t \in T} \sum_{\mathrm{v} \in \mathrm{V}} P_{\mathrm{d}jt} X_{t\mathrm{v}} I_{\mathrm{vd}} - Z_{\mathrm{d}j}^{P} = 1 \quad \text{for all } \mathrm{d} \in \mathrm{D}_{\mathrm{NE}}, \quad j = 1, 2, \ldots, n_{\mathrm{d}}, \tag{4.6}$$

$$\sum_{t \in T} \sum_{\mathrm{v} \in \mathrm{V}} R_{\mathrm{d}t} X_{t\mathrm{v}} I_{\mathrm{vd}} - Z_{\mathrm{d}}^{R} = 0 \quad \text{for all } \mathrm{d} \in \mathrm{D}_{\mathrm{NE}}, \tag{4.7}$$

$$X_{tv} \in \{0, 1\} \quad \text{for all } t \in T, \quad \mathrm{v} \in \mathrm{V}, \tag{4.8}$$

$$Z_{\mathrm{d}j}^{P} \in \mathrm{Z}^{+} \mathrm{V}\{0\} \quad \text{for all } \mathrm{d} \in \mathrm{D}_{\mathrm{NE}}, \quad j = 1, 2, \ldots, n_{\mathrm{d}}, \tag{4.9}$$

$$Z_{\mathrm{d}}^{R} \in \mathrm{Z}^{+} \mathrm{V}\{0\} \quad \text{for all } \mathrm{d} \in \mathrm{D}_{\mathrm{NE}}, \tag{4.10}$$

where sets T, D, D_{NE} and V, parameters $\{n_{\mathrm{d}}\}$, $\{I_{\mathrm{vd}}\}$, $\{P_{\mathrm{d}jt}\}$, and $\{R_{\mathrm{d}t}\}$, and variables $\{X_{t\mathrm{v}}\}$ are defined in VFSREP.

The objective function given by Equation 4.4 minimizes the total weighted amount of extraimmunization in the vaccine formulary subject to the dose requirements for each disease $\mathrm{d} \in \mathrm{D}$ and extraimmunization restrictions for each disease $d \in D_{NE}$. The objective function weights are subjective, and hence, allow the model user to weight extraimmunization differently for each disease $d \in D_{NE}$ or for periods when vaccination is permitted versus when vaccination is restricted. For example, a pediatrician may weigh those diseases that pose biological risks from extraimmunization more heavily. In the non-weighted case (i.e., $\rho_d = \gamma_d = 1$ for all disease $d \in D_{NE}$), the objective function minimizes the total number of extra doses administered for all diseases $d \in D_{NE}$. In any case, if the minimum total weighted amount of extraimmunization equals zero, then the answer to VFSREP is yes. Constraint given by Equation 4.5 ensures that for each disease $d \in D_E$, at least one vaccine that provides immunization for disease $d \in D$ is administered in some period when dose $j = 1, 2, \ldots, n_d$ may be administered. Similarly, Constraint given by Equation 4.6 ensures that for each disease $d \in D_{NE}$, exactly one vaccine that provides immunization for disease $d \in D$ is administered in some period when dose $j = 1, 2, \ldots, n_d$ may be administered, plus any extra doses that are administered in the periods when dose $j = 1, 2, \ldots, n_d$ is permitted. Finally, constraint given by Equation 4.7 ensures that the number of doses administered in periods when vaccination is restricted equals zero, plus any extra doses that are administered in the periods when vaccination is restricted. Constraints given by Equations 4.8 through 4.10 are the binary and integer constraints on the respective decision variables.

4.3 Computational Complexity, Algorithms, and Heuristics

This section briefly discusses the computational complexity of VFSLBP and VFSREP, and presents a description of an exact dynamic programming algorithm and two heuristics for solving the discrete optimization problems in Section 4.2.

VFSLBP and VFSREP are NP-complete in the strong sense [15–17]. Therefore, in the worst case, both these problems are intractable, which means it is likely that a significant amount of computing effort will be needed to find the optimal vaccine formulary for a given childhood immunization schedule. In fact, both VFSLBP and VFSREP remain NP-complete even when the sets T, D, and V, or when dose (n_d, $d \in D$) and cost (c_v, $v \in V$) parameters are significantly restricted (see Refs. [16,17]). However, there exist special cases of VFSLBP and VFSREP that are solvable in polynomial time [16,17]. These polynomial special cases occur when the valency of the vaccine set is restricted. For example, in the case of monovalent vaccines (i.e., $\text{Val}(v) = 1$ for all vaccines $v \in V$), the constraint matrix for VFSLBP(O) and VFSREP(O) is totally unimodular, which implies that VFSLBP(O) and VFSREP(O) may be solved using linear programming [1,16,17]. Other polynomial special cases occur when the flexibility in the childhood immunization schedule is removed. For example, a childhood immunization schedule is said to be tight if every required dose of vaccine for each disease d ∈ D may be administered in exactly one period (i.e., for dose $j = 1, 2, \ldots, n_{\mathrm{d}}$ and disease d ∈ D, $P_{\mathrm{d}jt} = 1$ for exactly one period $t \in T$), and hence, a tight childhood immunization schedule implies a less flexible childhood immunization schedule (see Refs. [16,17] for detailed special cases of VFSLBP and VFSREP).

In Section 4.2, VFSLBP(O) and VFSREP(O) are modeled as integer programming (IP) problems, and hence, may be solved using several well-known integer optimization techniques (such as branch and bound; see Ref. [27]). Another useful and robust exact algorithm is dynamic programming (DP). A DP algorithm for VFSLBP(O) and VFSREP(O) is now presented.

Given the stated set of inputs for VFSLBP(O) or VFSREP(O), the DP algorithm solves each respective problem one period at a time beginning at the first period (i.e., $t = 1$), and steps through each period in T until $t = \tau$. Therefore, the set T defines the stages of the DP algorithm. Define $m_{\mathrm{d}t}$ as the minimum number of vaccine doses required for disease d ∈ D through period $t \in T$ and $M_{\mathrm{d}t}$ as the maximum number of vaccine doses required for disease d ∈ D through period $t \in T$.

Define a state in the DP algorithm as the number of doses of a vaccine that have been administered for each disease through period $t \in T$. Formally, a state in period $t \in T$ is a δ-dimensional vector $\mathbf{S}_t = (S_{t1}, S_{t2}, \ldots, S_{t\delta})$, where $S_{t\mathrm{d}}$ is the number of doses of a vaccine that have been administered for disease $\mathrm{d} = 1, 2, \ldots, \delta$, in periods $1, 2, \ldots, t$. Therefore, the state space in period $t \in T$ is $\Omega_t = \{\mathrm{S}_t \in \mathrm{Z}^\delta : m_{\mathrm{d}t} \le S_{t\mathrm{d}} \le M_{\mathrm{d}t}$ for all $\mathrm{d} \in \mathrm{D}\}$, where Z denotes the set of all integers. The decision in period $t \in T$ is which vaccines to administer that immunize against the diseases requiring vaccination in this period (i.e., the binary decision variables $X_{t\mathrm{v}}$), and is represented by the δ-dimensional binary vector $\mathbf{Y}_t = (Y_{t1}, Y_{t2}, \ldots, Y_{t\delta})$, where $Y_{t\mathrm{d}} = 1$ implies $X_{t\mathrm{v}} = 1$ for some vaccine v ∈ V that immunizes against disease d ∈ D (i.e., $I_{\mathrm{vd}} = 1$). The decision space in period $t \in T$ is defined as $\mathbf{\Phi}_t = \{\mathbf{Y}_t \in \mathbf{B}^\delta : 0 \le Y_{t\mathrm{d}} \le M_{\mathrm{d}t} - m_{\mathrm{d}(t-1)}$ for all $\mathrm{d} \in \mathrm{D}\}$, where $\mathbf{B}$ denotes the binary set $\{0, 1\}$. These states and decisions define the DP algorithm system dynamics: $\mathbf{S}_t = \mathbf{S}_{t-1} + \mathbf{Y}_t$. Because $\mathbf{Y}_t \in \mathbf{\Phi}_t$ is a binary vector, a state $\mathbf{S}_t \in \Omega_t$ is accessible from state $\mathbf{S}_{t-1} \in \Omega_{t-1}$ only if $\mathbf{S}_t - \mathbf{S}_{t-1}$ is also a binary vector.

Given that $\mathbf{Y}_t = \mathbf{S}_t - \mathbf{S}_{t-1}$, then a transition from state $\mathbf{S}_{t-1} \in \Omega_{t-1}$ to state $\mathrm{S}_t \in \Omega_t$ requires that a dose of vaccine be administered in period $t \in T$ for each disease in the set $\mathrm{D}_t = \{\mathrm{d} \in \mathrm{D} : Y_{t\mathrm{d}} = 1\}$. The sets $\mathrm{V}_t = \{\mathrm{v} \in V : I_{\mathrm{vd}} = 1$ and $\mathrm{d} \in \mathrm{D}_t\}$ (i.e., the set of vaccines that immunize against any disease that requires vaccination in period $t \in T$) and D_t define a sub-instance of VFSLBP(O) for VFSREP(O).

For VFSLBP(O), each such sub-instance is a set-covering problem instance, termed SCP(Y_t), with base set D_t and the collection of subsets V_t (see Ref. [27] for a formal definition of the set-covering problem). The specific set-covering problem instance for period $t \in T$ and decision $\mathbf{Y}_t \in \Phi_t$ is given by SCP(Y_t)

$$\text{Min} \quad \sum_{v \in V_t} c_v X_{tv}$$

s.t.

$$\sum_{v\hat{I}V_t} X_{tv} I_{vd} \geq 1 \quad \text{for all } d \in D_t,$$

$$X_{tv} \in \{0, 1\} \qquad \text{for all } v \in V_t.$$

To characterize the cost of decision $\mathbf{Y}_t \in \mathbf{\Phi}_t$, which is the cost of transitioning from state $\mathbf{S}_{t-1} \in \Omega_{t-1}$ in period $(t-1) \in T$ to state $S_t \in \Omega_t$ in period $t \in T$, define the one-period cost function $C_t(\mathbf{S}_{t-1}, \mathbf{Y}_t)$ as the cost of vaccination in period $t \in T$given state $\mathbf{S}_{t-1} \in \Omega_{t-1}$ and decision $\mathbf{Y}_t \in \mathbf{\Phi}_t$. Note, however, that this one-period cost in period $t \in T$ depends only on decision $\mathbf{Y}_t \in \mathbf{\Phi}_t$, and hence, the optimal value of SCP($\mathbf{Y}_t$) $= C_t(\mathbf{S}_{t-1}, \mathbf{Y}_t) = C_t(\mathbf{Y}_t)$. Therefore, the optimal one-period cost over all possible decisions in period $t \in T$ is given by $\min_{\mathbf{Y}_t \in \Phi_t} C_t(\mathbf{Y}_t)$.

For VFSREP(O), each sub-instance is termed IP($\mathbf{Y}_t$). To describe IP($\mathbf{Y}_t$), the following definitions are needed. Let

$D_{Et} = D_E \cup D_t$ and $D_{NEt} = D_{NE} \cup D_t$ for any period $t \in T$

$Z^P_{dt} \in Z^+ \cap \{0\}$ be the number of extra doses of vaccine administered for disease $d \in D_{NE}$ in period $t \in T$ such that $Y_{td} = 1$

$Z^R_{dt} \in Z^+ \cap \{0\}$ be the number of extra doses of vaccine administered for disease $d \in D_{NE}$ in period $t \in T$ such that $Y_{td} = 0$ (i.e., for disease $d \in D_{NE} \backslash D_{NEt}$).

The specific sub-instance for VFSREP(O) for period $t \in T$ and decision $Y_t \in \mathbf{\Phi}_t$ is given by IP($\mathbf{Y}_t$)

$$\text{Min} \quad \sum_{d \in D_{NEt}} \rho_d Z^P_{dt} + \sum_{d \in D_{NE} \backslash D_{NEt}} \gamma_d Z^R_{dt}$$

s.t.

$$\sum_{v \in V} X_{tv} I_{vd} \geq 1 \qquad \text{for all d } \in D_{Et},$$

$$\sum_{v \in V_t} X_{tv} I_{vd} - Z^P_{dt} = 1 \quad \text{for all d } \in D_{NEt},$$

$$\sum_{v \in V_t} X_{tv} I_{vd} - Z^R_{dt} = 0 \quad \text{for all d } \in D_{NE} \backslash D_{NEt},$$

$$X_{tv} \in \{0, 1\} \qquad \text{for all v } \in V_t,$$

$$Z^P_{dt}, Z^R_{dt} \in Z^+ \text{ V}\{0\} \qquad \text{for all d } \in D_{NE}.$$

To characterize the cost of decision $\mathbf{Y}_t \in \mathbf{\Phi}_t$, which is the cost of transitioning from state $\mathbf{S}_{t-1} \in \Omega_{t-1}$ in period $(t-1) \in T$ to state $\mathbf{S}_t \in \mathbf{\Omega}_t$ in period $t \in T$, define the one-period cost function $C_t(\mathbf{S}_{t-1}, \mathbf{Y}_t)$ as the amount of extraimmunization in period $t \in T$ given state $\mathbf{S}_{t-1} \in \Omega_{t-1}$ and decision $\mathbf{Y}_t \in \mathbf{\Phi}_t$. Note, however, that this one-period cost in period $t \in T$ depends only on decision $\mathbf{Y}_t \in \mathbf{\Phi}_t$, and hence, the optimal value of IP($\mathbf{Y}_t$) $= C_t(\mathbf{S}_{t-1}, \mathbf{Y}_t) = C_t(\mathbf{Y}_t)$. Therefore, the optimal one-period value over all possible decisions in period $t \in T$ is given by $\underset{\mathbf{Y}_t \in \mathbf{\Phi}_t}{\text{Min}} C_t(\mathbf{Y}_t)$.

For VFSLBP(O), define $Z_t(S_t)$ as the minimum cost of a vaccine formulary that immunizes against all diseases through period $t \in T$ subject to the number of required doses at the end of time $t \in T$ being equal to $S_t \in \Omega_t$. Similarly, for VFSREP(O), define $Z_t(S_t)$ as the minimum weighted (as defined by ρ_d and γ_d for disease d $\in D_{NE}$) amount of extraimmunization of a vaccine formulary that immunizes against all diseases through the period $t \in T$ subject to the number of required doses at the end of period $t \in T$ being equal to $S_t \in \Omega_t$. Therefore, the DP optimality equation is given by the recurrence relation

$$Z_t(\mathbf{S}_t) = \underset{\mathbf{Y}_t \in \Phi_t, \mathbf{S}_{t-1} \in \Omega_{t-1} : \mathbf{S}_t = \mathbf{S}_{t-1} + \mathbf{Y}_t}{\text{Min}} \{C_t(\mathbf{Y}_t) + Z_{t-1}(\mathbf{S}_{t-1})\}.$$

Furthermore, the optimal value of the vaccine formulary that satisfies a given childhood immunization schedule is given by

$$z^* = \underset{\mathbf{S}_T \in \Omega_T}{\text{Min}} Z_t(\mathbf{S}_t),$$

where Ω_τ is the state space for the final period $\tau \in T$. The DP algorithm for VFSLBP(O) and VFSREP(O) is now formally given.

4.3.1 Dynamic Programming Algorithm for VFSLBP(O) and VFSREP(O)

Step 1 Initialize

a. Initial state, $\mathbf{S}_0 \leftarrow 0$ (the δ-dimensional zero vector)
b. Initial extraimmunization contribution, $Z_0(\mathbf{S}_0) \leftarrow 0$
c. Set $m_{\text{d}0}, M_{\text{d}0} \leftarrow 0$ for all d $\in$ D
d. Initial stage, $t \leftarrow 1$

Step 2 Compute

$$Z_t(\mathbf{S}_t) = \underset{\mathbf{Y}_t \in \Phi_t, \mathbf{S}_{t-1} \in \Omega_{t-1}: \mathbf{S}_t = \mathbf{S}_{t-1} + \mathbf{Y}_t}{\text{Min}} \{C_t(\mathbf{Y}_t) + Z_{t-1}(\mathbf{S}_{t-1})\}$$

for each state $\mathbf{S}_t \in \Omega_t$.

Step 3 If $t < \tau$, then $t \leftarrow t + 1$ and return to Step 2. Else, stop and return $z^* = \underset{\mathbf{S}_\tau \in \Omega_\tau}{\text{Min}} Z_t(\mathbf{S}_t)$.

The worst case complexity of the DP algorithm is exponential in the cardinality of the disease set D [16,17]. This is not surprising, as the DP algorithm solves VSFLBP(O) and VFSREP(O) to optimality. However, the DP algorithm offers several advantages, both theoretically and computationally. First, this algorithm is efficient in practice with the 2006 Recommended Childhood Immunization Schedule (see Section 4.4), as this schedule yields a reasonable state/decision space, and the sub-instances in each period $t \in T$ are small. Second, the DP algorithm offers insight into the theoretical structure of VFSLBP and VFSREP. For example, the fact that the DP algorithm for VFSLBP(O) yields SCP instances in each period allows one to exploit existing theory and algorithms for this problem. Third, the structure of the DP algorithm is ideally suited to solve partial childhood immunization schedules that arise when a child has already been partially immunized and then reenters the healthcare system to complete the immunization schedule (this problem is termed the schedule completion problem). Fourth, the structure of the DP algorithm makes it easier to impose restrictions that are schedule-specific into each sub-instance (see Ref. [31] for some of the restrictions that are specific to the Recommended Childhood Immunization Schedule). Last, the structure of the DP algorithm is well suited for solving VFSLBP(O) or VFSREP(O) related problems that include stochastic variations. For example, during a given time $t \in T$, a parent/guardian may refuse a particular dose of vaccine if the number of injections required is unreasonably high. Therefore, as each vaccine is administered, the probability that a parent/guardian refuses another injection increases (this problem is termed the balking problem).

Given the computational complexity of VFSLBP and VFSREP, it is useful (even necessary) to design heuristics that do not guarantee optimality but execute in time that is polynomial in the size of the inputs. Two heuristics for VFSLBP(O) and VFSREP(O) are now presented.

The first heuristic uses the solution from a linear program (LP) to construct a feasible integer solution to VFSLBP(O) or VFSREP(O). This technique has been applied to several other well-known discrete optimization problems [18]. Relaxing the binary constraint for VFSLBP(O) yields the LP relaxation

$$\text{Min} \quad \sum_{t \in T} \sum_{\mathrm{v} \in \mathrm{V}} c_{\mathrm{v}} X_{t\mathrm{v}}$$

s.t.

$$\sum_{t \in T} \sum_{\mathrm{v} \in \mathrm{V}} P_{\mathrm{d}jt} X_{t\mathrm{v}} I_{\mathrm{vd}} \geq 1 \quad \text{for all } \mathrm{d} \in \mathrm{D},\ j = 1, 2, \ldots, n_{\mathrm{d}}$$

$$X_{t\mathrm{v}} \geq 0 \quad \text{for all } t \in T,\ \mathrm{v} \in \mathrm{V}.$$

Similarly, relaxing the binary and integer constraints for VFSREP(O) yields the LP relaxation

$$\text{Min} \quad \sum_{\mathrm{d} \in \mathrm{D_{NE}}} \left[\rho_{\mathrm{d}} \left(\sum_{j=1}^{n_{\mathrm{d}}} Z_{\mathrm{d}j}^{P} \right) + \gamma_{\mathrm{d}} Z_{\mathrm{d}}^{R} \right]$$

s.t.

$$\sum_{t \in T} \sum_{\mathrm{v} \in \mathrm{V}} P_{\mathrm{d}jt} X_{t\mathrm{v}} I_{\mathrm{vd}} \geq 1 \quad \text{for all } \mathrm{d} \in \mathrm{D_E},\ j = 1, 2, \ldots, n_{\mathrm{d}},$$

$$\sum_{t \in T} \sum_{\mathrm{v} \in \mathrm{V}} P_{\mathrm{d}jt} X_{t\mathrm{v}} I_{\mathrm{vd}} - Z_{\mathrm{d}j}^{P} = 1 \quad \text{for all } \mathrm{d} \in D_{\mathrm{NE}},\ j = 1, 2, \ldots, n_{\mathrm{d}},$$

$$\sum_{t \in T} \sum_{\mathrm{v} \in \mathrm{V}} R_{\mathrm{d}t} X_{t\mathrm{v}} I_{\mathrm{vd}} - Z_{\mathrm{d}}^{R} = 0 \quad \text{for all } \mathrm{d} \in \mathrm{D_{NE}},$$

$$0 \leq X_{t\mathrm{v}} \leq 1 \quad \text{for all } t \in T,\ \mathrm{v} \in \mathrm{V},$$

$$Z_{\mathrm{d}j}^{P} \geq 0 \quad \text{for all } \mathrm{d} \in \mathrm{D_{NE}},\ j = 1, 2, \ldots, n_{\mathrm{d}},$$

$$Z_{\mathrm{d}}^{R} \geq 0 \quad \text{for all } \mathrm{d} \in \mathrm{D_{NE}}.$$

Let X^*_{LP} denote the optimal decision vector for the $X_{t\text{v}}$ variables in the respective LP relaxation. A simple heuristic would be to round each fractional variable in the decision vector X^*_{LP} that is greater than some threshold value $1/\alpha$. For example, a suitable choice is $\alpha = \max_{\text{d}\in\text{D}} \alpha_{\text{d}}$, where $\alpha_{\text{d}} = (\Sigma_{\text{v}\in\text{V}} I_{\text{vd}})\big(\max_{j=1,2,\ldots,n_{\text{d}}} \Sigma_{t\in T} P_{\text{d}jt}\big)$(see Refs. [16,17].). However, if X^*_{LP} contains several fractional variables, then this technique tends to round too many variables to one, thereby yielding a poor solution. Instead of rounding all variables greater than or equal to the $1/\alpha$ threshold, it seems reasonable to round only a few variables with fractional values close to one, because these variables are more likely to equal one in the optimal integer solution. The MAX Rounding heuristic limits the number of rounded variables by selecting the variables with large fractional values.

To present the MAX Rounding heuristic, some additional notation is required. Define $\mathbf{D} = \{(\text{d},j) : \text{d} \in \text{D}, j = 1, 2, \ldots, n_{\text{d}}\}$ to be the set of all diseases ordered by dose, where $|\mathbf{D}| = \Sigma^{\delta}_{\text{d}=1} n_{\text{d}}$. For all periods $t \in T$ and vaccines $\text{v} \in \text{V}$, define $C_{t\text{v}} = \{(\text{d},j) \in \mathbf{D} : I_{\text{vd}} = 1 \text{ and } P_{\text{d}jt} = 1\}$, which specifies the diseases and dose that vaccine $\text{v} \in \text{V}$ immunizes against in period $t \in T$. Therefore, $C_{t\text{v}} \subseteq \mathbf{D}$ for all periods $t \in T$ and vaccines $\text{v} \in \text{V}$. Furthermore, in the case when all diseases $\text{d} \in \text{D}$ have mutually exclusive time windows, at most one $(\text{d}, j) \in \mathbf{D}$ for all diseases $\text{d} \in \text{D}$ is contained in any set $C_{t\text{v}}$ because, for a given disease $\text{d} \in \text{D}$ and period $t \in T$, $P_{\text{d}jt} = 1$ for at most one dose $j = 1, 2, \ldots, n_{\text{d}}$, and hence, each set $C_{t\text{v}}$ does not contain multiple doses for any disease $\text{d} \in \text{D}$. Lastly, let $f_{t\text{v}} = X^*_{\text{LP}_{t\text{v}}}(0)$ if $X^*_{\text{LP}_{t\text{v}}} \geq 1/\alpha (< 1/\alpha)$ for all periods $t \in T$ and vaccines $\text{v} \in \text{V}$, which specifies the value of vaccine $\text{v} \in \text{V}$ in period $t \in T$. Therefore, the MAX Rounding heuristic limits the number of rounded variables by greedily selecting (at each iteration) the most valuable available vaccine $\text{v} \in \text{V}$ that immunizes against the most disease doses (not yet covered) in period $t \in T$ (i.e., rounds the variable $X^*_{\text{LP}_{t\text{v}}}$ that maximizes $f_{t\text{v}} \cdot |C_{t\text{v}}|$) until every disease dose $(\text{d}, j) \in \mathbf{D}$ is covered by some vaccine $\text{v} \in \text{V}$ in period $t \in T$. The MAX Rounding heuristic is now formally given.

4.3.2 MAX Rounding Heuristic for VFSLBP(O) and VFSREP(O)

Step 1 Initialize

a. Solve the respective LP relaxation of VFSLBP(O) or VFSREP(O)
b. $f_{t\text{v}} \leftarrow X^*_{\text{LP}_{t\text{v}}}(0)$ for all $t \in T$, $\text{v} \in \text{V}$ such that $X^*_{\text{LP}_{t\text{v}}} \geq (<) 1/\alpha$
c. $X_{t\text{v}} \leftarrow 0$ for all $t \in T$ and $\text{v} \in \text{V}$
d. $\hat{C}_{t\text{v}} \leftarrow C_{t\text{v}}$ for all $t \in T$ and $\text{v} \in \text{V}$

Step 2 While $C = \bigcup_{\{tv: X_{tv}=1\}} C_{tv} \neq \mathbf{D}$ do

a. $(t', \mathrm{v}') \leftarrow \arg\max_{t \in T, \mathrm{v} \in \mathrm{V}} f_{t\mathrm{v}} \cdot |\hat{C}_{t\mathrm{v}}|$ (select the nonempty set $\hat{C}_{t\mathrm{v}}$ with the largest fractional value times the number of disease doses covered by vaccine $\mathrm{v} \in \mathrm{V}$ in period $t \in T$)
b. $X_{t'\mathrm{v}'} \leftarrow 1$ (administer vaccine $\mathrm{v}' \in \mathrm{V}$ in period $t' \in T$)
c. $\hat{C}_{t\mathrm{v}} \leftarrow \hat{C}_{t\mathrm{v}} \backslash \hat{C}_{t'\mathrm{v}'}$ for all $t \in T$ and $\mathrm{v} \in \mathrm{V}$ (remove all the disease doses covered by vaccine $\mathrm{v}' \in \mathrm{V}$ in period $\mathrm{t}' \in T$ from all remaining sets)

Step 3 For all $\mathrm{d} \in \mathrm{D}_{\mathrm{NE}}$ (for VFSREP(O) only)

a. For all $j = 1,2,\ldots,n_{\mathrm{d}}$ $Z^P_{\mathrm{d}j} \leftarrow \sum_{t \in T} \sum_{\mathrm{v} \in \mathrm{V}} P_{\mathrm{d}jt} X_{t\mathrm{v}} I_{\mathrm{vd}} - 1$
b. $Z^R_{\mathrm{d}} \leftarrow \sum_{t \in T} \sum_{\mathrm{v} \in \mathrm{V}} R_{\mathrm{d}t} X_{t\mathrm{v}} I_{\mathrm{vd}}$

Step 4 Compute and return $\sum_{t \in T} \sum_{v \in V} c_{\mathrm{v}} X_{t\mathrm{v}}$ (for VFSLBP(O)) or $\Sigma_{\mathrm{d} \in \mathrm{D}_{\mathrm{NE}}} \left[\rho_{\mathrm{d}} \left(\sum_{j=1}^{n_{\mathrm{d}}} Z^P_{\mathrm{d}j} \right) + \gamma_{\mathrm{d}} Z^R_{\mathrm{d}} \right]$ (for VFSREP(O))

The MAX Rounding heuristic executes in $O(\mathbf{T}_{LP} + |\mathbf{D}|\tau\upsilon)$ time, where $\mathbf{T}_{LP}$ is the time required to solve the LP relaxation of VFSLBP(O). Furthermore, the MAX Rounding heuristic returns a feasible solution, because every iteration of the while loop (i.e., Step 2) administers a vaccine that satisfies at least one dose requirement for some disease $\mathrm{d} \in \mathrm{D}$ (i.e., every iteration covers at least one $(\mathrm{d}, j) \in \mathbf{D}$). Moreover, MAX Rounding heuristic is an α-approximation algorithm for VFSLBP(O) and VFSREP(O) (see Refs. [16,17]), which means that the algorithm runs in polynomial time and returns a solution no worse than $\alpha \cdot z^*$, where z^* is the optimal value of VFSLBP(O) or VFSREP(O).

The second heuristic for VFSLBP(O) and VFSREP(O) uses an intuitive greedy approach and does not require the solution of a linear program, which (in theory) should be more efficient. The Greedy heuristic for VFSLBP(O) iteratively selects the lowest cost available vaccine that immunizes against the most disease doses, while the Greedy heuristic for VFSREP(O) iteratively selects the vaccine that incurs the smallest penalty for extraimmunization and immunizes against the most disease doses. Specifically, the extraimmunization penalty for vaccine $\mathrm{v} \in \mathrm{V}$ in time $t \in T$ is $W_{t\mathrm{v}} = \sum_{\{\mathrm{d} \in \mathrm{D}: I_{\mathrm{vd}}=1\}} w_{\mathrm{d}t}$, where

$$w_{\mathrm{d}t} = \begin{cases} \rho_{\mathrm{d}} & \text{if } \mathrm{d} \in \mathrm{D}_{\mathrm{NE}}, (\mathrm{d},j) \in C_{t\mathrm{v}} \\ & \text{for some } j = 1, 2, \ldots, n_{\mathrm{d}}, \text{ and } (\mathrm{d},j) \in \mathbf{C} = \cup_{t\mathrm{v}: X_{t\mathrm{v}}=1} C_{t\mathrm{v}} \\ \gamma_{\mathrm{d}} & \text{if } \mathrm{d} \in \mathrm{D}_{\mathrm{NE}}, (\mathrm{d},j) \notin C_{t\mathrm{v}} \text{ for some } j = 1, 2, \ldots, n_{\mathrm{d}} \\ 0 & \text{otherwise,} \end{cases}$$

because the penalty for extraimmunization is ρ_{d} if dose requirement $j = 1, 2, \ldots, n_{\mathrm{d}}$ for disease $\mathrm{d} \in \mathrm{D}_{\mathrm{NE}}$ is satisfied by some vaccine in an earlier iteration, γ_{d} if vaccine $\mathrm{v} \in \mathrm{V}$ immunizes against disease $\mathrm{d} \in \mathrm{D}_{\mathrm{NE}}$ but does not satisfy some dose requirement in period $t \in T$, or zero for all diseases $\mathrm{d} \in \mathrm{D}_{\mathrm{NE}}$ such that vaccine $\mathrm{v} \in \mathrm{V}$ does not provide immunization, (i.e., $I_{\mathrm{vd}} = 0$) and for all diseases $\mathrm{d} \in \mathrm{D}_{\mathrm{E}}$. Using the same notation from above for $\mathbf{D} = \{(\mathrm{d}, j) : \mathrm{d} \in \mathrm{D}, j = 1, 2, \ldots, n_{\mathrm{d}}\}$ and $C_{t\mathrm{v}} = \{(\mathrm{d}, j) \in \mathbf{D} : I_{\mathrm{vd}} = 1$ and $P_{\mathrm{d}jt} = 1\}$ for all periods $t \in T$and vaccines $\mathrm{v} \in \mathrm{V}$, the Greedy heuristics are now formally given.

4.3.3 Greedy Heuristic for VFSLBP(O)

Step 1 Initialize

a. $X_{t\mathrm{v}} \leftarrow 0$ for all $t \in T$ and $\mathrm{v} \in \mathrm{V}$
b. $\hat{C}_{t\mathrm{v}} \leftarrow C_{t\mathrm{v}}$ for all $t \in T$ and $\mathrm{v} \in \mathrm{V}$

Step 2 While $C = \bigcup_{\{t\mathrm{v}:X_{t\mathrm{v}}=1\}} C_{t\mathrm{v}} \neq \mathbf{D}$ do

a. $(t', \mathrm{v}') \leftarrow \arg\min_{t \in T, v \in V} c_{\mathrm{v}}/|\hat{C}_{t\mathrm{v}}|$ (select the nonempty set $\hat{C}_{tv}$ with the smallest cost per disease doses covered by vaccine $\mathrm{v} \in \mathrm{V}$ in period $t \in T$)
b. $X_{t'\mathrm{v}'} \leftarrow 1$ (administer vaccine $\mathrm{v}' \in \mathrm{V}$ in period $t' \in T$)
c. $\hat{C}_{tv} \leftarrow \hat{C}_{tv} \backslash \hat{C}_{t'v'}$ for all $t \in T$ and $\mathrm{v} \in \mathrm{V}$ (remove all the disease doses covered by vaccine $\mathrm{v}' \in \mathrm{V}$ in period $t' \in T$ from all remaining sets)

Step 3 Compute and return $\Sigma_{t \in T} \Sigma_{\mathrm{v} \in \mathrm{V}} c_{\mathrm{v}} X_{t\mathrm{v}}$

4.3.4 Greedy Heuristic for VFSREP(O)

Step 1 Initialize

a. $X_{t\mathrm{v}} \leftarrow 0$ for all $t \in T$ and $\mathrm{v} \in \mathrm{V}$
b. $\hat{C}_{t\mathrm{v}} \leftarrow C_{t\mathrm{v}}$ for all $t \in T$ and $\mathrm{v} \in \mathrm{V}$

Step 2 While $C = \bigcup_{\{t\mathrm{v}:X_{t\mathrm{v}}=1\}} C_{t\mathrm{v}} \neq \mathbf{D}$ do

a. Compute $W_{t\mathrm{v}}$ for all $t \in T$ and $\mathrm{v} \in \mathrm{V}$ (compute extraimmunization penalty for vaccine $\mathrm{v} \in \mathrm{V}$ in period $t \in T$)
b. $(t', \mathrm{v}') \leftarrow \arg\min_{t \in T, \mathrm{v} \in \mathrm{V}} W_{t\mathrm{v}}/|\hat{C}_{t\mathrm{v}}|$ (select the nonempty set $\hat{C}_{t\mathrm{v}}$ with the smallest penalty per disease doses covered by vaccine $\mathrm{v} \in \mathrm{V}$ in

period $t \in T$. Break ties by selecting vaccine $\mathrm{v} \in \mathrm{V}$ that immunizes against the most diseases in period $t \in T$)

c. $X_{t'\mathrm{v}'} \leftarrow 1$ (administer vaccine $\mathrm{v}' \in \mathrm{V}$ in period $t' \in T$)
$\hat{C}_{t\mathrm{v}} \leftarrow \hat{C}_{t\mathrm{v}} \backslash \hat{C}_{t'\mathrm{v}'}$ for all $t \in T$ and $\mathrm{v} \in \mathrm{V}$ (remove all the disease doses covered by vaccine $\mathrm{v}' \in \mathrm{V}$ in period $t' \in T$ from all remaining sets)

Step 3 For all $\mathrm{d} \in \mathrm{D}_{\mathrm{NE}}$

a. For all $j = 1, 2, \ldots, n_{\mathrm{d}}$
$Z^P_{\mathrm{d}j} \leftarrow \sum_{t\in T} \sum_{\mathrm{v}\in \mathrm{V}} P_{\mathrm{d}jt} X_{t\mathrm{v}} I_{\mathrm{vd}} - 1$

b. $Z^R_{\mathrm{d}} \leftarrow \sum_{t\in T} \sum_{\mathrm{v}\in \mathrm{V}} R_{\mathrm{d}t} X_{t\mathrm{v}} I_{\mathrm{vd}}$

Step 4 Compute and return $\sum_{\mathrm{d}\in \mathrm{D}_{\mathrm{NE}}} \left(\rho_{\mathrm{d}} \left(\sum_{j=1}^{n_{\mathrm{d}}} Z^P_{\mathrm{d}j}\right) + \gamma_{\mathrm{d}} Z^R_{\mathrm{d}}\right)$.

Both Greedy heuristics execute in $O(|\mathbf{D}|\tau\upsilon)$ time, and return a feasible solution, because each iteration of the while loop (i.e., Step 2) administers a vaccine that satisfies at least one dose requirement for some disease $\mathrm{d} \in \mathrm{D}$ (i.e., every iteration covers at least one $(\mathrm{d}, j) \in \mathbf{D}$). Therefore, the Greedy heuristics should be more efficient than the MAX Rounding heuristics. Moreover, the authors show that the Greedy heuristic for VFSLBP(O) is an H_β-approximation algorithm, where $\beta \equiv \max_{t\in T, \mathrm{v}\in \mathrm{V}} |C_{t\mathrm{v}}|$ and $H_k \equiv \Sigma_{i=1}^{k} \frac{1}{i}$, the sum of the first k elements in the harmonic series [16].

4.4 Computational Comparison of Algorithms and Heuristics

This section reports computational results comparing the MAX Rounding and Greedy heuristics and the DP algorithm presented in Section 4.3 for both VFSLBP(O) and VFSREP(O). For comparative purposes, computational results are also reported for an IP branch and bound (IP B&B) algorithm. The MAX Rounding and Greedy heuristics and the DP and IP B&B algorithms were executed on two sets of test problems to demonstrate their computational effectiveness and limitations. The first test problem is the 2006 Recommended Childhood Immunization Schedule. The second set of test problems are randomly generated based on hypothetical future childhood immunization schedules. The size of these randomly generated childhood immunization schedules assumes that future Recommended Childhood Immunization Schedules will expand to include more diseases and periods, and hence, will require a larger number of both monovalent and combination vaccines. These assumptions are reasonable, given the recent trends in expanding the schedule. For example, four periods and three diseases have been added to the Recommended Childhood Immunization Schedule since 1995, and there are currently several vaccine products that are marketed and tested for use in children [8,10,19].

For VFSLBP(O), the solution quality effectiveness measure θ is reported for each heuristic, where $\theta = Z_{\text{Heuristic}}/Z^*$ and $Z_{\text{Heuristic}}$ is the objective function cost returned by the heuristic and Z^* is the optimal objective function cost (returned by the exact algorithms). The execution time (in CPU seconds) is also reported for each heuristic and exact algorithm, which is the efficiency-effectiveness measure. For VFSREP(O), the solution quality effectiveness measure Z is the value of the objective function. Furthermore, for all childhood immunization schedules, $\rho_{\text{d}} = \gamma_{\text{d}} = 1$ for all diseases $\text{d} \in \text{D}_{\text{NE}}$, and hence, Z specifies the number of extra vaccine doses administered. All heuristics and exact algorithms were coded and executed in MATLABv7.0 on a 2.4 MHz Pentium IV with 1 GB of RAM including the IP B&B algorithm (using default settings) from MATLAB's optimization toolbox. The IP B&B algorithm for VFSREP(O) used an open source mixed integer optimization routine (see Ref. [32]).

The first test problem is the 2006 Recommended Childhood Immunization Schedule displayed in Figure 4.1. Therefore, D = {1 = hepatitis B, 2 = diphtheria–tetanus–pertussis, 3 = *Haemophilus influenzae* type b, 4 = polio, 5 = measles–mumps–rubella, 6 = varicella, 7 = pneumococcus, 8 = influenza, 9 = hepatitis A} with dose vector $n = (3, 5, 4, 4, 2, 1, 4, 1, 2)$, because diphtheria, tetanus, and pertussis are considered one disease and measles, mumps, and rubella are also considered one disease, and $T = \{1, 2, \ldots, 10\}$. The schedule parameters $P_{\text{d}jt}$ for diseases $\text{d} \in \text{D}$, dose $j = 1, 2, \ldots, n_{\text{d}}$, and periods $t \in T$ are all obtained from Figure 4.1. For example, for disease d= 1 = hepatitis B and dose $j = 2$, $P_{\text{d}jt} = 1(0)$ for periods $t = 2, 3(1, 4, 5, 6, 7, 8, 9, 10)$.

For VFSLBP(O), the vaccine set is V = {1 = {1}, 2 = {2}, 3 = {3}, 4 = {4}, 5 = {5}, 6 = {6}, 7 = {7}, 8 = {8}, 9 = {9}, 10 = {2,3}, 11 = {1,3}, 12 = {1,2,4}}. The parameters I_{vd} are indicated by the set V. For example, vaccine 1 is the monovalent vaccine for disease 1 (hepatits B) and vaccine 12 is the combination vaccine Pediarix® that immunizes against diseases 1 (hepatitis B), 2 (diphtheria–tetanus–pertussis), and 4 (polio). Three different cost scenarios are evaluated. The first scenario only considers the actual purchase price of the vaccines. In particular, the cost vector $c = (9.00, 12.75, 7.66, 10.42, 16.67, 52.25, 54.12, 9.71, 12.10, 24.62, 24.50, 38.34)$, where $c_{\text{v}}, \text{v} = 1, 2, \ldots, 12$, is the federal contract purchase price (in U.S. dollars) for vaccine $\text{v} \in \text{V}$ [3]. The second scenario includes the purchase price of the vaccine and a fixed injection cost of \$10 per injection, and the final scenario includes the purchase price, the fixed injection cost, and a

Table 4.1 VFSLBP(O) Results for 2006 Recommended Childhood Immunization Schedule

	Scenario 1			Scenario 2			Scenario 3		
Algorithm	*Z*	Time	θ	*Z*	Time	θ	*Z*	Time	θ
MAX Rounding	499.05	0.13	1.00	736.77	0.13	1.02	796.77	0.13	1.02
Greedy	499.05	0.06	1.00	719.81	0.05	1.00	779.81	0.05	1.00
DP	499.05	0.32		719.81	0.30		779.81	0.31	
IP B&B	499.05	0.91		719.81	0.92		779.81	0.92	

preparation cost of $3 per injection. Table 4.1 reports the objective function cost Z and execution time (in CPU seconds) for each heuristic and exact algorithm and for each scenario. Table 4.1 also reports the solution quality effectiveness measure θ for each heuristic.

Hall et al. [16] show that VFSLBP(O) is solvable in polynomial time when all vaccines $v \in V$ are monovalent vaccines, and hence, the results reported in Table 4.1 for scenario 1 are not surprising, given that most vaccines $v \in V$ are monovalent. In fact, the combination vaccines are not competitively priced when considering purchase price alone. The fixed costs considered in scenarios 2 and 3 penalize the monovalent vaccines and make the combination vaccines more economical. For example, the purchase prices for monovalent vaccines 1, 2, and 4 sum to $32.17, which is less than the $38.34 purchase price for the combination vaccine Pediarix (i.e., $v = 12$). However, in scenario 2, the total cost of the combination vaccine Pediarix is $48.34, whereas the total costs for monovalent vaccines 1, 2, and 4 sum to $62.17. Observe the efficiency of the DP algorithm described in Section 4.3 compared to the IP B&B algorithm.

For VFSREP(O), two different sets of vaccines, V_1 and V_2, are evaluated on two different sets of diseases that restrict extraimmunization, D_{NE1} and D_{NE2}. The vaccine sets are $V_1 = \{1 = \{1\}, 2 = \{2\}, 3 = \{3\}, 4 = \{4\}, 5 = \{5\}, 6 = \{6\}, 7 = \{7\}, 8 = \{8\}, 9 = \{9\}, 10 = \{2,3\}, 11 = \{1,3\}, 12 = \{1,2,4\}, 13 = \{5,6\}\}$ and $V_2 = \{1 = \{1\}, 2 = \{2,3,4\}, 3 = \{1,9\}, 4 = \{4\}, 5 = \{5\}, 6 = \{6\}, 7 = \{7\}, 8 = \{8\}, 9 = \{9\}, 11 = \{1,3\}, 12 = \{1,2,4\}, 13 = \{5,6\}, 14 = \{1,2,3,4\}\}$, where V_1 represents a set of pediatric vaccines currently licensed for use in the United States and V_2 represents a set of pediatric vaccine with fewer monovalent vaccines and more combination vaccines, some of which are not yet licensed for use in the United States, but are projected to be in the future. The parameters I_{vd} are indicated within the sets V_1 and V_2, respectively. The disease sets are $D_{NE1} = \{1,2,3,4\}$ and $D_{NE2} = D = \{1,2,3,4,5,6,7,8,9\}$. Table 4.2 reports the solution quality and execution time (in CPU seconds) for each heuristic and exact algorithm and for each vaccine set and disease set combination.

Table 4.2 VFSREP(O) Results for 2006 Recommended Childhood Immunization Schedule

	V_1 and D_{NE1}		V_1 and D_{NE2}		V_2 and D_{NE1}		V_2 and D_{NE2}	
Algorithm	Z	Time	Z	Time	Z	Time	Z	Time
MAX Rounding	0	0.14	1	0.27	3	0.17	4	0.25
Greedy	0	0.09	0	0.11	1	0.09	1	0.11
DP	0	0.36	0	0.36	1	0.47	1	0.45
IP B&B		9.66		5.02		16.94		12.89

Hall et al. [17] show that VFSREP(O) is solvable in polynomial time when all vaccines v ∈ V are monovalent vaccines or when there exists a corresponding monovalent vaccine for every disease d ∈ D, and, hence, the results for the solution quality and execution time reported in Table 4.2 are not surprising, given that most diseases have a corresponding monovalent vaccine (particularly in vaccine set V_1). Furthermore, the MAX Rounding and Greedy heuristics for VFSREP(O) were both more efficient than the exact algorithms. In all cases, the Greedy heuristic returned the optimal solution.

Excluding the IP B&B algorithm for VFSREP(O), all heuristics and exact algorithms for VFSLBP(O) and VFSREP(O) executed in less than a second. However, as the next set of test problems will illustrate, this is unlikely to occur for future Recommended Childhood Immunization Schedules, as the schedule expands and more combination vaccines are licensed for use and enter the market.

The second set of test problems for VFSLBP(O) considers hypothetical future childhood immunization schedules. Each heuristic and exact algorithm were executed on 30 randomly generated childhood immunization schedules with 15 periods, 75 vaccines, and 11 diseases. Therefore, each random childhood immunization schedule reflects a gradual expansion in the sets D (from 9 to 11 diseases) and T (from 10 to 15 periods) and a significant increase in the number of available vaccines, particularly, combination vaccines. In each random childhood immunization schedule, $1 \leq n_{\mathrm{d}} \leq 5$ for all disease d ∈ D, $1 \leq \mathrm{Val(v)} \leq 6$ and $c_{\mathrm{v}} \sim U(10, 80)$ (uniformly distributed) for all vaccines v ∈ V, and $P_{\mathrm{d}jt} = 1$ for at most three periods $t \in T$ for every disease d ∈ D and dose $j = 1, 2, \ldots, n_{\mathrm{d}}$. Table 4.3 reports the average μ and the standard deviation σ for the execution time (in CPU seconds) and solution quality θ averaged over the 30 random childhood immunization schedules.

The solutions returned by the MAX Rounding and Greedy heuristics on average were within 7 percent of the optimal solution. Across all 30 randomly generated childhood immunization schedules, the optimal solution was returned 15 times by at least one of these heuristics. The exact algorithms required significantly more time to execute than the heuristics (i.e., the least efficient heuristic MAX Rounding was seven times

Table 4.3 VFSLBP(O) Results for Future Childhood Immunization Schedule

Algorithm	Time		θ	
	μ	σ	μ	σ
MAX Rounding	0.42	0.05	1.07	0.06
Greedy	0.24	0.03	1.06	0.07
DP	3.1	1.2		
IP B&B	40.6	46.5		

faster than the most efficient exact algorithm DP). Moreover, the IP B&B algorithm, on average, required significantly more execution time than the DP algorithm. Furthermore, the DP algorithm showed far less variability in its execution time.

The second set of test problems for VFSREP(O) considers hypothetical near-term future childhood immunization schedules. Each heuristic and exact algorithm were executed on 30 randomly generated childhood immunization schedules with 15 periods, 30 vaccines, and 11 diseases. In each random childhood immunization schedule, $1 \leq n_{\mathrm{d}} \leq 5$ for all diseases $\mathrm{d} \in \mathrm{D}, 1 \leq \mathrm{Val(v)} \leq 6$ for all vaccines $\mathrm{v} \in \mathrm{V}$, and $P_{\mathrm{d}jt} = 1$ for at most three periods $t \in T$ for every disease $\mathrm{d} \in \mathrm{D}$ and dose $j = 1, 2, \ldots, n_{\mathrm{d}}$. For each randomly generated childhood immunization schedule, each heuristic and exact algorithm was executed three times, where in execution 1, 2, and 3, $\delta_{\mathrm{NE}} = 4, 8$, and 11, respectively. Table 4.4 reports the solution quality and execution time (in CPU seconds) averaged over the 30 random childhood immunization schedules for each value of δ_{NE}. An additional measure λ that indicates the number of childhood immunization schedules that the respective heuristic or exact algorithm found the optimal solution is also reported. The IP B&B algorithm found the optimal solution for λ of the 30 random childhood immunization schedules, but exceeded the default execution time limit (two hours) or default iteration limit (10^7) for the remaining $(30 - \lambda)$ random childhood immunization schedules. The

Table 4.4 VFSREP(O) Results for Future Childhood Immunization Schedule

Algorithm	$\delta_{\mathrm{NE}} = 4$			$\delta_{\mathrm{NE}} = 8$			$\delta_{\mathrm{NE}} = 11$		
	Z	Time	λ	Z	Time	λ	Z	Time	λ
MAX Rounding	0.70	0.80	25	7.53	0.83	4	16.77	0.85	0
Greedy	1.03	0.23	18	7.87	0.29	2	13.97	0.35	0
DP	0.43	1.74	30	4.77	1.80	30	10.33	1.81	30
IP B&B	0.43	518	30	4.48	1095	27	10.07	1767	27

statistics reported in Table 4.4 are averaged over the λ random childhood immunization schedules for which the IP B&B algorithms found the optimal solution, which is why the average Z values for the IP B&B and DP algorithms differ when $\delta_{\mathrm{NE}} = 8$ and 11.

The data reported in Table 4.4 shows that for VFSREP(O), the MAX Rounding and Greedy heuristics found better solutions for $\delta_{\mathrm{NE}} \ll \delta$. For the smaller values of δ_{NE}, the MAX Rounding heuristic slightly outperformed the Greedy heuristic, while the Greedy heuristic outperformed the MAX Rounding heuristic for $\delta_{\mathrm{NE}} = \delta$. Across all values of δ_{NE}, the Greedy heuristic was the most efficient compared with MAX Rounding and the DP and IP B&B exact algorithms. The DP algorithm executed two to seven times slower than the heuristics; however, the IP-MIN algorithm, on average, executed approximately 300–1000 times slower than the DP algorithm. Furthermore, the DP algorithm found the optimal solution for all 90 instances of VFSREP(O) reported in Table 4.4, with little sensitivity to the value of δ_{NE}. Conversely, the IP B&B algorithm only found the optimal solution for 84 of the 90 instances of VFSREP(O) reported in Table 4.4, and the average execution time for IP B&B algorithm more than tripled when δ_{NE} went from four to eleven diseases.

The observed difference in execution time between the heuristics and exact algorithms reported in Tables 4.3 and 4.4 could be problematic for practical uses. For example, a Web page used to find a good vaccine formulary for a given childhood immunization schedule would require an algorithm to execute in real-time, because most Web users would terminate a Web application that required several seconds or minutes to execute. Moreover, the difference in execution time between the heuristics and exact algorithms will provide an efficient analysis of larger childhood immunization schedules that may involve Monte Carlo simulation (see Ref. [22]) or the balking problem (described in Section 4.3), where either of these may require the solution of hundreds of thousands of VFSLBP(O) or VFSREP(O) instances. Furthermore, the childhood immunization schedule may need to be solved for each child, on a case-by-case basis, and hence, efficient algorithms are needed to provide, in real-time, practical value for the public health community. See Refs. [16,17] for more extensive computational results for both VFSLBP(O) and VFSREP(O).

4.5 Conclusion

This chapter presented two discrete optimization models that are used to address the cost of immunizing a child and the issue of extraimmunization. As more combination vaccines come to market and the Recommended Childhood Immunization Schedule becomes more complex to include more diseases and cover more periods, VFSLBP(O) and VFSREP(O) will capture the combinatorial explosion of alternatives for public health policy-makers and administrators, vaccine manufacturers, pediatricians, and parents/guardians by identifying vaccine formularies and schedules that minimize the cost of fully immunizing a child and safely use combination vaccines, which will help address safety concerns in pediatric immunization, reduce costs, and reduce vaccine wastage associated with extraimmunization.

In general, VFSLBP(O) and VFSREP(O) are NP-hard unless the vaccines, schedule parameters, or disease set are significantly restricted. This chapter presented a DP algorithm that solves VFSLBP(O) and VFSREP(O) to optimality. In Section 4.4, this DP algorithm was compared computationally to an IP B&B algorithm, and these results showed that the DP algorithm was significantly more efficient (at least eight times faster). Furthermore, for VFSREP(O), the execution time of the DP algorithm was insensitive to the size of the set D_{NE} (the average execution time remained nearly constant at 1.74 seconds when $\delta_{NE} = 4$ to 1.81 seconds when $\delta_{NE} = 11$), whereas the execution time of IP B&B algorithm was sensitive to the size of the set D_{NE} (the average execution time tripled from 518 seconds when $\delta_{NE} = 4$ to 1767 seconds when $\delta_{NE} = 11$). However, for most of the randomly generated childhood immunization schedules, both the DP and IP B&B algorithms required at least twice as much time to execute when compared to the execution time of the heuristics presented in Section 4.3. Moreover, the average execution time for each heuristic was less sensitive to increases in the size of the childhood immunization schedule. These heuristics will allow more efficient analysis of larger childhood immunization schedules and practical analysis involving Monte Carlo simulation or finding an optimal vaccine formulary for each child on a case-by-case basis, which will require the solution of several unique VFSLBP(O) and VFSREP(O) instances.

Acknowledgments

This research has been supported in part by the National Science Foundation (DMI-0457176, DMI-0456945). The second author was supported in part by the Air Force Office of Scientific Research (FA9550-04-1-0110). The views expressed in this chapter are those of the authors and do not reflect the official policy or position of the United States Air Force, Department of Defense, National Science Foundation, or the United States Government.

References

1. R. K. Ahuja, T. L. Magnanti, and J. B. Orlin. *Network Flows*. Prentice-Hall: Upper Saddle River, New Jersey, 1993.
2. R. L. Burden and J. D. Faires. *Numerical Analysis*, 6th ed. Brookes-Cole: New York, 1997.

3. CDC Vaccine Price List. Center for Disease Control and Prevention, National Immunization Program Web site. November 21, 2005, http://www.cdc.gov/nip/vfc/cdc_ vac_price_list.htm, 2005.
4. Centers for Disease Control and Prevention. Recommended childhood and adolescent immunization schedule—United States. *Morbidity and Mortality Weekly Report*, 54(52):Q1–Q4, 2006.
5. Centers for Disease Control and Prevention. Recommended childhood and adolescent immunization schedule—United States. *Morbidity and Mortality Weekly Report*, 53:Q1–Q3, 2005.
6. Centers for Disease Control and Prevention. General recommendations on immunization. *Morbidity and Mortality Weekly Report*, 51(RR-2), 2002.
7. Centers for Disease Control and Prevention. Combination vaccines for childhood immunization. *Morbidity and Mortality Weekly Report*, 48(18)(RR-5), 1999.
8. Centers for Disease Control and Prevention. Recommended childhood and adolescent immunization schedule—United States. *Morbidity and Mortality Weekly Report*, 44(RR-5):1–9, 1995.
9. R. T. Chen, F. DeStefano, R. Pless, G. Mootrey, P. Kramarz, and B. Hibbs. Challenges and controversies in immunization safety. *Infectious Disease Clinics of North America*, 15(1):21–39, 2001.
10. S. L. Cochi. National immunization survey. Presentation to National Press Club, July 27, 2005.
11. M. L. Cohen. Changing patterns in infectious disease. *Nature*, 406:762–767, 2000.
12. D. S. Diekema. Responding to parental refusals of immunization of children. *Pediatrics*, 115(5):1428–1431, 2005.
13. K. M. Edwards and M. D. Decker. Combination vaccines. *Infectious Disease Clinics of North America*: *Vaccine Recommendations, Challenges and Controversies*, 15(1):209–230, 2001.
14. E. A. Flanagan-Klygis, L. Sharp, and J. E. Frader. Dismissing the family who refuses vaccines: A study of pediatrician attitudes. *Archives of Pediatrics and Adolescent Medicine*, 159(10):929–934, 2005.
15. M. R. Garey and D. S. Johnson. *Computers and Intractability*: *A Guide to the Theory of NP-Completeness*. W.H. Freeman: New York, 1979.
16. S. N. Hall, S. H. Jacobson, and E. C. Sewell. An analysis of the general minimum cost vaccine selection problem. Technical Report, University of Illinois, Urbana, Illinois, 2006 (Submitted).
17. S. N. Hall, S. H. Jacobson, and E. C. Sewell. Designing childhood vaccination schedules with limited extraimmunization: A complexity and computational analysis. Technical Report, University of Illinois, Urbana, Illinois, 2006.
18. D. S. Hochbaum, Ed. *Approximation Algorithms for NP-Hard Problems*. PWS Publishing Company: Boston, Massachusetts, 1997.
19. Infectious Diseases in Children. Almost 200 New Drugs in Development for Use in Children. 15(10):41–44, 2002.
20. S. H. Jacobson, T. Karnani, and E. C. Sewell. Analyzing the economic value of the hepatitis B-*Haemophilus influenzae* type b combination vaccine by reverse engineering a formulary selection algorithm. *Vaccine*, 21:2169–2177, 2003.
21. S. H. Jacobson, T. Karnani, and E. C. Sewell. Assessing the impact of wastage on pediatric vaccine immunization formulary costs using a vaccine selection algorithm. *Vaccine*, 22:2307–2315, 2004.

22. S. H. Jacobson and E. C. Sewell. Using Monte Carlo simulation to determine combination vaccine price distributions for childhood disease. *Health Care Management Science*, 5:135–145, 2002.
23. S. H. Jacobson, E. C. Sewell, D. A. Allwine, E. A. Medina, and B. G. Weniger. Designing pediatric vaccine formularies and pricing pediatric combination vaccines using operations research models and algorithms. *Expert Review of Vaccines*, 2(1):15–19, 2003.
24. S. H. Jacobson, E. C. Sewell, R. Deuson, and B. G. Weniger. An integer programming model for vaccine procurement and delivery for childhood immunization: A pilot study. *Health Care Management Science*, 2:1–9, 1999.
25. S. H. Jacobson, E. C. Sewell, and T. Karnani. Engineering the economic value of two pediatric combination vaccines. *Health Care Management Science*, 8(1):29–40, 2005.
26. S. H. Jacobson, E. C. Sewell, and R. A. Proano. An analysis of the pediatric vaccine supply shortage problem. *Health Care Management Science*, 9(4):371–389, 2006.
27. G. L. Nemhauser and L. A. Wolsey. *Integer and Combinatorial Optimization*. John Wiley & Sons: New York, 1999.
28. S. A. Plotkin and W. A. Orenstein. *Vaccines*, 4th ed. W.B. Saunders, Philadelphia, PA, 2004.
29. E. C. Sewell and S. H. Jacobson. Using an integer programming model to determine the price of combination vaccines for childhood immunization. *Annals of Operations Research*, 119:261–284, 2003.
30. E. C. Sewell, S. H. Jacobson, and B. G. Weniger. Reverse engineering a formulary selection algorithm to determine the economic value of pentavalent and hexavalent combination vaccines. *Pediatric Infectious Disease Journal*, 20(11):S45–S56, 2001.
31. E. C. Sewell, K. Stuart K., S. H. Jacobson, and S. N. Hall. Selecting optimal pediatric vaccine formularies using a dynamic programming algorithm. Technical Report, Southern Illinois University, Edwardsville, Illinois, 2005.
32. S. Tawfik. Mixed-integer LP. MATLAB Central File Exchange Web site. Retrieved Oct. 4, 2005, from http://www.mathworks.com/matlabcentral/fileexchange/loadCategory.do, 2005.
33. B. G. Weniger. Economic analysis to meet the challenges of new combination vaccines, Presented at Vaccine Economics: Planning a Research Agenda for the Challenge of New and Improved Vaccines, Atlanta, Georgia, 1996.
34. B. G. Weniger, R. T. Chen, S. H. Jacobson, E. C. Sewell, R. Deuson, J. R. Livengood, and W. A. Orenstein. Addressing the challenges to immunization practice with an economic algorithm for vaccine selection. *Vaccine*, 16(19):1885–1887, 1998.

Chapter 5

Optimal Spending on HIV Prevention and Treatment: A Framework for Evaluating Cost-Effectiveness with Example Application to the India AIDS Initiative

Margaret L. Brandeau, Elisa F. Long, David W. Hutton, and Douglas K. Owens

CONTENTS

5.1 Introduction 148
5.2 Model-Based Framework 149
5.3 Designing an Epidemic Model to Support Cost-Effectiveness Analysis 151
    5.3.1 Model Scope 151
    5.3.2 Population Groups 154
    5.3.3 Risk Factors 154
    5.3.4 Input Parameters 154
    5.3.5 Health and Economic Outcomes 155

5.3.6 Sensitivity Analysis 155
5.3.7 Evaluating Multiple Interventions 156
5.4 Example Model: Evaluating the India AIDS Initiative 156
5.4.1 HIV in India 156
5.4.2 Avahan Project 158
5.4.3 Framework for Cost-Effectiveness Analysis of the Avahan Project 158
5.4.4 Illustrative Model 159
5.4.4.1 Description 159
5.4.4.2 Data 161
5.4.4.3 Effect of Interventions 161
5.5 Developing a Tailored Model 167
5.5.1 Homogeneous Sexual Behavior 167
5.5.2 Random Mixing 167
5.5.3 Transmission Risk Factors 167
5.5.4 Injection Drug Use 168
5.5.5 Disease Progression and Treatment 168
5.5.6 Nonconstant Population Group Sizes 168
5.5.7 Spread of Other Diseases 168
5.5.8 Realism versus Tractability 169
5.6 Discussion 169
Acknowledgments 170
Appendix: Equations of Illustrative Model 170
References 171

Abstract Resources to control the spread of human immunodeficiency virus (HIV) are insufficient to meet all needs. To achieve the maximum health benefit with limited resources, information about program cost-effectiveness is needed. We present a model-based framework for evaluating the cost-effectiveness of HIV prevention and treatment programs, which we illustrate by suggesting how programs in the Gates Foundation's India AIDS Initiative could be assessed. A framework of this type can help identify the most cost-effective programs in a portfolio of HIV interventions, thus guiding the allocation (and possible reallocation) of scarce funds. Additionally, a model-based framework allows one to see how cost-effectiveness estimates and policy conclusions may change for different values of uncertain parameters, and helps identify areas where further data collection is most critically needed.

5.1 Introduction

In 2004, approximately 5 million people became infected with HIV, bringing the total number of people worldwide living with HIV/AIDS to an estimated 40 million [1]. Although the majority of cases thus far have occurred in sub-Saharan Africa, HIV is spreading rapidly into new regions. It has been projected that within the next five years, 50–75 million people in Nigeria, Ethiopia, Russia, China, and India will acquire HIV infection, which is almost twice the 30–35 million cases expected in sub-Saharan Africa [2].

The international community has committed significant resources, knowledge, and leadership toward slowing the spread of HIV. Major efforts have been funded by The Global Fund to Combat Malaria, Tuberculosis and HIV [3], the United Kingdom Department of International Development [4], the United States Agency for International Development [4,5], the World Bank [4], and the Bill and Melinda Gates Foundation [6]. In 2004, some US$6.1 billion was spent on HIV prevention and treatment worldwide; in 2005, this figure rose to $8.3 billion [7]. Despite these efforts, overall funding for HIV/AIDS has fallen significantly short of the estimated need. In 2005, total HIV/AIDS funding was $4 billion less than the estimated need [7].

Information about the cost-effectiveness of different HIV prevention and treatment programs can help policy makers make the best use of limited program funds. By assessing the health benefits achieved by a given investment, cost-effectiveness analysis identifies the programs that represent the most effective use of limited resources. This information can be used to guide the allocation (and possible reallocation) of HIV funds, thus maximizing the impact of limited resources.

This chapter presents a model-based framework, shown in Figure 5.1, for evaluating the cost-effectiveness of HIV prevention and treatment programs (Section 5.2). We describe some of the issues that are relevant in designing a model to support such a cost-effectiveness analysis (Section 5.3). We illustrate our ideas with a simple model framework to show how programs in the Gates Foundation's India AIDS Initiative [8] could be assessed (Section 5.4). We explain how one could develop a tailored model to evaluate specific interventions (Section 5.5). We conclude with discussion (Section 5.6).

5.2 Model-Based Framework

Evaluation of the cost-effectiveness of interventions is paramount to understanding how best to use limited HIV prevention and treatment resources. One program may be much more effective than another in changing HIV risk behavior but also much more expensive. Cost-effectiveness analysis (CEA) can help to determine the program that represents a more efficient expenditure of resources [9]. The incremental cost-effectiveness of a program is estimated as

$$\frac{\text{Costs}_{\text{With Program}} - \text{Costs}_{\text{Without Program}}}{\text{Health Benefits}_{\text{With Program}} - \text{Health Benefits}_{\text{Without Program}}}$$

1. Develop an epidemic model that appropriately captures HIV transmission and progression in the population—in particular, in the target population and among those whom they may contact—and that captures the cost and health consequences of an intervention. As appropriate, the model may also incorporate the spread of other diseases such as other sexually transmitted diseases or tuberculosis.

2. Use sensitivity analysis to help identify key elements needed in such a model and to identify the data for which the most accurate estimates are needed.

3. Collect necessary data, including data on
   (a) Risk groups and risk factors,
   (b) Demographic factors,
   (c) Baseline disease prevalence and incidence,
   (d) Baseline risk factors,
   (e) Effects of interventions, and
   (f) Direct and indirect costs of the intervention.

4. Calibrate the model so that its projections in the absence of any new interventions are broadly in keeping with known epidemic trends.

5. Apply the model to estimate the cost-effectiveness of the intervention. Include all costs and health consequences of the intervention, including those that accrue beyond the time horizon of the intervention. Use sensitivity analysis to determine how cost-effectiveness estimates change for different values of uncertain parameters.

Figure 5.1 Framework for cost-effectiveness analysis of HIV interventions.

Health benefits are typically measured in quality-adjusted life years (QALYs) experienced by individuals in the population, but could also be measured in disability-adjusted life years (DALYs) experienced, life years experienced, or HIV infections averted.

An HIV prevention or treatment program may change an individual's risk factors for acquiring or transmitting HIV. For example, a condom promotion program may increase condom use; treatment of HIV-infected individuals with antiretroviral drugs may reduce the infectivity of those individuals; a program to treat sexually transmitted diseases (STDs) may reduce the prevalence of such STDs, thereby reducing the chance of HIV transmission during sexual contact [10]. Such programs can affect not only an individual's risk of acquiring or transmitting HIV infection but also the spread of HIV in the target population and in the broader population as a whole. Thus, an appropriate HIV epidemic model is needed to translate the behavioral and biological changes due to an intervention into meaningful epidemiological outcomes (e.g., QALYs gained or HIV infections averted).

Understanding the effectiveness of an intervention in preventing the spread of HIV is, however, only half of the picture. CEA also requires an understanding of the costs incurred or saved as a result of the intervention. These include not only the direct costs of the intervention (e.g., program administration cost and cost per person reached) but also indirect costs that occur due to changes in health status of targeted individuals as well as others in the population (e.g., treatment for HIV among newly identified individuals and treatment for STDs). A model of the HIV epidemic, with costs for each health state, is needed to support accurate estimation of the total cost of an intervention.

Finally, a model should incorporate the relationship between the level of investment in an intervention and its effectiveness in changing risk factors (e.g., increase in condom use, reduction in needle sharing). Many analyses assume a linear relationship for these production functions (i.e., they assume that an intervention's effectiveness is proportional to its cost), but other functional forms may more accurately capture this relationship (e.g., decreasing returns to scale) [11,12].

A variety of HIV epidemic models have been developed. Table 5.1 summarizes a representative sample of such models. These range from very simple models that can be represented by a few equations to complex simulation models that incorporate detailed demographic and risk factors. Some of the models were designed primarily to project new HIV cases; other models also allow one to evaluate the effects of prevention programs.

5.3 Designing an Epidemic Model to Support Cost-Effectiveness Analysis

In this section, we discuss issues relevant to designing an epidemic model to support CEA of HIV prevention and treatment programs.

5.3.1 Model Scope

One consideration in model design is overall scope. A high-level model that captures general epidemic trends might allow for CEA of broad categories of interventions, without capturing specific details of disease transmission. Such a model may be easy to create and implement, but may ignore critical components of disease transmission, such as the effect of different risk groups. A lower-level model, on the other hand, might consider very specific details of disease transmission or progression (e.g., distinguishing circumcised versus noncircumcised males [13]), but will require more data. A key factor when deciding the level of detail to include is the availability of reliable data.

Table 5.1 Summary of Selected HIV Epidemic Models

Model Name	Purpose	Model Type	Transmission Modes	Transmission Factors	Key Population Groups	Mathematical Logic	Implementation
Anderson–May Model [41,42]	Epidemic projection	Dynamic compartmental model	Sexual	None	Age and gender stratification; susceptible, infected	Deterministic	Solution of differential equations
UNAIDS Model [43,44]	Estimate HIV prevalence, mortality	Hybrid model (compartmental/parametric curves)	Sexual	None	Adults, children	Deterministic	Solution of differential equations
WHO Model [45]	Project adult AIDS incidence	AIDS incidence projection model	None	None	HIV-positive, AIDS-diagnosed	Deterministic	Estimation and projection of AIDS cases
iwgAIDS Model [46]	Epidemic projection, given a set of interventions	Dynamic simulation tool	Sexual, mother-to-infant, blood transfusion, needle sharing	Condom use, STDs, anal intercourse	Age and gender stratification	Deterministic	Differential equation solvers
AVERT Model [47]	Estimate HIV incidence, given interventions	Static mathematical model	Sexual	Condom use, STDs (ulcerative, non-ulcerative)	Males, females	Deterministic	Mathematical calculations

SimulAIDS Model [14,48]	Simulate effect of interventions on HIV prevalence, incidence	Simulation tool	Sexual	STDs (ulcerative, non-ulcerative)	Age and gender stratification	Stochastic	Monte Carlo simulation
CMH Model [49,50]	Simulate effect of interventions on HIV prevalence, incidence	Dynamic compartmental simulation tool	Heterosexual, mother-to-infant	Condom use	Males, females; high-risk, low-risk; undergoing antiretroviral therapy	Deterministic	Numerical projection of equations
STDSIM Model [51]	Simulate effect of interventions on STD prevalence, incidence	Microsimulation tool	Homosexual, heterosexual, mother-to-infant	STDs	Age and gender stratification, high-risk females (CSW, pregnant women)	Stochastic	Monte Carlo simulation
Asian Epidemic Model [13,38]	Project new HIV infections by age and gender	Dynamic compartmental model	Heterosexual (CSW and nonpaid), needle sharing	Condom use, circumcision, STDs	Males (CSW client, nonclient, IDU sharing, IDU nonsharing), females (direct CSW, indirect CSW, non-CSW); age stratification	Deterministic	Numerical projection of equations

5.3.2 Population Groups

Another choice is whether individuals or cohorts of people are considered. When individuals are modeled, heterogeneity in risky behaviors and other risk factors can be included [14,15]. However, as the number of individuals increase, computational requirements increase dramatically, because such a model must consider the interactions among all individuals. A more tractable but less realistic approach is to model cohorts of individuals with similar risk behaviors (e.g., frequent injection drug user [IDU], client of commercial sex worker [CSW], etc.). This is typically done using a compartmental model. The choice of whether to model individuals or cohorts should depend on the heterogeneity of the population of interest and available data.

If cohorts of individuals are modeled, risk groups must be defined in enough detail so that the costs and effects of interventions can be captured, but not in so much detail that reliable data cannot be collected for each risk group. For example, to evaluate an HIV intervention in a country such as Russia or India, it might be important to distinguish not only high-risk populations (e.g., CSWs, IDUs) and low-risk populations (e.g., the general population) but also bridge populations (e.g., clients of CSWs, sex partners of IDUs) that provide a vector for transmission from high-risk groups to the general population.

5.3.3 Risk Factors

Relevant risk factors—that is, factors that affect an individual's risk of acquiring or transmitting HIV infection—must also be identified. Risk factors may include, for example, frequency of sex acts or frequency of drug injection, whether an individual is infected with an STD (other than HIV), or whether an individual is receiving antiretroviral therapy. The definitions of risk groups and risk factors, as well as other assumptions of the epidemic model, help to determine the data that must be collected.

It may be necessary to include the transmission of STDs, tuberculosis (TB), or other diseases if their spread is significantly affected by the intervention. For example, an intervention that treats STDs as part of an HIV education program is likely to reduce the prevalence and incidence of such STDs, and thus may reduce HIV transmission.

5.3.4 Input Parameters

The parameters in a transmission model may be static or dynamic. A model that does not allow parameters to change over time may require substantially less data than a model that uses time-varying parameters, because it does not consider how a parameter value evolves over time. However, the insights drawn from such a model may be limited because many behavioral and epidemiological parameters do vary over time. In the implementation of a model with constant parameters, it is reasonable to limit the time horizon to a period over which few parameters might change. Although more complex, a model with time-varying parameters may be able to capture transmission patterns more accurately.

The transmission model may incorporate deterministic or stochastic events. A stochastic model requires more data than a deterministic model, as one must estimate the probability distribution of a parameter's possible values. However, unlike a deterministic model, a stochastic model can provide insight into the variability of health and economic outcomes.

5.3.5 Health and Economic Outcomes

A comprehensive CEA must measure, at the minimum, discounted life years gained or HIV infections averted, as well as the net present value of all incremental costs or cost savings that accrue from the intervention. Health benefits and costs may accrue beyond the time horizon for which efficacy data is available; these future costs and benefits should be estimated and included. Total costs are typically estimated by summing the direct costs of the intervention (e.g., cost of a condom distribution program), the indirect cost for each group of individuals (e.g., annual healthcare cost for an HIV-infected CSW), and any incremental costs that occur after the end of the time horizon. The incremental cost of an intervention is the difference between the total discounted cost for the intervention and the total discounted cost for the base case. Incremental health benefits are measured analogously. All costs and health benefits should be discounted to the present. Examples of comprehensive CEAs of HIV prevention and treatment programs can be found elsewhere [16–20].

5.3.6 Sensitivity Analysis

An important advantage of a model-based approach to CEA is that it allows one to perform sensitivity analysis on model assumptions and parameter values. Sensitivity analysis can be used to help guide model structure. For example, results from a more detailed model (e.g., a model that distinguishes ulcerative versus nonulcerative STDs) can be compared to results from a simpler model (e.g., a model that does not distinguish these two types of STDs) to determine whether the additional level of model detail is required to obtain good cost-effectiveness estimates. Sensitivity analysis can show how cost-effectiveness estimates change for different values of uncertain parameters. Additionally, sensitivity analysis can identify those variables whose uncertainty has the greatest effect on the cost-effectiveness estimates, and thus guide data-collection efforts. If the value of a parameter (over its plausible range of values) has little effect on cost-effectiveness estimates, then a readily obtained estimate will suffice; otherwise, the value for the parameter should be carefully estimated, perhaps based on several sources.

5.3.7 Evaluating Multiple Interventions

Because data requirements for a detailed model are likely to be quite large, it may be feasible to perform very detailed CEAs of selected interventions only. To evaluate the effect of several independent interventions, results from detailed CEAs of the individual interventions could be combined; for example, by adding results of local interventions to estimate a regional effect, or by adding results of regional interventions to estimate a nationwide effect. However, interventions are often not independent; for example, a general education campaign may increase awareness of HIV and thus increase the effectiveness of other HIV prevention programs. To estimate the combined effect of interventions that are not independent, a high-level model may be the only feasible choice.

5.4 Example Model: Evaluating the India AIDS Initiative

We now illustrate our ideas with a simple model we developed to show how one could assess the cost-effectiveness of programs in the India AIDS Initiative, also known as the Avahan Project. As a preliminary, we briefly describe the HIV epidemic in India and the efforts of the Avahan Project.

5.4.1 HIV in India

An estimated 5.7 million people in India are currently HIV-infected, including almost 2 million women and 100,000 children [1,21]. HIV was initially confined to specific high-risk groups in southern and northeast India, but has now begun to spread to the general population. One-fifth of new infections now occur among monogamous married women [22], and almost 3 percent of new infections occur among newborns [23]. Although HIV prevalence among adults aged 15–49 in India is less than 1 percent, the potential for many new infections exists given India's population of 1.1 billion people. It has been estimated that unless prevention efforts have significant success, India may have as many as 20–25 million people living with HIV within five years—more than any other country [2].

The majority of HIV cases in India occur in six states: Tamil Nadu, Maharashtra, Karnataka, Andhra Pradesh, Nagaland, and Manipur [24]. These states account for 70 percent of all HIV cases in India, but only 30 percent of the total population. Transmission via heterosexual contact is the primary mode of HIV transmission in India (85 percent of cases), and is most common in the southern states of Tamil Nadu, Maharashtra, Karnataka, and Andhra Pradesh.

High-risk groups that contribute to spread of HIV in India include CSWs and IDUs. An estimated two million CSWs live in India [25]. HIV prevalence among CSWs varies significantly by region, from 11 percent in Kolkata to over 50 percent in Pune and Mumbai [26]. With such high HIV prevalence among CSWs, the low use of condoms is a serious concern.

Clients of CSWs—many of whom are married—provide the main transmission route from high-risk groups to the general population. Truck drivers are especially important in HIV transmission because they are highly mobile and often visit many CSWs along the highways. They can spread HIV to CSWs as well as to their regular and nonregular partners at home, thus introducing HIV to small communities in rural India. India has an estimated five million long-distance truck drivers, many of whom are sexually promiscuous. As few as 11 percent consistently use condoms with CSWs [21,27].

Coinfection with HIV and another STD is thought to increase the probability of transmission during sexual activity [10]. In a survey of the general population in India, almost 5 percent reported having symptoms of STDs in the past year [28]. Among groups at high risk for HIV, the prevalence of STD symptoms was much higher; almost half of CSWs and nearly one-third of their clients reported STD symptoms within the last year [28].

Injection drug use accounts for approximately 2.4 percent of HIV cases in India [29]. Injection drug use fuels the spread of HIV in the northeastern states of Nagaland and Manipur, where HIV prevalence among some groups of drug injectors is 40 percent or more. In addition to transmission via drug injection, IDUs can spread HIV to others via sexual contact. Up to 45 percent of wives of IDUs are also infected [30].

HIV infection significantly increases the chance that an individual will become infected with TB. The annual likelihood of developing active TB is 50 times higher in HIV-infected individuals than in uninfected individuals [31]. In India, 60 percent of HIV-infected individuals eventually acquire active TB [32]. Additionally, the presence of TB infection increases the progression of HIV in infected individuals. In many countries, and especially in India, the spread of HIV has exacerbated the spread of TB and vice versa. This is a serious problem in a country that already accounts for 20 percent of all TB cases worldwide, and where TB is a leading cause of death [33]. In 2004, an estimated 1.8 million new TB cases occurred in India and 330,000 people died of TB [33,34].

India faces a number of challenges in its fight against HIV/AIDS. Universal access to free or low-cost medical care does not exist. Many healthcare centers have limited resources and are overburdened with patients seeking basic medical treatment. Stigmatization and discrimination significantly hinder HIV prevention efforts [2,4], as does the low social status of women.

Perhaps, the most significant hindrance to an adequate HIV response in India is the difficulty of treating a disease that affects less than 1 percent of the adult population, in a country where other diseases dominate and resources are scarce. Although the national response to HIV in India is improving, current HIV treatment levels are inadequate. At the end of 2005, only 7 percent of the estimated 770,000 people needing treatment received antiretroviral treatment (only 7,000 people received free treatment through the public sector) [1,35]. The Indian government intends to expand treatment coverage to 188 centers, with the target of providing 100,000 people with free treatment by 2007 [1,35].

5.4.2 Avahan Project

The Avahan Project, funded by the Bill and Melinda Gates Foundation, is an ambitious five-year program (2004–2009) for HIV prevention in India [8]. The initiative supports a variety of programs that target high-risk groups, including CSWs and their clients, with programs for STD control, condom promotion, and education about HIV risk. By targeting high-risk populations, the project aims to reduce the spread of HIV into so-called bridge populations and into the general population. Programs are to be implemented in 100 high-prevalence districts in the most affected states, at 50–75 highway stops, and in other hotspots in low-prevalence states. District-level interventions aim to reach 250,000 CSWs not yet reached by other interventions, 4.5 million CSW clients, and 60,000 IDUs. Interventions at highway stops aim to stop transmission from clients of CSWs to their partners, and aim to reach 1.5 million truck drivers.

To help these initiatives succeed, the project aims to improve communication and cooperation among private, public, and government partners. A community advocacy program will aim to reduce stigmatization of HIV, and will include mass communication campaigns aimed at high-risk groups. Effects of the programs will be measured through repeated surveillance of HIV, STDs, and risk behaviors, and by surveys.

5.4.3 Framework for Cost-Effectiveness Analysis of the Avahan Project

To maximize the impact of the $200 million expenditure, it is important to accurately estimate the relative efficiency of each HIV prevention program in the Avahan Project. The foundation for such analysis should include an appropriate model of HIV transmission and disease progression that can translate the changes in risk factors due to an intervention (e.g., changes in number of sexual partners, or change in infectivity due to drug therapy) into meaningful epidemiological outcomes (e.g., QALYs), and that can estimate all the costs that accrue from an intervention. The model should be sophisticated enough to capture important characteristics of the epidemic (e.g., key risk groups and risk behaviors), yet simple enough so that data requirements are not excessive. The model should provide a broad representation of the epidemic but, more importantly, should capture the effects of the interventions on the epidemic.

A key factor to include in such a model is transmission from high-risk groups (e.g., CSWs) to bridge groups (e.g., clients of CSWs) and from bridge groups to the general population (e.g., wives of CSW clients). Such a model should include sexual transmission of HIV and, if appropriate, transmission via birth and via injection drug use. It may be desirable to include the transmission of STDs, TB, or other diseases if their spread is significantly affected by the intervention. The model should also capture relevant changes in health status (e.g., deaths, changes in HIV prevalence, changes in STD, or TB prevalence) and associated costs.

Although a variety of models that capture the spread of HIV and other STDs have been developed, none is readily applicable to evaluating programs in the Avahan Project. In the following section, we present an illustrative version of such a model.

5.4.4 Illustrative Model

5.4.4.1 Description

Figure 5.2 shows a schematic illustration of a simple model that could provide a foundation for CEA of programs in the Avahan Project. Such a model is not intended to exactly capture the growth of the epidemic, but instead to approximate the primary modes of heterosexual HIV transmission in India as a basis for evaluating the cost-effectiveness of interventions. In the model, the population under consideration (adults in India, and HIV-infected newborns) is divided into mutually exclusive, collectively exhaustive compartments. The adult population is divided into three groups: CSWs, clients of CSWs, and all other adults in the general population. Each group is subdivided into HIV-infected and uninfected subgroups. CSWs are assumed to be female and their clients are assumed to be male.

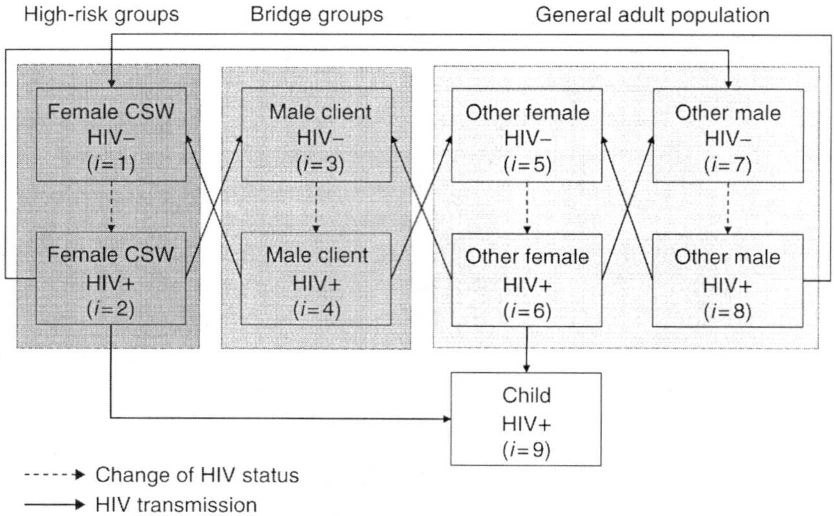

Figure 5.2 Schematic diagram of illustrative model.

A key feature of the model is the flow of the virus from high-risk groups to bridge groups to the general population. The high-risk individuals—CSWs—are the initial sources of the infection. They can infect the bridge population—their clients. In turn, individuals in the bridge population can infect CSWs and can infect their other partners, who are members of the general population. At the end of the chain, females and males in the general population can infect one another. Additionally, HIV-infected females can give birth to HIV-infected children. Equations of the model are shown in the Appendix.

Our illustrative model includes HIV transmission only via heterosexual contact and from mothers to newborns. To model sexual transmission, we considered number of sexual partners, frequency of sexual contact, probability of condom use, and the chance that either partner has an STD. Mother-to-child transmission depends on the birthrate, the chance a woman is HIV-infected, and the chance of infection transmission to the child.

The purpose of an epidemic model of this type is to create a baseline projection that captures salient aspects of HIV spread so that the effects of interventions can be measured. It may be reasonable to make various simplifying assumptions in the development of such a model (e.g., as we have done, one might ignore geographic isolation, age differences, or individual sexual behavior differences). However, the model should incorporate enough detail to demonstrate how the high-risk sexual behavior of CSWs and their clients—and changes in their risk factors due to the interventions—affect the overall HIV epidemic.

5.4.4.2 Data

The illustrative model requires the following epidemiological data: estimates of compartment sizes (number of CSWs, clients, etc.); baseline HIV prevalence in each population group (compartment); factors relating to HIV transmission, including the number of sex acts, the chance of transmission per unprotected sex act, as a function of whether an STD is present, the probability that an STD is present, and the probability a condom is used; and the death rate from HIV. On the basis of a review of the literature and available databases, as well as interviews conducted in India with HIV prevention and treatment experts, we estimated values for all model parameters (Table 5.2) to represent the situation in Mumbai.

5.4.4.3 Effect of Interventions

Using such a model, the effect of a program can be assessed by adjusting relevant model parameters as appropriate. For instance, to estimate the cost-effectiveness of a program that increases condom use among CSWs by 10 percent, one could increase by 10 percent the value of the parameter that specifies the probability of condom use by CSWs. One could then use the model to measure the resulting health outcomes and costs, and then compare these to health outcomes and costs in the base case (which uses the original parameter value for probability of condom use by CSWs) to determine the incremental cost-effectiveness of the program.

For the illustrative model, we measured costs and benefits over a 10-year time horizon. We considered three types of programs: a program aimed at increasing condom use between CSWs and clients; a program aimed at reducing STD prevalence among CSWs; and a program aimed at decreasing the number of sex acts between CSWs and clients. We assumed that interventions take place and immediately affect behavior at the beginning of the time horizon, but that direct intervention costs are incurred annually. For health outcomes, we measured the (discounted) number of HIV infections averted over the time horizon and (discounted) number of QALYs gained.

Figure 5.3 shows how each program would reduce the prevalence of HIV after five and ten years, as a function of its effectiveness. For example, Figure 5.3 shows that with zero program effectiveness (or no prevention programs), adult HIV prevalence in Mumbai would be almost 2.6 percent in

Table 5.2 Data Used in Illustrative Model

Parameter	Value	Source
Population group sizes		
Number of CSWs	50,000	[25,52]
Number of clients	325,000	[25,52]
Number of other female partners	3,990,000	[37]
Number of other male partners	3,570,000	[37]
Annual birth rate	22.01/1000	[37]
HIV prevalence		
CSWs	60 percent	[25,52]
Clients of CSWs	5 percent	Estimated [25]
Other female	0.51 percent	Estimated [29]
Other male	0.46 percent	Estimated [29]
HIV transmission		
Number of sex acts per month		
CSW (with clients)	65	[53,54]
CSW (with other males)	9	[53]
Client (with other females)	6	Estimated
Other males (with other females)	8	Estimated
Probability of transmission per unprotected sex act		
Female to male – No STD	0.000835	Estimated
Male to female – No STD	0.0025	Estimated
Female to male – STD	0.0025	Estimated
Male to female – STD	0.0075	Estimated
Probability of STD (assumed constant)		
CSW	0.30	[28,53]
Client	0.15	[28,53]
Other female partner	0.02	[28]
Other male partner	0.02	[28]
Probability of condom use		
CSW (with clients)	0.75	[28,53]
CSW (with other males)	0.20	[28,53]
Client (with other females)	0.20	[28,53]
Other males (with other females)	0.20	[28]
Probability of vertical transmission	0.30	[55]
HIV progression		
Annual death rate from HIV	0.0833	[55]
Life expectancy of HIV+ child	3 years	[56]
Life expectancy of HIV+ person at end of horizon	6 years	Estimated
Life expectancy of HIV– person at end of horizon	25 years	Estimated
Quality multipliers		
HIV–	1.00	Estimated
HIV+	0.84	Estimated
Costs (US$)		
Cost per CSW reached	19.21	[57]
General annual per capita healthcare costs for HIV–	27	[58]
General annual per capita healthcare costs for HIV+	32	Estimated
Discount rate	3 percent	[9]

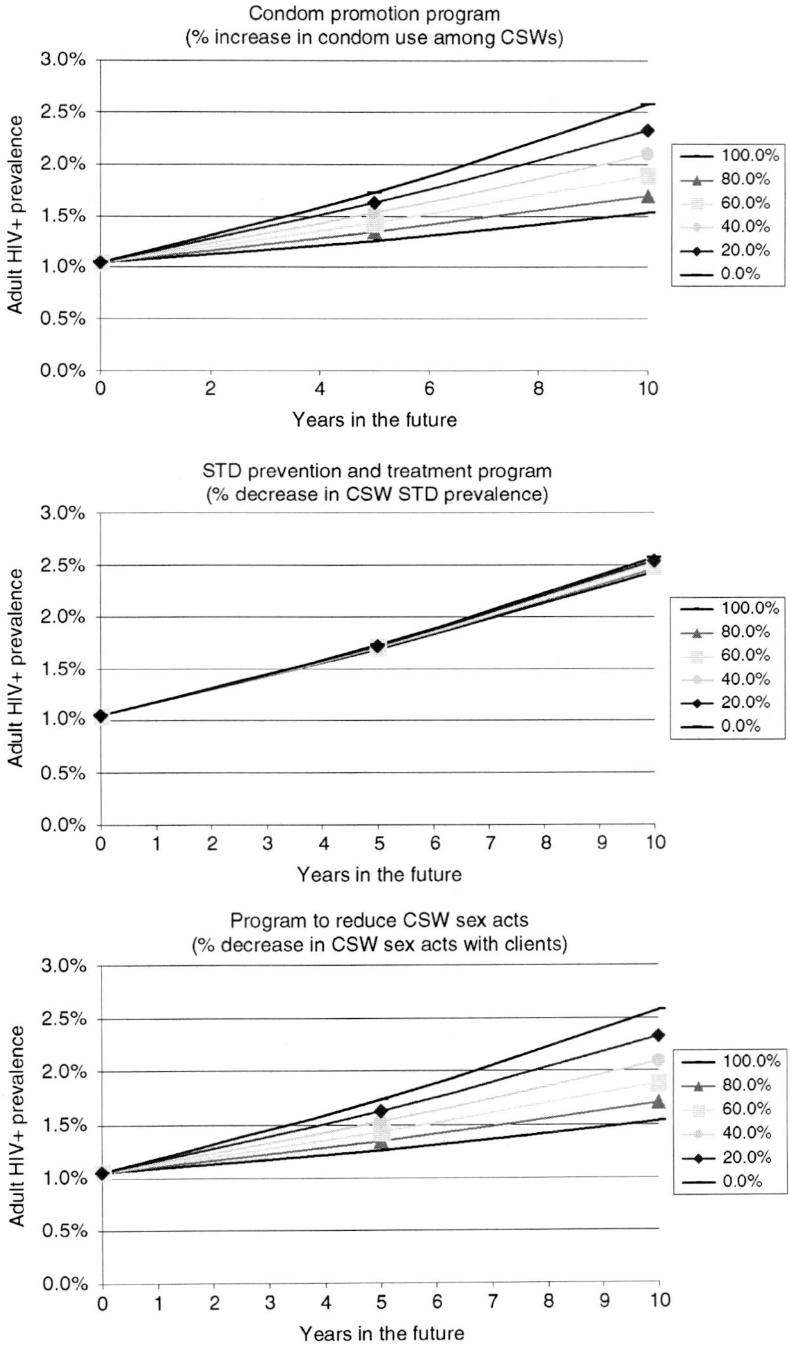

Figure 5.3 Example analyses using illustrative model: Impact of program effectiveness on HIV prevalence among adults after five and ten years.

ten years. If the condom promotion program were to reduce unprotected CSW sex acts by 20 percent, HIV prevalence among adults in ten years would be about 2.3 percent. Figure 5.3 shows that the condom promotion program and the program aimed at decreasing CSW sex acts would likely have a much greater impact on HIV prevalence than the STD prevention and treatment program.

Figures 5.4 and 5.5 show example cost-effectiveness analyses performed using the illustrative model. For each program, we considered different annual costs per targeted individual (ranging from \$5 to \$50) and different levels of program effectiveness (expressed as a percentage change from baseline). We then calculated iso-curves of constant cost per infection averted (Figure 5.4) and cost per QALY gained (Figure 5.5).

The three curves in each panel of Figure 5.4 correspond to a cost-effectiveness ratio of \$500 per HIV infection averted, \$1000 per HIV infection averted, and \$2000 per HIV infection averted, respectively. Thus, for example, Figure 5.4 shows that the condom promotion program will cost \$500 per HIV infection averted if it costs \$5 per CSW reached per year and leads to a 6 percent decrease in the number of unprotected sex acts; or if it costs \$15 per CSW per year and leads to a 19 percent decrease in the number of unprotected sex acts. Figure 5.4 shows that programs that increase condom usage or reduce CSW sex acts could have a low cost per HIV infection averted even with relatively minor changes in behavior. However, an STD reduction program would have to be very inexpensive and lead to very dramatic reductions in STDs among CSWs to cost \$500 per HIV infection averted.

In Figure 5.5, the three lines in each panel show cost-effectiveness ratios of \$100 per QALY gained, \$200 per QALY gained, and \$400 per QALY gained, respectively. Thus, for example, Figure 5.5 shows that if a condom promotion program costs \$20 per CSW reached and achieves a 10 percent decrease in the number of unprotected sex acts, the program has an incremental cost-effectiveness ratio of \$100 per QALY gained. If the same program achieves 5 percent decrease in the number of unprotected sex acts, the program has a cost-effectiveness ratio of about \$200 per QALY gained; if the decrease is only 2 percent, the program will cost about \$400 per QALY gained. According to the World Health Organization, a health intervention can be considered cost-effective if each DALY averted costs less than three times the gross domestic product (GDP) per capita [36]. In our analysis, we assumed that one QALY gained is equivalent to one DALY averted. For India, per capita GDP is approximately \$3300 [37]. Figure 5.5 shows that each of the three programs would be highly cost-effective according to WHO guidelines even if the programs induce only modest behavior change.

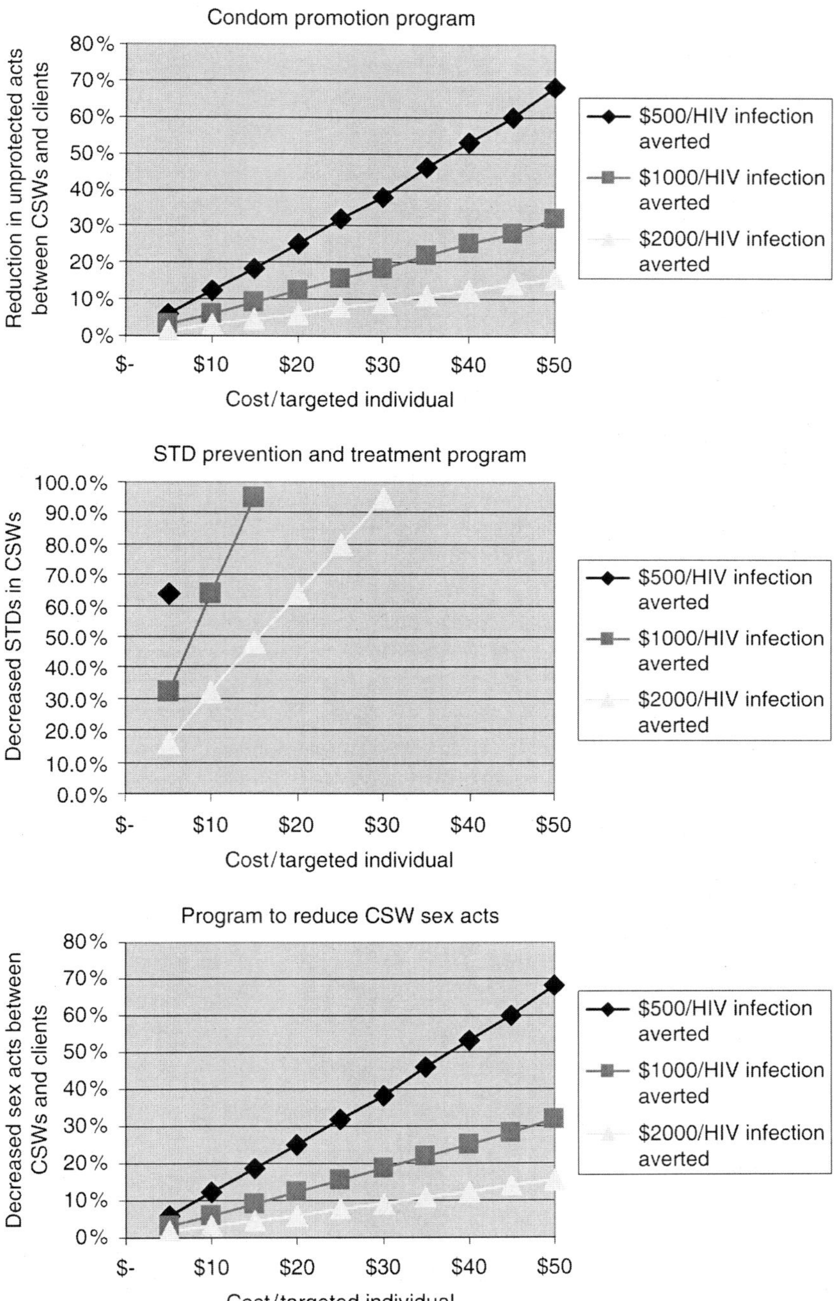

Figure 5.4 Example analyses using illustrative model: Cost per HIV infection averted as a function of annual program cost and effectiveness.

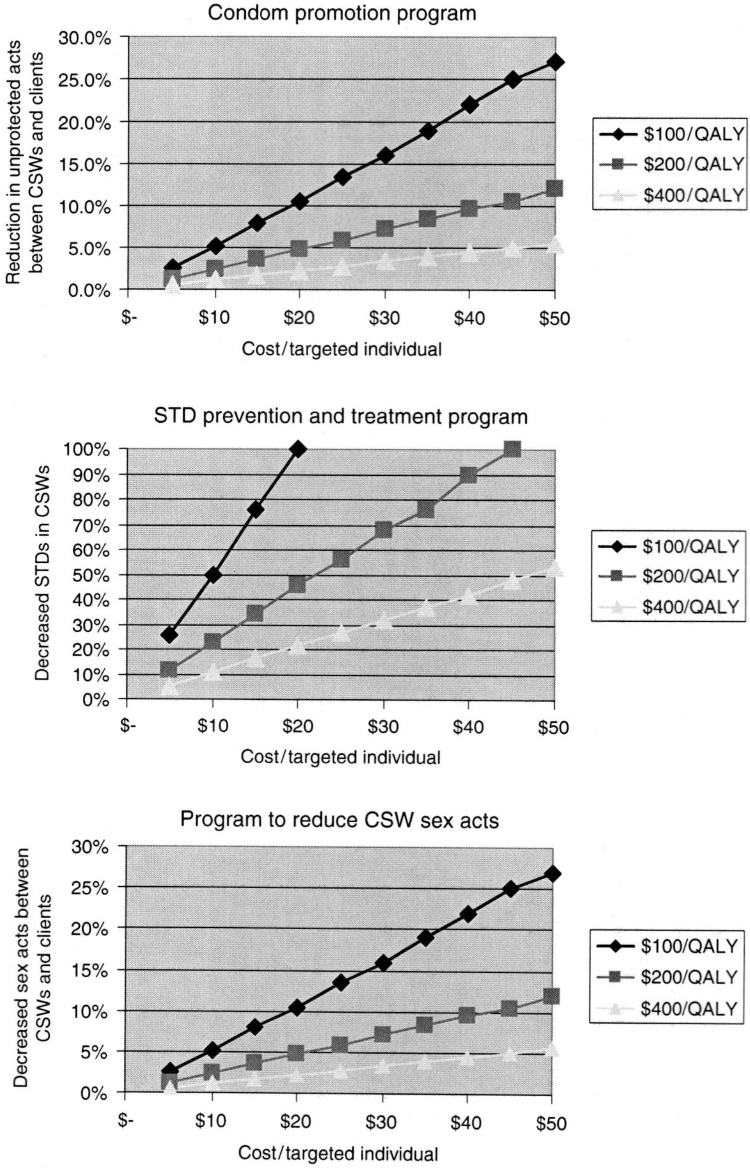

Figure 5.5 Example analyses using illustrative model: Cost per quality-adjusted life year (QALY) gained as a function of annual program cost and effectiveness.

5.5 Developing a Tailored Model

We have presented a very simple model to evaluate interventions in the Avahan project. In this section, we discuss factors that might be included in a more detailed model.

5.5.1 Homogeneous Sexual Behavior

Our illustrative model assumed that all individuals in a group have the same average number of sex acts per unit time. However, not all individuals in the groups we distinguished are the same; for example, some CSWs are brothel-based, while others are not, and some CSWs may have more sexual encounters than others. One way to capture this heterogeneity is to develop a model with more compartments than we have included. For example, one could include compartments for CSWs with many sexual encounters and compartments for CSWs with fewer sexual encounters, as in the Asian Epidemic Model [38].

5.5.2 Random Mixing

If a simple, high-level model is implemented, sexual contact between CSWs and clients could be modeled as a random mixing process. In reality, all CSWs do not mix randomly with all clients. This effect could be partially captured by an epidemic model with more compartments than we have shown; for example, one could subdivide CSW and client compartments according to mixing patterns. Another approach is to use models of social networks to capture mixing patterns and the resultant spread of HIV [15,39]. Social network models can capture very detailed mixing patterns, but have the disadvantage of being data- and computation-intensive.

5.5.3 Transmission Risk Factors

A detailed model might consider how sexual transmission of HIV is affected by the presence or absence of an STD, as in our illustrative model. Because the chance of HIV transmission may be higher in the presence of an ulcerative STD than in the presence of a nonulcerative STD, a further distinction could be made between these categories of STDs.

5.5.4 Injection Drug Use

Our illustrative model ignored HIV transmission via injection drug use. In some parts of India, injection drug use is a significant contributor to the epidemic. An intervention that reduces high-risk sexual behavior may affect HIV incidence less in a region where injection drug use is a significant driver of the epidemic. For interventions in regions with significant injection drug use, it may be important to consider the effects of injection drug use on the epidemic.

5.5.5 Disease Progression and Treatment

To accurately model the health benefit accruing from antiretroviral therapy, a detailed model of HIV progression and the effects of treatment on disease progression, along with estimates of quality-of-life multipliers for each disease stage, are needed [17]. Incorporating disease progression, and the associated changes in infectivity and quality of life, may be particularly important when evaluating interventions that include treatment for HIV. These effects can be captured by including model compartments for individuals in different stages of HIV infection and with different treatment status.

5.5.6 Nonconstant Population Group Sizes

To accurately capture population dynamics, deaths from HIV infection should be included, as well as deaths from other causes, births, and changes in the size of the overall population. Additionally, it may be appropriate to incorporate changes in the sizes of risk groups that occur for other reasons. For example, one goal of the Avahan Project is to reduce the demand for commercial sex, which could reduce the total number of CSWs.

5.5.7 Spread of Other Diseases

A detailed model could allow for the presence of STDs (other than HIV) via parameters that specify STD prevalence exogenously for each population group. Because the spread of STDs is a dynamic process (similar to that for HIV), an intervention that cures one individual's STD may prevent an STD infection in the individual's partners. Thus, it may be desirable to include the spread of STDs in the HIV epidemic model, particularly when evaluating the cost-effectiveness of interventions that aim to treat and eliminate STDs. Similarly, it may be desirable to include the effects of an intervention on the spread of TB. HIV-infected individuals in India have significant rates of TB coinfection. HIV-infected individuals have an increased chance of developing active TB, and thus can transmit TB to other people. Adding other diseases to an evaluation may substantially complicate model development and will increase the requirements for input data, but may be important in some cases.

5.5.8 Realism versus Tractability

We have discussed a number of features that could be included in an epidemic model to more realistically model HIV transmission in India. In developing such a model, a balance must be achieved between model realism and data requirements. If the model is too simple, it may not properly capture the effects of interventions on health and costs. If the model is too detailed, it may not be possible to collect reliable data, and the outputs of the model may not be credible. As highlighted above, the key goal is to develop a model that can evaluate the most important costs and health benefits of an intervention, and thus provide an accurate assessment of program cost-effectiveness.

5.6 Discussion

We have presented a model-based framework for evaluating the cost-effectiveness of HIV prevention and treatment programs, which we illustrated using the Gates Foundation's India AIDS Initiative as an example. A framework of this type can be used to develop accurate assessments of the cost-effectiveness of HIV interventions, thus helping policy makers maximize the impact of limited HIV prevention and treatment resources. For cost-effectiveness analysis of interventions in a specific geographic region, one must consider local disease dynamics, the behavioral and demographic factors of the population of interest, the availability of reliable data, and the specific intervention that is being evaluated.

A model-based approach has several important advantages. It allows one to capture the effects of an intervention on disease transmission and progression among targeted individuals, as well as among other individuals in the population. It allows one to capture the direct and indirect costs of an intervention that accrue over time. It allows one to perform sensitivity analysis on model assumptions and key parameter values.

In addition to HIV, other infectious diseases pose a serious threat to global health. Every year, two million people die from TB and another eight million become sick from TB [3]. However, 95 percent of TB deaths could be prevented with relatively inexpensive treatments. STDs are another serious health problem. In 1999, 340 million people aged 15–49 were newly infected with a curable STD (chlamydia, gonorrhea, syphilis, or trichomoniasis) [40]. Over 150 million of these infections occurred in South and Southeast Asia. STDs can lead to frequent illness, infertility, disability, or death. The model-based framework we suggest could be applied to evaluate the cost-effectiveness of interventions targeting these diseases.

Control of HIV and other infectious diseases is a critical global health concern. Limited prevention and treatment resources must be spent in the most effective manner. A model-based framework, such as the one presented in this chapter, can be used to assess the cost-effectiveness of such interventions, and thus improve health outcomes.

Acknowledgments

This work was supported by the Gates Foundation Policy Research Network and by a grant from the National Institute on Drug Abuse, National Institutes of Health (R-01-DA-15612). Ms. Long also received support from the INFORMS Seth Bonder Scholarship.

Appendix

Equations of Illustrative Model

Let $X_i(t)$ denote the number of individuals in compartment i at time t. The model compartments are indexed as follows: CSW, HIV− ($i = 1$), CSW, HIV+ ($i = 2$); Client, HIV− ($i = 3$); Client, HIV+ ($i = 4$); Other female, HIV− ($i = 5$); Other female, HIV+ ($i = 6$); Other male, HIV− ($i = 7$); Other male, HIV+ ($i = 8$); Child, HIV+ ($i = 9$). The sizes of the larger population groups (CSWs, Clients, Other female, and Other male) are assumed to be constant. Let μ_i be the death rate from HIV of individuals in compartment i, and $\lambda_i(t)$ be the rate at which individuals in (HIV uninfected) compartment i become infected (i.e., the HIV sufficient contact rate) at time t. Let ρ be the birthrate, and ν be the chance of HIV transmission from a mother to her newborn. The model is specified by the following differential equations:

$$\frac{dX_i(t)}{dt} = \mu_{i+1}X_{i+1}(t) - \lambda_i(t)X_i(t) \quad i = 1, 3, 5, 7$$

$$\frac{dX_i(t)}{dt} = -\mu_i X_i(t) + \lambda_{i-1}(t)X_{i-1}(t) \quad i = 2, 4, 6, 8$$

$$\frac{dX_i(t)}{dt} = \rho\nu[X_2(t) + X_6(t)] \quad i = 9$$

We now describe how the sufficient contact rates $\lambda_i(t)$ are calculated. For i = 1, 3, 5, 7, and j = 2, 4, 6, 8, let α_i be the transmission rate to an individual in compartment i per unprotected sex act with an individual in compartment j, let β_{ij} be the probability that a condom is used during sex acts between individuals in compartments i and j, and let δ_{ij} be the average number of sex acts per unit time that an individual in compartment i has with individuals in compartments $j - 1$ and j. Then the sufficient contact rates can be expressed as

$$\lambda_i(t) = 1 - \prod_{j=2,4,6,8} \left[\left(1 - \alpha_i(1 - \beta_{ij}) \left[\frac{X_j(t)}{X_{j-1}(t) + X_j(t)} \right] \right)^{\delta_{ij}} \right] \quad i = 1, 3, 5, 7$$

References

1. Joint United Nations Programme on HIV/AIDS (UNAIDS), 2006 Report on the Global HIV/AIDS Epidemic, Geneva, Switzerland: United Nations; 2006.
2. United States Central Intelligence Agency National Intelligence Council, The next wave of HIV/AIDS: Nigeria, Ethiopia, Russia, India, and China. United States Central Intelligence Agency National Intelligence Council, 2002 (accessed August 17, 2006, at http://www.dni.gov/nic/special_nextwaveHIV.html).
3. The Global Fund, The Global Fund to Fight AIDS, Tuberculosis and Malaria. 2004 (accessed August 17, 2006, at http://www.theglobalfund.org).

4. Avert.org, HIV and AIDS in India. Avert, 2006 (accessed August 17, 2006, at http://www.avert.org/aidsindia.htm).
5. United States Agency for International Development, HIV/AIDS Overview 2005 (accessed August 17, 2006, at http://www.usaid.gov).
6. Bill and Melinda Gates Foundation, HIV/AIDS in India: Time to Act. Bill and Melinda Gates Foundation, 2006 (accessed August 17, 2006, at http://www.gatesfoundation.org/nr/downloads/globalhealth/India/Backgrounder_050131.pdf).
7. Joint United Nations Programme on HIV/AIDS (UNAIDS), Resource needs for an expanded response to AIDS in low- and middle-income countries, 2005 (accessed August 17, 2006, at http://data.unaids.org/publications/irc-pub06/resourceneedsreport_en.pdf).
8. Bill and Melinda Gates Foundation, Avahan: India AIDS Initiative. Bill and Melinda Gates Foundation, 2006 (accessed August 17, 2006, at http://www.gatesfoundation.org/globalhealth/pri_diseases/hivaids/hivprogramspartnerships/avahan.htm).
9. M.R. Gold, J.E. Siegel, L.B. Russell, and M.C. Weinstein, Eds., *Cost-Effectiveness in Health and Medicine*, Oxford University Press, New York, 1996.
10. D.T. Fleming and J.N. Wasserheit, From epidemiological synergy to public health policy and practice: The contribution of other sexually transmitted diseases to sexual transmission of HIV infection, *Sexually Transmitted Infections* 75, 3–17, 1999.
11. M.L. Brandeau, G.S. Zaric, and V. de Angelis, Improved allocation of HIV prevention resources: Using information about program effectiveness, *Health Care Management Science* 8, 19–28, 2005.
12. E.H. Kaplan, Economic analysis of needle exchange, *AIDS* 9, 1113–1119, 1995.
13. T.J. Saidel, D. Des Jarlais, W. Peerapatanapokin, J. Dorabjee, S. Singh, and T. Brown, Potential impact of HIV among IDUs on heterosexual transmission in Asian settings: Scenarios from the Asian Epidemic Model, *International Journal of Drug Policy* 14, 63–74, 2003.
14. R.S. Bernstein, D.C. Sokal, S.T. Seitz, B. Auvert, J. Stover, and W. Naamara, Simulating the control of a heterosexual HIV epidemic in a severely affected east African city, *Interfaces* 28, 101–126, 1998.
15. M. Kretzschmar and L.G. Wiessing, Modeling the spread of HIV in social networks of injecting drug users, *AIDS* 12, 801–811, 1998.
16. A.D. Paltiel and K.A. Freedberg, The cost-effectiveness of preventing cytomegalovirus disease in AIDS patients, *Interfaces* 28, 34–51, 1998.
17. G.D. Sanders, A.M. Bayoumi, V. Sundaram, et al., Cost-effectiveness of screening for HIV in the era of highly active antiretroviral therapy, *New England Journal of Medicine* 352, 570–585, 2005.
18. G.S. Zaric, P.G. Barnett, and M.L. Brandeau, HIV transmission and the cost effectiveness of methadone maintenance, *American Journal of Public Health* 90, 2000, 1100–1111.
19. P.E. Sax, R. Islam, R.P. Walensky, et al., Should resistance testing be performed for treatment-naive HIV-infected patients? A cost-effectiveness analysis, *Clinical Infectious Diseases* 41, 1316–1323, 2005.
20. A.P. Johnson-Masotti, S.D. Pinkerton, K.J. Sikkema, J.A. Kelly, and D.A. Wagstaff, Cost-effectiveness of a community-level HIV risk reduction intervention for

women living in low-income housing developments, *Journal of Primary Prevention* 26, 345–362, 2005.
21. Joint United Nations Programme on HIV/AIDS (UNAIDS), 2004 Report on the Global AIDS Epidemic, Geneva, Switzerland: United Nations; 2004.
22. P. Chatterjee, Spreading the word about HIV/AIDS in India, *Lancet* 361, 1526–1527, 2003.
23. National AIDS Control Organization (NACO), Surveillance for AIDS Cases in India, 2005 (accessed August 17, 2006, at http://www.nacoonline.org/facts_rep-ortfeb.htm).
24. World Health Organization (WHO), Summary country profile for HIV/AIDS treatment scale-up: India, July 2004, Geneva, Switzerland: World Health Organization; 2004.
25. C.B.S. Venkataramana and P.V. Sarada, Extent and speed of spread of HIV infection in India through the commercial sex networks: A perspective, *Tropical Medicine and International Health* 6, 1040–1061, 2001.
26. J. Cohen, HIV/AIDS in India, *Science* 304, 504–513, 2004.
27. K.S. Rao, R.D. Pilli, A.S. Rao, and P.S. Chalam, Sexual lifestyle of long distance lorry drivers in India: Questionnaire survey, *British Medical Journal* 318, 162–163, 1999.
28. National AIDS Control Organization (NACO), Executive Summary-HRG Part-I (Female Sex Workers and Their Clients), 2004 (accessed August 17, 2006, at http://www.naco.nic.in/indianscene/executive1.htm).
29. Avert.org, India HIV & AIDS Statistics, 2006 (accessed August 17, 2006, at http://www.avert.org/indiaaids.htm).
30. S. Panda, A. Chatterjee, S.K. Bhattacharya, et al., Transmission of HIV from injecting drug users to their wives in India, *International Journal of Sexually Transmitted Diseases and AIDS* 11, 468–473, 2000.
31. World Health Organization, Frequently Asked Questions About TB and HIV, 2006 (accessed August 17, 2006, at http://www.who.int/tb/hiv/faq/en/index.html).
32. P. Chatterjee, India's battle against the tuberculosis-HIV dual epidemic, *Lancet Infectious Diseases* 4, 70, 2004.
33. C. Dye, India's leading role in tuberculosis epidemiology and control, *Indian Journal of Medical Research* 123, 481–484, 2006.
34. World Health Organization (WHO), Global tuberculosis control: Surveillance, planning, financing, Geneva, Switzerland: World Health Organization; 2006.
35. World Health Organization (WHO), Summary country profile for HIV/AIDS treatment scale-up: India, June 2005, Geneva, Switzerland: World Health Organization; 2005.
36. World Health Organization (WHO), The World Health Report 2002—reducing risks, promoting healthy life, Geneva, Switzerland: World Health Organization; 2002.
37. United States Central Intelligence Agency (CIA), The World Factbook: India 2006. United States Central Intelligence Agency (CIA), 2006 (accessed August 17, 2006, at https://www.cia.gov/cia/publications/factbook/geos/in.html).
38. Family Health International (FHI), The Asian Epidemic Model (AEM): Structure and Development of the Model, Research Triangle Park, North Carolina; 2004.

39. M. Morris and M. Kretzschmar, Concurrent partnerships and transmission dynamics in networks, *Social Networks* 17, 299–318, 1995.
40. Avert.org, Indian HIV and AIDS Statistics. Avert, 2005 (accessed August 17, 2006, at http://www.avert.org/indiaaids.htm).
41. R.M. May and R.M. Anderson, Transmission dynamics of HIV infection, *Nature* 326, 137–142, 1987.
42. R.M. Anderson, R.M. May, T.W. Ng, and J.T. Rowley, Age-dependent choice of sexual partners and the transmission dynamics of HIV in sub-Saharan Africa, *Philosophical Transactions of the Royal Society of London B—Biological Sciences* 336, 135–155, 1992.
43. UNAIDS Reference Group on Estimates Modeling and Projections, Improved methods and assumptions for estimation of the HIV/AIDS epidemic and its impact: Recommendations of the UNAIDS Reference Group on Estimates, Modeling and Projections, *AIDS* 16, W1–W14, 2002.
44. Joint United Nations Programme on HIV/AIDS (UNAIDS), Epidemiological software and tools. Joint United Nations Programme on HIV/AIDS (UNAIDS), 2006 (accessed August 17, 2006, at http://www.unaids.org/en/HIV_data/Epidemiology/episoftware.asp).
45. J. Chin and S.K. Lwanga, Estimation and projection of adult AIDS cases: A simple epidemiological model, *Bulletin of the World Health Organization* 69, 399–406, 1991.
46. S.T. Seitz and G.E. Mueller, Viral load and sexual risk: Economic and policy implications for HIV/AIDS, in: E.H. Kaplan and M.L. Brandeau, Eds. *Modeling the AIDS Epidemic: Planning, Policy and Prediction*. New York: Raven Press; 1994, 461–480.
47. T.M. Rehle, T.J. Saidel, S.E. Hassig, P.D. Bouey, E.M. Gaillard, and D.C. Sokal, AVERT: A user-friendly model to estimate the impact of HIV/sexually transmitted disease prevention interventions on HIV transmission, *AIDS* 12 (Suppl 2), S27–S35, 1998.
48. N.J. Robinson, D.W. Mulder, B. Auvert, and R.J. Hayes, Modeling the impact of alternative HIV intervention strategies in rural Uganda, *AIDS* 9, 1263–1270, 1995.
49. N.J.D. Nagelkerke, P. Jha, S.J. de Vlas, et al., Modeling HIV/AIDS epidemics in Botswana and India: Impact of interventions to prevent transmission, *Bulletin of the World Health Organization* 80, 89–96, 2002.
50. N.J.D. Nagelkerke, P. Jha, S.J. de Vlas, et al., Modeling the HIV/AIDS epidemics in India and Botswana: The effect of interventions. World Health Organization, Commission on Macroeconomics and Health Working Paper Series, Working Paper WG5:4, 2001 (accessed August 17, 2006, at http://www.cmhealth.org/docs/wg5_paper4.pdf).
51. C.P.B. Van der Ploeg, C. Van Vliet, S.J. De Vlas, et al., STDSIM: A microsimulation model for decision support in STD control, *Interfaces* 28, 84–100, 1998.
52. P. Chandrasekaran, G. Dallabetta, V. Loo, S. Rao, H. Gayle, and A. Alexander, Containing HIV/AIDS in India: the unfinished agenda, *Lancet Infectious Diseases* 6, 508–521, 2006.
53. National AIDS Control Organization (NACO), National Baseline General Population Behavioral Surveillance Survey—2001. National AIDS Control Organization (NACO), 2001 (accessed August 17, 2006, at http://www.nacoonline.org/publication/31.pdf).

54. Family Health International (FHI), Mapping of commercial sex access points and relevant service outlets in Maharashtra, 2001 (accessed August 17, 2006, at http://www.fhi.org/NR/rdonlyres/ejl2pf7uynw5bx276oqyhqtavj662j6n6r2vbdan3zqa7hzkdjhbvtzvmm5mx2ho4js3kkw5ddik2f/Mappavertprepstudyindia.pdf).
55. Joint United Nations Programme on HIV/AIDS (UNAIDS), Prevention of HIV transmission from mother to child: Strategic options. Joint United Nations Programme on HIV/AIDS (UNAIDS), 1999 (accessed August 17, 2006, at http://data.unaids.org/Publications/IRC-pub05/Prevention_en.pdf).
56. N. Walker, B. Schwartlander, and J. Bryce, Meeting international goals in child survival and HIV/AIDS, *Lancet* 360, 284–289, 2002.
57. L. Guinness, L. Kumaranayake, B. Rajaraman, et al., Does scale matter? The costs of HIV-prevention interventions for commercial sex workers in India, *Bulletin of the World Health Organization* 83, 747–755, 2005.
58. World Health Organization, Core Health Indicators, 2006 (accessed August 17, 2006, at http://www3.who.int/whosis/core/core_select.cfm).

Chapter 6

Optimization over Graphs for Kidney Paired Donation

Sommer E. Gentry

CONTENTS

6.1 Finding Edges 182
6.2 Designing the Objective 184
6.3 Solution Methods 186
6.4 Current Status of Paired Donation 189
6.5 Projecting the Impact of Kidney Paired Donation 190
Acknowledgments 192
References 193

Resource allocation problems are ubiquitous in the optimization literature, and perhaps no resource is more ethically and practically challenging to allocate than the limited number of suitable solid organs for transplantation. The most frequently transplanted solid organs are kidney and liver, in that order. Optimization models may consider the objectives of and decisions made by various stakeholders in transplantation, such as individual recipients (living donor livers) [1], or society in the form of the United Network for Organ Sharing (UNOS) (deceased donor kidneys) [2]. Su et al. demonstrated that individual autonomy can degrade the performance of a societal optimization mechanism [3]. This chapter will focus on the case of kidney paired donation, introducing both the medical background and

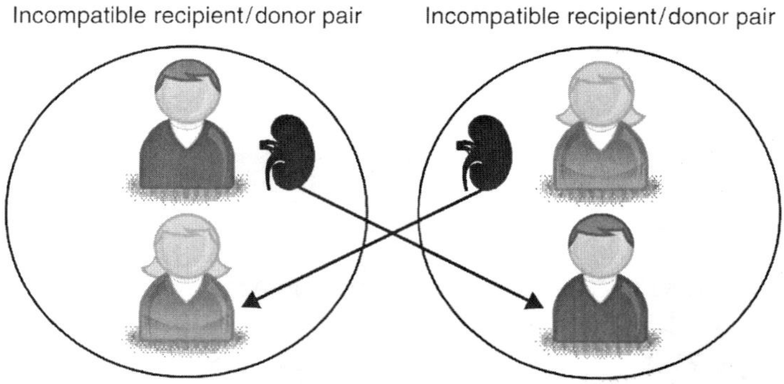

Figure 6.1 A kidney paired donation.

mathematical tools used to find a societally optimal allocation of organs for this type of transplant [4].

A kidney paired donation involves an exchange of living kidney donors between two recipient/donor pairs [5,6], as illustrated in Figure 6.1. The donor of each pair wants to give a kidney to the recipient of the pair, but cannot do so because of blood or tissue incompatibility. Instead, the donor of the first pair gives to the recipient of the second pair, and the donor of the second pair gives to the recipient of the first pair. Because neither donor will donate unless his own loved one can receive a kidney, these exchanges can only occur if a reciprocally compatible pair can be found. The donor operations are begun simultaneously to avoid the possibility of a donor reneging after his own recipient has gotten a kidney.

The UNOS is charged with coordinating transplantation activities in the United States. Over 70,000 hopeful recipients are on the UNOS waiting list for a kidney transplant [7]. Fewer than 11,000 kidneys became available from deceased donors in 2006, and there are limited prospects for increasing the level of deceased donation. However, over 6000 living donors gave one of their kidneys to either a relative or an unrelated recipient in 2006. Living donation is on the rise for many reasons, including the availability of a minimally invasive donor operation with a faster recovery time. However, about one-third of all recipient/donor pairings will be found to be incompatible. Presently, most of these incompatible donors are simply turned down and do not donate. UNOS has proposed a national kidney paired donation registry that could enable many of these donors to donate a kidney so that their loved one can receive one [8].

Every kidney paired donation constitutes a net benefit to the system because two people donate who otherwise would not have given a kidney. Further, if there are many recipient/donor pairs with various feasible

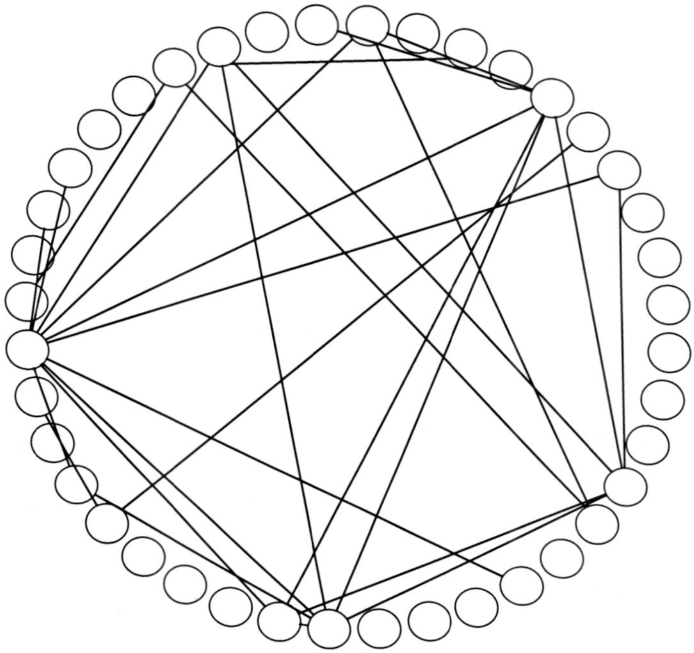

Figure 6.2 A typical kidney paired donation graph.

paired donation arrangements among them then even greater benefit can be achieved by using a graph-theoretic maximum matching algorithm to find the optimal allocation, selecting which foursomes should proceed with paired donation.

A graph consists of a set of nodes and a set of edges connecting them, as in Figure 6.2. A graph model for kidney paired donation is as follows: let each node on a graph represent an incompatible recipient/donor pair. Each undirected edge on the graph represents a reciprocally compatible foursome for paired donation, meaning that the donor of the first pair is compatible with the recipient of the second pair, and vice versa. Index the n nodes of the graph by i in $\{1, 2, \ldots, n\}$, and let edge ij refer to the edge connecting node i with node j.

A matching on a graph is any subset of the edges having the property that no two edges in the matching are incident on the same node. In paired donation terms, this means that in a matching, a single incompatible recipient/donor pair (node) cannot be involved in more than one paired donation (edge). A maximum matching, also called a maximum cardinality matching, includes the largest possible number of edges. In fact, a more nuanced model called maximum weighted matching, which maximizes the sum of positive weights on the edges in the matching, is appropriate for determining the best collection of exchanges on a kidney paired donation graph. Some papers on kidney paired donation refer to a match, meaning a single edge on the graph that has been selected to be in the matching. Also, the word "matching" is often used as a verb, to describe the process of selecting a preferred matching.

Algorithms for optimal matching operate on a fixed graph of incompatible pairs. In reality, incompatible pairs present to transplant centers sequentially. If each new arrival triggered a search for a paired donation opportunity then the optimization would be trivial, amounting to rank-ordering the pairs that are reciprocally compatible with the new one. Deceased donor kidneys are allocated as they arrive based on a rank-ordering of recipients. This is necessary because a kidney recovered from a deceased donor cannot be stored, whereas living donation can be scheduled for any convenient time. Higher transplantation rates will be achieved for paired donation if incompatible pairs accumulate in a registry before the preferred matching is selected [4]. Figure 6.3 shows a projection of the relationship between transplantation rates and the number of incompatible pairs in a registry. See Section 6.5 for more about how this data is generated. The transplant community has recognized the need to accumulate incompatible pairs, but it is thought that participants will demand that a matching be selected about every three months, even if the registries are initially small.

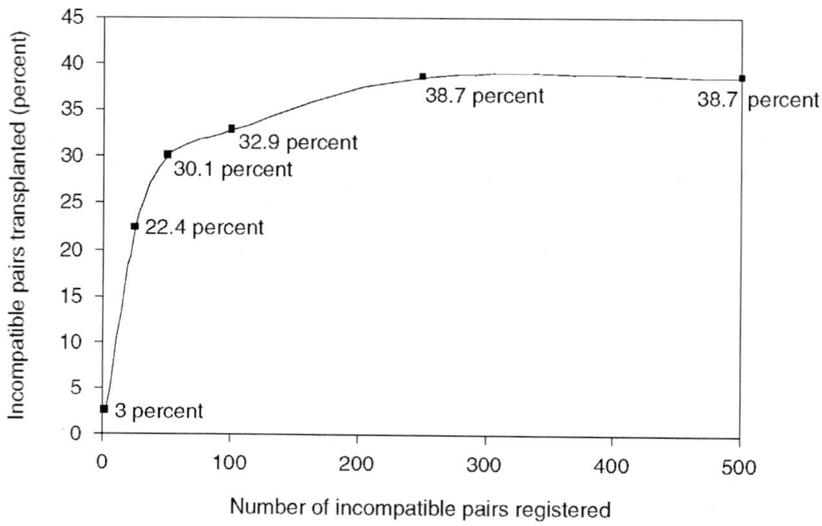

Figure 6.3 Percent of incompatible pairs transplanted for various numbers of incompatible pairs accumulated, using maximum weighted matching.

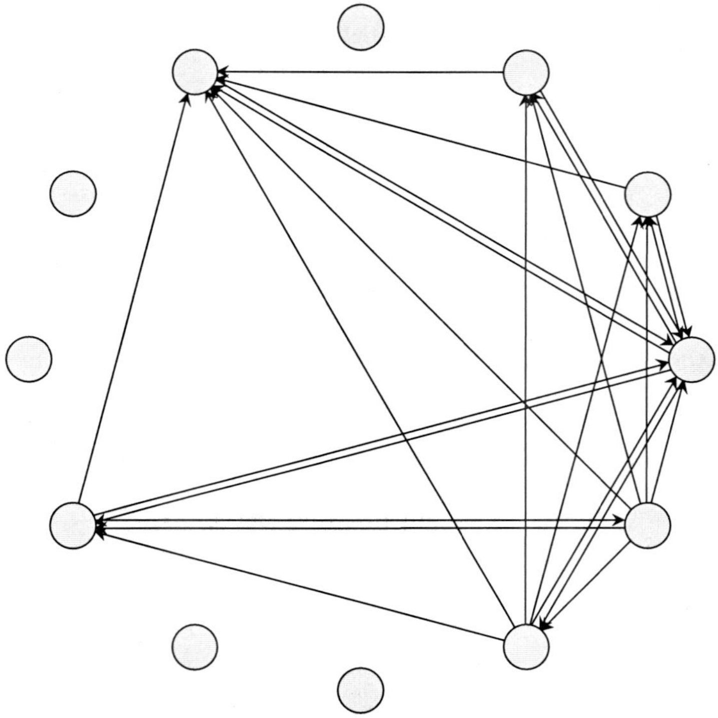

Figure 6.4 A typical directed kidney paired donation graph for m-way donations.

Significant generalizations of the basic kidney paired donation arrangement have been explored. Kidney donor exchanges involving more than two incompatible pairs are important in theory [9] and practice [10]. If m incompatible pairs are involved, the procedure is called an m-way paired donation. Representing these larger exchanges requires a directed graph, as in Figure 6.4, with directed edges where the donor at the source node is compatible with the recipient at the target node. An opportunity for an m-way paired donation corresponds to the presence of a cycle of length m. A cycle of length m is a directed path that touches m nodes before revisiting the first node on the path. More transplants would be possible with two- and three-way paired donations than with only two-way [11]. Still, three-way paired donations introduce logistical difficulties in practice and add algorithmic complexity in selecting a matching. Roth et al. have shown that, asymptotically and with strong assumptions about compatibility, allowing three-way and four-way paired donations would achieve the same number of transplants as allowing unrestricted m-way paired donations [12]. The development in this chapter is restricted to two-way paired donation, which is equivalent to matching in undirected graphs, unless explicitly stated.

Other generalizations include domino paired donation, list paired donation, and voluntary compatible pair participation. Domino paired donation begins with an altruistic, nondirected kidney donor, who offers to give a kidney to any stranger in need. Instead of the anonymous donor giving to the person at the top of the UNOS waiting list, the nondirected donor gives to the recipient of an incompatible pair, and the donor of the incompatible pair gives to the person at the top of the waiting list [13].

In list paired donation, the donor of an incompatible pair gives preemptively to the person at the top of the waiting list, and in return the recipient of the incompatible pair is moved to the top of the waiting list for the next available deceased donor organ [14]. List paired donation is less desirable than living paired donation because kidneys from living donors perform better than kidneys from deceased donors, making the exchange somewhat unfair. Also, list paired donation opportunities frequently involve giving a non-O blood group kidney to the list in return for a type O kidney, but blood group-O recipients already have long waiting times compared with other subgroups, so list paired donation exacerbates that disparity. Matching procedures are available which incorporate n-way paired, domino, and wait list paired donations [15]. Zenios proposed a dynamic controller that offers each incompatible pair either an immediate wait list paired donation or registration for a later living paired donation [16].

Compatible recipient/donor pairs have already participated in some paired donations. Voluntary participation by recipients who have compatible donors would greatly increase the probability of transplantation for incompatible pairs. Voluntary participation could also benefit the compatible pairs by helping them find younger or more immunologically favorable donors [17].

6.1 Finding Edges

In an operational kidney paired donation system, the incompatible recipient/donor pairs who are registered will be stored in a database that comprises the nodes of a graph. The first step in finding an optimal matching is to draw the edges of the graph. The reciprocal compatibility of every foursome, or every pair of nodes, must be checked in turn, so that in a database containing n incompatible pairs there are $[(n^2 - n)/2]$ potential edges to check.

If m-way donations with $m = 2, 3, \ldots, M$ are permitted then one approach is to find every cycle of length $m \leq M$ in the graph. The resulting cycles may be referred to as m-edges. A cycle-finding algorithm cannot

Table 6.1 Blood Group Incompatibilities

If the Recipient of the Pair Has Blood Group	If the Donor of the Pair Has Blood Group			
	O	A	B	AB
O	Compatible	Incompatible	Incompatible	Incompatible
A	Compatible	Compatible	Incompatible	Incompatible
B	Compatible	Incompatible	Compatible	Incompatible
AB	Compatible	Compatible	Compatible	Compatible

in general be polynomial-time in n and M, because the output could be on the order of $\binom{n}{M}$. However, logistical difficulties in m-way donation will usually require that M is a small constant that does not depend on n. The degree of a node is the number of different edges incident on it, that is, the number of different paired donation opportunities for that incompatible pair. Finding the edges in a paired donation graph does not solve the optimal allocation problem, unless the graph is so sparse that all nodes have degree zero or one. A procedure for finding two-way and three-way edges is described in Kaplan [18].

An undirected edge or m-edge represents an opportunity for every recipient on the edge to exchange to a compatible donor. There are two major categories of incompatibilities: blood group incompatibility and tissue incompatibility. The four blood groups are O, A, B, and AB, and their respective compatibilities are shown in Table 6.1.

Tissue incompatibility means that a recipient has preformed antibodies to some of the donor's HLA (human leukocyte antigens), which will cause the recipient to reject the kidney. Each person carries up to two HLA-A antigens, two HLA-B antigens, and two HLA-DR antigens, but some of these entries may be blank in which case the person has fewer than six total. There are about 25 common HLA-A antigens, about 50 common HLA-B antigens, and about 20 common HLA-DR antigens [19]. A recipient may have preformed antibodies to any number of antigens that are different from his own. A recipient with preformed antibodies is called sensitized to those antigens. A donor can still donate to a recipient even if the donor's antigens are not identical to the recipient's, provided that the recipient is not sensitized to the donor's antigens. A zero-mismatch transplant is one in which the donor does not carry any antigens that are not also antigens of the recipient. Opportunities for zero-mismatch transplants are fairly rare, but when they do occur, the donated kidney will have a longer expected graft survival time.

Some recipients are highly sensitized, meaning that they have preformed antibodies to many common HLA types. A panel-reactive antibody (PRA) test estimates what percentage of the population a particular recipient is sensitized to. For instance, a recipient with a PRA of 80 is predicted to be tissue incompatible with approximately 80 percent of the population. A better diagnostic than a PRA test is a screening that characterizes the recipient's antibodies, enabling a prediction of sensitivity to any particular donor's HLA. Finally, before any transplant, there must be a physical crossmatch test in which the compatibility of the donor and recipient are verified. If the donor and recipient are crossmatch-positive, then the transplant cannot proceed. If they are crossmatch-negative then the transplant proceeds.

6.2 Designing the Objective

When the reciprocally compatible two-way edges among a set of incompatible pair nodes have been found, the next step is to define a preferred, or optimal, matching on the resulting graph. A matching on the graph represents a decision about which foursomes should proceed with paired donation. The simplest objective would be to maximize the number of transplants accomplished, but a more useful objective is one that reflects the complexities of transplantation in general and paired donation in particular. For instance, prioritizing zero-mismatch transplant opportunities and pediatric recipients would mimic current practice in allocation of deceased donor kidneys [20].

It is tempting to believe that a simply stated utilitarian objective, say, maximize number of years of life gained, could be appropriate. However, because the benefit of each organ transplanted accrues only to the individual who receives it, the system as a whole has the responsibility to fairly apportion those benefits to different subgroups. It is well established that organ allocation decisions must balance utility with equity [2,20]. For instance, it would certainly be untenable to allocate all deceased donor organs only to Caucasian recipients, even though Caucasian recipients do have longer expected graft and patient survival times than other recipients.

In what follows, positive integer weights w_{ij} are assigned to the edge of the graph connecting node i with node j. These weights are higher for the higher priority organ exchanges, but the weights must be chosen carefully because seemingly reasonable choices of weights may lead to very poor matchings [21].

Some of the factors which UNOS proposes to consider in the objective include pediatric recipients, prior live donor recipients, zero-mismatches, travel requirements, waiting time, and the total number of transplants. Required travel is still seen as a major impediment to a national kidney paired donation registry [22]. If the two incompatible pairs in a paired donation are being treated at different hospitals, the donor of each pair will usually travel to the opposite hospital, which may be expensive and result in a difficult separation for the families involved. Other arrangements are possible; recently a team from Johns Hopkins transported a kidney from a live donor's operation in San Francisco to a recipient in Baltimore. In any of these arrangements, a shorter distance between the hospitals is better.

Each edge weight $w_{ij} = \sum_k f_{ijk}$. The f_{ijk} are positive integers denoting the points awarded for the presence or prevalence of the kth paired donation factor for edge ij. Let $k = 0$ denote the factor weighting for the total number of transplants, and note that the f_{ij0} should be the same for all i, j because every transplant is equally valuable. As another example, if $k = 1$ denotes the pediatric recipients factor weighting then f_{ij1} should be either 0, q, or $2q$ for some $q > 0$, according to whether nodes i and j include zero, one, or two pediatric recipients. If $k = 2$ denotes the travel requirements factor then f_{ij2} should be largest (most preferred) when node i and node j represent incompatible pairs who are both registered at the same hospital.

The size of the f_{ijk} terms relative to each w_{ij} specifies a trade-off between each of the factors being considered. In fact, the true kidney paired donation problem has multiple objectives, and the weights proposed here constitute a linear scalarization of the objective vector [23]. It is not the role of optimization experts to set these f_{ijk}. The transplant community, including donors, recipients, and their advocates, administrators, and physicians, should come to a consensus about the goals of a paired donation matching system. However, maximum weighted matching is an opaque algorithm, and thus it is a nontrivial task to translate a community's ethical judgments to numerical weightings. Transplant professionals have turned to simulations to guide the choice of the f_{ijk} for various factors [8], as will be discussed in Section 6.5.

As an alternative to simulation, a recent report offers an example of how theoretical analysis might inform the transplant community's weighting choices [21]. Suppose that there are K important factors, and so k ranges from 1 to K. The size of the f_{ij0} terms expresses the value placed on the total number of transplants performed. Because all f_{ij0} are the same, call this value f_0. Let 2μ be the maximum number of transplants that could be performed if no other factors were considered. It is not unreasonable that the most preferred matching may contain fewer than 2μ transplants. However, for arbitrary f_{ijk}, the preferred matching may in the worst case have only μ transplants. Using some simple inequalities that maximum matchings satisfy, it has been shown that at least a fraction $f_0/(\max_{i,j} \sum_k f_{ijk})$ of the maximum number of transplants possible will be performed in the maximum weighted matching. This gives a lower bound on the size of the maximum weighted matching. Graphs achieving this lower bound may easily be constructed. If the preferred matching must contain at least a fraction α of the maximum achievable transplants then f_0 can be set equal to $\alpha/(1-\alpha)\max_{i,j}\sum_{k=1}^{k=K} f_{ijk}$.

Further, even if physicians demand that the absolute largest number of transplants possible be performed, the other important factors need not be ignored. It is possible to express preferences among the matchings containing the largest number of transplants, using standard weighted matching with an appropriate edge weighting scheme. For a graph containing n incompatible recipient/donor pairs, if $\max_{i,j}\sum_{k=1}^{k=K} f_{ijk} = W$ then f_0 can be set at $(W/2)(n-1)$. Then the maximum weighted matching contains 2μ transplants, and considering the other factors imposes no penalty on the total number of transplants. The proofs of this and related propositions exploit the property that the size of any matching is an integer.

6.3 Solution Methods

Given an edge-weighted graph, a maximum weighted matching is a matching that maximizes the sum of the weights of edges in a matching. A maximum weighted matching is not necessarily unique. Solution algorithms for maximum weighted matching, sometimes referred to as maximum edge-weight matching or just weighted matching, have been extensively studied over the past half-century [24–26].

It is straightforward to write a maximum weighted matching problem directly as an integer program. Let x_{ij} denote a decision variable which takes value 1 if the edge connecting node i with node j is in the matching, and value 0 if that edge is not in the matching. The weight of any matching is $Z = \sum w_{ij}x_{ij}$. The optimal matching problem can be formulated as given by Equations 6.1 through 6.3:

$$\text{Max } Z = \sum_{i<j} w_{ij}x_{ij} \tag{6.1}$$

$$\text{s.t.: for all } j, \sum_{i<j} x_{ij} \leq 1 \tag{6.2}$$

$$x_{ij} \text{ in } \{0, 1\} \; (x_{ij} \text{ are binary}) \tag{6.3}$$

To obtain a linear programming formulation of the maximum matching problem, the integer constraint on the x_{ij} variables in Equation 6.3 must be relaxed to obtain Equation 6.4:

$$0 \leq x_{ij} \leq 1 \tag{6.4}$$

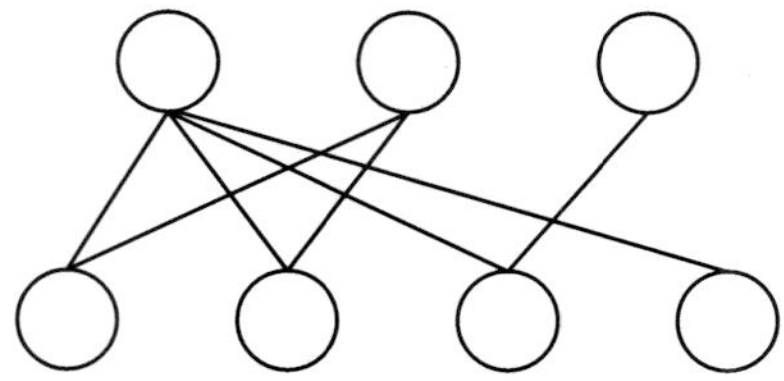

Figure 6.5 A bipartite graph.

Unfortunately, the relaxed problem might have fractional variables at the optimum, meaning that the solution does not correspond to a matching at all. The relaxation in Equation 6.4 is permissible under certain conditions: if the graph is bipartite, or if additional constraints given by Equation 6.5 are added.

A bipartite graph is one that can be partitioned into two classes, with all the edges of the graph between the two classes and no edges connecting members of the same class, as in Figure 6.5. Maximum edge-weight matching on a bipartite graph is also known as the assignment problem. Assignment problems can be solved with specialized algorithms, or by a direct application of the linear program in Equations 6.1, 6.2, and 6.4. This is because the basic feasible solutions to the linear program described by Equations 6.1, 6.2, and 6.4 have all x_{ij} equal to 0 or 1. In a graph containing separate nodes for donors and recipients, with edges connecting each donor to every compatible recipient, it would be an assignment problem to decide which donors should give to which recipients.

In kidney paired donation, the donors are not willing to donate unless their recipients can be transplanted, and the resulting graph is not bipartite, as is obvious from Figure 6.2. A non-bipartite graph is exactly a graph with odd-length cycles. Consider the L node subsets of odd cardinality, $\mathrm{S}_1, \mathrm{S}_2, \ldots \mathrm{S}_L$, in a non-bipartite graph, where the number of nodes in S_l is $2\mathrm{s}_l + 1$. It is certainly true that no matching contains more than s_l edges which are incident only on the nodes of S_l, so the following constraints are satisfied by any matching:

$$\text{For all } l \text{ in } 1, 2, \ldots, L, \quad \sum_{i,j \in \mathrm{S}_l} x_{ij} \leq \mathrm{s}_l \tag{6.5}$$

Edmonds showed that when these additional constraints given by Equation 6.5 are included, the linear programming formulation has integral basic feasible solutions that correspond to matchings. This formulation yields a primal-dual algorithm for solving the weighted matching problem [27]. Lawler provided a more efficient implementation of Edmonds' algorithm [24]. Galil, Micali, and Gabow used specialized data structures to further improve the algorithm's running time [25]. An implementation of the latter algorithm can be found in the graph theoretic C++ library LEDA [28]. A thorough exposition of the primal-dual algorithm for weighted matching in general graphs, as well as a more complete discussion than that summarized here, is contained in Papadimitriou [29].

One optimization method that will suffer combinatorial explosion as paired donation registries grow larger is to exhaustively catalog the feasible matchings to find the best one, as was proposed by Kim [30]. Although there are at most a number of edges proportional to n^2 on the graph, the number of matchings on a graph may be exponential in the number of nodes. There are estimated to be millions of different matchings on a typical paired donation graph of 100 nodes [4]. The criteria for selecting a preferred matching from a complete list of matchings in Ref. [30] could instead be expressed using appropriate choices for factor weights f_{ijk}.

Heuristic approaches that have been suggested for allocating kidney paired donation could have poor performance. The greedy algorithm begins with an empty matching, and first adds the edge with the largest weight to the matching, next adds the edge with the largest weight among those that could be added to the matching, and so on, until no more edges can be added. Variants of the greedy algorithm have been used in actual paired donation programs [31]. In one operational paired donation program, a greedy algorithm selected a matching with only two transplants, where actually four were possible. A medical review committee manually discovered the shortfall and decided to use the larger matching. In the example of Figure 6.6, the optimal matching would have matched all the recipients in the greedy matching and some other recipients also.

If up to M-way donations are permissible for $M > 2$ then the analogous optimization problem is necessarily an integer program. Let P be the number of cycles of length $m < M$ in the directed graph. Let x_i for $i = 1, 2, \ldots, P$ denote a decision variable which takes value 1, if the ith

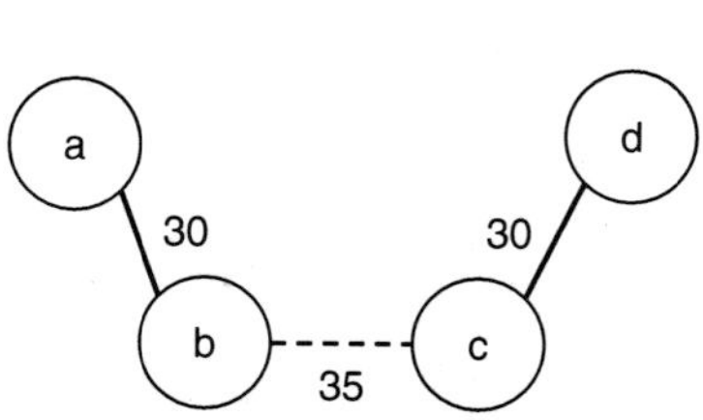

Figure 6.6 A largest-edge-first greedy algorithm would choose the dashed edge, and transplant only b and c. The maximum weighted matching includes the solid edges and transplants a, b, c, and d.

m-edge will be used in the allocation, and value 0 if that edge is not used in the allocation. Let w_i have the obvious meaning: the value of an m-way paired donation among the nodes of the ith m-edge. The weight of any allocation is $Z = \sum w_i x_i$. Let $I(j)$ denote the set of all m-edges incident on the node j. Then the optimization problem for up to M-way paired donation is

$$\text{Max } Z = \sum_i w_i x_i \tag{6.6}$$

$$\text{s.t.: for nodes } j = 1, 2, \ldots, n, \quad \sum_{i \in \mathrm{I}(j)} x_i \leq 1 \tag{6.7}$$

$$x_i \text{ in } \{0, 1\} \ (x_i \text{ are binary}) \tag{6.8}$$

Constraint given by Equation 6.7, analogous to constraint given by Equation 6.2, ensures that only one edge incident on each node j is included in the allocation.

Another allocation method is available which does not require finding all cycles in the graph, provided that paired donations involving any number of pairs are permissible. Roth used a modified top-trading-cycles algorithm for this case, based on an ordered preference list for each incompatible pair [9].

6.4 Current Status of Paired Donation

In 2006, 85 people received transplants through paired donation in the United States, which falls far short of the projected one to two thousand additional transplants per year. In the United States, the only paired donation registries are those run by individual transplant centers or coalitions of transplant centers. Although UNOS proposed a nationwide registry, that proposal was on hold because the legality of paired donation in the United States had been questioned. The National Organ Transplantation Act (NOTA) forbids receiving valuable consideration for donating an organ [32]. However, the U.S. Department of Justice recently ruled that paired donation does not constitute valuable consideration under NOTA. Also, bills to the same effect passed both houses of Congress in 2007, although in slightly different versions. The Canadian Council for Donation and Transplantation has also drafted a proposal for a registry to allocate paired donations in Canada. South Korea and the Netherlands already operate nationwide kidney paired donation registries [33,34].

6.5 Projecting the Impact of Kidney Paired Donation

United Network for Organ Sharing (UNOS) collects data about every potential transplant recipient and every living and deceased donor, but it does not collect data about incompatible donors who have come forward for particular recipients. Consequently, simulation models are used to generate nationwide projections of the impact of kidney paired donation. The model presented here has been used extensively to answer clinical questions concerning paired donation [35,36], and is based on a decision tree model by Zenios et al. [37]. Table 6.2 details the probability assumptions that are made and the sources of the data. Simulating incompatible

Table 6.2 Data Sources Used in Simulations, Including Values Where They Are Concise

	Probability (Percent)	Source of Data
Race of recipients	(Exclude other)	UNOS waiting list additions 2003
Caucasians	52.6 percent	
African-Americans	30.4 percent	
Hispanics	17.0 percent	
Blood group	Allele combination frequencies, by race	Zenios et al. [37]
HLA-A,-B,-DR	By race	Leffell et al. [19]
Donor relationship		UNOS live kidney donors, 2003
Parent	19.7 percent	
Child	16.8 percent	
Sibling	42.4 percent	
Spouse	10.0 percent	
Unrelated	11.2 percent	
PRA range		UNOS waiting list additions 2003
0–9	71.31 percent	
9–80	18.66 percent	
80–100	10.02 percent	
Positive crossmatch rate		Assumed from definition of panel-reactive antibody (PRA)
PRA 0–9	5 percent (25 percent maternal)	
PRA 10–80	45 percent (65 percent maternal)	
PRA 80–100	90 percent (95 percent maternal)	
Pediatric	3.07 percent	UNOS waiting list additions 2003
Prior live donor	0.052 percent	UNOS waiting list additions 2003
Transplant center	Of 242 centers	UNOS waiting list additions 2003

pairs is made more difficult by two factors: (1) the blood group and HLA are inherited traits although most people's donors are blood relatives and (2) the group of recipients with incompatible donors is biased in blood group and sensitization levels.

First, a recipient and his social network, including two parents, a sibling, a spouse, a friend, and two children are simulated. The race of the recipient's parents, sibling, and children is chosen according to the makeup of the kidney recipient population, and the races of the recipient's friend and recipient's spouse are the same as the recipient's race with 90 percent probability. The blood group allele combinations (like OO, AO, BB, AB) and HLA profiles of the parents, friend, and spouse are drawn from race-dependent population distributions. The blood group alleles and HLA profiles of the recipient and his sibling are based on probabilistic inheritance from their parents, and likewise the blood group alleles and HLA profiles of the recipient's children are based on probabilistic inheritance from the recipient and his spouse. Between two and four members of the social network are chosen to be the recipient's candidate donors, according to the prevalence of each donor relation in the UNOS database of actual living donors.

Next, the decision tree in Figure 6.6 classifies the recipient and donors into one of these categories: the recipient either has at least one donor who is compatible, has no viable donor, or has only an incompatible donor or donors. Donors are either medically or psychologically unsuitable with probability 56.7 percent, except spouses who are unsuitable with probability 25 percent because fewer spouses are unwilling to donate. The PRA of each recipient is assigned in low, medium, and high categories. According to the recipient's PRA, a simulated crossmatch test determines whether the recipient is sensitized to each candidate donor. Female recipients are more likely to be crossmatch-positive with their spouses and children due to sensitization during pregnancy.

If the recipient's only medically eligible donor(s) is incompatible by blood group or crossmatch then that recipient/donor pair is registered as a node in the kidney paired donation graph. To estimate the number of incompatible pairs who would present annually, new recipients and families are generated in the upper oval of Figure 6.6, until the number of simulated recipients with a compatible donor (dashed lower oval) equals the number of actual recipients with a compatible donor in a recent year. Assuming one candidate donor per recipient, about 4443 incompatible pairs would be projected to present annually; assuming two, about 3584 pairs annually; assuming four, about 2406.

Table 6.3 shows the blood groups of incompatible donor and recipient pairs. It is instructive to compare the blood groups of donors and recipients. The shortage of blood group-O donors and surplus of blood group-O recipients result in a low (less than 50 percent) projected match rate for

Table 6.3 Projected Blood Groups of Incompatible Pairs (Percent)

If the Recipient of the Pair Has Blood Group	If the Donor of the Pair Has Blood Group				
	O	A	B	AB	All Donors
O	20.39	27.89	11.20	1.57	61.06
A	6.48	10.44	4.93	2.45	24.31
B	2.94	4.97	2.94	2.04	12.88
AB	0.38	0.63	0.40	0.34	1.75
All recipients	30.20	43.93	19.48	6.40	

incompatible donors, but voluntary compatible donation could resolve the blood group imbalance. The numerical projections of paired donation's impact are very sensitive to the assumed recipient PRA levels, the assumed number of donors per person, and the likelihood of finding a compatible donor for highly sensitized recipients. Only the first of these elements is known with a high degree of confidence.

Using a collection of databases of simulated incompatible pairs, clinically important questions can be answered. For instance, placing restrictions on a recipient and donor's willingness to accept a faraway match will reduce the number of edges incident on that node, and simulations can demonstrate the effect of travel restrictions on match rates [35]. As another example, by repeated application of maximum weighted matching to the same set of graphs, one may compare the match rates overall and for various subgroups as the f_{ijk} are varied [8]. The curse of dimensionality will prevent a complete manual survey of f_{ijk} settings. For $K = 8$ (factors), and only five trial settings for each factor, there will be about 400,000 candidates matching objectives. Because there are a finite number of different matchings on any graph, there will be a finite number of meaningfully distinct objective designs. However, the simulation method is useful for projecting the effect of particular changes to the objective settings.

Kidney paired donation is poised to become a major part of the solution to the organ shortage crisis. Although the concept of paired donation dates to 1986 [6], recent progress towards a nationwide registry was spurred by a number of simulation and optimization papers that quantitatively established its potential to generate additional donations. Paired donation is an excellent example of how optimization can have a major and immediate impact on clinical medicine.

Acknowledgments

The developments presented here are based on an intensive collaboration with Dorry Segev, MD, and the Johns Hopkins Comprehensive

Transplant Center, led by Robert Montgomery, MD, DPhil. The Naval Academy Research Council also supported this research. Midshipman Kate Williams, U.S. Navy (USN), assembled the illustration in Figure 6.1.

References

1. Alagoz, O., Maillart, L.M., Schaefer, A.J., and Roberts, M.S. The optimal timing of living-donor liver transplantation. *Management Science*, 2004. **50**(10): 1420–1430.
2. Zenios, S.A., Wein, L.M., and Chertow, G.M., Evidence-based organ allocation. *American Journal of Medicine*, 1999. **107**(1): 52–61.
3. Su, X. and Zenios, S.A., Patient choice in kidney allocation: A sequential stochastic assignment model. *Operations Research*, 2005. **53**(3): 443–455.
4. Segev, D.L., Gentry, S.E., Warren, D.S., Reeb, B., and Montgomery, R.A. Kidney paired donation and optimizing the use of live donor organs. *JAMA*, 2005. **293**(15): 1883–1890.
5. Montgomery, R.A., Zachary, A.A., Ratner, L.E., Segev, D.L., Hiller, J.M., Houp, J., Cooper, M., Kavoussi, L., Jarrett, T., Burdick, J., Marey, W.R., Melancon, J.K., Kozlowski, T., Sinpkins, C.E., Phillips, M., Desai, A., Collins, V., Reeb, B., Kraus, E., Rabb, H., Leffell, M.S., and Warren, D.S. Clinical results from transplanting incompatible live kidney donor/recipient pairs using kidney paired donation. *JAMA*, 2005. **294**(13): 1655–1663.
6. Rapaport, F.T., The case for a living emotionally related international kidney donor exchange registry. *Transplantation Proceedings*, 1986. **18**(3), Suppl. (2): 5–9.
7. UNOS, United Network for Organ Sharing data as of May 1, 2007, obtained from and available at the UNOS website. 2007.
8. UNOS, Concept Proposal for a National Kidney Paired Donation Program through the Organ Procurement and Transplantation Network/United Network for Organ Sharing (Kidney and Pancreas Transplantation Committee). Presented August 28, 2006 for Public Comment after Presentation to the UNOS Board. 2006.
9. Roth, A.E., Sönmez, T., and Ünver, M.U., Kidney exchange. *Quarterly Journal of Economics*, 2004. (May): 457–488.
10. McLellan, F., US surgeons do first "triple-swap" kidney transplantation. *Lancet*, 2003. **362**(9382): 456.
11. Saidman, S.L., Roth, A.E., Sonmez, T., Unver, M.U., and Delmonico, F., Increasing the opportunity of live kidney donation by matching for two and three way exchanges. *Transplantation*, 2006. **81**(5): 773–782.
12. Roth, A.E., Sonmez, T., and Unver, M.U., Efficient kidney exchange: Coincidence of wants in markets with compatibility-based preferences. *American Economic Review*, 2007. **97**(3).
13. Montgomery, R.A., Gentry, S.E., Marks, W.A., Warren, D.S., Hiller, J., Houp, J., Zachary, A.A., Melancon, J.K., Maley, W.R., Rabb, H., Sinpkins, C., and Seger, D.L. Domino paired kidney donation: A strategy to make best use of live non-directed donation. *Lancet*, 2006. **368**(9533): 419–421.
14. Delmonico, F.L., Morrissey, P.E., Lipkowitz, G.S., Stoff, J.S., Himmelfarb, J., Harmon, W., Pavlakis, M., Mah, H., Goguen, J., Luskin, R., Milford, E.,

Basadonna, G., Chobanian, M., Bothout, B., Lorber, M., and Rohrer, R.J. Donor kidney exchanges. *American Journal of Transplantation*, 2004. **4**(10): 1628–1634.
15. Roth, A.E., Sönmez, T., Ünver, M.U., Delmonico, F.L., and Saidma, S.L. Utilizing list exchange and nondirected donation through 'chain' kidney paired donations. *American Journal of Transplantation*, 2006. **6**(11): 2694–2705.
16. Zenios, S., Optimal control of a paired-kidney exchange program. *Management Science*, 2002. **48**(3): 328.
17. Gentry, S.E., Montgomery, R., and Segev, D.L., Blood group O recipients and compatible kidney paired donation. *American Transplant Congress*, 2007.
18. Kaplan, I., Houp, J., Leffell, M.S., Hart, J.M., and Zachary, A.A. A computer match program for paired and unconventional kidney exchanges. *American Journal of Transplantation*, 2005. **5**(9): 2306.
19. Leffell, M.S., Steinberg, A.G., Bias, W.B., Machan, C.H., and Zachary, A.A. The distribution of HLA antigens and phenotypes among donors and patients in the UNOS registry. *Transplantation*, 1994. **58**(10): 1119–1130.
20. UNOS. Policy 3.5: Allocation of deceased donor kidneys. 2006 [cited; available from: http://www.optn.org/PoliciesandBylaws/policies/pdfs/policy_7.pdf.]
21. Gentry, S.E., Michael, T.S., and Segev, D.L., Maximum matchings for allocating kidney paired donation. 2006, United States Naval Academy Technical Report.
22. Woodle, E.S., The potential of paired donation programs: Modeling and reality. *American Journal of Transplantation*, 2005. **5**(8): 1787–1788.
23. Ehrgott, M. and Gandibleux, X., A survey and annotated bibliography of multi-objective combinatorial optimization. *OR Spectrum*, 2000. **22**: 425–460.
24. Lawler, E.L., *Combinatorial Optimization: Networks and Matroids*. 1976, New York: Holt, Rinehart & Winston.
25. Galil, Z., Micali, S., and Gabow, H., An O(EV log V) algorithm for finding a maximal weighted matching in general graphs. *SIAM Journal on Computing*, 1986. **15**(1): 120–130.
26. Edmonds, J., Paths, trees, and flowers. *Canadian Journal of Mathematics*, 1965. **17**: 449–467.
27. Edmonds, J., Matching and a polyhedron with 0–1 vertices. *Journal of Research of the National Bureau of Standards*, 1965. **69B**: 125–130.
28. LEDA 5.2. 2006, Algorithmic Solutions Software GmbH.
29. Papadimitriou, C.H. and Steiglitz, K., *Combinatorial Optimization: Algorithms and Complexity*. 1982, Mineola, New York: Dover Publications, Inc. pp. 218–270.
30. Kim, B.S., Kim, Y.S., Kim, S.I., Kim, M.S., Lee, H.Y., Kim, Y.-L., Kim, C.D., Yana, C.W., Kim, B.S., Han, D.J., Kim, Y.S., Kim, S.J., Lee, H.Y., and Kim, D.J. Outcome of multi-pair donor kidney exchange by a web-based algorithm. *Journal of the American Society of Nephrology*, 2007. **18**: 1000–1006.
31. Keizer, K.M., deklerk, M., Haase – Kromwijk, B.J., and Weimar, W. The Dutch algorithm for allocation in living donor kidney exchange. *Transplantation Proceedings*, 2005. **37**(2): 589–591.
32. NOTA, National Organ Transplantation Act Public Law Number 98–507. 1984.
33. Park, K., Park, J.W., Koo, Y.M., and Kim, J.H. Exchange donor program in kidney transplantation. *Transplantation*, 1999. **67**(2): 336–338.
34. de Klerk, M., Keizer, K., and Weimar, W., Donor exchange for renal transplantation. *New England Journal of Medicine*, 2004. **351**(9): 935–937.

35. Segev, D.L., Gentry, S.E., Melancon, J.K., and Montgomery, R.A. Characterization of waiting times in a simulation of kidney paired donation. *American Journal of Transplantation*, 2005. **5**(10): 2448–2455.
36. Gentry, S.E., Segev, D.L., and Montgomery, R.A., A comparison of populations served by kidney paired donation and list paired donation. *American Journal of Transplantation*, 2005. **5**(8): 1914–1921.
37. Zenios, S.A., Woodle, E.S., and Ross, L.F., Primum non nocere: Avoiding harm to vulnerable wait list candidates in an indirect kidney exchange. *Transplantation*, 2001. **72**(4): 648–654.

Chapter 7

Introduction to Radiation Therapy Planning Optimization

Gino J. Lim

CONTENTS

7.1 Introduction 198
7.2 Applications and Methods 200
7.2.1 Radiation Treatment Planning Procedure 200
7.2.2 Use of Optimization Techniques 201
7.2.3 Gamma Knife Radiosurgery 202
7.2.3.1 Introduction 202
7.2.3.2 Optimization Model Formulation 203
7.2.3.3 Solution Techniques 205
7.2.4 Three-Dimensional Conformal Radiation Therapy 205
7.2.4.1 Introduction 205
7.2.4.2 Optimization Model Formulation 208
7.2.5 Intensity Modulated Radiation Therapy 211
7.2.5.1 Introduction 211
7.2.5.2 Optimization Model Formulation 212
7.2.6 Other Radiation Therapy Devices 216
7.2.6.1 Tomotherapy 216
7.2.6.2 Proton Therapy 217
References 218

Abstract This chapter is designed to introduce recent advances in optimization of radiation therapy planning for cancer patients. Topics include

Gamma Knife, conventional three-dimensional conformal radiotherapy (3DCRT), intensity modulated radiation therapy (IMRT), tomotherapy, and proton therapy. A short problem description, model formulation, and solution techniques for each topic are presented.

7.1 Introduction

The National Cancer Institute estimates that approximately 1.4 million new cancer cases are expected to be diagnosed in the United States in 2006 [5]. They also estimate that over 570,000 Americans are expected to die of cancer this year. Cancer is the second leading cause of death in the United States, exceeded only by heart disease. Treatment options for cancer are determined by type and stage of the cancer and include surgery, radiation therapy, chemotherapy, immunotherapy, etc. Physicians often use a combination of those treatments to obtain the best results.

Our application is based on radiation therapy. Thanks to the continuous development of new treatment machines and technologies; it is now possible to have much greater control over the treatment delivery than was possible in the past. Researchers in the optimization community have made significant contributions in improving the quality of such treatment plans for cancer patients [4,7,16,15,27,29,30,33,44,48,51,57,58]. The common objective of radiotherapy planning is to achieve tumor control by planning a significant total dose of radiation to the cancerous region to sterilize the tumor without damaging the surrounding healthy tissues. One of the major difficulties in treatment planning is due to the presence of organs-at-risk (OARs). An OAR is a healthy organ located close to the target. The dose of radiation must be severely constrained to avoid reaching an OAR because an overdose in the OAR may lead to medical complications. OAR is also termed sensitive structure or critical structure in the literature. There are several survey articles that cover the essential elements of the radiation treatment planning problem, see Refs. [20,35,45,48].

Our aim in this chapter is to introduce recent advances in optimization models and solution techniques to improve the delivery of radiation for cancer patients. Two types of radiation therapy are the most common and include teletherapy (or external beam therapy) and brachytherapy. Radiation is delivered from outside the body and directed at the patient's tumor location using special radiation delivery machines in teletherapy, (see Figure 7.1). Different devices produce different types of radiation and they include Cobalt-60 machines (such as Gamma Knife radiosurgery), linear accelerators (such as intensity modulated radiation therapy), neutron beam machines, orthovoltage x-ray machines, and proton beam machines. In brachytherapy, radioactive substances are placed within the tumor region in the form of wires, seeds, or rods. Types of brachytherapy are categorized

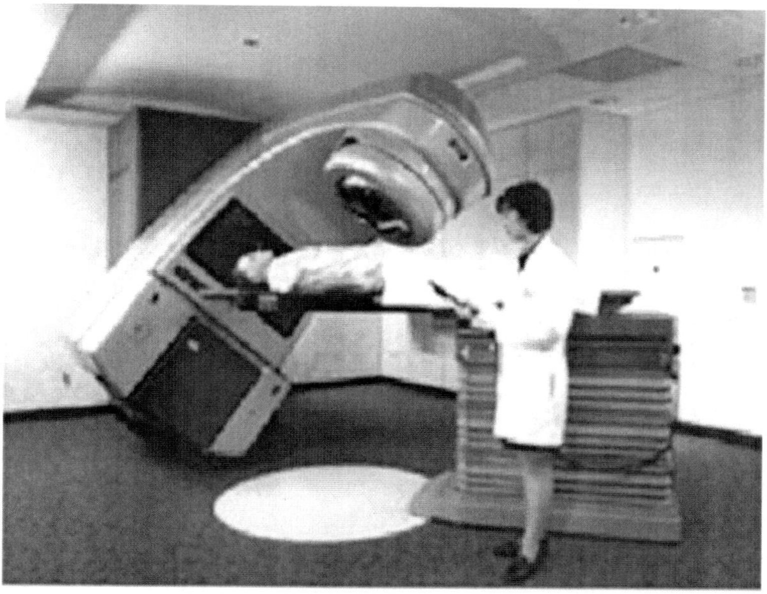

Figure 7.1 An external beam therapy machine.

depending on how the radioactive sources are placed inside the body such as interstitial brachytherapy, intracavitary brachytherapy, intraluminal radiation therapy, and radioactively tagged molecules given intravenously. There are two types of radiation treatment planning: forward planning and inverse planning. In forward planning, treatment plans are typically generated by a trial and error approach. Therefore, this process can be very tedious and time consuming, and does not necessarily produce high-quality treatment plans. On the other hand, there has been a significant move toward inverse treatment planning. Such a move is due to significant advances in modern technologies such as imaging technologies and computer control to aid the delivery of radiation. The inverse treatment planning procedure allows modeling highly complex treatment planning problems from brachytherapy to external beam therapy. Inverse planning is also called computer-based treatment planning.

In inverse treatment planning, an objective function is defined to measure the goodness (quality) of a treatment plan. Two types of objective functions are often used: dose-based models and biological (radiobiological) models. The biological model argues that optimization should be based on the biological effects resulting from the underlying radiation dose distributions. The treatment objective is usually to maximize the tumor control probability (TCP) while keeping the normal tissue complication probability (NTCP) within acceptable levels. In the dose-based model, achieving accurate radiation dose distributions on organs of interest is the main concern. The treatment objective is to minimize the deviation between the projected dose that the patient will receive and the prescribed dosage that the physician provides. This is the main approach we will describe in this chapter. The biological aspect is implicitly given in the physician's prescription.

7.2 Applications and Methods

7.2.1 Radiation Treatment Planning Procedure

When a patient comes in for a treatment, the doctor will choose what type of radiation beam to use for the treatment. The choice of radiation will depend on the type of cancer the patient has and how far into the body the radiation should penetrate to reach the tumor volume.

The next step is to identify the three-dimensional (3D) shapes of organs of interest in the patient's body. The location and the volume of organs are obtained by using (3D) imaging techniques such as computer tomography (CT) or magnetic resonance imaging (MRI). Based on (3D) images, a physician specifies the tumor region as gross tumor volume (GTV), clinical target volume (CTV), planning target volume (PTV), and OARs. GTV represents the volume of the known tumor. CTV represents the volume of the suspected microscopic spread. PTV is the marginal volume necessary to account for setup variations and organ and patient motion, i.e., PTV = GTV + marginal volume around the GTV. Typically, PTV is used in designing treatment plans and we call PTV a target in this chapter. Organ geometries are the key input data for designing a treatment plan.

A radiation physicist and a dosimetrist meet to decide what kind of radiation delivery machine to use and the number of treatments for the patient. Optimization algorithms are crucial to determine how much and where to deliver radiation in the patient's body. For most types of cancer, radiation therapy is administered five days each week for five–eight weeks. Using small radiation doses daily rather than a few large doses helps protect normal body tissues in the treatment area. Resting over the weekend will allow some time for normal cells to recover from the radiation damage.

In optimization, the (3D) volume is represented by a grid of voxels. There are several inputs required in optimization approaches in radiation treatment planning. The first input describes the machine that delivers radiation. The second and troublesome input is the dose distribution of a particular treatment problem. A dose distribution consists of dose contribution to each voxel of the region of interest from a radiation source. It can be expressed as a functional form or a set of data. However, a difficulty of using such distributions is either the functional form is highly nonlinear [16] or the amount of data which specifies the dose distribution is too large [30]. This problem needs to be overcome in a desirable automated treatment planning tool. The third common input is the set of organ geometries that are of interest to the treatment. Further common inputs are the desired dose levels for each organ of interest. These are typically provided by physicians. Other types of inputs can also be specified depending on the treatment planning problems. However, a desirable treatment planning system should be able to generate high-quality treatment plans with minimum additional inputs and human guidance.

7.2.2 Use of Optimization Techniques

Two major goals in treatment planning optimization are speed and quality. Solution quality of a treatment plan can be measured by uniformity, conformity, and avoidance [15,30,33]. Fast solution determination in a simple manner is another essential part of a clinically useful treatment planning procedure. Acceptable dose levels of these requirements are established by various professional and advisory groups.

It is important for a treatment plan to have uniform dose distributions on the target so that cold and hot spots can be minimized. A cold spot is a portion of an organ that receives below its required radiation dose level. On the other hand, a hot spot is a portion of an organ that receives more than the desired dose level. The uniformity requirement ensures that radiation delivered to the tumor volume will have a minimum number of hot spots and cold spots on the target. This requirement can be enforced using lower and upper bounds on the dose, or approximated using penalization. The conformity requirement is used to achieve the target dose control while minimizing the damage to OARs or healthy normal tissue. This is generally expressed as a ratio of cumulative dose on the target over total dose prescribed for the entire treatment. This ratio can be used to control conformity in optimization models. As we mentioned earlier, a great difficulty of producing radiation treatment plans is the proximity of the target to the OARs. An avoidance requirement can be used to limit the dose delivered to OARs. Finally, simplicity requirements state that a treatment plan should be as simple as possible. Simple treatment plans typically reduce the treatment time as well as implementation error. In this chapter, we introduce a few optimization models and solution techniques that are practically useful for radiation treatment modalities: Gamma Knife radiosurgery, conventional three-dimensional conformal therapy (3DCRT) [30], intensity modulated radiation therapy (IMRT) [3,22]. Many treatment planning models are also developed for proton therapy [59] and tomotherapy [14,25]. A brief overview of these new techniques are discussed at the end of this chapter.

7.2.3 Gamma Knife Radiosurgery

7.2.3.1 Introduction

The Gamma Knife is a highly specialized treatment unit that provides an advanced stereotactic approach to the treatment of tumor and vascular malformations within the head [17]. The Gamma Knife delivers a single, high dose of radiation emanating from 201 Cobalt-60 unit sources. All 201 beams simultaneously intersect at the same location in space to form an approximately spherical region that is typically termed a shot of radiation (see Figure 7.2).

Gamma Knife radiosurgery begins by finding the location and the size of the tumor. After administering local anesthesia, a stereotactic head frame is fixed to the patient's head using adjustable posts and fixation screws. This frame establishes a coordinate frame within which the target location is known precisely and serves to immobilize the patient's head within an attached focussing helmet during the treatment. An MRI or CT scan is used to determine the position of the treatment volume in relation to the coordinates determined by the head frame. Once the location and the volume of the tumor are identified, the neurosurgeon, the radiation oncologist, and the physicist work together to develop the patient's treatment plan. Multiple shots are often used in a treatment using a Gamma Knife due to the irregularity and size of tumor shapes and the fact that the focussing helmets are only available in four sizes (4, 8, 14, and 18 mm).

The plan aims to deliver a high dose of radiation to the intracranial target volume with minimum damage to the surrounding normal tissue.

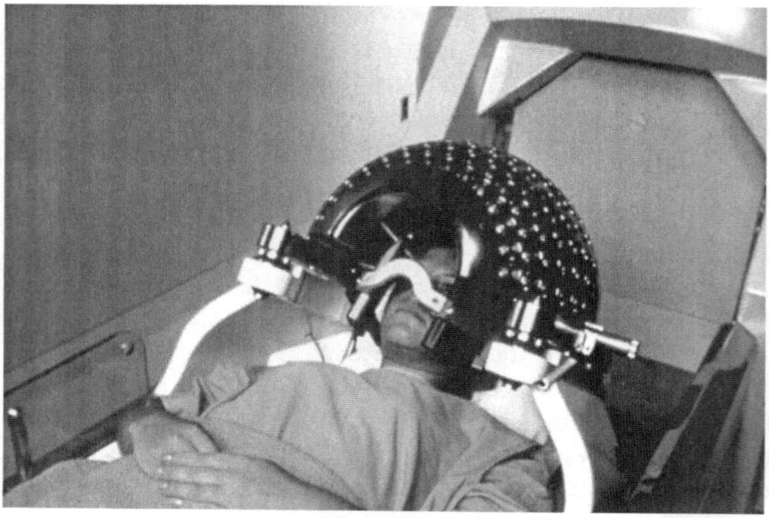

Figure 7.2 Radiation delivery: a collimator is positioned on patient's head.

The treatment goals can vary from one neurosurgeon to the next. But the following requirements are typical for a treatment plan, although the level of treatment and importance of each may vary.

1. A complete 50 percent isodose line coverage of the target volume. This means that the complete tumor volume must receive at least 50 percent of the maximum dosage delivered to the target. This can be thought of as a uniformity requirement.
2. Minimize the nontarget volume, which is covered by a shot or the series of delivered shots. This requirement is clear and can be thought of as a conformity requirement.
3. Limit the amount of dosage, which is delivered to organs at risk that are close to the target. Such requirements can be thought of as an avoidance requirement.

There are standard rules established by the Radiation Therapy Oncology Group (RTOG) that recommend the acceptable uniformity and conformity requirements. In addition to these requirements, it is also preferable to use a small number of shots to limit the treatment times and thus increase the number of patients that can be treated.

7.2.3.2 Optimization Model Formulation

Most commonly known optimization models include Mixed Integer Programming (MIP) Model and Mixed Integer Nonlinear Programming (MINLP) Model. MIP Models guarantee the global optimality, but they are not practically useful due to the long computation time. We discuss an MINLP Model that has shown to be practically useful [15]. A variant of this approach has been successfully implemented for planning treatments [47].

Suppose that the number of radiation shots for the treatment is given a priori. Adding the goal of minimizing this number is typically straightforward in the optimization model. However, solving such models can be extremely difficult.

Decision Variables: Consider a grid $\mathcal{G}$ of voxels. Let $\mathcal{T}$ denote the subset of voxels that are within the target and $\mathcal{N}$ represents the subset of voxels that are not in the target. Let $D_{i,j,k}$ denote the amount of radiation dose that a voxel (i, j, k) receives. In general, there are three types of decision variables.

1. *A set of discrete coordinates* (x_s, y_s, z_s): these are the target locations for the (ellipsoidal) shots.
2. *A discrete set of collimator sizes* w: currently four different sizes of focussing helmets are available (4, 8, 14, and 18 mm).

3. *Radiation exposure time* $t_{s,w}$: the amount of radiation to be delivered for each shot centered at location (x_s, y_s, z_s).

Constraints:

1. *Uniformity – isodose line coverage:* A treatment plan is normally considered acceptable if a $100 \cdot \theta$. percent isodose curve encompasses the tumor region. For example, 50 percent isodose curve is a curve that encompasses all voxels that receive at least 50 percent of the maximum radiation dose that is delivered to any voxels in the target volume.

$$\theta \leq D_{i,j,k} \leq 1, (i, j, k) \in \mathcal{T}. \tag{7.1}$$

2. *Choosing shot sizes:* The location of the shot center is chosen by a continuous optimization process. Choosing the particular shot width at each shot location is a discrete optimization problem that is treated by approximating the step function

$$H(t) = \begin{cases} 1 & \text{if } t > 0 \\ 0 & \text{if } t = 0 \end{cases}$$

by a nonlinear function,

$$H(t) \approx H_\alpha(t) = \frac{2\arctan(\alpha t)}{\pi}$$

For increasing values of α, H_α becomes a closer approximation to the step function H. This process is typically called smoothing.

An optimization model of Ref. [15] is described as follows:

$$\begin{aligned} \text{Min} \quad & \sum_{(i,j,k)\in\mathcal{N}} D_{i,j,k} \\ \text{s.t.} \quad & D_{i,j,k} = \sum_{(s,w)\in\mathcal{S}\times\mathcal{W}} t_{s,w} D_{w(S,i,j,k)} \\ & \theta \leq D_{i,j,k} \leq 1, \quad \forall (i,j,k) \in \mathcal{T} \\ & n = \sum_{(s,w)\in\{1,\ldots,n\}\times\mathcal{W}} H_\alpha(t_{s,w}) \\ & t_{s,w} \geq 0. \end{aligned} \tag{7.2}$$

7.2.3.3 Solution Techniques

The most critical problem for solving the optimization model Equation 7.2 is the large number of voxels that are needed when dealing with large irregular tumors (both within and outside of the target). This makes the optimization problem computationally intractable. One approach to overcoming this problem is to remove a large number of the nontarget voxels from the model. While this improves the computation time, this typically weakens the conformity of the dose to the target. Ferris et al. [15] proposed a sequential solution approach to speed up the time while maintaining conformality. First, a coarse grid problem is solved as a nonlinear programming (NLP) Model using a set of systematically reduced data points. Then the finer grid NLP problem with full data points is solved using the starting point that was obtained by the coarse grid model in the previous stage. Typically, the solution from this finer grid model is very close to a good local optimum for the MINLP. Using this solution, the full MINLP Model is finally solved to determine the values of the three sets of decision variables for this problem.

7.2.4 Three-Dimensional Conformal Radiation Therapy

7.2.4.1 Introduction

We learned from Section 7.2.3 that the Gamma Knife is specifically designed for treating diseases in the human brain. 3DCRT adds much greater flexibility to radiation treatments that can treat various cancer patients in the brain, breast, prostate, etc. One of the main strategies for minimizing morbidity in 3DCRT is to reduce the dose delivered to normal tissues that are spatially well separated from the tumor. This can be done by using multiple beams from different angles. A single radiation beam leads to a higher dose delivered to the tissues in front of the tumor than to the tumor itself. In consequence, if one were to give a dose sufficient to control the tumor with a reasonably high probability, the dose to the upstream tissues would likely lead to unacceptable morbidity. A single beam would only be used for very superficial tumors, where there is little upstream normal tissue to damage. For deeper tumors, one uses multiple cross-firing beams delivered within minutes of one another: all encompass the tumor, but successive beams are directed toward the patient from different directions to traverse different tissues outside the target volume. The delivery of cross-firing beams is greatly facilitated by mounting the radiation-producing equipment on a gantry: multileaf-collimator (MLC).

Several directed beams noticeably change the distribution of dose, as is illustrated in Figure 7.3. As a result, dose outside the target volume can often be quite tolerable even when dose levels within the target volume are high enough to provide a substantial probability of tumor control.

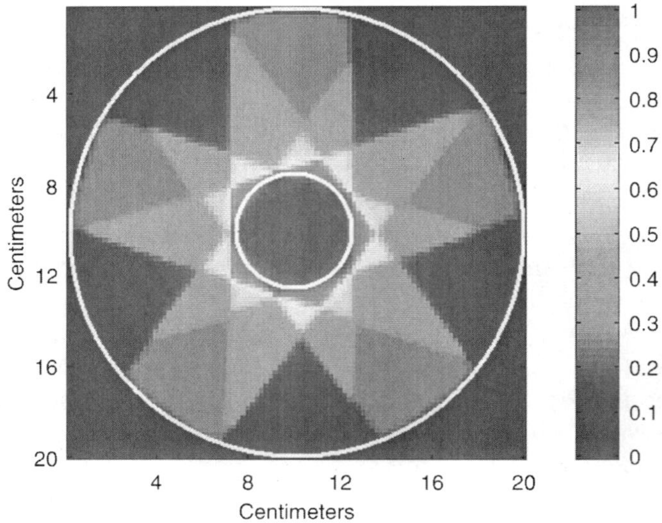

Figure 7.3 Effect of multiple beams: a hot spot is formed in the middle by five beams.

The leaves of the MLC are computer controlled and can be moved to the appropriate positions to create the desired beam shape. From each beam angle, (3D) anatomical information is used to shape the beam of radiation to match the shape of the tumor. Given a gantry angle, the view of the tumor that the beam source can see through the MLC is called the beam's-eye-view (BEV) of the target (see Figure 7.4) [18]. This BEV approach ensures adequate irradiation of the tumor while reducing the dose to normal tissue.

Wedge Filters: A wedge (also called a wedge filter) is a tapered metallic block with a thick side (the heel) and a thin edge (the toe) (see Figure 7.5). This metallic wedge varies the intensity of the radiation in a linear fashion from one side of the radiation field to the other. When the wedge is placed in front of the aperture, less radiation is transmitted through the heel of the wedge than through the toe. Figure 7.5b shows an external 45° wedge, so named because it produces isodose lines that are oriented at approximately 45°. The quality of the dose distribution can be improved by incorporating a wedge filter into one or more of the treatment beams. Wedge filters are particularly useful in compensating for a curved patient surface, which is common in breast cancer treatments.

Two different wedge systems are used in clinical practice. In the first system, four different wedges with angles 15°, 30°, 45°, and 60° are available, and the therapist is responsible for selecting one of these wedges

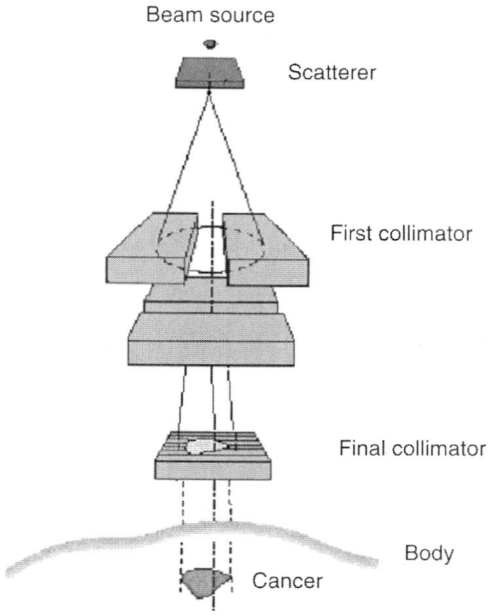

Figure 7.4 A beam's-eye-view (BEV) is a two-dimensional (2D) shape of a tumor viewed by the beam source at a fixed angle.

and inserting it with the correct orientation. In the second system, a single 60° wedge (the universal wedge) is permanently located on a motorized mount located within the head of the treatment unit. This wedge can be rotated to the desired orientation or removed altogether, as required by the treatment plan.

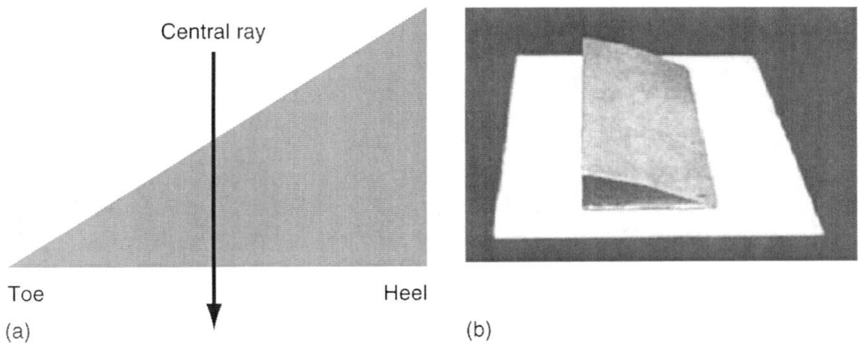

Figure 7.5 Wedges.

7.2.4.2 Optimization Model Formulation

Suppose that the data to the optimization models is given. Let $\mathcal{D}_{(i,j,k),A}$ be the dose contribution to voxel (i, j, k) from a beam of weight 1 from angle A, $\mathcal{S}$ be a collection of voxels on the sensitive structure(s), and $\mathcal{N}$ be a collection of voxels on the normal tissue. When wedges are allowed in the optimization, the data will be provided as $\mathcal{D}_{(i,j,k),A,F}$ that represents the dose contribution to voxel (i, j, k) from a beam of weight 1 from angle A, using wedge orientation F.

Beam Weight Optimization: The classical optimization problem in conformal radiation therapy is to choose the weights (or intensity levels) to be delivered from a given set of angles. Suppose w_A represents the beam weight delivered from angle A, $D_{(i,j,k)}$ for the total dose deposited to voxel (i, j, k) and λ represent the relative weighting factors in the objective function. Given a set $\Omega = \mathcal{T} \cup \mathcal{S} \cup \mathcal{N}$, a general optimization model that determines optimal radiation intensity is

$$\begin{aligned} \text{Min}_w \quad & \lambda_t f(\mathcal{D}_\mathcal{T}) + \lambda_s f(\mathcal{D}_\mathcal{S}) + \lambda_n f(\mathcal{D}_\mathcal{N}) \\ \text{s.t.} \quad & D_\Omega = \sum\nolimits_{A \in \mathcal{A}} \mathcal{D}_{\Omega,A} w_A, \\ & l \le D_\mathcal{T} \le u, \\ & 0 \le w_A, \quad \forall A \in \mathcal{A}. \end{aligned} \tag{7.3}$$

Hard upper and lower bound constraints are imposed on the target dose so that, in the worst case, the resulting solution will satisfy the minimum requirement for a treatment plan. Objective function $f(\mathcal{D})$ can be defined based on the planner's preference, but a general function can be written as

$$f(\mathcal{D}) = \|\mathcal{D}(\cdot) - \theta\|_p, p \in \{1, 2, \infty\}.$$

Note that θ is the desired dose level for an organ of interest. These problems can be cast as a quadratic programming (QP) problem ($p = 2$), minimizing the Euclidean distance between the dose delivered to each voxel and the prescribed dose [7,43,48,50]. Furthermore, linear programming (LP) has also been extensively used to improve conventional treatment planning techniques [2,27,38,45,48]. The strength of LP is its ability to control hot and cold spots or integral dose on the organs using constraints, and the presence of many state-of-the-art LP solvers. The LP Model replaces the Euclidean norm objective function of a QP with a polyhedral one, for which standard reformulations (see [30,34]) result in LP problems. While these techniques still suffer from large amounts of data in $\mathcal{D}_{(i,j,k),A}$, they are typically solved in acceptable time frames. These models tend to find optimal solutions more quickly than the corresponding QP formulations.

Another technique to convert the quadratic (or more generally convex) problem to a linear program is via a piecewise-linear approximation of the objective (see Ref. [44]). For a quadratic function, a uniform spacing for the breakpoints guarantees small approximation errors from the piecewise-linear interpolant [26]. Because the piecewise-linear interpolant is convex, standard techniques can be used to reformulate this as a linear program [19,26].

Equivalent Uniform Dose (EUD) [8]: Recently, some of the medical physics literature has been advocating the use of other forms of objective function in place of the ones outlined above. A popular alternative to those given above is that of generalized equivalent uniform dose (EUD). This is defined on a per structure basis as

$$\mathrm{EUD}_a(D,\Omega) = \left(\frac{1}{\mathrm{card}\,(\Omega)} \sum_{(i,j,k)\in\Omega} D^a_{(i,j,k)} \right)^{1/a}.$$

Note that EUD is a scaled version of the a-norm of the dose to the particular structure, and hence is known to be a convex function for any $a \geq 1$ and concave for $a \leq 1$. Thus the problem

$$\begin{aligned} \mathrm{Max}_w \quad & \mathrm{EUD}_a(D,\mathcal{T}) \\ \text{s.t.} \quad & D_\Omega = \textstyle\sum_{A\in\mathcal{A}} \mathcal{D}_{\Omega,A} w_A, \quad \Omega = \mathcal{T} \cup \mathcal{S} \cup \mathcal{N}, \\ & \mathrm{EUD}_b(D,\mathcal{S}) \leq \phi, \\ & \mathrm{EUD}_c(D,\mathcal{N}) \leq u, \\ & 0 \leq w_A, \qquad \forall A \in \mathcal{A}. \end{aligned}$$

is a convex optimization problem provided $a \leq 1$ and $b, c \geq 1$. As such, NLP algorithms will find global solutions to these problems.

Beam Angle Selection and Wedge Orientation Optimization: Optimization also lends itself to solving the more complex problem of selecting which angles and wedge orientations to use as well as their intensities. MIP is a straightforward technique for these types of problems. We describe an optimization model that simultaneously optimizes beam angles, wedge orientations, and beam intensities. Wedges are placed in front of the collimator to produce a gradient over the dose distribution and can be effective for reducing the radiation dose to organs at risk. This can be done by adding an extra dimension F to the variable w_A:

$$\begin{aligned} \text{Min}_{w,\psi} \; & \lambda_t f(\mathcal{D}_T) + \lambda_s f(\mathcal{D}_S) + \lambda_n f(\mathcal{D}_N) \\ \text{s.t.} \quad & D_\Omega = \textstyle\sum_{A\in\mathcal{A}} \mathcal{D}_{\Omega,A,F} w_{A,F}, \\ & w_{A,F} \le M \cdot \psi_A, \\ & l \le D_T \le u, \\ & \sum_{a\in\mathcal{A}} \psi_a \le K, \\ & \psi_A \in \{0,1\}, \quad \forall A \in \mathcal{A}. \end{aligned} \tag{7.4}$$

The variable ψ_A is used to determine whether or not to use an angle A for delivery. The choice of M plays a critical role in the speed of the optimization; further advice on its choice is given in Ref. [30]. Note that the data for this problem is considerably larger, increasing by a factor related to the number of wedge orientations allowed.

7.2.4.2.1 Solution Techniques

Simulated annealing (SA) has been well accepted in the medical community [39,55]. But the weakness of SA in the optimization point is its inability to verify the optimality. On the other hand, it is possible to find a global optimal solution for Equation 7.4. However, due to its slow convergence, using the MIP Model has not been very useful for designing a treatment plan in the hospital. Recently, Lim et al. [30] proposed an iterative solution approach that solves the MIP problem fast (within 20 minutes for two clinical case examples presented). It is termed a three-phase approach.

Three-phase approach is a multiphase technique that ramps up to the solution of the full problem via a sequence of models. Essentially, the models are solved in increasing order of difficulty, with the solution of one model providing a good starting point for the next. The models differ from each other in the selection of voxels included in the formulation, and in the number of beam angles allowed.

If the most promising beam angles can be identified in advance, the full problem can be solved with a small number of discrete variables. One simple approach for removing unpromising beam angles is to remove from consideration those that pass directly through any OAR [46]. A more elaborate approach [42] introduces a score function for each candidate angle, based on the ability of that angle to deliver a high dose to the PTV without exceeding the prescribed dose tolerance to OAR or to normal tissue located along its path. Only beam angles with the best scores are included in the model. We now describe the three-phase approach:

1. *Phase 1: Selection of Promising Beam Angles*: The aim in this phase is to construct a subset of beam angles $\mathcal{A}_1$ that are likely to appear in the final solution of Equation 7.4. We solve a collection of r MIPs, where each MIP is constructed from a reduced set of voxels consisting of the voxels in the PTV, a randomly sampled 10 percent of the OAR voxels ($\mathcal{S}$), and the voxels in $\mathcal{R}_\rho(\mathcal{T})$; that is

$$\Omega_1 = \{\mathcal{T} \cup \mathcal{S} \cup \mathcal{R}_\rho(\mathcal{T})\}.$$

We define $\mathcal{A}_1$ as the set of all angles $A \in \mathcal{A}$ for which $w_A > 0$ for at least one of these r-sampled problems.

2. *Phase 2: Treatment Beam Angle Determination*: In the next phase, we select K or fewer treatment beam angles from $\mathcal{A}_1$. We solve Equation 7.4 using $\mathcal{A}_1$ and a reduced set of voxels defined as follows:

$$\Omega_2 = \{\mathcal{T} \cup \mathcal{S} \cup \mathcal{R}_\rho(\mathcal{T}) \cup \mathcal{N}_1\}.$$

Note that $|\mathcal{A}_1|$ is typically greater than or equal to K, so the binary variables play a nontrivial role in this phase.

3. *Phase 3: Final Approximation*: In the final phase, we fix the K beam angles (by fixing $\psi_{\mathcal{A}_1} = 1$ for the angles selected in Phase 2 and $\psi_A = 0$ otherwise) and solve the resulting simplified optimization problem over the complete set of voxels. This final approximation typically takes much less time to solve than the full-scale model, because of both the smaller amount of data (due to fewer beam angles) and the absence of binary variables.

Although there is no guarantee that this technique will produce the same solution as the original full-scale model Equation 7.4, Lim et al. [30] have found that the quality of its approximate solution is close to optimal based on several numerical experiments.

7.2.5 Intensity Modulated Radiation Therapy

7.2.5.1 Introduction

A sophisticated form of treatment planning approach known as intensity modulated radiation therapy (IMRT) allows a number of differently shaped beams with different uniform radiation intensities to be delivered from each direction, which allows a high degree of flexibility in delivering radiation dose distribution from each beam angle [3,22]. In IMRT treatment planning, two-dimensional (2D) beams are divided into several hundred or thousand pencil beams to generate very precise dose distribution on the treatment volume.

Decision Variables: First, one needs to decide how many beam angles need to be coordinated for the treatment (beam angle optimization). For each beam angle, radiation is delivered using an MLC. In practice, an MLC is designed so that one leaf can only move one direction with a discrete distance. Therefore, we divide an MLC as an $M \times N$ grid of pixels. M is for the number of leaves in an MLC (note that this number can vary from one manufacturer to another), and N is for the number of discrete units that a leaf can move. Second, radiation intensity maps (fluence maps) for such beam angles need to be optimized to conform the (3D) radiation dose requirement to control the tumor (fluence map optimization). For a fixed beam angle, the fluence map contains real numbers in a set of (2D) discrete coordinates that are associated with the MLC. Because no machine can deliver such a nonuniform real intensity map, the intensity maps are first approximated as multiples of a physically deliverable minimum discrete unit (this number can be a fraction). For example, an approximated intensity map for a 3×4 MLC may look as follows (we assume that the minimum discrete value allowed is 0.5 in this case):

$$\mathcal{W} = \begin{pmatrix} 0 & 0.5 & 2.0 & 1.5 \\ 0.5 & 2.5 & 3.5 & 2.0 \\ 0 & 1.0 & 2.0 & 1.5 \end{pmatrix} = 0.5 \times \begin{pmatrix} 0 & 1 & 4 & 3 \\ 1 & 5 & 7 & 4 \\ 0 & 2 & 4 & 3 \end{pmatrix}. \tag{7.5}$$

Third, because we cannot deliver nonuniform radiation (see Equation 7.5) to the treatment volume with one open beam shape, an intensity map is decomposed into several unique shape matrices such that each matrix can contain zeros and uniform value. This is called beam segmentation problem.

7.2.5.2 Optimization Model Formulation

7.2.5.2.1 Modeling Dose Deposition

Dose deposition models specify the amount of radiation deposited at points in the irradiated region, relative to some known dose in a reference point [53]. There are numerous methods of calculating dose deposition, and many of them are highly complex, particularly in three dimensions. We describe here a simple 2D model as developed in Ref. [21], simply to illustrate the basics of the dose calculation process.

Assuming a finite number of angles a and sub-beams s, we wish to calculate the dose contribution of sub-beam (a, s) to the dose point p. In three dimensions, the dose points result from discretizing the irradiated area into small cubes called voxels. In two dimensions, the dose points may be visualized as squares. Because of scatter, dose point p may still receive radiation even if it is not directly in the field of sub-beam (a, s). We let δ be the depth of the dose point along the path of sub-beam (a, s) and let o be the off-axis distance (or distance from p to the sub-beam). If we let $x_{(a,s)}$ be the amount of energy being propagated along sub-beam (a, s) then dose point p receives

$$\mathrm{e}^{\eta o}\mathrm{e}^{\mu\delta}x_{(a,s)}$$

units of radiation from sub-beam (a, s). Here η and μ are parameters that vary with the beam's energy and model the effects of scatter and attenuation. This model assumes that as δ and o increase, the amount of dose deposited in p decreases exponentially. We let $d_{(p,a,s)} = \mathrm{e}^{\eta o}\mathrm{e}^{\mu\delta}$ be the rate in gray per second (Gy/t) at which radiation along sub-beam s in angle a is deposited into dose point p, where $d_{(p,a,s)} \geq 0, \forall(p, a, s)$. Then the total amount of dose deposited in a particular p is

$$\mathcal{D}_p = \sum_{(a,s)} d_{(p,a,s)}x_{(a,s)}.$$

We define the dose matrix A to be the collection of $d_{(p,a,s)}$ where the rows of A are indexed by p and the columns are indexed by (a, s). It is important to mention that this collection is patient specific. Because dose is directly proportional to exposure time (fluency) we can alternately define $x_{(a,s)}$ to be the fluence value for sub-beam (a, s). Then the linear transformation $x \mapsto Ax$ maps the fluency pattern x into the anatomy and determines the amount of radiation deposited at every dose point in the anatomy [12]. The linearity is not just an assumption of the model—it is experimentally validated that the dose is deposited linearly [21]. So-called phantoms are used to test the nature of radiation deposition in the human body. Phantoms are made from materials with the same radiation absorption properties as human tissue [53].

7.2.5.2.2 Beam Angle and Fluence Map Optimization

Because the optimal set of beam angles are intimately related to the optimal fluence map for each angle, these two problems must be dealt together in the problem formulation. Let a denote an angle, $a \in \mathcal{A}$, l denote the leaf index of the collimator, $l = 1, 2, \ldots, m$, and p represent the position of the leaf, $p = 1, 2, \ldots, n$. Then, formulating an optimization model for optimizing both beam angles and the fluence maps is a simple extension to Equation 7.4 that we discussed for the conventional 3DCRT and the model is given as

$$
\begin{aligned}
\text{Min}_{w,\psi} \ & \lambda_t f(\mathcal{D}_T) + \lambda_s f(\mathcal{D}_S) + \lambda_n f(\mathcal{D}_\mathcal{N}) \\
\text{s.t.} \quad & D_\Omega = \textstyle\sum_{a \in \mathcal{A}} \mathcal{D}_{\Omega,a,l,p} w_{a,l,p}, \\
& w_{a,l,p} \le M \cdot \psi_a, \\
& l \le D_T \le u, \\
& \sum_{a \in \mathcal{A}} \psi_a \le K, \\
& \psi_a \in \{0,1\}, \quad \forall a \in \mathcal{A}.
\end{aligned}
$$

See more details of this formulation and others in Refs. [29,32].

7.2.5.2.2.1 Solution Methods Solving this problem using any classical optimization techniques may take too long for any clinicians to use for their daily treatment planning. Lim et al. [32] propose a fast MIP solution approach and an LP-based iterative method that exploits score functions. However, due to the computational difficulties with large data in solving the optimization problem, heuristic methods are often used in practice [28].

7.2.5.2.3 Beam Segmentation Optimization

Consider a matrix

$$
\mathcal{W} = \begin{pmatrix} w_{1,1} & w_{1,2} & \cdots & w_{1,n} \\ \vdots & & & \vdots \\ w_{m,1} & w_{m,2} & \cdots & w_{m,n} \end{pmatrix},
$$

where $w_{i,j} \in \mathbb{Z}$, for $i = 1, \ldots, m$, $j = 1, \ldots, n$. Our objective is to decompose the matrix $\mathcal{W}$ into K binary matrices S^k such that

$$
\mathcal{W} = \sum_{k=1}^{K} \mu_k \cdot S^k, \tag{7.6}
$$

where

$$
\begin{aligned}
S^k &= \left[s_{i,j}^k\right], \\
s_{i,j}^k &\in \{0,1\}, \\
i &\in \{1,2,\ldots,m\}, \\
j &\in \{1,2,\ldots,n\}, \\
\mu_k &\in \mathbb{Z}, \\
k &\in \{1,2,\ldots,K\}.
\end{aligned}
$$

Solving Equation 7.6 is quite easy in general. However, this problem becomes extremely difficult to solve when we impose the following two objectives and a physical constraint.

Objectives:

1. Minimize the value of K.
2. Minimize T = the sum of the matrix multipliers, i.e., $T = \sum_{k=1}^{K} \mu_k$.

Consecutive One Constraint:
For each row of a binary matrix S_k, if there are more than one nonzero elements (1's), their sequence must be consecutive, i.e., zeros are not allowed to break the nonzero sequence. For example,

$$0\ 1\ 1\ 1\ 0$$

is a feasible sequence. But,

$$0\ 1\ 0\ 1\ 0$$

is not allowed because the sequence of ones is not continuous.

Note that there can be many more constraints to this problem depending on the machine that is used for radiation delivery. Some of the common constraints are overlap elimination constraint, interleaf collision constraint, and tongue-and-groove constraint. Details about these more elaborate constraints can be found in Refs. [1,23,24,54].

7.2.5.2.3.1 Solution Methods This is a combinatorial optimization problem that is proven to be strongly NP-hard [10]. Optimization formulations have been proposed including integer nonlinear program (INLP) and integer programming (IP) [32]. IP Models are easier to solve than INLP Models. IP Models with relatively small numbers of rows and columns can be solved within a reasonable amount of time using a branch-and-bound method [56]. However, as the matrix size increases (say, larger than 10) and the maximum value of the matrix $\mathcal{W}$ increases, finding global solutions for the IP Models can take too long for treatment planners to use. Therefore, both researchers and planners use various heuristics. Engel [13] proposed a heuristic that generates optimal T, but K is still not optimal. Lim et al. [31] proposed a two-stage integer programming approach that improved Engel's K. Both methods claimed that solving problems with a matrix size larger than 10 × 10 takes less than a few seconds. A genetic algorithm has also been used by other researchers [9].

7.2.6 Other Radiation Therapy Devices

7.2.6.1 Tomotherapy

The recent development of IMRT on conventional LINACs provided a major increment in radiation therapy dose delivery. A course of radiation treatment often consists of up to 40 daily treatment fractions. For each fraction, the patient must be repositioned. This repositioning has inherent uncertainties that relate not only to setting the patient up to external reference marks, usually laser alignment with marks positioned on the skin surface, but also to the movement of internal organs from day-to-day. To address the issues of highly conformal dose distributions as well as accounting for patient setup and organ motion uncertainties, a new technology, known as tomotherapy, has been developed. Theoretically, it provides better targeting with a corresponding reduction in the dose to normal tissues. This allows a higher dose to the tumor, which results in an increased probability of tumor control [11,36].

Tomotherapy, which literally means slice therapy, is a term derived from tomography. The first implementation of this concept was performed by NOMOS Corporation [6] and was provided as an add-on accessory to existing linear accelerators. The add-on feature consists of a set of MLCs that provide a narrow fan beam shape [11] projecting a maximum width at the patient of about 20 cm. The fan beam thickness can be either 0.8 or 1.6 cm and each leaf projects a shadow of about 1 cm width on the patient. When the leaves are in the beam, that portion of the beam is fully shielded except for a minor (0.5 percent) transmission component. Either the leaf is open or closed for that slice providing binary dose delivery, i.e., for that portion of the beam, the beam is either on or off. The open beam components are generally referred to as beamlets or pencil beams. Radiation delivery consists of a machine that rotates around the patient while the beam is on and the leaves rapidly move in and out depending on whether that beamlet is aimed at the target or at normal tissues. After two simultaneous slices have been delivered, the patient is translated by two slice thicknesses and the next two slices are delivered until the total treatment volume is covered. This is called serial tomotherapy.

7.2.6.1.1 Optimization Methods

Shepard et al. [49] introduced iterative approaches for optimizing dose in tomotherapy. Olivera et al. [41] described optimization techniques for large-scale helical tomotherapy. Helical tomotherapy is an integrated therapeutic technique, which includes planning, delivery, and verification capabilities. Helical tomotherapy allows for irradiation of a large number of targets and region at risk (RAR) over broad regions of the body; a large-scale optimization technique is necessary. Usually a tomotherapy treatment will have tens to hundreds of thousands of pencil beams for which intensity needs to be optimized. Moreover, the number of voxels where the dose needs to be computed is on the order of 1,000,000. The complexity and size of the optimization is the price paid in tomotherapy to obtain coplanar deliveries that are neither limited by the number of beam directions nor the degree of modulation used during delivery. They claim that a complex optimization plan will not lead to a complex delivery, thanks to the simplicity and capabilities of the tomotherapy concept. More recent publication on tomotherapy can be found in Ref. [40].

7.2.6.2 Proton Therapy

7.2.6.2.1 Introduction

Heavy charged particles, such as protons, produce energy deposition patterns in tissue which are superior to single beams of photons and electrons for treatment of cancer. The dose distribution from a monoenergetic beam of protons has an entrance region of slowly rising dose followed by a sharp increase, called Bragg peak, near the end of range. By superimposing beams of different energies, the Bragg peak can be spread to generate a dose distribution, which provides moderate entrance dose, uniform high dose within the target tissue, and almost zero dose beyond the target (see Figure 7.6). Highly conformal dose distributions with low integral dose can

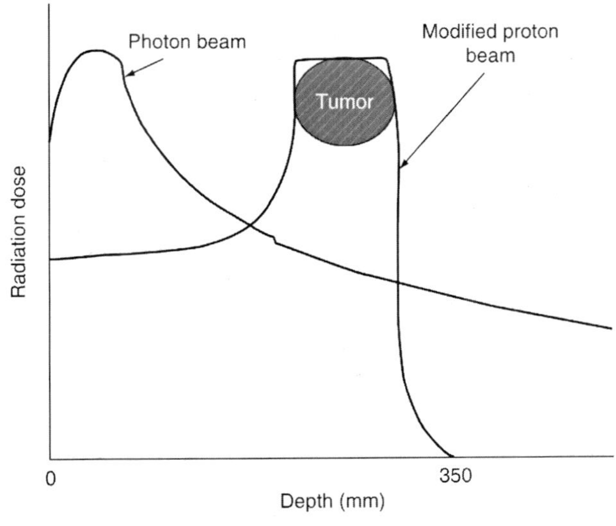

Figure 7.6 A comparison of the amount of radiation delivered with conventional photon beam radiation therapy versus proton therapy. Conventional therapy is distinguished by a relatively high entrance dose and an exit dose. By contrast, proton therapy has a much lower entrance dose and no exit dose.

be achieved with a small number of proton treatment fields convergent on a target region [37].

7.2.6.2.2 Intensity Modulated Proton Therapy

Intensity modulated proton therapy (IMPT) is a technique for radiation treatment of cancer, which allows one to deliver highly target-conformal dose distributions. The use of IMPT may potentially result in better sparing of the normal tissue, than is achievable with the proton passive scattering or the intensity modulated therapy with photons. Unlike the passive scattering technique, IMPT does not require the use of patient-specific hardware, such as field forming apertures and range compensators. Robust mathematical models and efficient solution techniques for optimizing proton treatment planning parameters will make this precise treatment technique more desirable for future cancer treatment [52].

References

1. R. Alfredo and C. Siochi. Minimizing static intensity modulation delivery time using an intensity solid paradigm. *International Journal of Radiation Oncology, Biology, Medicine*, 43(3):671–680, 1999.
2. G. K. Bahr, J. G. Kereiakes, H. Horwitz, R. Finney, J. Galvin, and K. Goode. The method of linear programming applied to radiation treatment planning. *Radiology*, 91:686–693, 1968.
3. T. Bortfeld, A. L. Boyer, W. Schlegel, D. L. Kahler, and T. J. Waldron. Realization and verification of three-dimensional conformal radiotherapy with modulated fields. *International Journal of Radiation Oncology: Biology, Physics*, 30(4):899–908, 1994.
4. T. Bortfeld and W. Schlegel. Optimization of beam orientations in radiation therapy: Some theoretical considerations. *Physics in Medicine and Biology*, 38(2):291–304, 1993.
5. American Cancer Society. Cancer facts and figures, 2006. http://www.cancer.org/downloads/STT/CAFF2006f4PWSecured.pdf.
6. M. P. Carol. Peacock: A system for planning and rotational delivery of intensity-modulated fields. *International Journal of Imaging Systems and Technology*, 6:56–61, 1995.
7. Y. Chen, D. Michalski, C. Houser, and J. M. Galvin. A deterministic iterative least-squares algorithm for beam weight optimization in conformal radiotherapy. *Physics in Medicine and Biology*, 47:1647–1658, 2002.
8. B. Choi and J. O. Deasy. The generalized equivalent uniform dose function as a basis for intensity-modulated treatment planning. *Physics in Medicine and Biology*, 47:3579–3589, 2002.
9. C. Cotrutz and L. Xing. Segment-based dose optimization using a genetic algorithm. *Physics in Medicine and Biology*, 48:2987–2998, 2003.
10. M. Ehrgott, D. Baatar, H. W. Hamacher, and G. J. Woeginer. Decomposition of integer matrices and multileaf collimator sequencing. Technical report, Fachbereich Mathematik Technische Universität Kaiserslautern, 2004.

11. J. V. Dyk, T. Kron, G. Bauman, and J. J. Battista. Tomotherapy: A revolution in radiation therapy. http://www.lrcc.on.ca/research/pdf/Tomotherapy.pdf, 2007.
12. M. Ehrgott, A. Holder, and J. Reese. Beam selection in radiotherapy design. Technical Report 95, Trinity University Mathematics, San Antonio, TX, 2005.
13. K. Engel. A new algorithm for optimal multileaf collimator field segmentation. *Discrete Applied Mathematics*, 152:35–51, 2005.
14. G. Fang, B. Geiser, and T. R. Mackie. Software system for UW/GE tomotherapy prototype. In D. D. Leavitt and G. Starkshall, editors, Proceedings of the 12th International Conference on the Use of Computers in Radiation Therapy, Salt Lake City, pp. 332–334, St. Louis, Missouri, Medical Physics Publishing, 1997.
15. M. C. Ferris, J. -H. Lim, and D. M. Shepard. Optimization approaches for treatment planning on a Gamma Knife. *SIAM Journal on Optimization*, 13:921–937, 2003.
16. M. C. Ferris, J. -H. Lim, and D. M. Shepard. Radiosurgery treatment planning via nonlinear programming. *Annals of Operations Research*, 119:247–260, 2003.
17. J. C. Ganz. *Gamma Knife Surgery*. Springer-Verlag, Wien, Austria, 1997.
18. M. Goitein, M. Abrams, S. Rowell, H. Pollari, and J. Wiles. Multi-dimensional treatment planning: II. Beam's eye-view, back projection, and projection through CT sections. *International Journal of Radiation Oncology: Biology, Physics*, 9:789–797, 1983.
19. J. K. Ho. Relationships among linear formulations of spearable convex piecewise linear programs. *Mathematical Programming Study*, 24:126–140, 1985.
20. A. Holder. Radiotherapy treatment design and linear programming. In M.L. Brandeau, F. Saintfort, and W.P. Pierskalla, editors, *Operations Research and Health Care: A Handbook of Methods and Applications*. Kluwer Academic Publishers, Boston, MA, 2004, pp. 741–774.
21. A. Holder and B. Salter. A tutorial on radiation oncology and optimization. In H. Greenberg, editor, *Tutorials on Emerging Methodologies and Applications in Operations Research*, Chapter 4. Kluwer Academic Publishers, Boston, MA, 2004.
22. Intensity Modulated Radiation Therapy Collaborative Working Group. Intensity-modulated radiotherapy: Current status and issues of interest. *International Journal of Radiation Oncology: Biology, Physics*, 51(4):880–914, 2001.
23. T. J. Jordan and P. C. Williams. The design and performance characteristics of a multileaf collimator. *Physics in Medicine and Biology*, 39:231–251, 1994.
24. T. Kalinowski. Realization of intensity modulated radiation fields using multileaf collimators. *General Theory of Information Transfer and Combinatorics*, Electronic Notes in Discrete Mathematics, Vol. 21, pp. 319–320, Springer, Berlin, 2005.
25. J. M. Kapatoes, G. H. Olivera, J. P. Balog, H. Keller, P. J. Reckwerdt, and T. R. Mackie. On the accuracy and effectiveness of dose reconstruction for tomotherapy. *Physics in Medicine and Biology*, 46:943–966, 2001.
26. S. Kontogiorgis. Practical piecewise-linear approximation for monotropic optimization. *INFORMS Journal on Computing*, 12(4):324–340, 2000.
27. M. Langer and J. Leong. Optimization of beam weights under dose-volume restriction. *International Journal of Radiation Oncology: Biology, Physics*, 13:1255–1260, 1987.
28. M. Langer, S. Morrill, R. Brown, O. Lee, and R. Lane. A comparison of mixed integer programming and fast simulated annealing for optimized beam weights in radiation therapy. *Medical Physics*, 23:957–964, 1996.

29. E. K. Lee, T. Fox, and I. Crocker. Simultaneous beam geometry and intensity map optimization in intensity-modulated radiation therapy. *International Journal of Radiation Oncology, Biology and Physics*, 64(1):301–320, 2006.
30. G. J. Lim, M. C. Ferris, S. J. Wright, D. M. Shepard, and M. A. Earl. An optimization framework for conformal radiation treatment planning. *INFORMS Journal on Computing*, 19(3):366–380, 2007.
31. G. J. Lim and J. Choi. A two-stage integer programming approach for optimizing leaf sequence in IMRT. IE Tech. Report, IE0707-01, University of Houston, 2007.
32. G. J. Lim, J. Choi, and R. Mohan. Iterative solution methods for beam angle and fluence map optimization in intensity modulated radiation therapy planning. *OR Spectrum*, 2007 (in press).
33. J. -H. Lim. Optimization in Radiation Treatment Planning. PhD thesis, University of Wisconsin, Madison, WI, December 2002.
34. J. -H. Lim, M. C. Ferris, and D. M. Shepard. Optimization tools for radiation treatment planning in matlab. In M. L. Brandeau, F. Saintfort, and W. P. Pierskalla, editors, *Operations Research and Health Care: A Handbook of Methods and Applications*, Kluwer Academic Publishers, Boston, MA, 2004, pp. 775–806.
35. W. Lodwick, S. McCourt, F. Newman, and S. Humphries. Optimization methods for radiation therapy plans. In C. Borgers and F. Natterer, editors, *Computational Radiology and Imaging: Therapy and Diagnosis*, IMA Series in Applied Mathematics. Springer-Verlag, 1998.
36. T. R. Mackie, P. Reckwerdt, J. O. Deasy, J. Yang, B. Paliwal, and T. Kinsella. A comparison of three inverse treatment planning algorithms. *Medical Physics, Biology*, 39:91–106, 1994.
37. D. W. Miller. A review of proton beam radiation therapy. *Medical Physics*, 22(11):1943–1954, 1995.
38. S. Morrill, R. Lane, J. Wong, and I. I. Rosen. Dose-volume considerations with linear programming. *Medical Physics*, 6(18):1201–1210, 1991.
39. S. M. Morrill, K. S. Lam, R. G. Lane, M. Langer, and I. I. Rosen. Very fast simulated annealing in radiation therapy treatment plan optimization. *International Journal of Radiation Oncology, Biology and Physics*, 31:179–188, 1995.
40. G. H. Olivera, T. R. Mackie, K. Ruchala, W. Lu, and J. Kapatoes. Adaptive radiation therapy (art) strategies using helical tomotherapy. In T. Bortfeld, R. Schmidt-Ullrich, W. De Neve and D. E. Wazer, editors, *Image-Guided IMRT*. Springer, Berlin Heidelberg, 2006, pp. 399–405.
41. G. H. Olivera, P. J. Reckwerdt, D. M. Shepard, and T. R. Mackie. Large-scale helical tomotherapy optimization: Four clinical casestudies. Proceedings of the 22nd Annual International Conference of the IEEE, 4:3090–3092, 2000.
42. A. Pugachev and L. Xing. Incorporating prior knowledge into beam orientation optimization in IMRT. *International Journal of Radiation Oncology, Biology and Physics*, 54:1565–1574, 2002.
43. A. T. Redpath, B. L. Vickery, and D. H. Wright. A new technique for radiotherapy planning using quadratic programming. *Physics in Medicine and Biology*, 21:781–791, 1976.
44. H. E. Romeijn, R. K. Ahuja, J. F. Dempsey, and A. Kumar. A new linear programming approach to radiation therapy treatment planning problems. *Operations Research*, 54(2):201–216, 2006.
45. I. I. Rosen, R. Lane, S. Morrill, and J. Belli. Treatment plan optimization using linear programming. *Medical Physics*, 18(2):141–152, 1990.

46. C. G. Rowbottom, V. S. Khoo, and S. Webb. Simultaneous optimization of beam orientations and beam weights in conformal radiotherapy. *Medical Physics*, 28(8):1696–1702, 2001.
47. D. M. Shepard, L. S. Chin, S. J. DiBiase, S. A. Naqvi, J. Lim, and M. C. Ferris. Clinical implementation of an automated planning system for Gamma Knife radiosurgery. *International Journal of Radiation Oncology, Biology, Physics*, 56:1488–1494, 2003.
48. D. M. Shepard, M. C. Ferris, G. Olivera, and T. R. Mackie. Optimizing the delivery of radiation to cancer patients. *SIAM Review*, 41:721–744, 1999.
49. D. M. Shepard, G. H. Olivera, P. J. Reckwerdt, and T. R. Mackie. Iterative approaches to dose optimization in tomotherapy. *Physics in Medicine and Biology*, 45:69–90, 2000.
50. G. Starkschall. A constrained least-squares optimization method for external beam radiation therapy treatment planning. *Medical Physics*, 11:659–665, 1984.
51. J. Tervo and P. Kolmonen. A model for the control of a multileaf collimator in radiation therapy treatment planning. *Inverse Problems*, 16:1875–1895, 2000.
52. A. Trofimov and T. Bortfeld. Optimization of beam parameters and treatment planning for intensity modulated proton therapy. *Technoloy in Cancer Research & Treatment*, 2(5):437–444, 2003.
53. L. J. Verhey. Principles of radiation physics. In S. A. Leibel and T. L. Phillips, editors, *Textbook of Radiation Oncology*. W. B. Saunders Company, Philadelphia, PA, 1998, pp. 91–114.
54. J. Kung W. Que and J. Dai. Tongue-and-groove effect in intensity modulated radiotherapy with static multileaf collimator fields. *Physics in Medicine and Biology*, 49:399–405, 2004.
55. S. Webb. Optimisation of conformal radiotherapy dose distributions by simulated annealing. *Physics in Medicine and Biology*, 34(10):1349–1370, 1989.
56. L. A. Wolsey. *Integer Programming*. John Wiley & Sons, New York, NY, 1998.
57. X. Wu and Y. Zhu. A global optimization method for three-dimensional conformal radiotherapy treatment planning. *Physics in Medicine and Biology*, 46:109–119, 2001.
58. Y. Xiao, Y. Censor, D. Michalski, and J. M. Galvin. The least-intensity feasible solution for aperture-based inverse planning in radiation therapy. *Annals of Operations Research*, 119:183–203, 2003.
59. K. Yoda, Y. Saito, and H. Sakamoto. Dose optimization of proton and heavy ion therapy using generalized sampled pattern matching. *Physics in Medicine and Biology*, 42:2411–2420, 1997.

Chapter 8

Beam Orientation Optimization Methods in Intensity Modulated Radiation Therapy Treatment Planning

Dionne M. Aleman, H. Edwin Romeijn, and James F. Dempsey

CONTENTS

8.1 Introduction 224
8.2 BOO Model 225
8.2.1 Beam Orientations 225
8.2.2 Feasible Beam Space 226
8.2.2.1 Beam Data Generation 228
8.2.2.2 Time and Space Considerations 229
8.2.3 Objective Function 229
8.2.3.1 Fluence Map Optimization 229
8.2.3.2 Beam's-Eye-View Approaches 231
8.2.3.3 Geometric Considerations 235
8.2.3.4 Other Scoring Criteria 236
8.3 Optimization Methods 239
8.3.1 Solving the BOO + FMO Integer Program 239
8.3.1.1 Branch-and-Bound Techniques 240
8.3.1.2 Commercial Solvers 241
8.3.2 Metaheuristics 241

8.3.2.1 Simulated Annealing 242
8.3.2.2 Evolutionary and Genetic Algorithms 243
8.3.2.3 Response Surface Method 245
8.3.3 Local Search Algorithms 245
8.3.4 Greedy Algorithms 246
8.3.5 Other Optimization Approaches 247
8.4 Conclusions 247
References 248

Abstract The intensity modulated radiation therapy (IMRT) treatment planning problem consists of several subproblems, which are typically solved sequentially. This chapter addresses the beam orientation optimization (BOO) problem, and to some extent, the fluence map optimization (FMO) problem. The BOO problem is the problem of selecting from which beam orientations to deliver radiation to the patient. The goal of beam orientation optimization is to select the best beams from which to deliver the radiation so that the treatment plan can deliver the prescribed amount of radiation dose to the target cells while simultaneously delivering a small enough amount of radiation to the surrounding tissue so that nearby organs will continue to function properly after radiation therapy. The solution to the FMO problem, the problem of determining the amount of radiation intensity (fluence) of each beamlet in each beam, is generally accepted as the measure of the quality of a beam solution. However, due to the cost of evaluating the FMO problem, a number of alternative approaches have been taken to model and to solve the BOO problem. This chapter summarizes many of the BOO methods presented in the literature.

8.1 Introduction

In intensity Modulated Radiation Therapy (IMRT), radiation is delivered to the patient via external beams. Beam orientation optimization (BOO) in IMRT is the problem of selecting from which beam orientations to irradiate the patient, the goal of which is to increase the quality of a treatment plan or decrease the time required to deliver the treatment plan. The quality of a treatment plan is determined by its ability to deliver the prescibed amount of radiation to the target structures—which typically consist of a gross tumor volume (GTV) where disease is evident, and a planning tumor volume (PTV), an area surrounding the GTV where microscopic cancer spread is suspected—while simultaneously delivering an acceptably low amount of radiation to the surrounding normal tissue, called critical structures or organs-at-risk (OARs). One may intuitively expect that the selection of the beam orientations influences the quality of the treatment plan. One may then also expect that it is possible for treatment plans using different numbers of beams to have the same quality, depending on which beams are used in each plan. From a clinical perspective, it is desirable to minimize the time required to administer the treatment plan, and one way of achieving faster delivery times is to decrease the number of beams in the treatment plan. BOO thus serves to achieve two goals in IMRT: (1) to improve the quality of the treatment plan or (2) to reduce the number of beams required to deliver a high-quality treatment plan.

This chapter describes the BOO problem and approaches that have been used to solve it, and is organized as follows. Section 8.2 describes the BOO formulation; Section 8.3 discusses methods of solving the BOO problem; and Section 8.4 summarizes the abilities of the methods presented.

8.2 BOO Model

To optimize the beam orientations, a quantitatve measure to assess the quality of each vector of beam solutions, $\boldsymbol{\theta}$, must be obtained. Let this measure be $\mathcal{F}(\boldsymbol{\theta})$. Let $\mathcal{B}$ be the set of candidate beam orientations from which radiation may be delivered. For a solution consisting of k beams, the vector $\boldsymbol{\theta}$ must lie in $\mathcal{B}^k = \mathcal{B} \times \cdots \times \mathcal{B}$.

Selecting an objective function such that the best solution is obtained at the function's minimum, a basic formulation of the BOO problem is then

$$\text{Min } \mathcal{F}(\boldsymbol{\theta})$$

$$\text{s.t. } \boldsymbol{\theta} \in \mathcal{B}^k$$

The above formulation determines not only the beam orientations, but the number of beams as well. Although it is common to choose k to be a fixed value before solving the BOO problem, studies where k is a decision variable can be found in Schreibman et al. [46], Söderstrom and Brahme [51], and Stein et al. [52].

8.2.1 *Beam Orientations*

Several components of a linear accelerator, the machine used to deliver the radiation, can be translated and rotated so that radiation beams may originate from anywhere within a vast space around the patient. Despite the availability of an immeasurable number of beam orientations, for most cancer sites, typical IMRT treatment plans in practice use equi-spaced coplanar beams. Coplanar beams are those beams obtained from the rotation of only the gantry of the linear accelerator. Figure 8.1 shows a linear accelerator with arrows indicating the movements available to its components; the gantry rotation is highlighted. If all other components of the linear

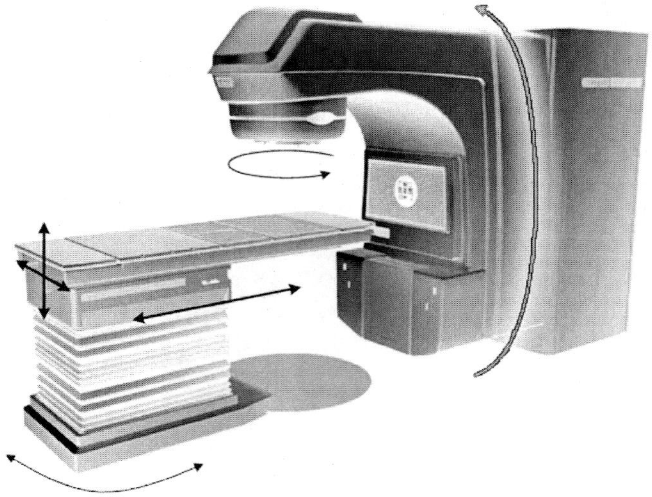

Figure 8.1 A linear accelerator and available movements; the gantry rotation is highlighted.

accelerator are fixed, the rotation of the gantry sweeps out a set of coplanar beams in a circular disc perpendicular to the couch. If the couch is also permitted to rotate, the rotation of the gantry and the couch create a spherical set of beams. The couch can also translate up, down, left, right, forward and backward, and the head of the gantry can rotate independently, creating an even larger set of beams. Beams obtained from the movement of more than just the gantry are known as *non-coplanar* beams.

The set of candidate beams $\mathcal{B}$ can be selected according to any user-specified criteria. Although the linear accelerator is able to deliver a continuous set of beam orientations from nearly any orientation in three-dimensional (3D) space by rotating/translating the various components indicated in Figure 8.1, it is common to only consider a discretized set of beam orientations due to limitations in machine tolerances. Because many degrees of freedom in a linear accelerator result in a very large set of available beam orientations, the size of the solution space $\mathcal{B}^k$ can be intractably large despite discretization. It is therefore also common to further restrict the set of candidate beams by considering only a subset of the discretized candidate beams. The most common restriction is to allow only coplanar beams.

8.2.2 Feasible Beam Space

Although there are many approaches to quantifying the quality of a beam solution and formulating the objective function to the BOO Model presented in Section 8.2, the constraints implemented do not vary as widely. Aside from the constraint that beam solutions be members of the candidate beam set $\mathcal{B}$, the only other common constraints placed upon the beam solutions in the literature are constraints enforcing geometric considerations. The geometric constraints typically considered involve the exclusion of parallel-opposed beams (beams that are 180° apart) and a minimum separation between angles in the beam solution.

The exclusion of parallel-opposed beams arises from the fact that the radiation deposited in a patient by a beam does not stop simply because it reaches the target structure; rather, it may continue on and deposit radiation in cells on the other side of the target. If parallel-opposed beams are used, it is clear that there is potential for the radiation delivered by the beams to completely overlap, and possibly create a situation in which the targets or critical structures become overdosed. Instances of the exclusion of parallel-opposed beams can be found explicitly in the methods presented in Rowbottom et al. [44], or implicitly in Das et al. [11], where parallel-opposed beams are avoided by selecting $\mathcal{B}$ in such a way that no beam has its parallel-opposed beam also in $\mathcal{B}$. Without having to restrict $\mathcal{B}$, a simple set of constraints to the BOO Model in Section 8.2 that exclude parallel-opposed beams is

$$\theta_h \neq \theta_j + 180, \quad h = 1, \ldots, k, \quad j = 1, \ldots, k, \quad j \neq h$$

$$\theta_h \neq \theta_j - 180, \quad h = 1, \ldots, k, \quad j = 1, \ldots, k, \quad j \neq h.$$

Note that the above constraints assume that angles will be coplanar and within the range $[0, 360)$.

The other common geometric constraint is that a user-specified minimum angle distance (δ) be observed (Das et al. [11], Rowbottom et al. [44]). Within the BOO Model presented in Section 8.2, such a requirement for coplanar angles could be expressed by the following set of constraints:

$$\theta_{h+1} - \theta_h \geq \delta, \quad h = 1, \ldots, k-1 \tag{8.1}$$

$$(360 - \theta_k) - \theta_1 \geq \delta \tag{8.2}$$

$$\theta_1 \leq \cdots \leq \theta_k \tag{8.3}$$

where

- constraint in Equation 8.3 requires that the angles are sorted in nondecreasing order, thus simplifying the other constraints;
- constraint in Equation 8.1 ensures that there are at least δ degrees between adjacent angles;

- constraint in Equation 8.2 ensures that there are also δ degrees between the first and last angles, which require special consideration due to the cyclic nature of angles, that is, $0° = 360°$.

Note that the above constraints assume that the candidate beam set $\mathcal{B}$ consists of angles in the range $[0, 360)$. Also, the above constraints only restrict the entry angles of the beams.

To incorporate both the entry and exit angles of the beams, the angle separation constraint could be formulated using the angle separation scoring mechanism S_{θ}^{sep} described in Section 8.2.3.3 to obtain

$$S_{\theta}^{\text{sep}} \geq \delta',$$

where δ' is a user-specified value indicating the desired amount of angle separation.

8.2.2.1 Beam Data Generation

For each beam orientation that is considered in $\mathcal{B}$, lengthy calculations must be made to determine the beam's effect on the patient's tissue and organs. This includes determining in which structure each voxel lies, which voxels are hit by which beamlets, and the amount of intensity of each beamlet that is deposited in each voxel through which it passes.

Beamlet dose computation models used in IMRT rely heavily on ray-tracing algorithms for voxel classification and determination of the radiological path (Fox et al. [17]). Voxel classification (Siddon [49]) establishes whether voxels are inside or outside the path of a radiation beam and classifies voxel centers as inside or outside of segmented targets and critical structures. The radiological path is the effective distance traveled by a beamlet when the effect of traveling through tissues of different densities is considered. The exact radiological path of a beamlet through the patient is required to correct for tissue heterogeneities in determining the dose deposition coefficients (Siddon [48]).

Siddon's ray-tracing algorithms (Siddon [48,49]) have been the standard methods used for ray-tracing in radiotherapy since the 1980s. In Siddon's polygon and voxel ray-tracing algorithms for voxel classification (point-in-polygon testing), structures are represented as 3D polygonal objects, known as Siddon Prisms, and the signs of cross products of rays passing through the polygons are used to determine whether a voxel lies inside or outside a structure. Despite its overwhelming use, Siddon's algorithm for polygon ray-tracing is very costly because the number of voxels in a patient. In 2005, Fox et al. [17] developed a novel approach to polygon ray-tracing that circumvents the need for cross products by translating the polygon structure onto a coordinate system, replacing the need for a cross product by the sign of the second coordinate of each voxel in the coordinate system.

In Siddon's algorithm for determining radiological paths (Siddon [48]), the radiological path must be determined for each voxel for every beamlet. This involves computations for millions of beamlet–voxel combinations. As reported by Jacobs et al. [24], a significant amount of computational time is required for these repeated calculations. Fox et al. [17] combined the incremental voxel ray-tracing algorithm presented by Jacobs et al. [24] with a method of virtual stereographic projection to significantly reduce the computational cost of obtaining radiological path lengths.

Using their polygon translation and incremental ray-tracing algorithms, Fox et al. [17] achieved a 100–300 fold improvement in computation time over Siddon's point-in-polygon algorithm. Although these methods provide a significant reduction in computation time, the consideration of non-coplanar beam orientations still poses difficulty in the BOO problem.

8.2.2.2 Time and Space Considerations

Because beam data calculations must be performed for each of millions of beamlet–voxel combinations, beam data generation is a lengthy process, requiring several minutes per beam using the algorithms described by Fox et al. [17]. For methods requiring a priori knowledge of the beam data for each beam in $\mathcal{B}$, the computation time and hard drive space required to store the beam data can render the consideration of a large number of candidate beams impractical. This issue is typically addressed by simply restricting the number of candidate beams in $\mathcal{B}$ to a manageable size (see Lee et al. [29]).

8.2.3 Objective Function

The objective function $\mathcal{F}(\boldsymbol{\theta})$ that measures the quality of beam vector $\boldsymbol{\theta}$ can be chosen to express a number of different criteria. For a given choice of $\boldsymbol{\theta}$, a patient will be treated using an optimal treatment plan obtained by solving the FMO problem—the problem of determining the optimal beamlets intensities given by $\boldsymbol{\theta}$. There are essentially two classes of BOO approaches: one that defines $\mathcal{F}$ to be the optimal solution value of the FMO problem and other that defines $\mathcal{F}$ as an approximation thereof based on intuition.

8.2.3.1 Fluence Map Optimization

In IMRT, each beam can be modeled as a collection of hundreds of small beamlets, the fluences of which can be controlled individually. These fluence values are known as a fluence map, and given a fixed set of beams, the optimization of these fluences is known as fluence map optimization.

Because the quality of a treatment plan can vary greatly depending on the fluence maps, the optimal solution value of the FMO problem is generally accepted as a quantitative measure of the quality of the treatment plan. Thus, the quality of a set of beams can be measured by the optimal solution of the FMO problem performed with these beams.

In the FMO problem, the patient's organs—both targets and critical structures—are irradiated using a predetermined set of beam angles, $\boldsymbol{\theta} \in \mathcal{B}^k$, where k is the number of beams in $\boldsymbol{\theta}$. Each beam is decomposed into a rectangular grid of beamlets with m rows and n columns, yielding typically 100–400 beamlets per beam. The position and intensity of all beamlets in a beam can be represented by a vector of values representing the beamlet intensities, called bixels. The set of all bixels in beam $\boldsymbol{\theta}$ is denoted by B_θ. The core task in IMRT treatment planning, and clearly the FMO problem in general, is finding radiation intensities for all beamlets.

Denote the total number of structures by S and say the targets are indexed as $s = 1, \ldots, T$ and the critical structures are $s = T + 1, \ldots, S$. Each structure s is discretized into a finite number v_s of volume cubes, known as voxels. Typically, around 350,000 voxels are required to accurately represent the targets and surrounding structures of a head-and-neck cancer site.

Because a beamlet must pass through a certain amount of tissue to reach a voxel, the dose received in a voxel from a beamlet may not be the full delivered intensity. Denote D_{ijs} as the dose received by voxel j in structure s from beamlet i at unit intensity. The D_{ijs} values are known as dose deposition coefficients, which are obtained from the radiological paths discussed in Section 8.2.2.1. Let x_i denote the intensity of bixel i. The dose z_{js} received by voxel j in structure s can be expressed by

$$z_{js} = \sum_{h=1}^{k} \sum_{i \in B_{\theta_h}} D_{ijs} x_i, \quad j = 1, \ldots, v_s, \quad s = 1, \ldots, S.$$

Say the objective function for the FMO problem given a vector of doses $\mathbf{z}$ is $F(\mathbf{z})$. A basic formulation of the FMO problem is then

$$\begin{aligned} \text{Min} \quad & F(\mathbf{z}) \\ \text{s.t.} \quad & z_{js} = \sum_{h=1}^{k} \sum_{i \in B_{\theta_h}} D_{ijs} x_i, \quad j = 1, \ldots, v_s, \quad s = 1, \ldots, S \\ & x_i \geq 0, \qquad\qquad i \in B_{\theta_h}, \quad h = 1, \ldots, k. \end{aligned}$$

In addition to the constraint that beamlet fluences be nonnegative, it is also possible to incorporate other requirements into the FMO Model. Other constraints employed in the literature span a myriad of criteria, including target dose homogeneity, tumor control probability (TCP) (Agren et al. [2], Lind et al. [34]), normal tissue complication probability (NTCP) (Grigorov et al. [20], Thomas et al. [53]), dose–volume histogram (DVH) constraints (Hou et al. [23], Romeijn et al. [42]), equivalent uniform dose (EUD) (Chapet et al. [8], Mavroidis et al. [36], Niemierko [37], Thomas et al. [53], Wu et al. [57]), and more. It is also possible to include many of these criteria in the FMO objective function.

As an alternative to the BOO Model given in Section 8.2, if the set of beam orientations $\mathcal{B}$ is finite, the BOO and FMO problems can be formulated together and solved simultaneously as a mixed-integer linear or nonlinear program (D'Souza et al. [56], Ehrgott and Johnston [14], Ferris et al. [16], Lee et al. [29,30], Lim et al. [33], Shepard et al. [47], Wang et al. [55]). The FMO formulation can be combined with BOO in the following model. Let y_θ be a binary variable indicating whether or not beam $\theta \in \mathcal{B}$ is used. If beam θ is used in the treatment plan then all the beamlets in θ, B_θ, are turned on, that is, they can have positive fluences up to some predetermined maximum intensity M. The simultaneous BOO + FMO MIP Model is then

$$\text{Min} \quad F(\mathbf{z})$$

$$\text{s.t.} \quad z_{js} = \sum_{h=1}^{k} \sum_{i \in B_{\theta_k}} D_{ijs} x_i, \quad j = 1, \ldots, v_s, \quad s = 1, \ldots, S$$

$$x_i \le M y_\theta, \quad i \in B_\theta, \ \theta \in \mathcal{B}$$

$$\sum_{\theta \in \mathcal{B}} y_\theta \le k$$

$$x_i \ge 0, \quad i \in B_\theta, \ \theta \in \mathcal{B}$$

$$y_\theta \in \{0, 1\}, \quad \theta \in \mathcal{B}.$$

8.2.3.2 Beam's-Eye-View Approaches

The concept of a beam's-eye-view (BEV) has been popular in BOO studies. A BEV is similar in concept to a bird's-eye-view, where the object being viewed is a patient as seen from a beam. The more of a target structure that is seen by the beam, the better candidate the beam is to be used in the treatment plan. Whether or not a beamlet intersects a voxel in a particular structure is contained within the beam data discussed in Section 8.2.2.1. By using beam solutions that have favorable BEVs, that is, that have the ability to irradiate a large amount of the target structures, a treatment plan will intuitively be capable of delivering a high amount of dose to the target structures while avoiding the critical structures. According to a study by Vijayakumar et al. [54], the advantages of BEV include, among other things, the availability of dose volume histogram profiles which may allow better definition of normal tissue tolerance and higher tumor control probability. Chen et al. [9], Cho et al. [10], Goitein et al. [18], Lu et al. [35], and Pugachev and Xing [39–41] have all used BEV or extensions thereof to select beam orientations.

8.2.3.2.1 Beam's-Eye-View

For each beamlet–voxel combination, the dose deposition coefficient D_{ijs} represents the amount of dose deposited in voxel j in structure s by beamlet i. In the BEV measure, it is only important to consider whether a beamlet in a beam intersects a voxel, not whether the dose deposition coefficient allows for a large amount of dose to be delivered. To this end, define r_{ijs} as

$$r_{ijs} = \begin{cases} 1 & D_{ijs} > \varepsilon, \\ 0 & \text{otherwise}, \end{cases} \tag{8.4}$$

where ϵ is a very small value. The beam's-eye-view score of beam θ, S_{θ}^{BEV}, is then simply the number of target voxels that can be reached by the beamlets in θ:

$$S_{\theta}^{\text{BEV}} = \sum_{s=1}^{T} \sum_{j=1}^{v_s} \min \left\{ 1, \sum_{i \in B_\theta} r_{ijs} \right\}.$$

The BOO objective function for the BEV approach is

$$\mathcal{F}(\boldsymbol{\theta}) = -\sum_{h=1}^{k} S_{\theta_h}^{\text{BEV}}.$$

8.2.3.2.2 Pseudo Beam's-Eye-View

Pugachev and Xing [39–41] developed a BEV variation called pseudo beam's-eye-view (pBEV) that assigns a score to each beam that accounts for the maximum intensity deliverable from the beam without exceeding the tolerances doses of the OARs. The score S_θ^{pBEV} associated with beam θ is given by

$$S_\theta^{\text{pBEV}} = \sum_{s=1}^{T} \frac{1}{v_s} \sum_{j=1}^{v_s} \left(\frac{d_{\theta js}}{P_s} \right)^2,$$

where

$d_{\theta js}$ is a measure of the total dose delivered to voxel j in structure s by beam θ

P_s is the prescribed radiation dose for the target s.

In Pugachev and Xing [39], the value $d_{\theta js}$ is calculated by $d_{\theta js} = d^0_{\theta js} \exp(-\mu \ell_{\theta js})$, where d_i^0 is the dose at zero depth delivered by the beamlet in θ that intersects with voxel j in structure s, $\ell_{\theta js}$ is the distance traveled by the beamlet before reaching the voxel and μ is the attenuation coefficient. The value $d_{\theta js}$ is calculated according to the following procedure (Pugachev and Xing [40]):

1. Determine the maximum beam intensity profile of beam θ as follows. For all beamlets $i \in B_\theta$:
   (a) Find the voxels crossed by beamlet i.
   (b) Assign x_i an intensity that could deliver a dose equal to or higher than the prescription dose in every target voxel crossed.
   (c) Calculate the minimum ratio by which x_i must be reduced to ensure that the tolerance dose for all critical structures and normal tissue crossed by beamlet i is not exceeded.
   (d) Reduce the value of x_i according to the minimum ratio. This value represents the maximum usable intensity of the beamlet.
2. Perform a forward dose calculation using the maximum beam intensity profile to obtain $d_{\theta js}$.

The BOO objective function for the pBEV approach is

$$\mathcal{F}(\boldsymbol{\theta}) = -\sum_{h=1}^{k} S_{\theta_h}^{\text{pBEV}}.$$

8.2.3.2.3 Target-Eye-View

Cho et al. [10] used an extension to BEV called target-eye-view (TEV), wherein the overlap of targets and critical structures in the BEV is accounted for. A nonoverlapping area is called a miss. Any overlap is called a hit; more specifically, a distal hit identifies a critical structure that lies behind a target in the BEV, whereas a proximal hit identifies a critical structure in front of the target. Marginal hits occur when the distance between a critical structure and a target is less than the beam's penumbra, the region at the edge of a

Table 8.1 Collision and Criticality Scores Implemented by Cho et al. [10] in Calculating Target-Eye-View (TEV)

Type of Overlap	Collision Score
Proximal hit	1.0
Marginal hit	0.8
Distal hit	0.7
Miss	0.0
Critical Structure	*Criticality Score*
Chiasm	4
Optic nerve	4
Optic global	2
Cavernous sinus	1
Brainstem	1

Source: From Cho, B.C.J., Roa, H.W., Robinson, D., and Murray, B., *Int. J. Radiat. Oncol. Biol. Phys.*, 23, 153, 1992.

radiation beam over which the dose rate changes rapidly as a function of distance from the beam axis (*The Physics of Radiation Therapy* [27]).

For every critical structure, the overlaps with every candidate beam are tabulated in a matrix. The TEV score for a beam θ, S_θ^{TEV}, is determined from these matrices. Each critical structure s is arbitrarily assigned a relative criticality score, c_s, weighting the importance of sparing one critical structure over another. Additionally, each type of overlap g is given a collision score, g_m. Table 8.1 displays the collision and criticality scores given by Cho et al. [10].

The TEV score is calculated by combining the scores of every critical structure for a particular beam orientation. Let G be the number of types of overlaps and let $y_{ijs} = \{1, \ldots, G\}$ be the type of hit occurring from the intersection of beamlet i and voxel j in structure s. Note that G must include a type of hit representing no hit, and the collision score for the no-hit type must be zero.

$$S_\theta^{\mathrm{TEV}} = \sum_{i \in B_\theta} \sum_{s=T+1}^{S} \sum_{j=1}^{v_S} \sum_{m=1}^{G} y_{ijs} c_s g_m.$$

From the definition, it is clear that a high TEV score may suggest a significant amount of overlap between a beam and the critical structures, thus decreasing the potential benefit of that beam in a treatment plan.

The BOO objective function for the TEV approach is

$$\mathcal{F}(\boldsymbol{\theta}) = \sum_{h=1}^{k} S_{\theta_h}^{\text{TEV}}.$$

8.2.3.3 Geometric Considerations

In addition to each beam's relation to the targets and critical structures, the relationship of each beam with the other beams in the plan has also been considered in the selection of beam orientations. One common geometric consideration is that the beams should be spaced far apart from each other.

A large separation between the angles is intuitively desirable because the motivation for delivering the radiation from several beam orientations is that the targets, at the intersection of these radiation beams, will receive a very high dose, whereas surrounding tissue will receive radiation from some, but not all the beams. It is hoped that by receiving substantially less radiation than the targets, the critical structures will receive a low enough amount of dose that they will continue function after the treatment. If two or more beams in a treatment plan are very close together, it is possible that the cumulative dose received by an organ in their path will exceed that organ's dose tolerance, and the organ will fail. Thus, the concept of maximum angle separation has been explored in the BOO literature to force the beams of a treatment plan to be sufficiently far apart (Das et al. [11], Das and Marks [12], Rowbottom et al. [44]).

A simple method of accounting for angle separation in the BOO Model is to add a constraint requiring that a user-specified minimum angle distance (δ) be observed (Das et al. [11], Rowbottom et al. [44]).

Alternatively, the angle separation can be maximized as part of the optimization procedure (Das and Marks [12]). In this scenario, Das and Marks [12] mathematically represented the angle separation between beams in a beam vector, $S_{\boldsymbol{\theta}}^{\text{sep}}$ by the geometric mean of the sines of the angles between beams:

$$S_{\boldsymbol{\theta}}^{\text{sep}} = \left(\prod_{j=1, j\neq i}^{k} \prod_{i=1}^{k} \sin(\alpha_{\theta_i,\theta_j}) \right)^{\frac{1}{k(k-1)}},$$

where $\alpha_{\theta_i,\theta_j}$ is the angle between beams θ_i and θ_j. Note that this formulation accounts for both entrance and exit beams, and therefore, no beams may have more than 90° separation; in other words, $\alpha_{\theta_i,\theta_j} \leq 90°$. By using the geometric mean rather than the arithmetic mean, if any two beams are equal, then $S_{\boldsymbol{\theta}}^{\text{sep}} = 0$. With this scoring method, it is desirable to have as large a value of $S_{\boldsymbol{\theta}}^{\text{sep}}$ as possible, thus achieving maximum angle separation.

Because it is desirable to maximize the angle separation and the basic BOO Model is formulated as a minimization, the BOO objective function for this approach is

$$\mathcal{F}(\boldsymbol{\theta}) = -\sum_{h=1}^{k} S_{\theta_h}^{\text{sep}}.$$

Other forms of geometric considerations have also been considered. Haas et al. [21] formulate the BOO problem in such a way that the objective function reflects the need for the edges of the combined beams to conform to the PTV. This value is measured as the intersection of area hit by the beams minus the area of the PTV. Define $r'_{\boldsymbol{\theta}js}$ as an indicator as to whether a voxel j in structure s is hit by any beamlet in any beam in $\boldsymbol{\theta}$:

$$r'_{\boldsymbol{\theta}js} = \min\left\{1, \sum_{h=1}^{k} \sum_{i \in B_{\theta_h}} r_{ijs}\right\}.$$

The value $S_{\boldsymbol{\theta}}^{\text{area}}$ representing the area intersected by all the beams in $\boldsymbol{\theta}$ less the area of the PTV, which is denoted by structure index s', is

$$S_{\boldsymbol{\theta}}^{\text{area}} = \sum_{s=1}^{S} \sum_{j=1}^{v_s} r'_{\boldsymbol{\theta}js} - v_{s'}.$$

With this measure, it is desirable to have smaller values, indicating that more of the PTV and less of the surrounding tissue is intersected.

The BOO objective function for this approach is

$$\mathcal{F}(\boldsymbol{\theta}) = \sum_{h=1}^{k} S_{\theta_h}^{\text{area}}.$$

8.2.3.4 *Other Scoring Criteria*

A wide variety of scoring criteria beyond just the previously described BEV and geometric considerations have been considered. Several of these alternative scoring methods are discussed below.

8.2.3.4.1 Mean Organ-at-Risk Data

D'Souza et al. [56] incorporated a scoring mechanism named mean organ-at-risk data (MOD) into their integer programming BOO formulation. To obtain the MOD score value for a beam, an initial treatment plan is delivered from the single unmodulated beam, which is shaped to the BEV of the PTV. In this treatment plan, the prescribed dose to the PTV is 1.8 Gy. After the beamlet intensities **x** for this treatment plan are obtained, the plan is normalized so that the mean dose in the PTV is 2.0 Gy, yielding a vector of beamlet intensities $\hat{\mathbf{x}}$. For a beam θ, the MOD score S_θ^{MOD} is the mean critical structure dose obtained from the normalized single-beam treatment plan:

$$S_\theta^{\text{MOD}} = \sum_{i \in B_\theta} \frac{1}{\sum_{s=T+1}^{S} v_s} \sum_{s=T+1}^{S} \sum_{j=1}^{v_s} D_{ijs} \hat{x}_i.$$

Clearly, a lower score indicates a beam with potentially low impact on critical structures.

The BOO objective function for the MOD approach is

$$\mathcal{F}(\boldsymbol{\theta}) = \sum_{b=1}^{k} S_{\theta_b}^{\text{MOD}}.$$

8.2.3.4.2 Entropy

Söderstrom et al. [50] used the concepts of entropy and Fourier transforms to gauge the effective of a beam by analyzing the optimal fluence maps to determine the amount of beam structure and information.

From statistical mechanics, a high entropy value corresponds to a high amount of disorder or a near-equilibrium state. Conversly, a low entropy corresponds to a low amount of disorder, in other words, a high amount of structure. To determine which beams are the most influential in creating a dose distribution, the beams with the highest structure (lowest entropy) are sought. Entropy for a beam θ is measured as

$$e_\theta = -\sum_{i \in B_\theta} x_i \log x_i.$$

To make comparisons between beams easier so that a high value corresponds to good beam directions, $S_\theta^{\text{entropy}}$ is used to score the beams:

$$S_\theta^{\text{entropy}} = 1 + \frac{\min_{\gamma \in \mathcal{B}} e_\gamma - e_\theta}{\max_{\gamma \in \mathcal{B}} e_\gamma}.$$

Noting that it is desirable to have larger entropy values, the BOO objective function for the entropy approach is

$$\mathcal{F}(\boldsymbol{\theta}) = -\sum_{b=1}^{k} S_{\theta_b}^{\text{entropy}}.$$

8.2.3.4.3 Fourier Transforms

In their scoring method based on Fourier transforms, Söderstrom et al. [50] examined the frequency contain of the beams' fluence maps to locate beams with large amounts of gross structure, which indicate beams in which a near-uniform dose may be delivered. To obtain a measure of the gross structure of a beam, the absolute value of the Fourier transform of the beam's fluence map is taken ($\bar{F}_\theta$). Because the low frequencies represent gross structure in the frequency space, the low frequency part of the Fourier spectra is integrated to measure the gross structure of a beam, $S_\theta^{\text{Fourier}}$:

$$S_\theta^{\text{Fourier}} = \int_0^{u_{\max}} \left|\bar{F}_\theta(u)\right| du,$$

where $u_{\max}$ is a frequency corresponding to the smallest structure of interest. Söderstrom et al. [50] choose the value for $u_{\max}$ as half the width of a collimator leaf or half the width of a voxel.

Because beams with large amount of gross structure are sought, the BOO objective function for the Fourier transform approach is:

$$\mathcal{F}(\boldsymbol{\theta}) = -\sum_{h=1}^{k} S_{\theta_h}^{\text{Fourier}}.$$

8.2.3.4.4 Path of Least Resistance

Gokhale et al. [19] proposed a scoring technique that evaluates beam orientations according to the path of least resistance to radiation from the tumor site to the patient surface. To determine the path of least resistance for all beam angles, a fictitious radiation source is placed at the target structure and the amount of radiation dose received at the surface of the patient is calculated using radiation transport calculations. The higher the dose received at the surface, the better the corresponding beam orientation. The path of least resistance can also be obtained from the radiological paths calculated while generating beam data as discussed in Section 8.2.2.1.

Letting S_θ^{path} denote the path length for a beam θ, the BOO objective function for the path of least resistance approach is

$$\mathcal{F}(\boldsymbol{\theta}) = \sum_{h=1}^{k} S_{\theta_h}^{\text{path}}.$$

8.2.3.4.5 Single- and Multi-Beam cost Functions

Oldham et al. [38] and Rowbottom et al. [44,45] used a single-beam cost function to assess the value of a single beam, and then combined the single-beam costs of the beams in a beam vector to obtain a multi-beam cost for that beam solution. Due to the complexity of the cost functions, the reader is referred to the original source for detailed information on the scoring mechanism.

Define S_θ^{single} to be the value of the single-beam cost function for a single beam θ. The BOO objective function for the single-beam cost function approach is

$$\mathcal{F}(\boldsymbol{\theta}) = \sum_{h=1}^{k} S_{\theta_h}^{\text{single}}.$$

Define $S_{\boldsymbol{\theta}}^{\text{multi}}$ to be the value of the multi-beam cost function for a beam vector $\boldsymbol{\theta}$. The BOO objective function for the multi-beam cost function approach is

$$\mathcal{F}(\boldsymbol{\theta}) = S_{\boldsymbol{\theta}}^{\text{multi}}.$$

8.3 Optimization Methods

Although the FMO Model can be formulated in such a way that its objective function is quadratic (Aleman et al. [3–5]) or even linear (Romeijn et al. [42,43]) with linear constraints, the FMO problem is fundamentally nonlinear with respect to the beam vector used as input. This inherent nonlinearity arises from the fact that the physics of dose deposition change with direction, that is, as a particular beam changes its angle, the effects of the dose deposited in the patient can change drastically. For example, if a beam solution contains a beam that passes near the spinal cord, the spinal cord could receive a sufficiently low amount of dose such that it survives the treatment. However, if that one beam is moved only a few degrees so that it passes through, or even just too close, to the spinal cord, the spinal cord could effectively be severed due to an excessive amount of radiation received. Thus, two solution vectors very close to one another would yield very different objective function values.

This nonlinearity is largely responsible for the emergence of so many different scoring criteria for a beam solution. As with the numerous methods of describing the quality of a beam solution, there have been numerous approaches to solve the many BOO formulations that have been presented.

8.3.1 *Solving the BOO + FMO Integer Program*

Solving the BOO+FMO mixed integer program described in Section 8.2.3.1 poses a particularly difficult problem due to the complex model formulation. The use of specialized cuts and column generation within branch-and-bound schemes has been employed, as has the use of commercial solvers.

8.3.1.1 Branch-and-Bound Techniques

Branch-and-bound is a common method of solving programming problems involving integer variables. The branch-and-bound algorithm recursively partitions the solution space into a relaxation of the original problem. More specifically, for an integer program, the relaxation of the original problem is the linearization of the binary variables. For the BOO+FMO Model presented in Section 8.2.3.1, this relaxation allows y_θ, $\theta \in \mathcal{B}$, to be a real-valued number in the interval [0,1]. This results in the following linear program (LP) relaxation of the original MIP:

$$\text{Min} \quad F(\mathbf{z})$$

$$\text{s.t.} \quad z_{js} = \sum_{h=1}^{k} \sum_{i \in B_{\theta_k}} D_{ijs} x_i, \quad j = 1, \ldots, v_s, \; s = 1, \ldots, S$$

$$x_i \le M y_\theta, \quad i \in B_\theta, \; \theta \in \mathcal{B}$$

$$\sum_{\theta \in \mathcal{B}} y_\theta \le k$$

$$x_i \ge 0, \quad i \in B_\theta, \; \theta \in \mathcal{B}$$

$$0 \le y_\theta \le 1, \quad \theta \in \mathcal{B}.$$

Because the above relaxation contains only linear constraints, it can be easily solved. If the solution to the relaxed problem, $(\hat{\mathbf{x}}, \hat{\mathbf{y}})$, where $\hat{\mathbf{x}}$ corresponds to the beamlet intensities of all beamlets in all beams in $\mathcal{B}$ and $\hat{\mathbf{y}}$ indicates whether or not each beam in $\mathcal{B}$ is used, contains an integer-valued $\hat{\mathbf{y}}$ then $\hat{\mathbf{y}}$ solves the original problem. Otherwise, $\hat{\mathbf{y}}$ contains at least one fractional value, and the branching process occurs. For a fractional variable $\hat{y}_{\theta'}$, two new problems are created by fixing the value of $\hat{y}_{\theta'}$ to 0 in one problem and to 1 in another problem. Each of these problems is solved, and if there are any fractional values in the solutions, the branching process is repeated. The process continues until an integer solution is found. As the algorithm proceeds, branches can be eliminated from further exploration if the relaxed solution does not improve on integer solutions obtained from another branch. This is the bound step of the branch-and-cut algorithm.

Lee et al. [29] use a branch-and-bound algorithm to solve the BOO+FMO mixed integer problem. In the branching process, Lee et al. [29] elect to branch not on individual variables, but on sets of variables.

Let $\hat{\mathbf{y}}$ be the solution to an LP relaxation at some node in the branch-and-bound tree. The branching scheme partitions $\mathcal{B}$ into $\mathcal{B}_1 \cup \mathcal{B}_2$ such that $\sum_{\theta\in\mathcal{B}_1} \hat{y}_\theta \approx \sum_{\theta\in\mathcal{B}_2} \hat{y}_\theta$. The sets $\mathcal{B}_1$ and $\mathcal{B}_2$ are also chosen so that the beams in each set are roughly in the neighborhood of each other. From these two sets, the LP problem is split into two new nodes, each containing one of the following constraints:

$$\sum_{\theta\in\mathcal{B}_1} y_\theta \leq \left\lfloor \frac{k}{2} \right\rfloor \quad \sum_{\theta\in\mathcal{B}_2} y_\theta \leq \left\lfloor \frac{k}{2} \right\rfloor.$$

At each node of the branch-and-bound tree, an LP relaxation of the original MIP must be solved. To keep this LP relaxation tractable, Lee et al. [29] imposed a column generation technique, which generates an initial master problem containing only about half of the original voxels. The voxels included in the master problem are selected so as to maintain a realistic representation of the original problem. Additional constraints and corresponding voxels (columns) are added to the problem as the column generation method proceeds. Voxels that are not incorporated after a certain number of iterations, are removed from the master problem to maintain a reasonable size for the master problem.

Cutting plane techniques may also be applied to the original MIP to remove fractional values from the solution space (Lee et al. [29]).

8.3.1.2 Commercial Solvers

CPLEX (ILOG, Inc., California) is a commercial software package, which can solve many types of optimization problems efficiently. In the BOO literature, CPLEX has been commonly used to solve the BOO+FMO mixed integer programming formulation (D'Souza et al. [56], Ehrgott and Johnston [14], Lee et al. [29], Wang et al. [55]).

Other commercial software systems have also been used. Ferris et al. [16] used GAMS in addition to CPLEX, and Shepard et al. [47] use MATLAB in conjunction with GAMS. Through GAMS, several commercial solvers may be accessed, including OSL, CPLEX, CONOPT, and MINOS.

8.3.2 Metaheuristics

Owing to the nonlinearity of the BOO problem when formulated based upon the optimal solution to the FMO problem, obtaining a globally optimal solution can be problematic. Many global optimization techniques have been applied to the BOO problem to search the solution without becoming trapped in local optima. The most common of these algorithms, simulated annealing and evolutionary/genetic algorithms, is discussed. In addition, the response surface method presented by Aleman et al. [4,5]

is also discussed because of its ability to include non-coplanar beam solutions although circumventing storage issues associated with a large $\mathcal{B}$.

8.3.2.1 Simulated Annealing

The simulated annealing algorithm (Kirkpatrick et al. [28]) is a stochastic neighborhood search method which allows for the escape from local optima. The simulated annealing algorithm is based on the Metropolis algorithm, wherein a neighboring solution to the current iterate is generated according to some probability distribution, and if it is an improving point, it becomes the current iterate. Otherwise, it becomes the current iterate with probability $\exp\{\Delta\mathcal{F}/T\}$ (Boltzmann probability), where $\Delta\mathcal{F}$ is the difference in objective function value between the current iterate and the newly generated point and T is the *temperature*, a measure of the randomness of the algorithm. If $T = 0$ then only improving points are selected. If T is very large then any move is accepted, which is essentially a random search. By allowing the algorithm to move to points that do not improve upon the best solution, the algorithm can avoid becoming trapped in a local optimum.

The simulated annealing algorithm starts with an initial temperature T_0 and performs a number of iterations of the Metropolis algorithm using $T = T_0$. Then, the temperature is decreased according to some cooling schedule such that $\{T_i\} \rightarrow 0$.

In selecting a neighboring point in the simulated annealing algorithm, one or more beams in the current beam vector may be changed. Let $\mathcal{N}_b(\boldsymbol{\theta})$ be the neighborhood of a single beam at index b in $\boldsymbol{\theta}$, defined for some neighborhood size δ as

$$\mathcal{N}_b(\boldsymbol{\theta}) = \{(\theta_1, \ldots, \theta_{b-1}, \theta \mod 360, \theta_{b+1}, \ldots, \theta_k) \in \mathcal{B}^k : \theta \in [\theta_b - \delta, \theta_b + \delta]\}.$$

Say there is a set of indices $H \subseteq \{1, \ldots, k\}$ to be changed to obtain a neighboring beam. A neighbor for beam $\boldsymbol{\theta}$ then lies in the set $\cup_{b\in H}\mathcal{N}_b(\boldsymbol{\theta})$.

The simulated annealing algorithm is as follows:

- *Initialization*
  - Choose an initial beam set $\boldsymbol{\theta}^{(0)}$ and calculate $F_0 = \mathcal{F}(\boldsymbol{\theta}^{(0)})$.
  - Set $\hat{\boldsymbol{\theta}} = \boldsymbol{\theta}^{(0)}$, $\hat{F} = F_0$, $i = 0$.
- *Iteration*
  1. Select $H \subseteq \{1, \ldots, k\}$, generate $\boldsymbol{\theta} \in \cup_{b\in H}\mathcal{N}_b(\boldsymbol{\theta}^{(i)})$, and calculate $F = \mathcal{F}(\boldsymbol{\theta})$.
  2. If $F < \hat{F}$, set $\hat{F} = F$, $F_{i+1} = F$, $\boldsymbol{\theta}^{(i+1)} = \boldsymbol{\theta}$ and $\hat{\boldsymbol{\theta}} = \boldsymbol{\theta}$. Otherwise, set $F_{i+1} = F$ and $\boldsymbol{\theta}^{(i+1)} = \boldsymbol{\theta}$ with probability $\exp\{(F_i - F)/T_i\}$.
  3. If stopping criteria are met, stop with solution $\hat{\boldsymbol{\theta}}$ and objective function value $\hat{F}$. Otherwise, set $i \leftarrow i + 1$ and repeat step 1.

Because of its ability to escape local optima, simulated annealing has been a popular method of solving the BOO problem. Bortfeld and Schlegel [7] used the fast simulated annealing algorithm described by Szu and Hartley [6], which employs a Cauchy distribution in generating neighboring points. Stein et al. [52], Rowbottom [44], and Djajaputra et al. [13] also used a Cauchy distribution in generating neighboring solutions. Lu et al. [35] randomly selected new points satisfying BEV and conventional wisdom criteria, and Pugachev and Xing [41] randomly generate new points and then vary them according to an exponential distribution. Each of the aforementioned implemenations of the simulated annealing algorithm accept improving solutions, and with the exception of Rowbottom et al. [44], who accepted only improving solutions (essentially $T_i = 0$ for all i), all accept nonimproving solutions with a Boltzmann probability.

Aleman et al. [3] used a geometric distribution in generating neighboring solutions, as well as employed a unique neighborhood structure based on the parallel-opposed beams of the current iterate. This neighborhood is called a flip neighborhood, and is defined for beam θ_b as the union of the neighborhoods of θ_b and its parallel-opposed beam (180° away). The parallel-opposed beam can be described by $\theta_b' = (\theta_b + 180) \bmod 360$, and the size of its neighborhood, δ^F, may be different than that of the neighborhood around θ_b, δ. Specifically, the flip neighborhood can be defined as

$$\mathcal{N}_b^F(\boldsymbol{\theta}) = \{(\theta_1, \ldots, \theta_{b-1}, \theta \text{ mod } 360, \theta_{b+1}, \ldots, \theta_k)$$
$$\in \mathcal{B}^k : \theta \in [\theta_b - \delta, \theta_b + \delta] \cup [\theta_b' - \delta^F, \theta_b' + \delta^F]\}.$$

Aleman et al. [3] showed that the flip neighborhood can lead to improved convergence times in the simulated annealing algorithm.

8.3.2.2 Evolutionary and Genetic Algorithms

Evolutionary/genetic algorithms have also been popular in the BOO literature. Ezzell [15], Haas et al. [21], Li et al. [31,32], and Schreibmann [46] have all used evolutionary/genetic algorithms or variants thereof to solve the BOO problem. These algorithms are popular due to their ability to search for solutions in multiple areas of the solution space simultaneously.

8.3.2.2.1 Generic Evolutionary/Genetic Algorithms

Ezzell [15], Haas et al. [21], Li et al. [31], and Schreibmann [46] base the optimization of the BOO problem on a genetic algorithm. The concept behind genetic algorithms is evolution guided by natural selection. Each solution, called an individual, is a member of a population, which is a collection of individuals. The fitness of each individual is based on the objective function value it provides. In the genetic algorithm, individuals are allowed to combine and reproduce new individuals through reproduction and mutation mechanisms akin to those occurring in nature. Through these mechanisms, current solutions are combined and perturbed to obtain new solutions, providing the means of searching through the solution space. To promote the survival of individuals with a high level of fitness, that is, solutions with good objective function values, the fertility of an individual depends on its fitness.

In the genetic algorithm implemented by Ezzell [15], an initial population is first chosen. Each individual in the population corresponds to a beam solution, and the beams in the solution are selected randomly from $\mathcal{B}$. Additionally, each beam in the solution is assigned a random weighting value, which is then normalized so that the sum of the weights in a particular beam solution is one.

In the reproduction scheme, crossover, mutation, and cloning mechanisms are considered. Firstly, when two individuals (parents) combine, crossover involves selecting which elements of the child (the new individual) will be taken from which of the parent individuals. If there is no crossover, which may occur according to some probability measure, then the child becomes a duplicate of the parent with superior fitness. Ezzell [15] used a single-point crossover, meaning that one segment of the child comes from one parent and the rest comes from the other parent. Secondly, mutation is the mechanism by which new genetic information is introduced into the population. Mutation allows beams that do not exist in any individual in the population to be explored in the solution space. Mutations are implemented by incorporating a small probability that each beam in a solution may be increased or decreased (mutation rate). Lastly, when each new generation is created, the best individual in the population is cloned so that it may continue to influence the search of the solution space. A mutation may be applied to this individual; if so, then another unaltered duplicate of the individual is also added to the population.

Additionally, operations for spontaneous generation and deletion are included in method presented by Ezzell [15]. Spontaneous generation is the insertion of completely new individuals into the population, although the uniform deletion policy employed randomly selects individuals to be removed from the population.

The methods presented by Haas et al. [21] and Schreibmann [46] differ from the above method in that instead of optimizing a particular single objective function, a multi-objective formulation of the BOO problem is considered. Using the concept of Pareto optimality, nondominated solutions are given higher fitness levels so that solutions near the Pareto efficient frontier will prosper.

8.3.2.2.2 Particle Swarm Algorithm

Li et al. [32] used a particle swarm algorithm, a variant of the evolutionary algorithm, to solve the BOO problem. Similarly to evolutionary algorithms, in the particle swarm algorithm, a population of solutions is used to explore the solution space. Rather than refer to each solution in the population as an individual, solutions in a population are called particles. Each particle moves through the solution space according to its velocity, which consists of both speed and direction. The velocity of a particle is determined by its own flying experience and the flying experience of its neighbors, with the goal of improving its position. As each particle retains a memory of its best position ever visited, the movement of each particle is an aggregated acceleration toward its best previously visited position and toward the best individual of a neighborhood.

Li et al. [32] applied the particle swarm optimization method using a global neighborhood and also using a local neighborhood.

8.3.2.3 Response Surface Method

Aleman et al. [4,5] used a response surface based approach to the BOO problem to allow for the consideration of non-coplanar beam orientations. The response surface method, first described by Jones [25] and Jones et al. [26], identifies promising solutions based on the performance of previous solutions. The function value and expected improvement over the current best solution of a certain point are estimated based on the function behavior learned from previously sampled points and their calculated objective function values. The function values of points are related by correlation functions that depend on each point's distance from the previously sampled points. From the correlation functions, the algorithm predicts the probability that the best solution will improve at unexplored points in the solution space. Using this probability, a promising solution is identified. For the BOO problem, beam data only needs to be generated for these promising solutions, thus saving both computation time and storage space.

8.3.3 Local Search Algorithms

Although it would seem that local search heuristics would be commonly employed to obtain locally optimal solutions to the BOO problem, in fact, very few studies using local search algorithms have been published. Aleman et al. [3] developed a deterministic neighborhood search algorithm called Add/Drop to identify locally optimal beam solutions.

The Add/Drop algorithm is essentially a deterministic version of the simulated annealing algorithm. As with their implementation of the simulated annealing algorithm, Aleman et al. [3] employed a unique neighborhood structure based on parallel-opposed beam positions. In the Add/Drop algorithm, each beam in an initial starting solution is replaced with the most improving beam in its neighborhood according to the objective function. Once each beam has been examined, the process repeats. The algorithm continues until there can be no further improvement in any beam in the beam solution. The resulting solution is locally optimal.

Using the same neighborhood definition as the simulated annealing algorithm in Section 8.3.2.1, the Add/Drop algorithm is as follows:

- *Initialization*
  - Choose an initial starting solution $\boldsymbol{\theta}^{(0)}$.
  - Set $\boldsymbol{\theta}^* = \boldsymbol{\theta}^{(0)}$ and $i = 0$.
- *Iteration*
  1. Select $h \in \{1, \ldots, k\}$, then generate $\bar{\boldsymbol{\theta}} \in \mathcal{N}_h(\boldsymbol{\theta}^{(i)})$.
  2. If $F(\bar{\boldsymbol{\theta}}) < F(\boldsymbol{\theta}^*)$, set $\boldsymbol{\theta}^* = \boldsymbol{\theta}^{(i+1)} = \bar{\boldsymbol{\theta}}$ and set $i \leftarrow i + 1$.
  3. If all points in $\cup_{h=1}^{k} \mathcal{N}_h(\boldsymbol{\theta}^{(i)})$ have been sampled without improvement, stop with $\boldsymbol{\theta}^*$ as a local minimum. Otherwise, repeat step 1.

Aleman et al. [3] also employed the flip neighborhood structure $\mathcal{N}_h^F(\boldsymbol{\theta})$, described in Section 8.3.2.1, in the Add/Drop algorithm and achieved results indicating improved convergence times over the regular neighborhood definition $\mathcal{N}_h(\boldsymbol{\theta})$.

8.3.4 Greedy Algorithms

For many of the scoring methods to estimate the quality of a beam solution described in Section 8.2.3, greedy algorithms are a fast and effective method of selecting beam solutions. A greedy algorithm is an algorithm that follows the problem-solving metaheuristic of making a locally optimal choice at each opportunity with the hope of finding the global optimum. Although greedy algorithms are not guaranteed to locate globally optimal solutions, they can still provide good solutions.

The scoring methods, which assign value to individual beams rather than to an entire beam solution, particularly benefit from greedy algorithms, as the optimal solution can be found easily with a greedy algorithm. Consider a scoring method that assigns a value S_θ to each $\theta \in \mathcal{B}$. The objective function in the BOO Model presented in Section 8.2 would be

$$\mathcal{F}(\boldsymbol{\theta}) = -\sum_{h=1}^{k} S_{\theta_h},$$

where the negative sum of scores is taken assuming the scoring method assigns high scores to quality beams, whereas the BOO Model is formulated as a minimization problem.

Given a fixed k, it is clear that the optimal solution to such an objective function can be found by simply selecting the k highest-scoring (or lowest-scoring if a low score is desirable) beams, resulting in the smallest obtainable value of $\mathcal{F}(\boldsymbol{\theta})$. The scoring approaches whose globally optimal solutions can be found in this manner are S_θ^{BEV}, S_θ^{pBEV}, S_θ^{TEV}, S_θ^{MOD}, $S_\theta^{\text{entropy}}$, and $S_\theta^{\text{Fourier}}$.

8.3.5 Other Optimization Approaches

In addition to the previously described optimization techniques, there are still other methods that have been employed in obtaining solutions to the BOO problem. Acosta et al. [1] and Holder and Salter [22] used a method based on the image compression technique called vector quantization. Rowbottom et al. [45] used artificial neural networks, and Das and Marks [12] used a quasi-Newton method to solve the BOO problem.

8.4 Conclusions

Just as there are numerous approaches to formulating and solving the FMO problem, there are many varied techniques applied to the BOO problem. Each of these methods has its own benefits and detriments.

The global optimization methods presented have the benefit of obtaining globally optimal solutions, but depending on the problem formulation, particularly in the case of a MIP formulation, may require extensive computing to reach the optimal solution. The metaheuristic approaches also have the ability to obtain globally optimal solutions, but again, at the expense of computation time. The local search and greedy algorithms, however, can obtain a locally optimal solution quickly, but the solution will not necessarily be globally optimal. For some of the scoring methods, the greedy algorithm can obtain the globally optimal solution, however, the quality of those objective functions compared to the other objective function options is unknown.

The use of scoring methods can reduce the complexity of the BOO formulation and save computation time, although it is widely believed that the optimal FMO solution is the most suitable objective function and thus provides the most accurate measure of a beam solution's quality. However, the optimal FMO solution can be difficult to obtain depending on the complexity of the FMO Model. Even with an FMO Model that can be easily solved, the time required to obtain the solution will likely still be more than the time required to calculate the scoring methods discussed earlier.

It is unclear whether any scoring methods can achieve the same quality of beam solution as the optimal FMO solution for all patients. Likewise, it is unknown whether the optimal FMO solution is required for a BOO Model to obtain a treatment plan that meets minimum clinical requirements for all patients.

References

1. R. Acosta, M. Ehrgott, A. Holder, D. Nevin, J. Reese, and B. Salter. Comparing beam selection strategies in radiotherapy treatment design: The influence of dose point resolution. Submitted for publication, 2005.
2. A. Agren, A. Brahme, and I. Turesson. Optimization of uncomplicated control for head and neck tumors. *Int. J. Radiat. Oncol. Biol. Phys.*, 19:1077–1085, 1990.
3. D.M. Aleman, A. Kumar, R.K. Ahuja, H.E. Romeijn, and J.F. Dempsey. Neighborhood search approaches to beam orientation optimization in intensity modulated radiation therapy treatment planning. Submitted for publication, 2007.
4. D.M. Aleman, H.E. Romeijn, and J.F. Dempsey. Beam orientation optimization methods in intensity modulated radiation therapy treatment planning. In: Proceedings of the IE Research Conference, Orlando, Florida, 2006.
5. D.M. Aleman, H.E. Romeijn, and J.F. Dempsey. A response surface approach to beam orientation optimization in intensity modulated radiation therapy treatment planning. Submitted for publication, 2006.
6. I. Bomze. Fast simulated annealing. *Phys. Lett.*, 122A:157–162, 1987.
7. T. Bortfeld and W. Schlegel. Optimization of beam orientations in radiation therapy: Some theoretical considerations. *Phys. Med. Biol.*, 38:291–304, 1993.
8. O. Chapet, E. Thomas, M.L. Kessler, B.A. Fraass, and R.K. Ten Haken. Esophagus sparing with imrt in lung tumor irradiation: An EUD-based optimization technique. *Int. J. Radiat. Oncol. Biol. Phys.*, 63(1):179–187, 2005.
9. G.T. Chen, D.R. Spelbring, C.A. Pelizzari, J.M. Balter, L.C. Myrianthopoulos, S. Vijayakumar, and H. Halpern. The use of beam's eye view volumetrics in the selection of non-coplanar radiation portals. *Int. J. Radiat. Oncol. Biol. Phys.*, 23:153–163, 1992.
10. B.C.J. Cho, W.H. Roa, D. Robinson, and B. Murray. The development of target-eye-view maps for selection of coplanar or non-coplanar beams in conformal radiotherapy treatment planning. *Med. Phys.*, 26:2367–2372, 1999.
11. S. Das, T. Cullip, G. Tracton, S. Chang, L. Marks, M. Anscher, and J. Rosenman. Beam orientation selection for intensity-modulated radiation therapy based on target equivalent uniform dose maximization. *Int. J. Radiat. Oncol. Biol. Phys.*, 55:215–224, 2003.
12. S.K. Das and L.B. Marks. Selection of coplanar or non-coplanar beams using three-dimensional optimization based on maximum beam separation and minimized nontarget irradiation. *Int. J. Radiat. Oncol. Biol. Phys.*, 38:643–655, 1997.
13. D. Djajaputra, Q. Wu, Y. Wu, and R. Mohan. Algorithm and performance of a clinical IMRT beam-angle optimization system. *Phys. Med. Biol.*, 48:3191–3212, 2003.
14. M. Ehrgott and R. Johnston. Optimisation of beam directions in intensity modulated radiation therapy planning. *OR Spectr.*, 25:251–264, 2003.

15. G.A. Ezzell. Genetic and geometric optimization of three-dimensional radiation therapy treatment planning. *Med. Phys.*, 23:293–305, 1996.
16. M.C. Ferris, R.R. Meyer, and W. D'Souza. Radiation treatment planning: Mixed integer programming formulations and approaches. In G. Appa, L. Pitsoulis, and H.P. Williams, editors, *Handbook on Modelling for Discrete Optimization*, Springer-Verlag, New York, 2006, pp. 317–340.
17. C. Fox, H.E. Romeijn, and J.F. Dempsey. Fast voxel and polygon ray-tracing algorithms for IMRT treatment planning. *Med. Phys.*, 33:1364–1371, 2006.
18. M. Goitein, M. Abrams, D. Rowell, H. Pollari, and J. Wiles. Multi-dimensional treatment planning: II. Beam's eye-view, back projection, and projection through CT sections. *Int. J. Radiat. Oncol. Biol. Phys.*, 9:789–797, 1983.
19. P. Gokhale, E.M. Hussein, and N. Kulkarni. The use of beam's eye view volumetrics in the selection of non-coplanar radiation portals. *Med. Phys.*, 23:153–163, 1994.
20. G.N. Grigorov, J.C. Chow, L. Grigorov, R. Jiang, and R.B. Barnett. IMRT: Improvement in treatment planning efficiency using NTCP calculation independent of the dose–volume-histogram. *Med. Phys.*, 33(5):1250–1258, 2006.
21. O.C. Haas, K.J. Burnham, and J. Mills. Optimization of beam orientation in radiotherapy using planar geometry. *Phys. Med. Biol.*, 43:2179–2193, 1998.
22. A. Holder and B. Salter. A tutorial on radiation oncology and optimization. In H. Greenberg, editor, *Tutorials on Emerging Methodologies and Applications in Operations Research*. Kluwer Academic Press, Boston, MA, 2004.
23. Q. Hou, J. Wang, Y. Chen, and J.M. Galvin. An optimization algorithm for intensity modulated radiotherapy—the simulated dynamics with dose–volume constraints. *Med. Phys.*, 30(1):61–68, 2003.
24. F. Jacobs, E. Sundermann, B. De Sutter, M. Christiaens, and I. Lemahieu. A fast algorithm to calculate the exact radiological path through a pixel or voxel space. *J. Comput. Inf. Technol. (CIT)*, 6:89–94, 1998.
25. D.R. Jones. A taxonomy of global optimization methods based on response surfaces. *J. Glob. Optim.*, 21:345–383, 2001.
26. D.R. Jones, M. Schonlau, and W.J. Welch. Efficient global optimization of expensive black-box functions. *J. Glob. Optim.*, 13:455–492, 1998.
27. Faiz M. Khan. *The Physics of Radiation Therapy*. Lippincott Williams and Wilkins, Philadelphia, PA, 1994.
28. S. Kirkpatrick and C.D. Gelatt. Optimization by simulated annealing. *Science*, 220:671–680, 1983.
29. E.K. Lee, T. Fox, and I. Crocker. Simultaneous beam geometry and intensity map optimization in intensity-modulated radiation therapy treatment planning. *Ann. Oper. Res.*, 119:165–181, 2003.
30. E.K. Lee, T. Fox, and I. Crocker. Integer programming applied to intensity-modulated radiation therapy treatment planning. *Int. J. Radiat. Oncol. Biol. Phys.*, 64:301–320, 2006.
31. Y. Li, J. Yao, and D. Yao. Automatic beam angle selection in IMRT planning using genetic algorithm. *Phys. Med. Biol.*, 49:1915–1932, 2004.
32. Y. Li, J. Yao, D. Yao, and W. Chen. A particle swarm optimization algorithm for beam angle selection in intensity-modulated radiotherapy planning. *Phys. Med. Biol.*, 50:3491–3514, 2005.
33. J. Lim, M. Ferris, D. Shepard, S. Wright, and M. Earl. An optimization framework for conformal radiation treatment planning. *INFORMS J. Comput.*, 19:366–380, 2007.

34. B.K. Lind, P. Mavroidis, S. Hyodynmaa, and C. Kappas. Optimization of the dose level for a given treatment plan to maximize the complication-free tumor cure. *Acta. Oncol.*, 38:787–798, 1999.
35. H.M. Lu, H.M. Kooy, Z.H. Leber, and R.J. Ledoux. Optimized beam planning for linear accelerator-based stereotactic radiosurgery. *Int. J. Radiat. Oncol. Biol. Phys.*, 39:1183–1189, 1997.
36. P. Mavroidis, B.K. Lind, and A. Brahme. Biologically effective uniform dose for specification, report and comparison of dose response relations and treatment plans. *Phys. Med. Biol.*, 46:2607–2630, 2001.
37. A. Niemierko. Reporting and analyzing dose distributions: A concept of equivalent uniform dose. *Med. Phys.*, 24:103–110, 1997.
38. M. Oldham, V. Khoo, C.G. Rowbottom, J. Bedford J, and S. Webb. A case study comparing the relative benefit of optimising beam-weights, wedge-angles, beam orientations and tomotherapy in stereotactic radiotherapy of the brain. *Phys. Med. Biol.*, 43:2123–2146, 1998.
39. A. Pugachev and L. Xing. Computer-assisted selection of coplanar beam orientations in intensity-modulated radiation therapy. *Phys. Med. Biol.*, 46:2467–2476, 2001.
40. A. Pugachev and L. Xing. Pseudo beam's-eye-view as applied to beam orientation selection in intensity-modulated radiation therapy. *Int. J. Radiat. Oncol. Biol. Phys.*, 51:1361–1370, 2001.
41. A. Pugachev and L. Xing. Incorporating prior knowledge into beam orientation optimization in IMRT. *Int. J. Radiat. Oncol. Biol. Phys.*, 54:1565–1574, 2002.
42. H.E. Romeijn, R.K. Ahuja, J.F. Dempsey, and A. Kumar. A new linear programming approach to radiation therapy treatment planning problems. *Oper. Res.*, 54:201–216, 2006.
43. H.E. Romeijn, R.K. Ahuja, J.F. Dempsey, A. Kumar, and J.G. Li. A novel linear programming approach to fluence map optimization for intensity modulated radiation therapy treatment planning. *Phys. Med. Biol.*, 38:3521–3542, 2003.
44. C.G. Rowbottom, M. Oldham, and S. Webb. Constrained customization of non-coplanar beam orientations in radiotherapy of brain tumours. *Phys. Med. Biol.*, 44:383–399, 1999a.
45. C.G. Rowbottom, S. Webb, and M. Oldham. Beam-orientation customization using an artificial neural network. *Phys. Med. Biol.*, 44:2251–2262, 1999b.
46. E. Schreibmann, M. Lahanas, L. Xing, and D. Baltas. Multiobjective evolutionary optimization of the number of beams, their orientations and weights for intensity-modulated radiation therapy. *Phys. Med. Biol.*, 49:747–770, 2004.
47. D.M. Shepard, M.C. Ferris, G.H. Olivera, and T.R. Mackie. Optimizing the delivery of radiation therapy to cancer patients. *SIAM Rev.*, 41:721–744, 1999.
48. R.L. Siddon. Fast calculation of the exact radiological path for a three-dimensional CT array. *Med. Phys.*, 12:252–255, 1985.
49. R.L. Siddon. Prism representation: A 3d ray-tracing algorithm for radiotherapy applications. *Phys. Med. Biol.*, 8:817–824, 1985.
50. S. Söderstrom and A. Brahme. Selection of suitable beam orientations in radiation therapy using entropy and fourier transform measures. *Phys. Med. Biol.*, 37:911–924, 1992.
51. S. Söderstrom and A. Brahme. Which is the most suitable number of photon beam portals in coplanar radiation therapy? *Int. J. Radiat. Oncol. Biol. Phys.*, 33:151–159, 1995.

52. J. Stein, R. Mohan, X.H. Wang, T. Bortfeld, Q. Wu, K. Preiser, C.C. Ling, and W. Schlegel. Number and orientations of beams in intensity-modulated radiation treatments. *Med. Phys.*, 24:149–160, 1997.
53. E. Thomas, O. Chapet, M.L. Kessler, T.S. Lawrence, and R.K. Ten Haken. Benefit of using biologic parameters (EUD and NTCP) in IMRT optimization for treatment of intrahepatic tumors. *Int. J. Radiat. Oncol. Biol. Phys.*, 62(2):571–578, 2005.
54. S. Vijayakumar, N. Low, G.T. Chen, L. Myrianthopoulos, H. Culbert, P. Chiru, D. Spelbring, A. Awan, I. Rosenberg, H. Halpern, and R.R. Weichselbaum. Beams eye view-based photon radiotherapy I. *Int. J. Radiat. Oncol. Biol. Phys.*, 21:1575–1586, 1991.
55. C. Wang, J. Dai, and Y. Hu. Optimization of beam orientations and beam weights for conformal radiotherapy using mixed integer programming. *Phys. Med. Biol.*, 48:4065–4076, 2003.
56. W.D. D'Souza, R.R. Meyer, and L. Shi. Selection of beam orientations in intensity-modulated radiation therapy using single-beam indices and integer programming. *Phys. Med. Biol.*, 49:3465–3481, 2004.
57. Q.W. Wu, R. Mohan, A. Niemierko, and R. Schmidt-Ullrich. Optimization of intensity-modulated radiotherapy plans based on the equivalent uniform dose. *Int. J. Radiat. Oncol. Biol. Phys.*, 52:224–235, 2002.

Chapter 9

Multileaf Collimator Shape Matrix Decomposition

Thomas Kalinowski

CONTENTS

9.1 Introduction 253
9.2 Mathematical Model 255
9.3 Decomposition Time Problem 259
9.3.1 Interleaf Collision Constraint 261
9.3.1.1 Linear Programming Approach 262
9.3.1.2 Network Flow Approach 268
9.3.1.3 Duality Approach 273
9.3.2 Tongue-and-Groove Constraint 276
9.4 Decomposition Cardinality Problem 279
9.4.1 Computational Complexity of the DC-Problem 279
9.4.2 Heuristics for the DC-Problem 280
9.4.2.1 Unconstrained Case 281
9.4.2.2 Interleaf Collision Constraint 283
References 285

9.1 Introduction

An important method in cancer treatment is the use of high energetic radiation. To kill tumor cells, the patient is exposed to radiation that is delivered by a linear accelerator whose beam head can be rotated about the treatment couch. Inevitably, the healthy tissue surrounding the tumor is also exposed to some radiation. So the problem arises to arrange the treatment such that the tumor receives a sufficiently high uniform dose while the damage to the normal tissue is as small as possible. The standard approach to this problem is as follows. First the patient body is discretized into the so-called voxels. The set of voxels is then partitioned into three sets: the clinical target volume, the critical structures, and the remaining tissue. There are certain dose constraints for each of these parts. Basically, the dose in the target volume has to be sufficient to kill the cancerous cells and the dose in the critical structures must not destroy the functionality of the corresponding organs. The determination of a combination of radiation fields is usually done by inverse methods based on certain physical models of how the radiation passes through a body. In the early 1990s, the method of intensity modulated radiation therapy (IMRT) was developed to obtain additional flexibility. Using a multileaf collimator (MLC) it is possible to form homogeneous fields of different shapes. By superimposing some homogeneous fields an intensity modulated field is delivered. An MLC consists of two banks of metal leaves that block the radiation and can be shifted to form irregularly shaped beams (Figure 9.1).

The most common approach in treatment planning is to divide the optimization into two phases. At first step, a set of beam angles and corresponding fluence matrices are determined. In a second step, a sequence of leaf positions for the MLC for each of the angles is determined that yields the desired fluence distribution. Very recently, there have been attempts to combine both steps into one optimization routine [8,20].

In this chapter we concentrate on the second step, the shape matrix decomposition problem. Suppose we have fixed the beam angles from which the radiation is released, and for each of the beam angles we are

Figure 9.1 The leaf pairs of a multileaf collimator (MLC).

given a fluence distribution that we want the patient to be exposed to. After discretizing the beam into bixels, we can assume that the fluence distribution is given as a nonnegative integer matrix A. Each row of the matrix corresponds to a pair of leaves of the MLC.

There are two methods in IMRT using MLCs that differ essentially in their technical realization, but the mathematical methods used to determine optimal treatment plans are quite similar. In the step-and-shoot mode the radiation is switched off whenever the leaves are moving, so the intensity modulation is the result of superimposing a finite number of homogeneous fields. In the dynamic mode the radiation is switched on during the whole treatment and the modulation is achieved by moving the leaves with varying speed. Clearly, in this setup the fluence at a particular point is proportional to the amount of time in which the point is exposed to radiation, i.e., not blocked by one of the leaves. Here, we consider only the step-and-shoot mode. Essentially, the most common approach to the dynamic mode can be seen as an imitation of this case (see Ref. [14] and the references therein).

9.2 Mathematical Model

The principle of the MLC in step-and-shoot mode is illustrated in Figure 9.2. Our aim is to determine a sequence of leaf positions and corresponding irradiation times such that the given fluence distribution is realized. Suppose the given matrix has size $m \times n$, i.e., we consider m leaf pairs, and for each leaf there are $n + 1$ possible positions. Then the leaf positions can be described by certain 0–1-matrices of size $m \times n$ called shape matrices,

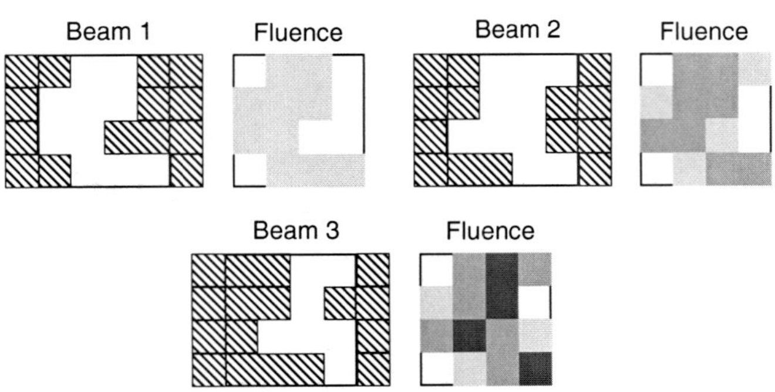

Figure 9.2 Intensity modulation by superimposing three beams of different shapes. In each step, the left figure shows a leaf position and in the right figure the gray scale indicates the total fluence.

where a 0-entry means the radiation is blocked and a 1-entry means that the radiation goes through.

For example, the first leaf position in Figure 9.2 corresponds to the shape matrix

$$\begin{pmatrix} 0 & 1 & 1 & 0 \\ 1 & 1 & 1 & 0 \\ 1 & 1 & 0 & 0 \\ 0 & 1 & 1 & 1 \end{pmatrix}.$$

Clearly, the superposition of differently shaped beams corresponds to positive linear combinations of shape matrices, where the coefficient of a shape matrix measures how long the corresponding field is applied. So, any representation of the given fluence matrix A as a positive integer linear combination of shape matrices is a feasible solution to our decomposition problem. For instance

$$A = \begin{pmatrix} 1 & 3 & 3 & 0 \\ 0 & 2 & 4 & 1 \\ 1 & 1 & 4 & 4 \\ 3 & 3 & 1 & 0 \end{pmatrix} = 2\begin{pmatrix} 0 & 1 & 1 & 0 \\ 0 & 0 & 1 & 0 \\ 0 & 0 & 1 & 1 \\ 1 & 1 & 0 & 0 \end{pmatrix} + \begin{pmatrix} 0 & 1 & 1 & 0 \\ 0 & 1 & 1 & 0 \\ 1 & 1 & 1 & 1 \\ 0 & 0 & 0 & 0 \end{pmatrix} + \begin{pmatrix} 1 & 0 & 0 & 0 \\ 0 & 1 & 1 & 1 \\ 0 & 0 & 1 & 1 \\ 1 & 1 & 1 & 0 \end{pmatrix}. \tag{9.1}$$

We denote the set of shape matrices by $\mathcal{S}$, and consider decompositions of the form $A = \sum_{S\in\mathcal{S}} u_S S$ with $u_S \in \mathbb{N}$ for all $S \in \mathcal{S}$. There are two quantities influencing the quality of a decomposition: the total irradiation time (proportional to the sum of the coefficients) and the number of necessary beams (the number of nonzero coefficients). Let $\mathcal{S}_0$ denote the set of matrices with nonzero coefficient. We can now formulate two different optimization problems, the decomposition time (DT) problem and the decomposition cardinality (DC) problem

$$\text{(DT)} \qquad \min\left\{\sum_{S\in\mathcal{S}} u_S \;\middle|\; A = \sum_{S\in\mathcal{S}} u_S S,\; u_S \in \mathbb{N}\right\}, \tag{9.2}$$

$$\text{(DC)} \qquad \min\left\{|\mathcal{S}_0| \;\middle|\; \mathcal{S}_0 \subseteq \mathcal{S},\; A = \sum_{S\in\mathcal{S}_0} u_S S,\; u_S \in \mathbb{N}\right\}. \tag{9.3}$$

Of course, one could also minimize some weighted sum of decomposition time and decomposition cardinality, i.e., an objective function

$$\sum_{S\in\mathcal{S}_0} u_S + \alpha|\mathcal{S}_0|,$$

where α is some positive constant. This objective function can be considered as total treatment time, where the parameter α depends on the used MLC and measures the average setup time, i.e., the time needed to move the leaves and check the setting. In a still more sophisticated model, one can include the possibility that the setup time between two different leaf positions depends on the amount of required leaf motion. Consequently, the order in which the beams are delivered becomes relevant and the corresponding objective function is

$$\sum_{S \in \mathcal{S}_0} u_S + \sum_{i=1}^{|\mathcal{S}_0|-1} \mu(S^{(i)}, S^{(i+1)}),$$

where $S^{(1)}, S^{(2)}, \ldots, S^{(|\mathcal{S}_0|)}$ is an ordering of the set of used shape matrices $\mathcal{S}_0$, and for two shape matrices S and S', $\mu(S, S')$ is proportional to the time necessary to change the setup of the MLC from the beam corresponding to S to the beam corresponding to S'.

The optimal value of Equation 9.2 can be computed very efficiently while the problem Equation 9.3 is computationally very hard (see Section 9.4). So the most common approach is to first compute the minimal DT, and then heuristically search for a decomposition that realizes this DT and also has a small DC.

Our model is still quite flexible: certain properties of the used MLC can be included in the definition of the shape matrices. From the design of the MLC it is clear that any shape matrix must have the consecutive ones property: in every row is a (possibly empty) interval of consecutive 1-entries and the remaining entries are 0. In addition, in some of the commercially available MLCs leaf overtravel is forbidden. That means, the left leaf of row i and the right leaf of row $i \pm 1$ must not overlap. In this case a shape matrix cannot contain two consecutive rows as follows:

$$\begin{pmatrix} 0 & 1 & 1 & 0 & 0 & 0 & 0 \\ 0 & 0 & 0 & 0 & 0 & 1 & 1 \end{pmatrix}.$$

Also some MLCs have a minimum leaf distance. That means if a row is not completely covered by either the right or the left leaf, a minimum distance δ between the two leaves in this row is present. In other words, the number of ones in a row is either 0 or at least δ. Another feature of most of the MLCs is the tongue-and-groove design. To prevent radiation from going through the gap between two adjacent leaves a design similar to the one indicated in Figure 9.3 is used. The small overlap between the regions that are covered by adjacent leaves causes underdosage effects as illustrated in Figure 9.4. To prevent such underdosage effects one has

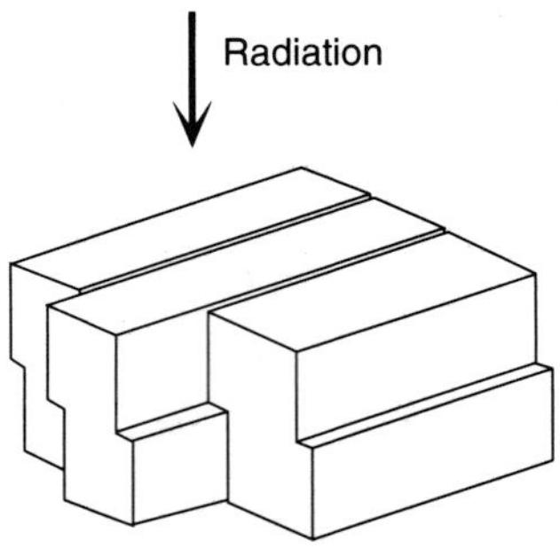

Figure 9.3 The tongue-and-groove design of the leaves of an MLC.

to require that $a_{i,j} \leq a_{i\pm1,j}$ implies that in each of the used beams, bixel $(i \pm 1, j)$ is open whenever bixel (i, j) is open, or in terms of the shape matrices:

$$a_{i,j} \leq a_{i-1,j} \wedge s_{i,j} = 1 \Longrightarrow s_{i-1,j} = 1,$$

$$a_{i,j} \geq a_{i-1,j} \wedge s_{i-1,j} = 1 \Longrightarrow s_{i,j} = 1$$

for $i = 2, \ldots, m$ and $j = 1, \ldots, n$.

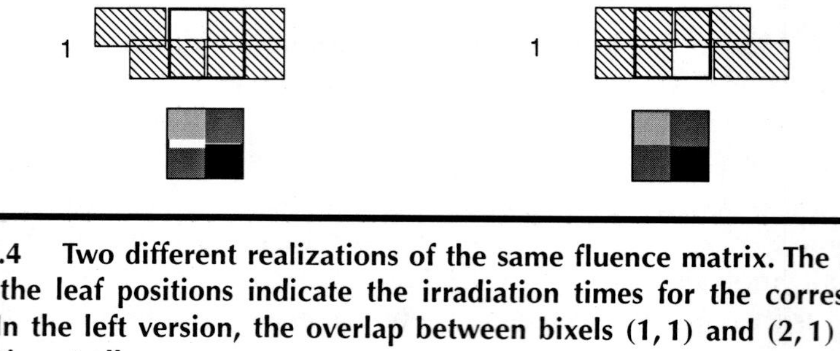

Figure 9.4 Two different realizations of the same fluence matrix. The numbers next to the leaf positions indicate the irradiation times for the corresponding beams. In the left version, the overlap between bixels (1, 1) and (2, 1) receives no radiation at all.

9.3 Decomposition Time Problem

Starting with Refs. [5,9] a number of different algorithms for the shape matrix decomposition problem have been proposed [7,13,15,21,22], some of them providing the optimal DT while others use heuristic methods for both objectives DT and DC. In this section, we concentrate on the DT-problem, and thus, without loss of generality, put all the nonzero coefficients u_s to 1, but allow the same shape matrix S to occur several times in the decomposition $A = \sum_{t=1}^{k} S^{(k)}$. First, we consider the version without additional constraints, i.e., the leaves in different rows move independently, and we neglect the tongue-and-groove underdosage. Then we can solve the decomposition problem for each row independently, and the optimal DT for the whole matrix is just the maximum of the optimal DTs of the single rows. All the algorithms that yield the exact optimum are essentially based on (disguised versions of) the following characterization of the minimal DT. For simplicity of notation, we add a 0-th and a $(n+1)$-th column to the matrix A by setting

$$a_{i,0} = a_{i,n+1} = 0 \quad \text{for } i = 1, 2, \ldots, m.$$

We define the i-th row complexity of A to be

$$c_i(A) = \sum_{j=1}^{n} \max\{0, a_{i,j} - a_{i,j-1}\},$$

and the complexity of A, $c(A) = \max_{1 \le i \le m} c_i(A)$.

THEOREM 1 [7]. *The minimal DT for a matrix A equals $c(A)$.*

PROOF. Let $\boldsymbol{b} = \begin{pmatrix} b_1 & b_2 & \cdots & b_n \end{pmatrix}$ denote the i-th row of the matrix A, i.e., $b_j = a_{i,j}$. First, we show that any representation of $\boldsymbol{b}$ as a sum of vectors with the consecutive ones property contains at least $c_i(A)$ terms. Suppose the vectors $\boldsymbol{s}^{(t)} \in \{0,1\}^n$ $(t = 1, 2, \ldots, k)$ define a representation

$$\boldsymbol{b} = \boldsymbol{s}^{(1)} + \cdots + \boldsymbol{s}^{(k)},$$

where each vector $\boldsymbol{s}^{(t)}$ has the consecutive ones property. For $t = 1, 2, \ldots, k$ let $\boldsymbol{b}^{(t)} = \boldsymbol{s}^{(1)} + \cdots + \boldsymbol{s}^{(t)}$ be the sum of the first t terms and put

$$c^{(t)} = \sum_{j=1}^{n} \max\{0, b_j^{(t)} - b_{j-1}^{(t)}\}.$$

Now let l_t and r_t denote the positions of the leaves corresponding to $\mathbf{s}^{(t)}$, i.e.,

$$s_j^{(t)} = \begin{cases} 1 & \text{if } l_t < j < r_t, \\ 0 & \text{otherwise.} \end{cases}$$

For $t > 1$ we obtain

$$\max\{0, b_j^{(t-1)} - b_{j-1}^{(t-1)}\} = \max\{0, b_j^{(t)} - b_{j-1}^{(t)}\} \qquad \text{for } j \notin \{l_t + 1, r_t\} \tag{9.4}$$

$$\max\{0, b_{l_t+1}^{(t-1)} - b_{l_t}^{(t-1)}\} = \begin{cases} \max\{0, b_{l_t+1}^{(t)} - b_{l_t}^{(t)}\} - 1 & \text{if } b_{l_t+1}^{(t)} > b_{l_t}^{(t)}, \\ \max\{0, b_{l_t+1}^{(t)} - b_{l_t}^{(t)}\} & \text{otherwise.} \end{cases} \tag{9.5}$$

$$\max\{0, b_{r_t}^{(t-1)} - b_{r_t-1}^{(t-1)}\} = \begin{cases} \max\{0, b_{r_t}^{(t)} - b_{r_t-1}^{(t)}\} + 1 & \text{if } b_{r_t}^{(t)} \geq b_{r_t-1}^{(t)}, \\ \max\{0, b_{r_t}^{(t)} - b_{r_t-r}^{(t)}\} & \text{otherwise.} \end{cases} \tag{9.6}$$

Consequently, $c^{(t-1)} \geq c^{(t)} - 1$ with equality if and only if $b_{l_t+1}^{(t)} > b_{l_t}^{(t)}$ and $b_{r_t}^{(t)} < b_{r_t-1}^{(t)}$. Summing up these inequalities for $2 \leq t \leq k$ and using $c^{(1)} = 1$ and $c^{(k)} = c_i(A)$, we obtain

$$1 + c^{(2)} + \cdots + c^{(k-1)} \geq c^{(2)} + c^{(3)} + \cdots + c^{(k-1)} + c_i(A) - (k-1),$$

or $k \geq c_i(A)$. To show that there is a decomposition of the i-th row with $c_i(A)$ terms we use induction on $k := c_i(A)$. If $k = 0$, then $\boldsymbol{b} = \mathbf{0}$ and there is nothing to prove. If $k > 0$, we put $b_0 = b_{n+1} = 0$ and define

$$l = \min\{j \mid 0 \leq j \leq n-1,\ b_j < b_{j+1}\},$$

$$r = \min\{j > l \mid l < r \leq n+1,\ b_j < b_{j-1}\},$$

$$s_j^{(k)} = \begin{cases} 1 & \text{if } l < j < r \\ 0 & \text{otherwise} \end{cases} \qquad (j = 1, \ldots, n)$$

Then by Equations 9.4 through 9.6, for $\boldsymbol{b}' := \boldsymbol{b} - \mathbf{s}^{(k)}$, we have

$$\sum_{j=1}^{n} \max\{0, b_j' - b_{j-1}'\} = \sum_{j=1}^{n} \max\{0, b_j - b_{j-1}\} - 1 = k - 1.$$

By induction, there is a decomposition $\boldsymbol{b}' = \boldsymbol{s}^{(1)} + \cdots + \boldsymbol{s}^{(k-1)}$. Together with $\boldsymbol{s}^{(k)}$ this yields the required decomposition of $\boldsymbol{b}$, and this concludes the proof. □

From the proof we can immediately derive an algorithm for the construction of a DT-optimal decomposition (see Algorithm 1). Of course the

Algorithm 1 (DT-optimal leaf sequence in the unconstrained case).
$k := 0$
while $A \neq 0$ **do**
  $k := k + 1$
  for $i = 1, 2, \ldots, m$ **do**
    if $a_{i,j} = 0$ for all $j = 1, 2, \ldots, n$ **then** $l_i := n,\ r_i := n + 1$
    else
      $l_i := \min\{j \mid 0 \leq j \leq n,\ a_{i,j} < a_{i,j+1}\}$
      $r_i := \min\{j \mid l_i < j \leq n + 1,\ a_{i,j-1} > a_{i,j}\}$
    for $j = 1, \ldots, n$ **do**
      if $l_i < j < r_i$ **then** $s_{i,j}^{(k)} := 1$ **else** $s_{i,j}^{(k)} := 0$
  $A := A - S^{(k)}$
return $S^{(1)}, \ldots, S^{(k)}$

choice of the l_i and r_i is not unique. In Ref. [5] this particular one is called sweep technique, because the leaves always move from left to right. As an example, consider the following decomposition of a matrix from Ref. [22].

$$\begin{pmatrix} 4&5&0&1&4&5\\ 2&4&1&3&1&4\\ 2&3&2&1&2&4\\ 5&3&3&2&5&3 \end{pmatrix} = \begin{pmatrix} 1&1&0&0&0&0\\ 1&1&0&0&0&0\\ 1&1&0&0&0&0\\ 1&0&0&0&0&0 \end{pmatrix} + \begin{pmatrix} 1&1&0&0&0&0\\ 1&1&0&0&0&0\\ 1&1&1&0&0&0\\ 1&0&0&0&0&0 \end{pmatrix} + \begin{pmatrix} 1&1&0&0&0&0\\ 0&1&0&0&0&0\\ 0&1&1&1&1&1\\ 1&1&1&0&0&0 \end{pmatrix} + \begin{pmatrix} 1&1&0&0&0&0\\ 0&1&1&1&0&0\\ 0&0&0&0&1&1\\ 1&1&1&1&1&0 \end{pmatrix}$$

$$+ \begin{pmatrix} 0&1&0&0&0&0\\ 0&0&0&1&0&0\\ 0&0&0&0&0&1\\ 1&1&1&1&1&0 \end{pmatrix} + \begin{pmatrix} 0&0&0&1&1&1\\ 0&0&0&1&1&1\\ 0&0&0&0&0&1\\ 0&0&0&0&1&1 \end{pmatrix} + \begin{pmatrix} 0&0&0&0&1&1\\ 0&0&0&0&0&1\\ 0&0&0&0&0&0\\ 0&0&0&0&1&1 \end{pmatrix} + \begin{pmatrix} 0&0&0&0&1&1\\ 0&0&0&0&0&1\\ 0&0&0&0&0&0\\ 0&0&0&0&1&1 \end{pmatrix}$$

$$+ \begin{pmatrix} 0&0&0&0&1&1\\ 0&0&0&0&0&1\\ 0&0&0&0&0&0\\ 0&0&0&0&0&0 \end{pmatrix} + \begin{pmatrix} 0&0&0&0&0&1\\ 0&0&0&0&0&0\\ 0&0&0&0&0&0\\ 0&0&0&0&0&0 \end{pmatrix}.$$

9.3.1 Interleaf Collision Constraint

The interleaf collision constraint (ICC) is present in many of the commercially available MLCs and forbids an overlap of the left leaf in row i and the right leaf in row $i \pm 1$. If l_i and r_i denote the leaf positions in row i $(i = 1, \ldots, m)$ this amounts to

$$\text{(ICC)} \qquad l_i < r_{i-1} \quad \text{and} \quad r_i > l_{i-1} \qquad (i = 2, \ldots, m)$$

9.3.1.1 Linear Programming Approach

An important conclusion from the following algorithm is that we can always construct a DT-optimal decomposition with unidirectional leaf movement. That means the leaves move only from left to right, or in other words, if $l_i^{(t)}$ and $r_i^{(t)}$ denote the leaf positions in row i corresponding to the t-th shape matrix, then $l_i^{(t)} \leq l_i^{(t+1)}$ and $r_i^{(t)} \leq r_i^{(t+1)}$ for all i and t. Such a decomposition $A = \sum_{t=1}^{k} S^{(t)}$ is completely determined once we know for each i and j, how often the leaves in row i are at position j, i.e., we have to know the numbers

$$\gamma_{i,j}^L = \left|\left\{t \mid l_i^{(t)} = j - 1\right\}\right|, \qquad \gamma_{i,j}^R = \left|\left\{t \mid r_i^{(t)} = j\right\}\right|. \tag{9.7}$$

The numbers $\gamma_{i,j}^L$ and $\gamma_{i,j}^R$ can be translated back into the shape matrices via

$$s_{i,j}^{(t)} = 1 \iff \sum_{j'=1}^{j} \gamma_{i,j'}^R < t \leq \sum_{j'=1}^{j} \gamma_{i,j'}^L \qquad (t = 1, \ldots, k). \tag{9.8}$$

This definition makes sense for any nonnegative $\gamma_{i,j}^L$ and $\gamma_{i,j}^R$. Now we formulate additional requirements for these values.

Lemma 1. *The matrices defined by Equation 9.8 sum up to A if and only if*

$$\gamma_{i,j}^L - \gamma_{i,j}^R = a_{i,j} - a_{i,j-1} \qquad (i = 1, \ldots, m;\ j = 1, \ldots, n+1). \tag{9.9}$$

Proof. "⇒": By hypothesis and Equation 9.8 we have

$$a_{i,j} = \sum_{j'=1}^{j} \gamma_{i,j}^L - \sum_{j'=1}^{j} \gamma_{i,j}^R.$$

For $j = 1$ we obtain

$$a_{i,1} - a_{i,0} = a_{i,1} = \gamma_{i,1}^L - \gamma_{i,1}^R.$$

And for $j > 1$,

$$a_{i,j} - a_{i,j-1} = \left(\sum_{j'=1}^{j} \gamma_{i,j}^{L} - \sum_{j'=1}^{j} \gamma_{i,j}^{R}\right) - \left(\sum_{j'=1}^{j-1} \gamma_{i,j}^{L} - \sum_{j'=1}^{j-1} \gamma_{i,j}^{R}\right) = \gamma_{i,j}^{L} - \gamma_{i,j}^{R}.$$

"$\Leftarrow$": Assume that Equation 9.9 is true and let $B = (b_{i,j})$ be the sum of the matrices defined by Equation 9.8. Then

$$b_{i,j} = \sum_{j'=1}^{j} \gamma_{i,j}^{L} - \sum_{j'=1}^{j} \gamma_{i,j}^{R} = \sum_{j'=1}^{j} (a_{i,j'} - a_{i,j'-1}) = a_{i,j} - a_{i,0} = a_{i,j}. \quad \square$$

The next lemma formulates the ICC in terms of the $\gamma_{i,j}^{L}$ and $\gamma_{i,j}^{R}$.

LEMMA 2. *If $\gamma_{i,j}^{L}$ and $\gamma_{i,j}^{R}$ encode a decomposition (not necessarily unidirectional) $A = \sum_{t=1}^{k} S^{(t)}$ with ICC as in Equation 9.7 then*

$$\sum_{j'=1}^{j} \gamma_{i-1,j'}^{L} \geq \sum_{j'=1}^{j} \gamma_{i,j'}^{R} \quad (i = 2, \ldots, m;\ j = 1, \ldots, n+1), \tag{9.10}$$

$$\sum_{j'=1}^{j} \gamma_{i,j'}^{L} \geq \sum_{j'=1}^{j} \gamma_{i-1,j'}^{R} \quad (i = 2, \ldots, m;\ j = 1, \ldots, n+1). \tag{9.11}$$

PROOF. We have

$$\sum_{j'=1}^{j} \gamma_{i-1,j'}^{L} = \left|\left\{t \ :\ l_{i-1}^{(t)} < j\right\}\right|, \quad \sum_{j'=1}^{j} \gamma_{i,j'}^{R} = \left|\left\{t \ :\ r_{i}^{(t)} \leq j\right\}\right|.$$

The ICC implies $\left\{t : r_i^{(t)} \leq j\right\} \subseteq \left\{t : l_{i-1}^{(t)} < j\right\}$, and this gives Equation 9.10. The statement in Equation 9.11 is proved similarly. $\square$

Of course, the DT equals the sum of all $\gamma_{i,j}^{L}$ (or equivalently all $\gamma_{i,j}^{R}$) along any row:

$$k = \sum_{j=1}^{n+1} \gamma_{i,j}^{L} = \sum_{j=1}^{n+1} \gamma_{i,j}^{R} \quad (i = 1, \ldots, m) \tag{9.12}$$

We can formulate the DT-problem with ICC as a linear program.

THEOREM 2. *The DT-problem with ICC is equivalent to*

$$\min\left\{k \mid \text{Equations 9.9 through 9.12},\ \gamma_{i,j}^L, \gamma_{i,j}^R \in \mathbb{N}\right\}. \tag{9.13}$$

PROOF. The above argument shows that every decomposition with unidirectional leaf movement gives rise to a feasible solution of Equation 9.13. Conversely, from every feasible solution of Equation 9.13 we obtain a (unidirectional) decomposition with k shape matrices (defined according to Equation 9.8). We show that the unidirectional leaf movement is no restriction: every decomposition $A = \sum_{t=1}^{k} S^{(t)}$ with ICC yields a feasible solution of Equation 9.13 with objective value k. Define $\gamma_{i,j}^L$ and $\gamma_{i,j}^R$ by Equation 9.7. It is clear that Equation 9.12 holds. By Lemma 2 we have Equations 9.10 and 9.11. From

$$a_{i,j} = \left|\left\{t \mid l_i^{(t)} < j < r_i^{(t)}\right\}\right| \quad \text{and} \quad a_{i,j-1} = \left|\left\{t \mid l_i^{(t)} < j-1 < r_i^{(t)}\right\}\right|,$$

it follows that

$$\begin{aligned} a_{i,j} - a_{i,j-1} &= \left|\left\{t \mid l_i^{(t)} = j-1,\ r_i^{(t)} > j\right\}\right| - \left|\left\{t \mid l_i^{(t)} < j-1,\ r_i^{(t)} = j\right\}\right| \\ &= \left|\left\{t \mid l_i^{(t)} = j-1,\ r_i^{(t)} \geq j\right\}\right| - \left|\left\{t \mid l_i^{(t)} \leq j-1,\ r_i^{(t)} = j\right\}\right| \\ &= \gamma_{i,j}^L - \gamma_{i,j}^R. \end{aligned}$$

So Equation 9.9 holds and this concludes the proof. □

Note that this also shows how an arbitrary leaf sequence can be transformed to an unidirectional one with the same DT: define the $\gamma_{i,j}^L$ and $\gamma_{i,j}^R$ according to Equation 9.7 and the new shape matrices with Equation 9.8. Obviously, the values

$$\tilde{\gamma}_{i,j}^L = \max\{0, a_{i,j} - a_{i,j-1}\}, \qquad \tilde{\gamma}_{i,j}^R = \max\{0, a_{i,j-1} - a_{i,j}\}$$

satisfy the conditions given in Equation 9.9 and these conditions imply $\gamma_{i,j}^L \geq \tilde{\gamma}_{i,j}^L$ and $\gamma_{i,j}^R \geq \tilde{\gamma}_{i,j}^R$ for all pairs (i, j). The $\tilde{\gamma}_{i,j}^L$ and $\tilde{\gamma}_{i,j}^R$ correspond to the sweep solution for the unconstrained case coming from Algorithm 1. Because

$$\gamma_{i,j}^L - \gamma_{i,j}^R = \tilde{\gamma}_{i,j}^L - \tilde{\gamma}_{i,j}^R = a_{i,j} - a_{i,j-1},$$

we can represent $\gamma_{i,j}^L$ and $\gamma_{i,j}^R$ with a single nonnegative variable $w_{i,j}$ via

$$\gamma_{i,j}^L = \tilde{\gamma}_{i,j}^L + w_{i,j}, \quad \gamma_{i,j}^R = \tilde{\gamma}_{i,j}^R + w_{i,j}.$$

With

$$T_i = \sum_{j=1}^{n+1} \tilde{\gamma}^L_{i,j} = \sum_{j=1}^{n+1} \tilde{\gamma}^R_{i,j} \quad (i = 1, \dots, m)$$

the constraints given by Equations 9.10 through 9.12 become

$$\sum_{j'=1}^{j} \tilde{\gamma}^L_{i-1,j'} + \sum_{j'=1}^{j} w_{i-1,j'} \geq \sum_{j'=1}^{j} \tilde{\gamma}^R_{i,j'} + \sum_{j'=1}^{j} w_{i,j'}$$
$$(i = 2, \dots, m;\ j = 1, \dots, n+1), \tag{9.14}$$

$$\sum_{j'=1}^{j} \tilde{\gamma}^L_{i,j'} + \sum_{j'=1}^{j} w_{i,j'} \geq \sum_{j'=1}^{j} \tilde{\gamma}^R_{i-1,j'} + \sum_{j'=1}^{j} w_{i-1,j'}$$
$$(i = 2, \dots, m;\ j = 1, \dots, n+1), \tag{9.15}$$

$$k = T_i + \sum_{j=1}^{n+1} w_{i,j} \quad (i = 1, \dots, m), \tag{9.16}$$

$$w_{i,j} \geq 0,\ w_{i,j} \in \mathbb{Z} \quad (i = 1, \dots, m;\ j = 1, \dots, n+1). \tag{9.17}$$

Observe that Equations 9.14 and 9.15 with $j = n + 1$ yield

$$T_{i-1} + \sum_{j=1}^{n+1} w_{i-1,j} \overset{\text{Equations 9.14}}{\geq} T_i + \sum_{j=1}^{n+1} w_{i,j} \overset{\text{Equations 9.15}}{\geq} T_{i-1} + \sum_{j=1}^{n+1} w_{i-1,j}.$$

Consequently, we have equality and thus Equation 9.16 follows from Equations 9.14 and 9.15. This simplifies the problem, because now we just have to minimize $\sum_{j=1}^{n+1} w_{i,j}$ for any row i, and the problem becomes, e.g.,

$$\min \left\{ \sum_{j=1}^{n+1} w_{1,j} \mid \text{Equations 9.14, 9.15, and 9.17} \right\}. \tag{9.18}$$

For a feasible solution $W = (w_{i,j})$ we denote the maximal index of a shape matrix having the left (right) leaf in row i to the left of column j by $I_L^{(i)}(j)$ $(I_R^{(i)}(j))$. In other words,

$$I_L^{(i)}(j) := \max\{t \mid l_i^{(t)} < j\} = \sum_{j'=1}^{j} \tilde{\gamma}_{i,j'}^L + \sum_{j'=1}^{j} w_{i,j'}$$

$$I_R^{(i)}(j) := \max\{t \mid r_i^{(t)} \leq j\} = \sum_{j'=1}^{j} \tilde{\gamma}_{i,j'}^R + \sum_{j'=1}^{j} w_{i,j'}$$

for $i = 1, \ldots, m, j = 1, \ldots, n+1$. In addition we put $I_L^{(i)}(0) = I_R^{(i)}(0) = 0$ for $i = 1, \ldots, m$. For the shape matrices $S^{(1)}, \ldots, S^{(k)}$ in the corresponding decomposition we have

$$s_{i,j}^{(t)} = 1 \iff I_R^{(i)}(j) < t \leq I_L^{(i)}(j),$$

hence for a feasible solution,

$$I_L^{(i)}(j) - I_R^{(i)}(j) = a_{i,j}.$$

Observe that the $I_L^{(i)}(j)$ and $I_R^{(i)}(j)$ are exactly the terms that occur in the constraints given by Equations 9.14 and 9.15. So these constraints can be rewritten as

$$I_L^{(i-1)}(j) \geq I_R^{(i)}(j), \quad I_L^{(i)}(j) \geq I_R^{(i-1)}(j) \tag{9.19}$$

for $i = 2, \ldots, m$, $j = 1, \ldots, n+1$. For convenience, we formulate the algorithm for the solution of Equation 9.18 in terms of the $I_L^{(i)}(j)$ and $I_R^{(i)}(j)$. Clearly, knowing these values is equivalent to knowing the $w_{i,j}$, and minimizing $\sum_{j=1}^{n+1} w_{i,j}$ is the same as minimizing $I_L^{(i)}(n+1)$. The idea is to determine the values in column j depending on the values in column $j-1$. We start with the lower bounds

$$I_L^{(i)}(j) := I_L^{(i)}(j-1) + \tilde{\gamma}_{i,j}^L, \qquad I_R^{(i)}(j) := I_R^{(i)}(j-1) + \tilde{\gamma}_{i,j}^R,$$

and then we run through the rows and eliminate violations by increasing the relevant values as little as possible. Note that by increasing the values in row $i-1$ we might create a new violation between row $i-1$ and row $i-2$. The recursive call of the function `Update` takes care of this.

Algorithm 2 (DT-optimal leaf sequence with ICC).
for $i = 1, \ldots, m$ **do** $I_L^{(i)}(0) := 0$; $I_R^{(i)}(0) := 0$
for $j = 1, \ldots, n+1$ **do**
  for $i = 1, \ldots, m$ **do**
    $I_L^{(i)}(j) := I_L^{(i)}(j-1) + \tilde{\gamma}_{i,j}^L$
    $I_R^{(i)}(j) := I_R^{(i)}(j-1) + \tilde{\gamma}_{i,j}^R$
  for $i = 2, \ldots, m$ **do**
    if $I_L^{(i)}(j) < I_R^{(i-1)}(j)$ **then**
      $I_L^{(i)}(j) := I_R^{(i-1)}(j)$; $I_R^{(i)}(j) := I_L^{(i)}(j) - a_{i,j}$
    if $I_R^{(i)}(j) > I_L^{(i-1)}(j)$ **then** `Update` $(i-1)$

Function `Update` (k)
$I_L^{(k)}(j) := I_R^{(k+1)}(j)$; $I_R^{(k)}(j) := I_L^{(k)}(j) - a_{i,j}$
if $\left(k \geq 2 \text{ and } I_R^{(k)}(j) > I_L^{(k-1)}(j)\right)$ **then** `Update(k` $-$ `1)`

THEOREM 3. *Algorithm 2 solves the DT-problem with ICC in time $O(m^2 n)$.*

PROOF. It is easy to see that after termination of the algorithm $I_L^{(i)}(j) - I_R^{(i)}(j) = a_{i,j}$ for $i = 1, \ldots, m$ and $j = 1, \ldots, n$. So, we indeed obtain a decomposition of matrix S. Also Equation 9.19 and hence Equations 9.14 and 9.15 hold, and the result corresponds to a feasible solution $W = (w_{i,j})$ of Equation 9.18. Let $\hat{W} = (\hat{w}_{i,j})$ be an optimal solution corresponding to $\hat{I}_L^{(i)}(j)$ and $\hat{I}_R^{(i)}(j)$.

Claim. At any time $I_L^{(i)}(j) \leq \hat{I}_L^{(i)}(j)$ and $I_R^{(i)}(j) \leq \hat{I}_R^{(i)}(j)$.

We prove this claim by induction on j. For $j = 1$, the initialization in the first inner loop gives $I_L^{(i)}(1) = a_{i,1}$ and $I_R^{(i)}(1) = 0$. The conditions for changing these values in the second inner loop are never satisfied, so our claim follows from

$$\hat{I}_L^{(i)}(1) = \tilde{\gamma}_{i,j}^L + \hat{w}_{i,1} \geq \tilde{\gamma}_{i,j}^L, \qquad \hat{I}_R^{(i)}(1) = \tilde{\gamma}_{i,j}^R + \hat{w}_{i,1} \geq \tilde{\gamma}_{i,j}^R = 0.$$

For $j > 1$, with induction the initialization in the first inner loop yields

$$\begin{aligned} I_L^{(i)}(j) = I_L^{(i)}(j-1) + \tilde{\gamma}_{i,j}^L \leq \hat{I}_L^{(i)}(j-1) + \tilde{\gamma}_{i,j}^L \leq \hat{I}_L^{(i)}(j-1) \\ + \tilde{\gamma}_{i,j}^L + \hat{w}_{i,j} = \hat{I}_L^{(i)}(j), \end{aligned}$$

$$\begin{aligned} I_R^{(i)}(j) = I_R^{(i)}(j-1) + \tilde{\gamma}_{i,j}^R \leq \hat{I}_R^{(i)}(j-1) + \tilde{\gamma}_{i,j}^R \leq \hat{I}_R^{(i)}(j-1) \\ + \tilde{\gamma}_{i,j}^R + \hat{w}_{i,j} = \hat{I}_R^{(i)}(j). \end{aligned}$$

Now suppose our claim is false and consider the step of the algorithm where we get $I_L^{(i)}(j) > \hat{I}_L^{(i)}(j)$ or $I_R^{(i)}(j) > \hat{I}_R^{(i)}(j)$ for the first time.

Case 1. The first condition for changing $I_L^{(i)}(j)$ is satisfied. Then for the values after the update we obtain

$$I_L^{(i)}(j) := I_R^{(i-1)}(j) \le \hat{I}_R^{(i-1)}(j) \overset{\text{ICC}}{\le} \hat{I}_L^{(i)}(j),$$

$$I_R^{(i)}(j) := I_L^{(i)}(j) - a_{i,j} \le \hat{I}_L^{(i)}(j) - a_{i,j} = \hat{I}_R^{(i)}(j),$$

contradicting the assumption.

Case 2. We are in the function `Update` (k). Then

$$I_L^{(i)}(j) := I_R^{(i+1)}(j) \le \hat{I}_R^{(i+1)}(j) \overset{\text{ICC}}{\le} \hat{I}_L^{(i)}(j),$$

$$I_R^{(i)}(j) := I_L^{(i)}(j) - a_{i,j} \le \hat{I}_L^{(i)}(j) - a_{i,j} = \hat{I}_R^{(i)}(j),$$

contradicting the assumption.

This proves the claim, and from $I_L^{(i)}(n+1) \le \hat{I}_L^{(i)}(n+1)$ and the optimality of $\hat{W}$ the optimality of W follows. Now let's consider the complexity. There are $m-1$ passes through the second inner for-loop, and in the worst case each of these calls the function `Update` which calls itself at most m times. So, the complexity of the second inner loop is $O(m^2)$, and because we have to run $n+1$ times through the outer loop, the total complexity of $O(m^2n)$ follows. □

Variants of this algorithm were presented in Refs. [2,15]. The proof given here is a mixture of these two references. The algorithm can also be adapted very easily to ensure minimum distances δ_0 and δ_1 between opposite leaves in the same row and in adjacent rows, respectively, if this is possible at all [15].

9.3.1.2 Network Flow Approach

A first network flow algorithm for DT-problem without ICC was proposed in Ref. [1]. Here we present a network flow formulation from Ref. [4], which also includes the ICC. The set of shape matrices is identified with the set of paths from D to D' in the layered directed graph (digraph) $G = (V, E)$, constructed as follows. The vertices in the i-th layer correspond to the possible pairs (l_i, r_i) $(1 \le i \le m)$, and two additional vertices D and D' are added:

$$V = \{(i,l,r) : i = 1,\dots,m;\ l = 1,\dots,n;\ r = l+1,\dots,n+1\} \cup \{D, D'\}.$$

Between two vertices (i,l,r) and $(i+1,l',r')$ is an arc if the corresponding leaf positions are consistent with the ICC, i.e., if $l' \le r-1$ and $r' \ge l+1$. In addition, E contains all arcs from D to the first layer, all arcs from the last layer m to D', and the arc (D', D), so

$$E = E_+(D) \cup E_-(D') \cup \bigcup_{i=1}^{m-1} E(i) \cup \{(D', D)\}, \text{ where}$$

$$E_+(D) = \{(D,(1,l,r)) \ :\ (1,l,r) \in V\},$$

$$E_-(D') = \{((m,l,r), D') \ :\ (m,l,r) \in V\},$$

$$E(i) = \{((i,l,r),(i+1,l',r')) \ :\ l' \le r-1,\ r' \ge l+1\}.$$

There is a bijection between the possible leaf positions and the cycles in G. This is illustrated in Figure 9.5 which shows two cycles in G for $m = 4$, $n = 2$, corresponding to the shape matrices

$$\begin{pmatrix} 1 & 0 \\ 0 & 1 \\ 1 & 1 \\ 1 & 0 \end{pmatrix} \text{ (straight lines) and } \begin{pmatrix} 0 & 1 \\ 1 & 1 \\ 1 & 0 \\ 0 & 1 \end{pmatrix} \text{ (dashed lines).}$$

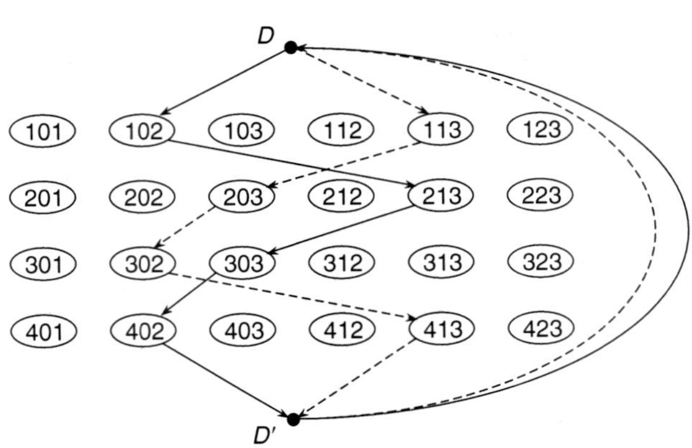

Figure 9.5 The vertices of G for $m = 4$, $n = 2$, and two cycles.

With a shape matrix S, given by $(l_1, r_1), (l_2, r_2), \ldots, (l_m, r_m)$, we associate a unit flow on the cycle

$$D, (1, l_1, r_1), (2, l_2, r_2), \ldots, (m, l_m, r_m), D', D.$$

Then any positive combination of shape matrices defines a circulation $\phi : E \to \mathbb{R}_+$ on G. For instance,

$$3\begin{pmatrix}1 & 0\\0 & 1\\1 & 1\\1 & 0\end{pmatrix} + 2\begin{pmatrix}0 & 1\\1 & 1\\1 & 0\\0 & 1\end{pmatrix} = \begin{pmatrix}3 & 2\\2 & 5\\5 & 3\\3 & 2\end{pmatrix},$$

corresponds to 3 units of flow on $(D, (1, 0, 2), (2, 1, 3), (3, 0, 3), (4, 0, 2), D')$, 2 units of flow on $(D, (1, 1, 3), (2, 0, 3), (3, 0, 2), (4, 1, 3), D')$, and 5 units of flow on (D', D). The amount of radiation that is released at bixel (i, j) equals the sum of the flows going through the vertices (i, l, r) with $l < j < r$, hence the conditions that must be satisfied by the circulation to correspond to a decomposition of A are

$$\sum_{l=1}^{j-1} \sum_{r=j+1}^{n+1} \sum_{l'=1}^{r-1} \sum_{r'=\max\{l,l'\}-1}^{n} \phi((i, l, r), (i + 1, l', r')) = a_{i,j}, \tag{9.20}$$

for $1 \le i \le m - 1,\ 1 \le j \le n$, and

$$\sum_{l=1}^{j-1} \sum_{r=j+1}^{n+1} \phi((m, l, r), D') = a_{mj}, \tag{9.21}$$

for $1 \le j \le n$. Because all of the flow must go through the arc (D', D), the DT of the decomposition corresponding to ϕ equals $\phi(D', D)$. Thus the DT-problem can be solved by finding a circulation satisfying conditions given by Equations 9.20 and 9.21 and having minimal cost with respect to the cost function $\alpha : E \to \mathbb{R}_+$,

$$\alpha(e) = \begin{cases} 1 & \text{if } e = (D, D'), \\ 0 & \text{otherwise.} \end{cases}$$

The graph G can be expanded to a graph $\hat{G} = (\hat{V}, \hat{E})$ so that, instead of the constraints given by Equations 9.20 and 9.21, the structure of $\hat{G}$

together with a capacity function on $\hat{E}$ forces the circulation to represent a decomposition of A.

$$\hat{V} = \{(i,l,r)_1, (i,l,r)_2 \mid 1 \le i \le m,\ 0 \le l < r \le n+1\}$$

$$\cup \{(i,j) \mid 1 \le i \le m,\ 0 \le j \le n\} \cup \{D, D'\}.$$

The arc set of $\hat{G}$ is $\hat{E} = \hat{E}^{\text{old}} \cup \hat{E}^1 \cup \hat{E}^2$, where

$$\hat{E}^{\text{old}} = \{((i,l,r)_2, (i+1,l',r')_1) \ :\ ((i,l,r),(i+1,l',r')) \in E\}$$

$$\cup \{(D,(1,l,r)_1) \ :\ (1,l,r)_1 \in \hat{V}\} \cup \{((m,l,r)_2, D') \ :$$

$$(m,l,r)_2 \in \hat{V}\} \cup \{(D',D)\},$$

$$\hat{E}^1 = \{((i,l,r)_1,(i,l)) \ :\ (i,l,r)_1 \in \hat{V}\} \cup \{((i,r-1),(i,l,r)_2) \ :$$

$$(i,l,r)_2 \in \hat{V}\},$$

$$\hat{E}^2 = \{((i,j-1),(i,j)) \ :\ i = 1,\dots,m;\ j = 1,\dots n\}.$$

Now a shape matrix with parameters l_i, r_i $(i = 1,\dots,m)$ corresponds to the cycle

$$D,(1,l_1,r_1)_1, (1,l_1), (1,l_1+1), \dots, (1,r_1-1), (1,l_1,r_1)_2,$$

$$(2,l_2,r_2)_1, (2,l_2), (2,l_2+1), \dots, (2,r_2-1), (2,l_2,r_2)_2,$$

$$\dots$$

$$(m,l_m,r_m), (m,l_m), (m,l_m+1), \dots, (m,r_m-1), (m,l_m,r_m)_2, D', D$$

Figure 9.6 shows the cycles in $\hat{G}$ corresponding to the cycles in Figure 9.5. Now the flow on the arc $((i,j-1),(i,j))$ equals the amount of radiation released at bixel (i,j) in the corresponding decomposition. Introducing lower and upper capacities $\underline{u}$ and $\overline{u}$ on the arcs of $\hat{G}$ by

$$\underline{u}(e) = \begin{cases} 0 & \text{if } e \in \hat{E}^{\text{old}} \cup \hat{E}^1, \\ a_{i,j} & \text{if } e = ((i,j-1),(i,j)) \in \hat{E}^2, \end{cases} \tag{9.22}$$

$$\overline{u}(e) = \begin{cases} \infty & \text{if } e \in \hat{E}^{\text{old}} \cup \hat{E}^1, \\ a_{i,j} & \text{if } e = ((i,j-1),(i,j)) \in \hat{E}^2, \end{cases} \tag{9.23}$$

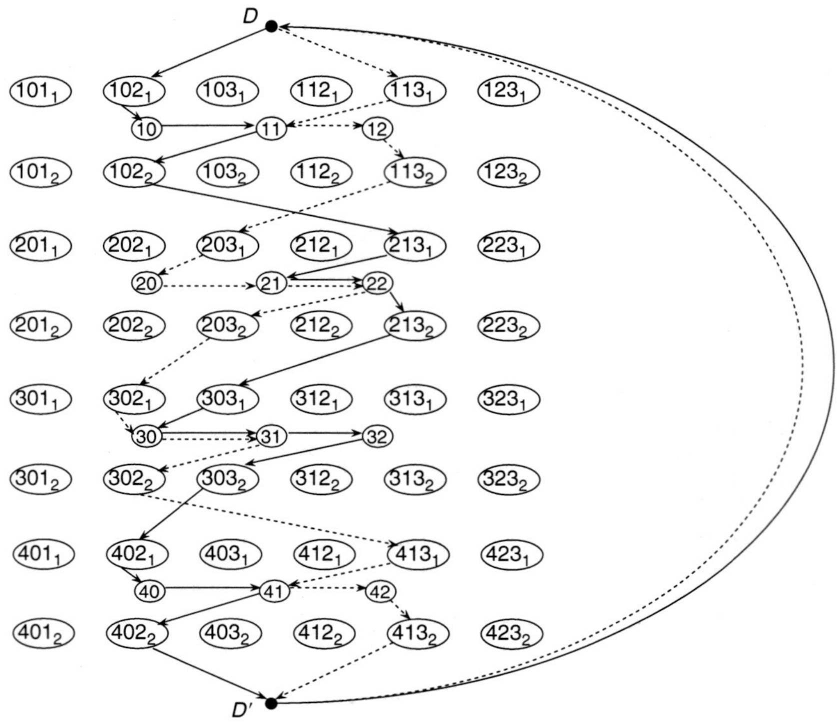

Figure 9.6 The vertices of $\hat{G}$ for $m = 4$, $n = 2$, and two cycles.

we make sure that the fluence matrix A is realized. Now to obtain another reformulation of the DT-problem we just have to require that the flow on the arc $((i, l, r)_1, (i, l))$ equals the flow on the arc $((i, r-1), (i, l, r)_2)$, because both of these correspond to the total amount of radiation that is released while $l_i = l$ and $r_i = r$.

THEOREM 4 [4]. *The DT-minimization problem is equivalent to the network flow problem*

$$\text{Min } \phi(D', D)$$

s.t. ϕ *a circulation in* $\hat{G} = (\hat{V}, \hat{E})$ *with lower and upper capacities* $\underline{u}$ *and* $\overline{u}$, *defined by Equations 9.22 and 9.23, and satisfying, for all* $(i, l, r)^{1,2} \in \hat{V}$,

$$\phi((i, l, r)_1, (i, l)) = \phi((i, r-1), (i, l, r)_2). \tag{9.24}$$

This formulation is quite close to a pure Min-Cost-Network-Flow problem. But the standard algorithms for this type of problem have to be adjusted to include the side constraint given by Equation 9.24. Doing this one obtains a polynomial time algorithm for the DT-problem with ICC (see Refs. [4,17]).

9.3.1.3 Duality Approach

Here we present another approach from Ref. [13] to the ICC, because it yields a nice characterization of the minimal DT, which can be modified to deal with the tongue-and-groove effect and also allows to derive a heuristic for the DC-problem in the next section. We only consider the problem without a minimum separation constraint, i.e., with $\delta_0 = \delta_1 = 0$ (introduced in the end of Section 9.3.1.1). Let the DT-ICC-graph $G = (V, E)$ be a digraph with vertex set V and arc set E defined as follows:

$$V = \{D, D'\} \cup \{(i, j) \mid 1 \leq i \leq m,\ 0 \leq j \leq n+1\}$$

$$E = \{(D, (i, 0)) \mid 1 \leq i \leq m\} \cup \{((i, n+1), D') \mid 1 \leq i \leq m\}$$

$$\cup \{((i, j), (i, j+1)) \mid 1 \leq i \leq m,\ 0 \leq j \leq n\}$$

$$\cup \{((i, j), (i+1, j)) \mid 1 \leq i \leq m-1,\ 1 \leq j \leq n-1\}$$

$$\cup \{((i, j), (i-1, j)) \mid 2 \leq i \leq m,\ 1 \leq j \leq n-1\}.$$

We define a weight function $w : E \to \mathbb{Z}$ by (recall that $a_{i,0} = a_{i,n+1} = 0$ for all i)

$$w(D, (i, 0)) = w((i, n+1), D') = 0 \quad (i = 1, \ldots, m)$$

$$w((i, j-1), (i, j)) = \max\{0, a_{i,j} - a_{i,j-1}\} \quad (i = 1, \ldots, m;\ j = 1, \ldots, n+1)$$

$$w((i, j), (i+1, j)) = -a_{i,j} \quad (i = 1, \ldots, m-1;\ j = 1, \ldots, n-1)$$

$$w((i, j), (i-1, j)) = -a_{i,j} \quad (i = 2, \ldots, m;\ j = 1, \ldots, n-1).$$

Figure 9.7 shows the digraph G for the matrix

$$A = \begin{pmatrix} 4 & 5 & 0 & 1 & 4 & 5 \\ 2 & 4 & 1 & 3 & 1 & 4 \\ 2 & 3 & 2 & 1 & 2 & 4 \\ 5 & 3 & 3 & 2 & 5 & 3 \end{pmatrix}.$$

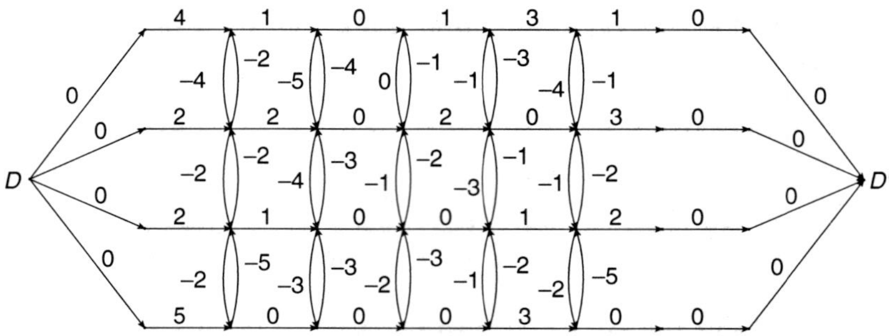

Figure 9.7 The DT-ICC-graph with arc weights corresponding to matrix A.

As usual, the weight $w(P)$ of a path P is just the sum of the weights of the arcs contained in P. Now we can formulate the theoretical result underlying the next decomposition algorithm.

Theorem 5. *The minimal DT of a decomposition of A satisfying the ICC equals the maximal weight of a path from D to D′ in G.*

In analogy with the unconstrained case we denote this maximal weight by $c(A)$.

Sketch of Proof. For the proof of this theorem we need to dualize the DT-problem (Equation 9.2). The LP–dual is

$$\text{Max}\left\{\sum_{i=1}^{m}\sum_{j=1}^{n} a_{i,j}y_{i,j} \mid \sum_{i=1}^{m}\sum_{j=1}^{n} s_{i,j}y_{i,j} \le 1 \text{ for all } S \in \mathcal{S}\right\}. \tag{9.25}$$

The basic idea of the proof is to associate with every (D, D')-path P a dual feasible solution $\boldsymbol{y}^{(P)}$ with objective value equal to the weight of P. By duality, this gives the lower bound for DT. In a second step, we will determine a shape matrix S that the maximal weight of a path with respect to $A' := A - S$ is strictly less than the maximal weight with respect to A, i.e., $c(A') < c(A)$. The value of $y_{i,j}^{(P)}$ depends on how the path P passes through the vertex (i, j).

$$y_{i,j}^{(P)} = \begin{cases} 1 & \text{if } (i, j-1), (i, j), (i, j+1) \in P \text{ and } a_{i,j-1} \le a_{i,j} > a_{i,j+1}, \\ -1 & \text{if } (i, j-1), (i, j), (i, j+1) \in P \text{ and } a_{i,j-1} > a_{i,j} \le a_{i,j+1}, \\ -1 & \text{if } (i, j-1), (i, j), (i \pm 1, j) \in P \text{ and } a_{i,j} < a_{i,j-1}, \\ -1 & \text{if } (i \pm 1, j), (i, j), (i, j+1) \in P \text{ and } a_{i,j+1} \ge a_{i,j}, \\ -1 & \text{if } (i \mp 1, j), (i, j), (i \pm 1, j) \in P, \\ 0 & \text{otherwise.} \end{cases}$$

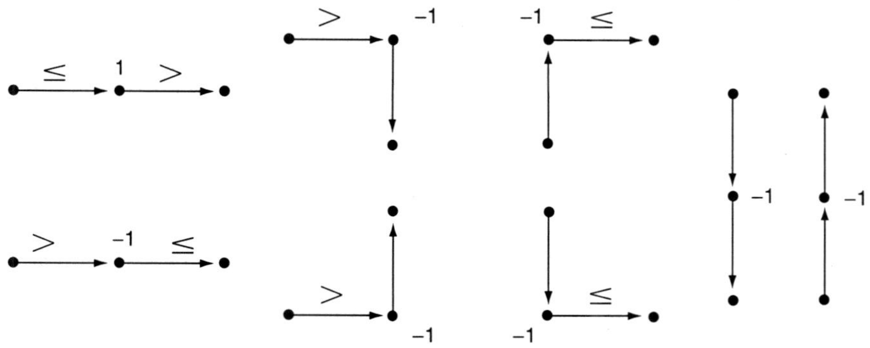

Figure 9.8 Illustration of the dual solution $y^{(P)}$. The labels of the arcs indicate the relation of $a_{i,j}$ to its neighbors on P.

Figure 9.8 illustrates this definition by showing the nonzero values of $y_{i,j}^{(P)}$ depending on the two neighbors of (i, j) in P. The following two lemmas establish the lower bound part of the theorem.

Lemma 3. *For every (D, D')-path P, $\mathbf{y}^{(P)}$ is a feasible solution for the problem given in Equation 9.25.*

Lemma 4. *For every (D, D')-path P, we have*

$$\sum_{i=1}^{m}\sum_{j=1}^{n} y_{i,j}^{(P)} a_{i,j} = w(P).$$

To construct a shape matrix reducing the maximal path weight we consider the following quantities

$$\alpha_1(i, j) := \max\{w(P) \mid P \text{ is a path from } D \text{ to } (i, j)\},$$

$$\alpha_2(i, j) := \max\{w(P) \mid P \text{ is a path from } (i, j) \text{ to } D\},$$

$$\alpha(i, j) := \alpha_1(i, j) + \alpha_2(i, j).$$

In Figure 9.9 we show the necessary information to determine the shape matrix. Observe that $c(A) = \max_{(i,j)} \alpha(i, j)$. We define a 0–1-matrix by

$$s_{i,j} = 1 \iff \alpha(i, j) = c(A), \quad \alpha_1(i, j) = a_{i,j} \text{ and } a_{i,j} > 0. \qquad (9.26)$$

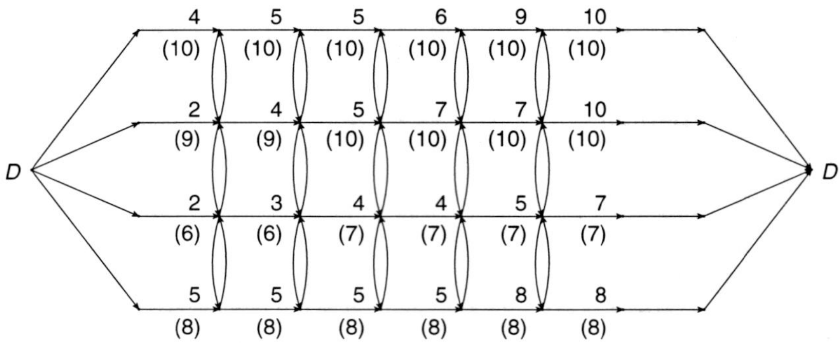

Figure 9.9 The values for α_1 and α (in parentheses) corresponding to the weights in Figure 9.7.

Lemma 5. *The matrix defined by Equation 9.26 is a shape matrix satisfying the ICC.*

Lemma 6. *For the shape matrix defined by Equation 9.26, the matrix $A' = A - S$ is still nonnegative and we have $c(A') = c(A) - 1$.*

Iterating this construction, we obtain a decomposition of A into $c(A)$ shape matrices and this concludes the proof. □

As a consequence of the proof we obtain Algorithm 3. Note that this algorithm yields unidirectional decompositions, i.e., the leaves move only from left to right.

Algorithm 3 (DT-optimal decomposition based on the DT-ICC-graph).
Determine the values of α_1 and α
while $A \neq 0$ **do**
  Determine S according to Equation 9.26
  $A := A - S$
  Update α_1 and α

9.3.2 Tongue-and-Groove Constraint

Recall that to prevent underdosage effects due to the tongue-and-groove design of the leaves we have to require

$$a_{i,j} \leq a_{i-1,j} \wedge s_{i,j} = 1 \implies s_{i-1,j} = 1, \tag{9.27}$$

$$a_{i,j} \geq a_{i-1,j} \wedge s_{i-1,j} = 1 \implies s_{i,j} = 1, \tag{9.28}$$

for $i = 2, \ldots, m$ and $j = 1, \ldots, n$. We call these the tongue-and-groove constraints (TGC). Here, we construct a decomposition with unidirectional leaf movement satisfying the TGC. Recall from Section 9.3.1.1 that such a decomposition is uniquely determined by the numbers $I_L^{(i)}(j)$, $I_R^{(i)}(j)$ ($i = 1, \ldots, m$, $j = 1, \ldots, n+1$). The following lemmas characterize decompositions satisfying TGC (respectively ICC and TGC) in terms of the $I_R^{(i)}(j)$ and $I_L^{(i)}(j)$.

Lemma 7. *The TGC are satisfied if and only if for* $i = 2, \ldots, m, j = 1, \ldots, n$,

(a) $a_{i,j} = 0$ *or* $a_{i-1,j} = 0$, *or*

(b) $I_R^{(i-1)}(j) \leq I_R^{(i)}(j) \leq I_L^{(i)}(j) \leq I_L^{(i-1)}(j)$, *or*

(c) $I_R^{(i)}(j) \leq I_R^{(i-1)}(j) \leq I_L^{(i-1)}(j) \leq I_L^{(i)}(j)$.

Proof. Assume the TGC are satisfied and $\min\{a_{i,j}, a_{i-1,j}\} > 0$. For the t-th shape matrix $S^{(t)} = (s_{i,j}^{(t)})$ we have

$$s_{i,j}^{(t)} = 1 \iff I_R^{(i)}(j) < t \leq I_L^{(i)}(j) \quad (i = 1, \ldots, m;\ j = 1, \ldots, n).$$

From this, we derive that $a_{i,j} \leq a_{i-1,j}$ and Equation 9.27 imply condition (b), while $a_{i,j} \geq a_{i-1,j}$ and Equation 9.28 imply condition (c). Conversely, assume $a_{i,j} \leq a_{i-1,j}$ and $s_{i,j}^{(t)} = 1$. It follows that condition (b) is true, and consequently $s_{i-1,j}^{(t)} = 1$. Similarly, from $a_{i,j} \geq a_{i-1,j}$ and $s_{i-1,j}^{(t)} = 1$ it follows that $s_{i,j}^{(t)} = 1$. □

Lemma 8. *The ICC and TGC are satisfied if and only if for* $i = 2, \ldots, m$, $j = 1, \ldots, n$,

(a) $I_R^{(i-1)}(j) \leq I_R^{(i)}(j) \leq I_L^{(i)}(j) \leq I_L^{(i-1)}(j)$, *or*

(b) $I_R^{(i)}(j) \leq I_R^{(i-1)}(j) \leq I_L^{(i-1)}(j) \leq I_L^{(i)}(j)$.

Proof. If $\min\{a_{i,j}, a_{i-1,j}\} > 0$ the proof is the same as for Lemma 7. If $a_{i,j} = 0$ we have $I_L^{(i)}(j) = I_R^{(i)}(j)$ and the ICC is equivalent to $I_R^{(i)}(j) \leq I_L^{(i-1)}(j)$ and $I_L^{(i)}(j) \geq I_R^{(i-1)}(j)$, so condition (a) follows. Similarly, condition (b) follows from $a_{i-1,j} = 0$. □

Algorithm 4 can be used to obtain leaf sequences satisfying TGC and ICC. The basic idea is similar to the one in Algorithm 2. We construct the $I_L^{(i)}(j)$ and $I_R^{(i)}(j)$ columnwise. In column j we start with the lower bounds $I_L^{(i)}(j) := I_L^{(i)}(j-1) + \max\{0, a_{i,j} - a_{i,j-1}\}$, $I_R^{(i)}(j) := I_R^{(i)}(j-1) + \max\{0, a_{i,j-1} - a_{i,j}\}$,

Algorithm 4 (DT-optimal leaf sequence with TGC and ICC).
for $i = 1, \ldots, m$ **do** $I_L^{(i)}(0) := 0$; $I_R^{(i)}(0) := 0$
for $j = 1, \ldots, n+1$ **do**
  for $i = 1, \ldots, m$ **do**
    $I_L^{(i)}(j) := I_L^{(i)}(j-1) + \max\{0, a_{i,j} - a_{i,j-1}\}$
    $I_R^{(i)}(j) := I_R^{(i)}(j-1) + \max\{0, a_{i,j-1} - a_{i,j}\}$
  for $i = 2, \ldots, m$ **do**
    if $a_{i,j} \leq a_{i-1,j}$ **then**
      if $I_R^{(i)}(j) < I_R^{(i-1)}(j-1)$ **then**
        $\Delta := I_R^{(i-1)}(j) - I_R^{(i)}(j)$
        $I_R^{(i)}(j) := I_R^{(i)}(j) + \Delta$; $I_L^{(i)}(j) := I_L^{(i)}(j) + \Delta$
      if $I_L^{(i)}(j) > I_L^{(i-1)}(j)$ **then** `Update` $(i-1)$
    else //the case $a_{i,j} > a_{i-1,j}$
      if $I_L^{(i)}(j) < I_L^{(i-1)}(j)$ **do**
        $\Delta := I_L^{(i-1)}(j) - I_L^{(i)}(j)$
        $I_R^{(i)}(j) := I_R^{(i)}(j) + \Delta$; $I_L^{(i)}(j) := I_L^{(i)}(j) + \Delta$
      if $I_R^{(i)}(j) > I_R^{(i-1)}(j)$ **then** `Update` $(i-1)$

Function `Update` (k)
if $a_{k,j} \leq a_{k+1,j}$ **then**
  $\Delta := I_R^{(k+1)}(j) - I_R^{(k)}(j)$
  $I_R^{(k)}(j) := I_R^{(k)}(j) + \Delta$; $I_L^{(k)}(j) := I_L^{(k)}(j) + \Delta$
else //the case $a_{k,j} > a_{k+1,j}$
  $\Delta := I_L^{(k+1)}(j) - I_L^{(k)}(j)$
  $I_R^{(k)}(j) := I_R^{(k)}(j) + \Delta$; $I_L^{(k)}(j) := I_L^{(k)}(j) + \Delta$
if $k \geq 2$, $a_{k,j} \leq a_{k-1,j}$ and $I_L^{(k)}(j) > I_L^{(k-1)}(j)$ **then** `Update` $(k-1)$
if $k \geq 2$, $a_{k,j} > a_{k-1,j}$ and $I_R^{(k)}(j) > I_R^{(k-1)}(j)$ **then** `Update` $(k-1)$

and eliminate the violations of Lemma 8. If $a_{i,j} \leq a_{i-1,j}$, condition (a) in Lemma 8 must be satisfied. This can be violated if $I_R^{(i)}(j) < I_R^{(i-1)}(j-1)$ or $I_L^{(i)}(j) > I_L^{(i-1)}(j)$. In the first case, we increase $I_L^{(i)}(j)$ and $I_R^{(i)}(j)$ by the minimum amount such that the condition holds, and in the second case we increase $I_L^{(i-1)}(j)$ and $I_R^{(i-1)}(j)$. In this second case, there might be a new violation between row $i-1$ and row $i-2$. The recursive call of the function `Update` (k) takes care of this. If there is no ICC, we just have to add the condition $\min\{a_{i,j}, a_{i-1,j}\} > 0$ to the conditions for changing the values $I_L^{(i)}(j)$ and $I_R^{(i)}(j)$. Let Algorithm 4′ denote the result of this modification.

Theorem 6 [16]. *Algorithm 4 yields DT-optimal decompositions with ICC and TGC under the additional condition of unidirectional leaf movement in* *time $O(m^2n)$. Algorithm 4′ yields DT-optimal decompositions with TGC and without ICC under the additional condition of unidirectional leaf movement in time $O(m^2n)$.*

The proof of this theorem is essentially the same as the proof of Theorem 3. For the problem with ICC and TGC, the condition on the unidirectional leaf movement can be dropped: using the duality based Algorithm 3 with a modified weight function yields a decomposition with unidirectional leaf movement that is optimal among all decompositions with ICC and TGC [12].

9.4 Decomposition Cardinality Problem

In this section, we consider the DC-problem (Equation 9.3). In the first subsection, we show that this problem is very hard, and in the second, we give heuristic approaches.

9.4.1 *Computational Complexity of the DC-Problem*

The fact that the DC-problem is NP-hard already for a single-row matrix was proved first by Burkart [6] who gave a reduction from 2-partition. A similar idea was used in Ref. [2] to reduce 3-partition showing the strong NP-hardness, i.e., the nonexistence of a pseudopolynomial algorithm unless $P = \text{NP}$.

Theorem 7 [2]. *The problem Equation 9.3 is strongly NP-hard, even for matrices with a single row.*

Proof. The decision version of the single-row DC-problem is as follows:

Instance: A vector $\boldsymbol{a} = (a_1, \ldots, a_n)$ with $a_i \in \mathbb{N}$, $K \in \mathbb{N}$

Question: Does a decomposition of $\boldsymbol{a}$ into at most K shape matrices exist?

Note that a shape matrix in this case is nothing else than a row vector with the consecutive ones property. We use reduction from the problem 3-partition, which is well known to be strongly NP-hard [10].

Instance: $B, Q \in \mathbb{N}$, $b_1, b_2, \ldots, b_{3Q} \in \mathbb{N}$ with $\sum_{j=1}^{3Q} b_j = QB$ and $\frac{B}{4} < b_j < \frac{B}{2}$ for all j

Question: Does a partitioning of $\{b_1, \ldots, b_{3Q}\}$ into triples $T_1, \ldots, T_Q$ such that $\sum_{b \in T_q} b = B$ for all $q = 1, \ldots, Q$ exist?

We define an instance of the DC-problem as follows.

$$n = 4Q,$$

$$a_j = \begin{cases} \sum_{k=1}^{j} b_k & \text{for } j = 1, \ldots, 3Q, \\ (4Q - j + 1)B & \text{for } j = 3Q + 1, \ldots, 4Q, \end{cases}$$

$$K = 3Q.$$

Now it is not difficult to see that the instance of 3-partition has answer YES if and only if the instance of the DC-problem has answer YES. □

We want to mention that this reduction proves even more, namely that it is already hard to find an approximate solution of the DC-problem. To be precise, the DC-problem is APX-hard even for single-row matrices with entries polynomially bounded in n. That means there is some $\varepsilon > 0$ such that, unless $P = \text{NP}$, there is no polynomial algorithm that decides whether the necessary number of shape matrices is K or at least $(1+\varepsilon)K$. This was shown in Ref. [3] using a result on the APX-hardness of 3-partition from Ref. [19].

The strong NP-hardness of 3-partition means that the problem remains NP-hard even if the input numbers are bounded by some constant. But in the reduction to the DC-problem, we produce a vector with very large entries, because we have to sum up all the numbers from the 3-partition instance. So, we can still hope for an efficient algorithm if we bound the entries of the matrix by some constant L, i.e., we require $a_{i,j} \leq L$ for all (i, j). Observe that for the case $L = 1$ any optimal solution to the DT-problem is also optimal for the DC-problem. And indeed, there is a result in this direction: for constant L, the DC-problem without ICC and TGC can be solved in time $O(mn^{2L+2})$ [11]. In Ref. [18], the algorithm was extended to find the exact minimum of the DC without the restriction that the DT has to be minimal. But these pseudopolynomial algorithms are of very limited practical value, not only because of the L in the exponent but also because the constant in the O-notation grows very fast with L. So, it is natural to require heuristic approaches to the DC-problem.

9.4.2 Heuristics for the DC-Problem

Most of the algorithms in the literature look for a decomposition with minimum DC among all decompositions with minimum DT. So the problem (which we also call DC-problem in the following) is

$$\textbf{(DC}'\textbf{)} \quad \min\left\{ |\mathcal{S}_0| \;\mid\; \mathcal{S}_0 \subseteq \mathcal{S},\; A = \sum_{S\in\mathcal{S}_0} u_S S,\; u_S \in \mathbb{N},\; \sum_{S\in\mathcal{S}_0} u_S \text{ is minimal.} \right\}.$$

Note that in general it is not possible to minimize both quantities simultaneously, as can be seen by the following example (from Ref. [14]):

$$\begin{pmatrix} 2 & 6 & 3 \\ 4 & 5 & 6 \end{pmatrix} = 3\begin{pmatrix} 0 & 1 & 1 \\ 1 & 1 & 1 \end{pmatrix} + \begin{pmatrix} 1 & 1 & 0 \\ 1 & 1 & 1 \end{pmatrix} + \begin{pmatrix} 1 & 1 & 0 \\ 0 & 1 & 1 \end{pmatrix} + \begin{pmatrix} 0 & 1 & 0 \\ 0 & 1 & 0 \end{pmatrix}.$$

This is a decomposition with DT = 6, which cannot be achieved with three shape matrices. But allowing DT = 7, three shape matrices are sufficient:

$$\begin{pmatrix} 2 & 6 & 3 \\ 4 & 5 & 6 \end{pmatrix} = 4\begin{pmatrix} 0 & 1 & 0 \\ 1 & 1 & 1 \end{pmatrix} + 2\begin{pmatrix} 1 & 1 & 1 \\ 0 & 0 & 1 \end{pmatrix} + \begin{pmatrix} 0 & 0 & 1 \\ 0 & 1 & 0 \end{pmatrix}.$$

So the problem (DC′) is really different from Equation 9.3. As before, let $c(A)$ denote the minimal DT for matrix A. A very natural greedy strategy is to look for a shape matrix S that can be extracted with a large coefficient u, such that $c(A - uS) = c(A) - u$, i.e., uS can be extended to a DT-optimal decomposition.

9.4.2.1 Unconstrained Case

The following greedy heuristic for the unconstrained case was proposed in Ref. [7]. We are looking for a pair (u, S) of a positive integer u and a shape matrix S, such that $A - uS$ is still nonnegative, $c(A - uS) = c(A) - u$ and u is maximal under these conditions. Let $u_{\max}$ be this maximal possible value. Using the notation introduced before Theorem 1, $c(A - uS) = c(A) - u$ is equivalent to

$$c_i(A - uS) \leq c(A) - u \quad (i = 1, \ldots, m).$$

Define the complexity gap of row i to be $g_i(A) := c(A) - c_i(A)$. As before, we describe the shape matrix by the parameters l_i and r_i $(i = 1, \ldots, m)$.

Lemma 9. *There is a pair (u, S) with $u = u_{max}$ and, for all i, either $l_i = r_i - 1$ or ($a_{i,l_i} < a_{i,l_i+1}$ and $a_{i,r_i-1} > a_{i,r_i}$).*

We put $d_{i,j} = a_{i,j} - a_{i,j-1}$ for $i = 1, \dots, m$, $j = 1, \dots, n$, and define

$$v_i(l, r) = \begin{cases} g_i(A) & \text{if } l = r - 1, \\ g_i(A) + \min\{d_{i,l+1}, -d_{i,r}\} & \text{if } l < r - 1 \text{ and } g_i(A) \leq \\ & \quad |d_{i,l+1} + d_{i,r}|, \\ \left(d_{i,l+1} - d_{i,r} + g_i(A)\right)/2 & \text{if } l \leq r \text{ and } g_i(A) > |d_{i,l+1} + d_{i,r}|. \end{cases}$$

LEMMA 10. *$c_i(A - uS) \leq c(A) - u$ if and only if $u \leq v_i(l_i, r_i)$.*

For convenience, we denote the set of pairs (l, r) to which we restrict our search in row i by $\mathcal{I}_i$, that is we put

$$\mathcal{I}_i := \{(l, r) \ : \ 0 \leq l \leq r - 1 \leq n \text{ and either } l = r - 1 \text{ or } (d_{i,l+1} > 0 \text{ and } d_{i,r} < 0)\}.$$

Clearly the nonnegativity of $A - uS$ is equivalent to $u \leq w_i(l_i, r_i)$ for all i, where

$$w_i(l, r) = \begin{cases} \infty & \text{if } l = r - 1, \\ \min\limits_{l<j<r} a_{i,j} & \text{if } l < r - 1. \end{cases}$$

Now we put, for $1 \leq i \leq m$ and $(l, r) \in \mathcal{I}_i$, $\hat{u}_i(l, r) = \min\{v_i(l, r), w_i(l, r)\}$, and for $i = 1, \dots, m$,

$$\tilde{u}_i = \max_{(l,r)\in\mathcal{I}_i} \hat{u}_i(l, r).$$

Then

$$u_{\max} = \min_{1\leq i\leq m} \tilde{u}_i.$$

To construct a shape matrix S such that, for $u = u_{\max}$, $A - uS$ is nonnegative and $c(A - uS) = c(A) - u$, we just have to find, for every $i \in [m]$, a pair $(l_i, r_i) \in \mathcal{I}_i$ with $\hat{u}_i(l_i, r_i) \geq u_{\max}$. A trivial way of doing this is to take a pair (l_i, r_i) where the maximum in the definition of $\tilde{u}_i$ is attained, i.e., with $\hat{u}_i(l_i, r_i) = \tilde{u}_i$. These (l_i, r_i) can be computed simultaneously with the calculation of $u_{\max}$ and this method yields $mn + n - 1$ as an upper bound for the DC of the decomposition. But there are better constructions for S after the determination of $u_{\max}$. We put

$$q(A) = \left|\{(i, j) \in [m] \times [n] \ : \ d_{i,j} \neq 0\}\right|,$$

and choose a shape matrix S so that $q(A - uS)$ is minimized. To make this precise, for $1 \leq i \leq m$ and $(l, r) \in \mathcal{I}_i$, we put

$$p_i(l,r) = \begin{cases} 2 & \text{if } d_{i,l+1} = -d_{i,r} = u_{\max}, \\ 1 & \text{if } d_{i,l+1} = u_{\max} \neq -d_{i,r} \text{ or } d_{i,l+1} \neq u_{\max} = -d_{i,r}, \\ 0 & \text{if } l = r+1 \text{ or } d_{i,l+1} \neq u_{\max} \text{ and } -d_{i,r} \neq u_{\max}. \end{cases}$$

Now, for (l_i, r_i) we choose among the pairs $(l, r) \in \mathcal{I}_i$ with $\hat{u}_i(l, r) \geq u_{\max}$ one with maximal value of $p_i(l, r)$, and if there are several of these we choose one with maximal value of $r - l$. As an example, we obtain the following decomposition:

$$\begin{pmatrix} 4&5&0&1&4&5 \\ 2&4&1&3&1&4 \\ 2&3&2&1&2&4 \\ 5&3&3&2&5&3 \end{pmatrix} = 4\begin{pmatrix} 1&1&0&0&0&0 \\ 0&0&0&0&0&1 \\ 0&0&0&0&0&1 \\ 1&0&0&0&0&0 \end{pmatrix} + 2\begin{pmatrix} 0&0&0&0&1&1 \\ 0&0&0&1&0&0 \\ 1&1&1&0&0&0 \\ 0&1&1&1&1&0 \end{pmatrix} + \begin{pmatrix} 0&0&0&1&1&1 \\ 1&1&1&1&0&0 \\ 0&0&0&1&1&0 \\ 1&1&1&0&0&0 \end{pmatrix}$$

$$+ \begin{pmatrix} 0&0&0&0&1&1 \\ 1&1&0&0&0&0 \\ 0&1&0&0&0&0 \\ 0&0&0&0&1&1 \end{pmatrix} + \begin{pmatrix} 0&1&0&0&0&0 \\ 0&1&0&0&0&0 \\ 0&0&0&0&1&0 \\ 0&0&0&0&1&1 \end{pmatrix} + \begin{pmatrix} 0&0&0&0&0&1 \\ 0&1&0&0&0&0 \\ 0&0&0&0&0&0 \\ 0&0&0&0&1&1 \end{pmatrix}.$$

9.4.2.2 Interleaf Collision Constraint

Using the min–max characterization of the minimal DT from Theorem 5, we can use the same strategy as for the unconstrained case: denoting by $c(A)$ the maximal weight with respect to A of a (D, D')-path in the DT-ICC-graph, we are looking for a pair (u, S) with maximal u such that $A - uS$ is nonnegative and $c(A - uS) = c(A) - u$. If this is the case, we call the pair (u, S) admissible. An additional difficulty comes from the fact that the influence of the extraction of uS on the weight of a path through a vertex (i, j) does not depend only on the i-th row of S so the determination of the values

$$\hat{u}_i(l,r) := \max\{u \mid \exists \text{ shape matrix } S \text{ with } l_i = l,\ r_i = r \text{ such that}$$
$$A - uS \text{ is nonnegative and } c(A - uS) = c(A) - u\}$$

becomes much harder. But suppose these values, or at least some upper bounds $u_0(i, l, r)$ for them, are given. Then for given u, every shape matrix S, such that (u, S) is admissible, corresponds to a path

$$D, (1, l_1, r_1), (2, l_2, r_2), \ldots, (m, l_m, r_m), D'$$

in the digraph defined in the beginning of Section 9.3.1.2 and illustrated in Figure 9.5, such that $u_0(i, l_i, r_i) \geq u$ for $i = 1, \ldots, m$. We put

$$\hat{u} = \max\{u \ : \ \text{There is a path } D, (1, l_1, r_1), \ldots, (m, l_m, r_m), D' \\ \text{with } u_0(i, l_i, r_i) \geq u \text{ for } i = 1, \ldots, m\}.$$

Clearly, $\hat{u}$ is an upper bound for the coefficient u in an admissible pair (u, S). The backtracking described in Algorithm 5 constructs an admissible pair (u, S) with maximal u. Starting with $u = \hat{u}$, the algorithm searches for a shape matrix S such that (u, S) is admissible, and if this is not possible, the value is decreased by one. The shape matrix is build up row by row, and the stopping criterion in row i is that after extracting the current candidates for the first i rows with coefficient u leads to a path P with all its vertices in the first i rows and $w(P) > c(A) - u$. The maximal weight of such a path is denoted by `MaxWeight(`i`)`. Iterating Algorithm 5 with $A' = A - uS$ we obtain a decomposition of A.

Algorithm 5 (Greedy step in the heuristic for the DC-problem with ICC).
Function `Construct Shape Matrix`
$u := \hat{u}$
finished := false
$l_0 := 0$; $r_0 := n + 1$
while not finished **do**
  `Complete Shape Matrix(1)`
  if not finished **then** $u := u - 1$

Function `Complete Shape Matrix(`i`)`
for (l_i, r_i) with $0 \leq l_i \leq r_{i-1} - 1$, $\max\{l_i, l_{i-1}\} + 1 \leq r_i \leq n + 1$
    and $u_0(i, l_i, r_i) \geq u$ **do**
  if `MaxWeight(`i`)` $\leq c(A) - u$ **then**
    if $i < m$ **then** `Complete Shape Matrix(`$i + 1$`)` **else** finished := true

The performance of this backtracking depends very much on the quality of the bounds $u_0(i, l, r)$. In Ref. [12] some bounds that work quite well in practice are described. A drawback of this method is that we have almost no control of the running time. Experiments with randomly generated matrices show that the algorithm is fast for the vast majority of matrices but there are some examples where it is extremely slow.

References

1. R.K. Ahuja and H.W. Hamacher. A network flow algorithm to minimize beam-on time for unconstrained multileaf collimator problems in cancer radiation therapy. *Networks*, 45(1):36–41, 2005.
2. D. Baatar, H.W. Hamacher, M. Ehrgott, and G.J. Woeginger. Decomposition of integer matrices and multileaf collimator sequencing. *Discrete Appl. Math.*, 152(1–3):6–34, 2005.
3. N. Bansal, D. Coppersmith, and B. Schieber. Minimizing setup and beam-on times in radiation therapy. In J. Diaz et al., editor, *Approximation, Randomization, and Combinatorial Optimization. Algorithms and Techniques*, volume 4110 of LNCS, Springer-Verlag, Heidelberg, 2006, pp. 27–38.
4. N. Boland, H.W. Hamacher, and F. Lenzen. Minimizing beam-on time in cancer radiation treatment using multileaf collimators. *Networks*, 43(4):226–240, 2004.
5. T.R. Bortfeld, D.L. Kahler, T.J. Waldron, and A.L. Boyer. X–ray field compensation with multileaf collimators. *Int. J. Radiat. Oncol. Biol. Phys.*, 28:723–730, 1994.
6. R. Burkart. Open problem session, Oberwolfach Conference on Combinatorial Optimization, November 24–29, 2002.
7. K. Engel. A new algorithm for optimal multileaf collimator field segmentation. *Discrete Appl. Math.*, 152(1–3):35–51, 2005.
8. K. Engel and E. Tabbert. Fast simultaneous angle, wedge, and beam intensity optimization in inverse radiotherapy planning. *Optimization Eng.*, 6(4):393–419, 2005.
9. J.M. Galvin, X.G. Chen, and R.M. Smith. Combining multileaf fields to modulate fluence distributions. *Int. J. Radiat. Oncol. Biol. Phys.*, 27:697–705, 1993.
10. M.R. Garey and D.S. Johnson. *Computers and Intractability, a Guide to the Theory of NP–Completeness.* W.H. Freeman, New York, 1979.
11. T. Kalinowski. The algorithmic complexity of the minimization of the number of segments in multileaf collimator field segmentation. Preprint 04/1, Institut für Mathematik, Universität Rostock, 2004.
12. T. Kalinowski. Reducing the tongue-and-groove underdosage in MLC segmentation. Preprint 04/3, Institut für Mathematik, Universität Rostock, 2004.
13. T. Kalinowski. A duality based algorithm for multileaf collimator field segmentation with interleaf collision constraint. *Discrete Appl. Math.*, 152(1–3):52–88, 2005.
14. T. Kalinowski. Realization of intensity modulated radiation fields using multileaf collimators. In R. Ahlswede et al., editor, *Information Transfer and Combinatorics*, volume 4123 of LNCS, Springer-Verlag, Heidelberg, 2006, pp. 1010–1055.
15. S. Kamath, S. Sahni, J. Li, J. Palta, and S. Ranka. Leaf sequencing algorithms for segmented multileaf collimation. *Phys. Med. Biol.*, 48(3):307–324, 2003.
16. S. Kamath, S. Sartaj, J. Palta, S. Ranka, and J. Li. Optimal leaf sequencing with elimination of tongue-and-groove underdosage. *Phys. Med. Biol.*, 49:N7–N19, 2004.
17. F. Lenzen. An integer programming approach to the multileaf collimator problem. Master's thesis, Department of Mathematics, University of Kaiserslautern, 2000.

18. M. Nußbaum. Min cardinality C1 decomposition of integer matrices. Master's thesis, Department of Mathematics, University of Kaiserslautern, TU Kaiserslautern, 2006.
19. E. Petrank. The hardness of approximation: Gap location. *Comput. Complex.*, 4:133–157, 1994.
20. H.E. Romeijn, R.K. Ahuja, J.F. Dempsey, and A. Kumar. A column generation approach to radiation therapy treatment planning using aperture modulation. *SIAM J. Optimization*, 15(3):838–862, 2005.
21. R.A.C. Siochi. Minimizing static intensity modulation delivery time using an intensity solid paradigm. *Int. J. Radiat. Oncol. Biol. Phys.*, 43:671–680, 1999.
22. P. Xia and L. Verhey. Multileaf collimator leaf sequencing algorithm for intensity modulated beams with multiple static segments. *Med. Phys.*, 25:1424–1434, 1998.

Chapter 10

Optimal Planning for Radiation Therapy

Mark Langer, Ronald Rardin, and Ali Tuncel

CONTENTS

10.1 Introduction 288
10.2 Treatment Planning of Dose Distributions 289
  10.2.1 Treatment Planning Decisions 291
  10.2.2 Planning Outputs 291
10.3 Treatment Optimization Modeling 292
  10.3.1 Assumptions 292
  10.3.2 Optimization Model Overview 294
  10.3.3 Dose Coefficients 294
  10.3.4 Dose and Homogeneity Constraints 296
  10.3.5 Scale of Models 296
    10.3.5.1 Feasibility Problems and Objectives 297
    10.3.5.2 Modeling Dose–Volume Constraints 298
10.4 Introducing Optimization of Orientations 299
  10.4.1 Optimization Approaches 299
    10.4.1.1 Mixed-Integer Linear Programming 300
    10.4.1.2 Linear Programming 300
    10.4.1.3 Conjugate Gradient Nonlinear Approaches 300
    10.4.1.4 Metaheuristics Including Simulated Annealing 301
    10.4.1.5 Direct Inverse Methods 301
    10.4.1.6 Biologic Function Formulations 301
10.5 Intensity Map Delivery 302
10.6 Four-Dimensional and Fractionated Planning 303
10.7 Conclusion 304
References 304

10.1 Introduction

Radiation therapy is used with over half of cancer patients. It is used as an adjuvant after surgery, either alone or combined with systemic therapies to increase cure rates for many tumor sites. It is also employed as the primary agent of local control to render patients disease free at the primary tumor site and associated lymph node basin while preserving function in surrounding organs. The words optimization and radiotherapy have long been intertwined. A citation search reveals the terms optimization or optimal, linked with radiation or radiotherapy[1] in publications dating to 1959.[2] A number of publications provide overviews of this field.[3–6]

Much of the early work concerned the optimal division of a course of radiation into individual sessions, called fractions, using patterns that could be generalized to wide populations. The work employed models of the differential response of tumor and normal tissues to the number of sessions into which the total prescribed dose is divided and the spacing of these sessions over time. It was recognized early that response and risk depended on irradiated volume and tumor volume, number of fractions and over time of the treatment course.[7] Volume ranges over a continuum between approximately 10^0 and 10^3 cm^3; the number of fractions ranges over integer values between roughly 3 and 40, and the overall treatment time is expressed in days in the range 3–60. The development of improved fractionation schedules based on predictive formulae remains an active area of research, whose findings might ultimately change the prescribed rules for radiation administration.[8–10] (see also Section 10.5). For now, treatment is arranged so that a cumulative plan distributing doses across various tissues can be decomposed into separate additive sessions, whose tissue dose levels all fall within their allowed ranges in each session and cumulatively.

This chapter concentrates on a different problem that is characteristic of dose optimization in radiotherapy. The problem is to optimize the spatial distribution of radiation dose in the individual patient. The optimization is performed over some function, which maps delivery parameter values to the patient's dose distribution. Radiation-absorbed dose, like temperature, is a quantity that can be defined at each point in the body. A treatment is evaluated by the distribution of dose values that it produces.

The concept of radiation dose is different from the familiar notion of drug dose, which refers to how much is given to the patient by some route, and how much is not to be taken up at some point. A precise description of the radiation dose distribution at every point in the body can be made from a specified delivery scheme. The basic planning challenge is to produce a distribution of dose values across all points of the body that meet the competing goals of safety in healthy tissue and eradication of disease in target. The radiation may arise from surgically implanted sources, in what

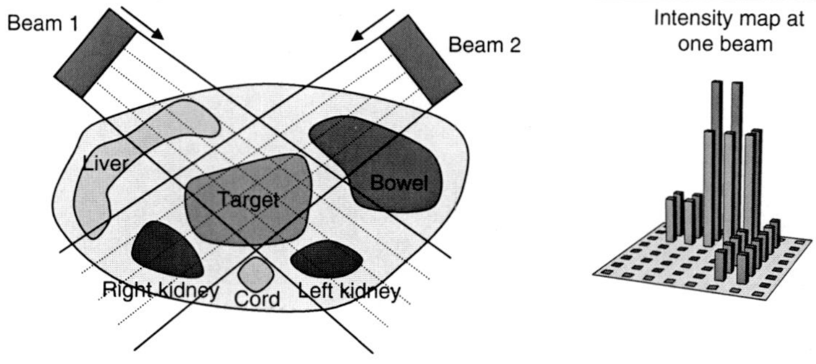

Box 10.1 Schematic of the external beam planning problem. Anatomic contours are displayed on a planar slice through the body, along with two beams from a larger set. Beams penetrate through the body in the directions shown by solid arrows (→). Beam borders, depicted by solid lines (–) diverge slightly outwards with distance. Use of multiple beams allows combined dose on the target tumor while distributing dose to surrounding tissues. A beam may be divided into smaller component units (illustrated by dotted lines), typically projected as a regularly spaced grid along the beam front. An intensity map at each beam angle determines how radiation is divided across components or beamlets of each grid.

is known as brachytherapy. Still, most is external beam therapy where radiation is generated from an accelerator that can be positioned around the patient at any angle (see Box 10.1). Beam angles, energy levels, time-on, and shape across the face of the beam then become the decision variables subject to planner control.

The remainder of this chapter focuses on those issues in external beam therapy planning. Still, other layers of complexity enter into consideration (see Box 10.2). These include treatment costs, and the risk that imprecision in realizing the expected dose distribution will produce a catastrophic failure (e.g., severing of the spinal cord function). The expected distribution is typically developed under a framework that is designed to limit treatment costs or session time and avoid catastrophic failure. The framework may limit the number of beams or their placement, or add safety margins to the boundaries of critical structures.

10.2 Treatment Planning of Dose Distributions

As already noted, radiation dose has a physical meaning that is analogous to tissue concentration in pharmacology, or temperature in engineering. Dose, in units of gray (1 Gy = 1 J/kg) denotes the energy per unit mass taken up in ionization within an infinitesimally small volume about a point. This ionization can produce chemical and ultimately biological effects.

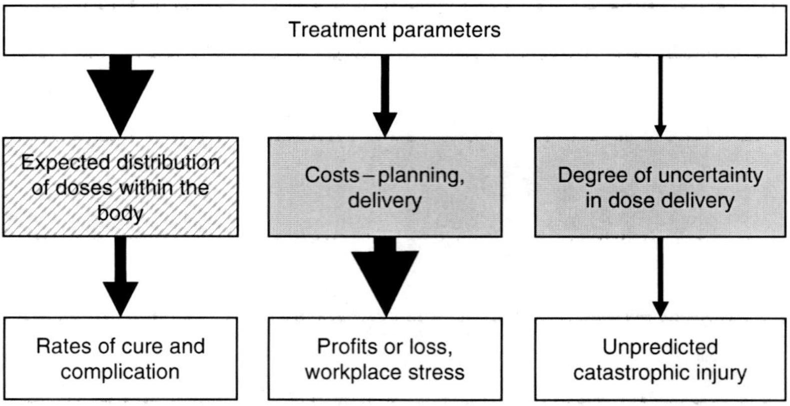

Box 10.2 Sketch of the treatment planning problem.
Shaded boxes indicate distinct endpoints considered for the individual. The central problem—dose formation—is shown in hatch. Unfilled boxes exhibit relations most reliably demonstrated in a population. The relative degrees of certainty in the relations are suggested by arrow widths.

A characteristic of radiation therapy is that the dose distribution can be fairly accurately determined for the individual patient from a given treatment scheme. The basic template is cast by the choice of beam angles, beam shape, and beam energy. An appealing property of radiotherapy is that the dose distribution can be refined, spatially and potentially over a rapid timescale, by manipulating the intensity profiles of the radiation beams. In cooking, one alters the temperature distribution by exchanging heat sources usually once or twice (fire vs electric; pot vs pan), reorienting food position a few times (flipping, stirring), choosing cooking time, and in a final refinement tuning the strengths in individual heating elements.

In analogy, the energy source in radiotherapy treatment is the incoming beam. Treatment design (refer to Box 10.1) entails the selection of beam energy from a limited number of choices, posing the source relative to the body at a limited number of positions, selecting intensity (on also known as fluence) at each position which will be proportional to time-on, and in a final refinement, varying intensity across different parts of the source. Either the intensity of an entire beam is uniformly changed or in a modern approach, the beam can be divided into smaller elements each transmitting a different intensity, a technique known as intensity-modulated radiotherapy.

Modulation of intensities can be accomplished by varying the exposure of each beam element, typically by moving thin blocking leaves in and out of the beam for varying lengths of time.[11] An alternative is to fabricate attenuating solids that project to different depths along the face of the beam.[12] As a first approximation, the exposure time or path attenuation of a beam element determines its intensity. However, the leaves or attenuators introduce perturbations into the intended intensity maps due to effects such as scatter, beam divergence, beam leakage, and nonuniformities in material, and choices may be limited by considerations of delivery time. These second-order effects force design choices that yield trade-offs in plan cost and quality.[13] The sequence of leaf positions (or attenuator geometry) that is taken to transmit a desired profile of intensities represents another layer of input that determines the final dose distribution.[14]

10.2.1 Treatment Planning Decisions

The decisions to be made in planning can be divided into those that set the dose constraints and goals, and those that determine the delivery of radiation, given the dose specifications. The dose specifications largely emerge from the medical environment, although sociological, economical, statistical, and mathematical forces shape their statements. Typically, the number of allowed beam positions is modest and only one or two beam energies are available, with acceptable choices in these areas made by human judgment. Once beam energy and placement choices are made, the planning problem devolves into the selection of intensities and the provision of a method to realize their delivery. When beams are divided into hundreds of smaller elements this intensity assignment problem is too large to be handled without automated assistance.

Other decisions that enter into planning, and which are susceptible to varying degrees of optimization, include the choice of dose constraints and the identification of anatomical compartments on which the constraints are to be placed. A statistical problem is to derive an accurate geometric distribution of the target given snapshots of its position, orientation and if deformable, its shape, over time.[15] In current 3D practice, a target identified on an image series considered to be instantaneously acquired is expanded to form a planning target volume that acknowledges motion over the several minutes needed to deliver treatment each session.[16] A problem in combinatorial geometry is to determine a minimum volume region that envelopes the shifting target with specified probability.[17,18] These problems are the subject of theoretic and empiric research under the rubric of 4D radiotherapy.[19]

10.2.2 Planning Outputs

The dose at each point in the body is determined by the treatment scheme that is setup. Each person's anatomy presents its own geometry, and as is too well known, there are secular and stochastic changes in one's own body over time (e.g., breathing). For the most part, work has concerned optimization over a fixed space, but improvements in radiologic technology have sparked interest in optimizing delivery over a space that can vary over time.

Exhibiting the dose distribution is itself a problem that is not straightforward. Physics-based simulations are typically employed after each iteration of planning to predict the dose distribution that would result from given parameter choices. Points are chosen along a fine grid in the structures of interest, and the dose at each point is estimated.[20]

Dose reports examined by the physician are graphical or tabular summaries of the dose distribution over discrete point sets into which structures of interest are resolved. Values for sample points are used to determine 2D curves of constant dose, called isodose lines (analogous to isotherm plots on weather charts). They are viewed superimposed over anatomy as a visual check on safety and efficacy. Small areas of over- or under-dose may be missed by this review, and estimates of the volume distribution of dose are almost impossible to make. Consequently, dose–volume histograms are generated to display the cumulative distribution of dose over any structure volume of interest (see Box 10.1).[21]

10.3 Treatment Optimization Modeling

Optimization in radiation therapy, as applied to intensity modulation, is often taken to mean the process of generating a computer-based assignment of intensity/fluence values for the beam elements, together with supplying a method for their delivery that in combination yield an acceptable plan. The planning process has been termed inverse plan optimization.[22]

Given the complexity of the joint planning and delivery problem, it is convenient if inexact to first consider separately one problem of producing an intensity map and the other of supplying a process for its delivery, and then to consider how the two problems interact.[23] This section will review the constraints and objectives that enter into intensity map generation, and the next section will explore solution methods. The problem of intensity map delivery will be discussed in Section 10.5.

10.3.1 Assumptions

An important simplification for those posed with optimizing treatment plans is that dose at any point can be taken as a linear combination of the intensities of the rays impinging on the patient. The method of intensity-modulated radiotherapy divides a beam into a panel of small unit elements, each of which is considered to carry a separate intensity value, and whose linear combination determines the dose distribution.[24] Strictly, the actual intensity need not be constant across a small beam element mainly because the tongue and groove construction of the metal leaves that slide into and out of the field to differentially block beam elements can lead to non uniform effects along the boundary lines, as can mismatch between the leaf cut and the divergence line of the abutting beam ray.[25,26] An accepted if imperfect approach is to consider the intensity across an element to be constant given certain leaf setup constraints taken to attenuate the boundary effects.

A much more difficult question is how dose affects treatment outcomes.[27] Although effects of dose on outcome are certainly nonlinear, they are also widely uncertain. Even the form of the relation is unknown other than an assumed monotonicity: increase in dose at a point should not decrease the observed effect. The relations of dose to outcome that give rise to constraints need not be linear; the number of intensity arguments and dose constraints may be too large to admit exact linear methods, and the effects on different tissues of how administered dose is divided over time are nonuniform and nonlinear. The complexity of optimization is further clouded when delivery effects are considered. Intensity perturbations at the leaf edge degrades delivery of the produced intensity map, and delivery conditions may argue to limit the complexity of the produced intensity maps, measured as local or global fluctuations in intensity values. These limits may be described by nonlinear conditions on intensity variation.[28]

Nonlinear parameterized formulae of complication or cure probabilities have been proposed. In particular, one may seek to optimize constructed biologic functions of tumor control probability (TCP) and normal tissue complication probability (NTCP), both nonlinear.[29,30] At present, however, they have not been shown to provide better measures of the safety or efficacy profile than those given by standard dose or dose–volume indicators.

These uncertainties have led physicians to standardize prescription directives, the constraints and objectives of the problem, by a compact set of dose limits. However expressed, radiation problems demand tight satisfaction of stated constraints and their solutions should have small optimization gaps. Violations of about 3 percent in strict absolute dose limits match allowed errors in dose reporting, forming one kind of acceptance criterion. Violations in constrained volume are less established, but a 3 percent tolerance can be taken to match the stated dose tolerance. Constraint violations are readily detected using dose–volume histograms or visual inspection of isodose lines, but optimization shortfalls are hidden. Optimization shortfalls of about 3 percent could produce effects detectable in populations of a size that can be accrued in clinical trials, amounting to about a 5 percent difference in the end point.

10.3.2 Optimization Model Overview

The classic optimization problem in radiotherapy is to produce beams along with their intensity maps given a specification of dose constraints and objectives. A beam in intensity-modulated therapy is defined by its energy, the orientation of its central ray, and its size and shape. For convenience, size and shape are usually derived from the projection of the target along with some specified margin onto the beam face (also termed beam front), a plane normal to the central ray.

Mathematical optimization in daily practice has been largely reserved for the intensity map generation of each beam. The choice of beam energy and placement is made through the experience of the planner. Beam energy is selected from one of the limited number of choices (two, for e.g.) the machinery makes available. Beam orientations are set in some fashion accepted by the planner so as to spread out the entrance directions of a manageable number of beams, allowing adjustments to limit projections over sensitive structures.

The remaining task, too large to manually handle, is to generate the intensity profiles of the selected beams. That problem can be stated (see Box 10.3) by considering a set of beams individually labeled $j \in B$, each one of which in turn is divided into a finer panel of beam elements (also known as beamlets or pencil beams), distinguished by labels $k \in U(j)$. Then dose at any point i can be viewed as the output of decision variables x_{jk}, the beamlet intensity for beam j, beamlet k. The expression of dose d_i at point i as a linear combination of beamlet intensities becomes

$$d_i = \sum_{j \in B} \sum_{k \in U(j)} a_{ijk} x_{jk}$$

where a_{ijk} is the dose delivered at i per unit intensity in beamlet k of beam j.

10.3.3 Dose Coefficients

Because the solution must meet dose limits over a sample of points, and verified using a different sampling than that used in the optimization, the model introduces the problem of stochastic sampling. An open question is whether geometric or algebraic features can be used to direct the sampling. The need to satisfy a strict limit (e.g., ±5 percent) on dose homogeneity within tumor suggests a dense sampling within that structure. On the other hand, estimates of the volume distribution of dose may potentially make use of the gradual, roughly log linear fall of dose with distance to create spatially nonuniform sampling schemes. These may preferentially sample

Dose is expressed as a linear combination of the component units of the beam:

(1.1) $d_i = \sum_{j \in B, k \in U(j)} a_{ijk} x_{jk}$

where d_i is the dose to point i, x_{jk} is the intensity contributed by component k within the grid panel $U(j)$ that divides beam j of the treatment set B, and a_{ijk} is the dose contribution to point i from a unit intensity in component k of beam j.

Upper and lower bounds on dose within any collection of points s representing organ or target structures take the form:

(1.2) $\underline{d}_i \leq d_i \leq \bar{d}_i \quad i \in s$

A dose–volume limit is of the form:

(1.3) $|\mathbf{I}| \geq \mathrm{f}|\boldsymbol{s}|$, where $i \in \mathbf{I} \subset \boldsymbol{s} \Leftrightarrow d_i \leq d^T$

where $\mathbf{I}$ is the set of points in $\boldsymbol{s}$ whose dose does not exceed the threshold value d^T, $|\mathbf{I}|$ is the count of $\mathbf{I}$, and $|\boldsymbol{s}|$, and f the fraction of elements in $\boldsymbol{s}$ that must not exceed the threshold dose bound d^T.

It is possible to present condition (1.3) in a mixed integer linear form through relations (1.4–5):

(1.4) $d_i - \delta_i (\bar{d}_i - d^T) \leq d^T, \ \delta_i \in \{0,1\}$

(1.5) $\sum_i \delta_i \leq (1 - \mathrm{f})|\boldsymbol{s}|$

A relative dose bound of H on the ratio of the minimum to maximum dose levels in tumor can be expressed through:

(1.6) $m \leq d_i \leq M$ *for* $i \in \mathbf{T}$, where $\mathbf{T}$ is the set of points in tumor.

(1.7) $m/M \geq \mathrm{H}$,

Box 10.3 Formulation of the independent intensity map planning problem. Relations (1.1)–(1.2) describe dose bounds that are linear in the selected intensities. The addition of dose volume limits (1.3) introduces combinatorial conditions. Relations (1.4)–(1.5) convert the combinatorial conditions into conditions on binary variables. Relations (1.6)–(1.7) complete a set of clinical conditions that can be combined with a linear objective such as maximizing minimum tumor dose m to obtain a mixed-integer linear programming formulation.

over regions where small variations in position are associated with dose differences that straddle critical values for constraint violations.

The matrix of coefficients a_{ijk} is very large and because it is difficult to force sparseness into the coefficient matrix, it slows down most computational methods. Although each beam unit, whose intensity is to be assigned traverses through only a tiny slab of the entire body, it generates dose beyond the borders of the slab, through effects, such as radiation scatter and leakage through the blocking leaves defining the slab. Although the effect at a distance may be only 1–3 percent of the dose along the central ray of the beamlet, the sum total of these distance effects from all hundreds of beamlets cannot be neglected. How these effects should be handled in the optimization, by iterative corrections or by introducing a constant leakage factor forms its own research question.

10.3.4 Dose and Homogeneity Constraints

The most widely employed constraints consist of absolute dose limits that are bounds on the maximum or minimum doses received in a structure, relative doses limits that restrict ratios of dose values (typically the maximum to minimum dose within the target or global maximum dose to target minimum), and dose–volume limits that restrict the volume distribution of dose. A dose–volume limit is a constraint on the fractional volume of a structure that can exceed (or fall below) a stated dose level, equivalent to a limit on a rank percentile distribution. Such constraints are useful because redundancies in the body mean that normal tissue failure or complication may be avoided if a sufficient fraction of organ volume is protected.

It is possible to pose dose limits in absolute terms or relative to some derived value, such as minimum or maximum dose. Typically, one finds dose or dose volume limits on healthy tissues are given as upper bounds to make treatment acceptable. Lower bounds are placed on targets of varying tumor cell concentration to make treatment worthwhile.

A homogeneity bound is also often placed over the relative values of the maximum and minimum doses in tumor (and sometimes over the maximum dose in the whole body). A homogeneity limit acts to align the costs of treatment with its benefits, and serves also to constrain the potential deviations of the dose distribution from historical experience.

Box 10.3 collects all these constraint forms assuming the linear relationship between beamlet intensity and dose. All constraints are linear if dose–volume considerations (1.3) are discounted, but become mixed-integer linear when binary variables are introduced to model dose–volume conditions as in (1.4)–(1.5).

10.3.5 Scale of Models

The scale of the planning problem is very large. Isodose lines are constructed from dose values determined on a fine spacing of points, e.g., every at $0.5 \times 0.5 \times 0.5\,\text{cm}^3$ intervals over volumes that are roughly $10 \times 10 \times 10$, or about 10,000 points in total. Not all these points are used to produce dose–volume histograms against which satisfaction of dose volume limits are verified, but even a quarter sampling of 2500 points yields a substantial sized constraint set. Intensities are assigned to beam components that spatially divide the beam into smaller units. Intensities assignments are to be distributed among seven or so beams, arrayed on a planar grid placed over each beam front, with grid spacing at $0.5 \times 0.5\,\text{cm}^2$ intervals. If each beam measures $10 \times 10\,\text{cm}^2$, the number of intensity arguments is roughly 400 per beam × 9 beams, or about 3600.

Forms in Box 10.3 require one decision variable for each beamlet intensity, one constraint for each tissue point, and two for target points. There are also binary variables for each dose–volume constrained point. Thus, the scale of the optimization to be solved easily ranges to thousands of decision variables and tens of thousands of constraints.

10.3.5.1 Feasibility Problems and Objectives

The planning task can be posed either as a feasibility task, such as meeting the set of bounds,[31,32] or as an optimization exercise by introducing one (or more) criterion functions.[33,34] Feasibility approaches begin by specifying absolute lower (and sometimes upper) limits on tumor dose at any point (variables m and M in Box 10.3). Then, the planning task is to come as close as possible for finding beamlet intensities that together satisfy both tumor and normal tissue limits. Often, requirements are classified as requirements that must be met vs others (including the minimum tumor dose) that can be viewed only as targets. Then some form of goal programming or weighted penalty function of under-satisfied goals drives the search for best beamlet intensities x_{jk}.

Where an explicit criterion function is to be maximized, a common objective is maximizing the minimum tumor dose. Common alternatives are to maximize the dose at some central point in tumor or the average tumor dose, given a tight homogeneity level. By raising the smallest dose received in some region of tumor, these objectives serve to diminish the likelihood that a clonogenic cell will survive, and lead to disease recurrence. The same principle can be found in cooking a hamburger to avoid cold regions in any of its volume. It has been estimated that each 1 percent point increase in dose is associated with a 1.5 percent increase in tumor control for tumors controlled about 50 percent of the time, a quantity called the γ_{50}.[35]

Each of these criterion functions, and indeed the goal programming form of infeasibility minimization, is easily expressed in terms of a linear objective function and added constraints. Thus, in the presence of only absolute or relative dose limits on points and tumor dose homogenity, the optimization problem can be modeled as a large linear program. When dose–volume limits are introduced, the problem becomes a mixed-integer linear program. Penalty function objectives, typically second order, make the models nonlinear.[36]

10.3.5.2 Modeling Dose–Volume Constraints

It is clear that dose–volume constraints pose interesting challenges in obtaining a tractable model for approximate optimization of treatment planning problems. In addition to the mixed-integer approach of Box 10.3, several other schemes have been employed to enforce the required effect, albeit approximately.

Box 10.4 illustrates the distinction between dose–volume limits, and a variant known as Conditional Value at Risk (CVaR), which is a concept derived from portfolio analysis. Both formats consider a specific percentile of the dose distribution. Dose–volume requires that percentile not to exceed the threshold dose. It can be considered a limit on the minimum of doses at or above the threshold. CVaR, on the other hand, constrains the average (or total) dose for points receiving the threshold or more.

The CVaR approximation is foreign to clinicians who have developed the protocols that produce dose–volume constraints—especially because it focuses on the higher dose points that have been planned for

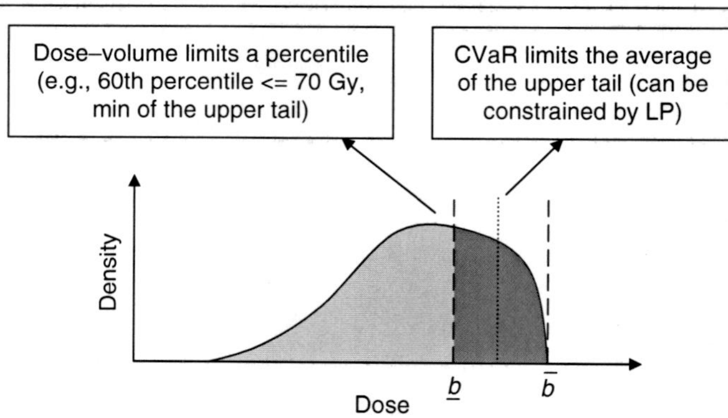

Linear constraints to implement CVaR are as follows:

$$\text{CVaR} = \phi + [(1-f)|S|]^{-1} \sum_{i \in S} w_i$$

$$d_i - \phi \le w_i \qquad i \in S$$

$$w_i \ge 0 \qquad i \in S$$

where ϕ is an auxiliary variable reflecting the $100 \times f$th percentile of the doses for $i \in S$ of a dose–volume tissue and auxiliary variables w_i compute any excess dose of point i over ϕ

Box 10.4 Dose–volume constraints.
These difficult-to-model constraint forms are combinatorial if represented exactly, as in Box 10.3, but admit a related CVaR form that can be modeled linearly.

sacrifice. Still, it has a significant computational advantage because it can be modeled with auxiliary variables and linear constraints as shown in Box 10.4.

Another approximate approach is derived from tissue geometry. Given coordinates of each target and protected tissue point, the distance from any protected point to the nearest target point can be computed in advance of the optimization.[37] Then the fraction $(1-f)$ of tissue points allowed higher than the threshold may be taken as those closest to points in the target. With those sacrificed tissue points preselected, (linear) maximum dose limits can be set to an upper limit $\bar{d}$ on the high-dose points, and threshold d^T for the rest. The intuition is that points in the target will receive higher does than any others, so that normal tissue points nearest to them are the ones most likely to be sacrificed.

10.4 Introducing Optimization of Orientations

A necessary preliminary step to treatment planning is to decide on a set of beam orientations. Not all orientations are available. Even when restricted to a common plane, the time needed to setup each new angle limits the number of orientations that can be realistically taken for each individual. Current practice is to preselect five–nine angles based on clinical experience with similar cases. An ideal would be to imbed angle selection into formulations like those of Box 10.3. Although this can be done in principle, the added computational burden to an already large-scale optimization tends to leave the model almost intractable. Nevertheless, models for simultaneous optimization of beam orientations and beam intensities have been suggested.[38,39]

Another promising class of approaches falling between these two extremes is to apply heuristics based on earlier experience and details of the case at hand to choose angles to use earlier or along with optimizing corresponding beamlet intensities.[40–42] Among proposed solution methods for such models are simulated annealing,[43] genetic algorithms,[44] particle swarm optimization,[45] and mixed-integer programming.[46]

10.4.1 Optimization Approaches

Although generally following the paradigm of Box 10.3, there are many variations in actual optimization approaches and even formulation strategies employed. Each has advantages in validity and convenience of use, but mathematical limits of the tools employed also introduce drawbacks.

10.4.1.1 Mixed-Integer Linear Programming

Mixed-integer linear programming approaches like those outlined above are exact formulations for maximizing one of the linear measures of tumor dose or minimizing goal shortfall when dose constraints are limited to specific points and dose–volume restrictions, and the dose at every point is a linear function of the beam weights.[47–49] In practice, however, the MIP search would normally be terminated with some provable bound on the remaining gap between the best known solution and the best possible. Linear programming relaxations often yield bounds within 5 percent in many cases. But in some, especially when the threshold dose $d^T \ll$ upper limit $\bar{d}$, the gaps can be large. When the LP relaxation is close to meeting the MIP constraints, good feasible solutions can be obtained by forcing into the protected set those points found with the lowest doses in the LP-relaxation solution. A lower bound to the MIP can be found by adding to the protected the minimum number of points required, constraining their doses to the lower threshold level d^T, and the solving to optimality the resulting LP. Intermediate approaches that iteratively assemble the protected set by adding to it some number of points with lowest received doses in the LP relaxation of the previous step can also be entertained.

10.4.1.2 Linear Programming

Linear programming formulations under similar constraint and objective function assumptions can deal with many of the issues in treatment planning optimization. Still, their use is limited because tolerance rules such as dose–volume limits do not reduce to linear constraints on specific points. Still, success has been reported by researchers using the CVaR measure of dose–volume discussed above, which can be modeled with linear programming at the cost of deviating from accepted treatment planning clinical standards.[50]

10.4.1.3 Conjugate Gradient Nonlinear Approaches

Conjugate gradient techniques, which are widely used in commercial and research systems, accept more general functions of the beam weights,[51] but do not guarantee optimal results for objectives or constraints that are not convex, including dose–volume limits.[52,53] Furthermore, the introduction of dose–volume and other constraints into an unconstrained optimization format to maximize a single optimization function using gradient or Newton's methods means that the correct weighting factors for summing the constraints and objective to yield the optimization function must be chosen.[54] Finding the correct and nonintuitive weighting factors put additional demands on the planner's time, and can lead to improved solutions being missed. It begs the question of how the correct weights are to be found because the problem of optimizing an objective subject to constraints is equivalent to maximizing a correctly weighted sum of the constraint and objective functions in the first place; it is the solution of the optimization problem that gives the correct weights.[55]

10.4.1.4 Metaheuristics Including Simulated Annealing

Simulated annealing and related metaheuristics are iterative randomized searches that have formed the basis of some commercial treatment planning algorithms.[56] Being unconstrained in their natural form, they have been extended to the problem of maximization of an objective function subject to constraints.[57] Simulated annealing algorithms have no natural criterion for termination, which can lead to inconsistent results, and their implementation may demand the assignment of penalty weights, raising the challenges seen with conjugate gradient methods. Comparative testing against mixed-integer programming has shown simulated annealing sometimes returns a minimum tumor dose that is as much as 12 Gy short of the highest value that could be delivered under the constraints.[47] Furthermore, plans obtained can fail to satisfy target dose homogeneity requirements because of the difficulty of enforcing hard constraints in the model.

10.4.1.5 Direct Inverse Methods

Although all approaches of this section may be termed inverse methods because they begin with planning requirements and seek beam and intensity parameters to satisfy them, there is a more direct alternative that begins with a fully specified dose distribution. Intensity options are then sought to minimize the variation from the given ideal. The approach's obvious drawback is that the best possible dose distributions are not known ahead of time. Still, the method has been successful in constructing plans that place high isodose lines tightly around irregularly shaped volumes. The method breaks down when treatment tolerance depends on the volumes of critical organs such as lung or kidney that receive low dose (e.g., $d^T = 20\,\text{Gy}$). It can return a plan, which is inferior in terms of both tumor control and risk of normal tissue injury than that given by manual planning.[58]

10.4.1.6 Biologic Function Formulations

Biologic function-based formulations based on treatment and complication probabilities mentioned above can also be employed to represent the dose optimization challenges. Still, the methods have not proved their ability to produce a clinically acceptable plan by themselves, and more standard criteria, such as a homogeneity limit, are used as an alternative.[59] Mathematical optimization of biologic functions remains incomplete, and surrogate functions are sometimes introduced to facilitate the generation of solutions.[60] Nevertheless, the solutions may fail to be optimal because the underlying functions and their valid surrogates are nonconvex, and can possess multiple local minima which are not enumerated.[61]

10.5 Intensity Map Delivery

Devices like the multileaf collimators (MLC) affect beamlet-by-beamlet variation by changing the beam aperture in a number of segments. As illustrated in Box 10.5, most implementations choose the leaf set up conditions convert continuous intensity values to whole number multiples, and search for a sequence of leaf positions which will deliver these values. The choice of leaf positions that can be assumed in delivery may be constrained by conditions designed to limit edge effects or other perturbations. More than one set of leaf positions may be feasible, and the problem is to find some set that optimizes one or more desired delivery properties. These include the number of times that the beam is turned off to reposition at least one leaf, or reducing the total beam time.[13,14,62]

Box 10.5 also details several of the dependencies that make aperture effects somewhat different than the sum of open beamlets. Some beamlet combinations cannot be formed by leafs on two sides, leaf collisions need to be avoided, boundary effects multiply the effect of beamlets according to their positions within the aperture, and closed beamlets still have some leaked radiation. Column generation techniques are natural to transform the Box 10.3 optimization over beamlets to one over apertures generated to address MLC limitations.

When MLC limitations are not considered directly in the planning optimization, e.g., by column generation, a post-optimization process is needed to convert computed beamlet intensities into segments. Relatively quick heuristics are known for this process and some aim to add conditions that limit the degradation of the selected intensity map by the delivery process.[13,14,62] These heuristics may aim to limit the total monitor units or number of times at least one leaf is moved that is needed to realize the delivery. Still, they can only approximately incorporate all the four considerations of Box 10.5, and neglect the effect of delivery on intensity optimization. Methods that integrate in some fashion intensity optimization with delivery have recently been considered.[26,47,63,64] These methods include optimization over preformed apertures rather than free intensity maps,[47,65] column generation to represent the segments at each iteration[63,64] and iterative optimization with correction for aperture perturbation at each step.[26]

For implementation, an intensity map must be decomposed into segments with blocked and open sets of beamlets exposed to a common intensity value.

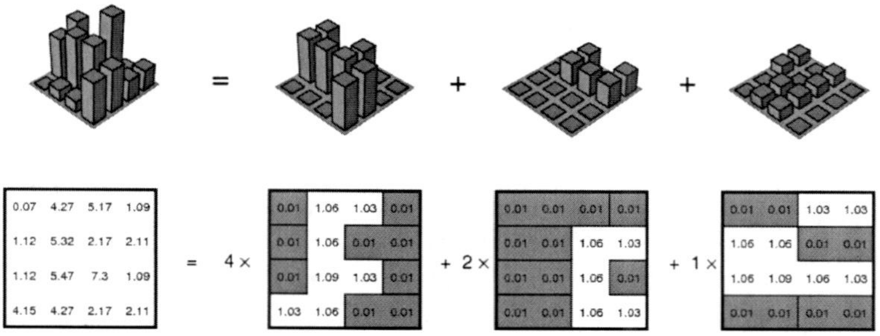

Not all segments can be formed into physically realizable structures, called apertures. Among the limitations that must be accounted for in estimating effects on segments are the following:

1. Apertures with blocked beamlets in holes not attached to either side cannot be implemented.
2. As a safety factor against *l*eaf collisions, leaves in adjacent rows should not be allowed to overlap.
3. Boundary effects in the open parts of apertures tend to inflate the presumed intensity of beamlets as they become further from boundaries.
4. Leakage may occur through leaves assumed closed in the aperture.

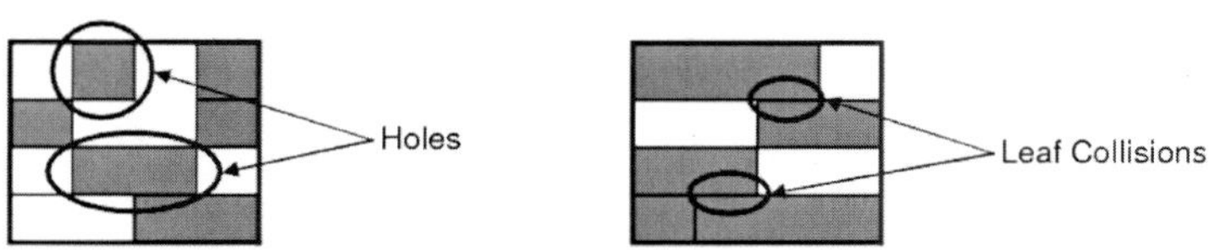

Box 10.5. Multileaf collimators.
A column generation approach may be employed to optimize over apertures rather than beamlets in the mixed-integer linear programming case of Box 3. A collection *C* of beamlets in *U(j)* for angle *j* is chosen to form an aperture with *C*-column coefficient in tissue point row *i*

$$a_{iC} = \sum_{k \in C} e_{kC} a_{ijk} + \sum_{k \in U,(j) \setminus C} L a_{ijk}$$

where e_{kC} is the boundary effect multiplier of beamlet *k* in aperture *C*, and *L* is the leakage multiplier for closed beamlets.

10.6 Four-Dimensional and Fractionated Planning

Most treatment planning optimization methods choose a plan once for the entire course of the optimization. Clinical practice, however, is to implement plans in a series of 25–50 daily fractions. Thus, fluence times for the desired segments of the chosen overall plan must actually be optimized over 3D and 4D. More recently, techniques have evolved to account for target motion during a session by tailoring delivery to a particular phase of the respiratory cycle.[66] In theory, matched delivery over more than one phase is possible.

Effectiveness of the treatment can be threatened if per-fraction limits are ignored in implementing a chosen overall plan. For example, a minimum dose of 2 Gy per-fraction may be required for sufficient impact on the target, but not more that 1.8–2 Gy may be allowed on healthy tissues. This raises the problem of maintaining overall and per-fraction dose limits in each structure when the treatment course is decomposed into fractions and phases within fractions.[67]

Recent advances in imaging provide updated images during the sequence of fractions and demand an adaptive reoptimization over the course of treatment.[68] In principle, adaptive reoptimization poses the challenge of accounting for both overall dose delivered and the dose per-fraction under the original and revised tissue geometry.[67]

A final challenge is tissue movement during a single fraction. For example, lungs deform from any fixed image as the patient breathes. Leaves can be made to track in real time, but this presents the problem of finding an optimal trajectory within each session to delivery the desired intensity map without unnecessarily lengthening session time.[69] As with all 4D issues, concepts remain the same as overall plan optimization, but the scale of the model tends to explode.

10.7 Conclusion

Radiation therapy planning touches on many modern topics in optimization: large-scale theory, sensitivity analysis, multiobjective optimization, combinatorial geometry, fractional set constraints, stochastic optimization, and discrete choice optimization. It presents practical issues of display of sensitivities across a multidimensional space, solution time limits, and bounds on shortfalls from optimality or constraint violations.[70] A particular feature of the problem is that the multiple problem expressions, derived from individual anatomic configurations are readily generated and in principle do not encode proprietary or identifiable data. This opens a wide field for testing optimization solutions for many classes of problems that transcend medicine or biology. The signature of the radiation planning problem is that its study enhances the sciences of the both optimization and medicine, and provides practical benefits and testable ideas to both fields.

References

1. Medline Ovid accessed 2/21/2007 using search (optimization or optimal.mp) and (radiation or radiotherapy.mp).

2. Carlson WD and Morkovin D. Time–dose relations for optimal radiation treatment of malignancy with minimal damage to normal tissues. *Miss. Valley Med. J.*, 81: 287–288, 1959.
3. Holder A and Salter B. A tutorial on radiation oncology and optimization. In H. Greenberg, (Ed.), *Tutorials on Emerging Methodologies and Applications in Operations Research*, chapter 4, Dordrecht: Kluwer Academic Publishers, 2004.
4. Bortfeld T. IMRT: A review and preview. *Phys. Med. Biol.*, 51(13): R363–R379, 2006.
5. Censor Y. Mathematical optimization for the inverse problem of intensity modulated radiation therapy. In Palta JR and Mackie TR (Eds.), *Intensity-Modulated Radiation Therapy: The State of The Art*, Madison, WI: Medical Physics Publishing, 2003, pp. 25–49.
6. Brahme A. Treatment optimization using physical and radiological objective functions. In Smith AR (Ed.), *Medical Radiology: Radiation Therapy Physics*, Berlin: Springer-Verlag, 1995, pp. 209–246.
7. Cohen L. The statistical prognosis in radiation therapy: A study of optimal dosage in relation to physical and biological parameters for epidermoid cancer. *Am. J. Roentgenol.*, 84: 741–753, 1960.
8. Fowler JF. Is there an optimum overall time for head and neck radiotherapy? A review, with new modelling. *Clinical Oncology*, 19: 8–22, 2007.
9. Bourhis J, Overgaard J, and Audry H. Hyperfractionated or accelerated radio therapy in head and neck cancer: A meta-analysis. *Lancet*, 368: 843–854, 2006.
10. Fenwick JD. Delay differential equations and the dose–time dependence of early radiotherapy reactions. *Med. Phy.*, 33: 3526–3540, 2006.
11. Bortfeld TR, Kahler DL, Waldron TJ, and Boyer AL. X-ray field compensation with multileaf collimators. *Int. J. Radiat. Oncol. Biol. Phys.*, 28: 723–730, 1994.
12. Chang SX, Cullip TJ, Deschesne KM, Miller EP, and Rosenman JG. Compensators: An alternative IMRT delivery technique. *J. Appl. Clin. Med. Phys.*, 5: 15–36, 2004.
13. Siochi RA. Minimizing static intensity modulation delivery time using an intensity solid paradigm. *Int. J. Radiat. Oncol. Biol. Phys.*, 43: 671–680, 1999.
14. Langer MP, Thai V, and Papiez L. Improved leaf sequencing reduces segments or monitor units needed to deliver IMRT using multileaf collimators. *Med. Phys.*, 28: 2450–2458, 2001.
15. Remeijer P, Rasch C, Lebesque JV, and van Herk M. Margins for translational and rotational uncertainties: A probability based approach. *Int. J. Radiat. Oncol. Biol. Phys.*, 53: 464–474, 2002.
16. Wambersie A and Landberg T. ICRU Report 62: Prescribing, recording and reporting photon beam therapy (supplement to ICRU report 50). International Commission on Radiation Units and Measurements, Inc., 1999, Bethesda, MD.
17. Papiez L and Langer M. On probablistically defined margins in radiation therapy. *Phys. Med. Biol.*, 51: 3921–3939, 2006.
18. Stroom J, De Boer H, Huizenga H, and Visser A. Inclusion of geometrical uncertainties in radiotherapy treatment planning by means of coverage probability. *Int. J. Radiat. Oncol. Biol. Phys.* 43: 905–919, 1999.
19. Langer MP, Papiez L, Spirydovich S, and Thai V. The need for rotational margins in intensity-modulated radiotherapy and a new method for planning target volume design. *Int. J. Radiat. Oncol. Biol. Phys.*, 63: 1592–1603, 2005.
20. Niemierko A and Goitein M. Random sampling for evaluating treatment plans. *Med. Phys.*, 17: 753–762, 1990.

21. Drzymala RE, Mohan R, Brewster L, Chu J, Goitein M, Harms W, and Urie M. Dose–volume histograms. *Int. J. Radiat. Oncol. Biol. Phys.*, 21: 71–78, 1991.
22. American Society for Therapeutic Radiology and Oncology (ASTRO) News. July/August 2001, Fairfax, VA, American Society for Therapeutic Radiology and Oncology.
23. Boyer AL and Yu CY. Intensity-modulated radiation therapy with dynamic multileaf collimators. *Semin. Radiat. Oncol.*, 9: 48–59, 1999.
24. Michalski D, Xiao Y, Censor Y, and Galvin JM. The dose–volume constraint satisfaction problem for inverse treatment planning with field segments. *Phys. Med. Biol.*, 49: 601–616, 2004.
25. Mohan R, Arnfield M, Tong S, Wu Q, and Siebers J. The impact of fluctuations in intensity patterns on the number of monitor units and the quality and accuracy of intensity modulated radiotherapy. *Med. Phys.*, 27: 1226–1237, 2000.
26. Siebers JV, Lauterbach M, Keall PJ, and Mohan R. Incorporating multi-leaf collimator leaf sequencing into iterative IMRT optimization. *Med. Phys.*, 29: 952–959, 2002.
27. Shephard DM, Olivera GH, Reckwerdt PJ, and Mackie TR. Iterative approaches to dose optimization in tomotherapy. *Phys. Med. Biol.*, 45: 69–90, 2000.
28. Alber M and Nušslin F. Intensity modulated photon beams subject to a minimal surface smoothing constraint. *Phys. Med. Biol.*, 45: 49–52, 2000.
29. Kutcher GJ and Burman C. Calculation of complication probability factors for nonuniform normal tissue irradiation: The effective volume method. *Int. J. Radiat. Oncol. Biol. Phys.*, 16: 1623–1630, 1989.
30. Niemierko A and Goitein M. Implementation of a model for estimating tumor control probability for an inhomogeneously irradiated tumor. *Radiother. Oncol.*, 29: 140–147, 1993.
31. Starkschall G, Pollack A, and Stevens CW. Treatment planning using a dose-volume feasibility search algorithm. *Int. J. Radiat. Oncol. Biol. Phys.*, 49: 1419–1427 2001.
32. Xiao Y, Censor Y, Michalski D, and Galvin JM. The least-intensity feasible solution for aperture-based inverse planning in radiation therapy. *Ann. Oper. Res.*, 119: 183–203, 2003.
33. Langer M, Brown R, Leong J, Urie M, Stracher M, and Shapiro JS. Large scale optimization of beam weights under dose–volume restrictions. *Int. J. Radiat. Oncol. Biol. Phys.*, 18: 887–893, 1990.
34. Cho P, Lee S, Marks R, Oh S, Sutlief S, and Phillips M. Optimization of intensity modulated beams with volume constraints using two methods: Cost function minimization and projection onto convex sets. *Med. Phys.*, 25: 435–443, 1998.
35. Chappell R and Fowler JF. Steepness of dose-response curve for larynx cancer (letter). *Radiother. Oncol.*, 30, 1994.
36. Spirou SV and Chui CS. A gradient inverse planning algorithm with dose–volume constraints. *Med. Phys.*, 25: 321–333, 1998.
37. Morrill SM, Rosen II, Lane RG, Belli JA. The influence of dose constraint point placement on optimized radiation therapy treatment planning. *Int. J. Radiat. Oncol. Biol. Phys.*, 19: 129–141, 1990.
38. Lee EK, Fox T, and Crocker I. Simultaneous beam angle and intensity map optimization in intensity modulated radiation therapy. *Int. J. Radiat. Oncol. Biol. Phys.*, 64: 301–320, 2006.

39. Yang RJ, Dai JR, Yang Y, and Hu YM. Beam orientation optimization for intensity-modulated radiation therapy using mixed integer programming. *Phys. Med. Biol.*, 51(15): 3653–3666, 2006.
40. Woudstra E, Heijman BJM, and Storchi PRM. Automated selection of beam orientations and segmented intensity-modulated radiotherapy (IMRT) for treatment of oesophagus tumors. *Radiother. Oncol.*, 77: 254–261, 2006.
41. Pugachev A and Xing L. Pseudo beam's eye view as applied to beam orientation selection in intensity modulated radiation therapy. *Int. J. Radiat. Oncol. Biol. Phys.*, 51: 1361–1370, 2001.
42. Gaede S, Wong E, and Rasmussen H. An algorithm for systematic selection of beam directions for IMRT. *Med. Phys.*, 31(2): 376–388, 2004.
43. Bortfeld T and Schlegel W. Optimization of beam orientations in radiation-therapy—some theoretical considerations. *Phys. Med. Biol.*, 38(2): 291–304, 1993.
44. Hou Q, Wang J, Chen Y, and Galvin JM. Beam orientation optimization for IMRT by a hybrid method of the genetic algorithm and the simulated dynamics. *Med. Phys.*, 30(9): 2360–2367, 2003.
45. Li YJ, Yao DZ, and Yao J. A particle swarm optimization algorithm for beam angle selection in intensity-modulated radiotherapy planning. *Phys. Med. Biol.*, 50(15): 3491–3514, 2005.
46. Wang C, Dai JR, and Hu YM. Optimization of beam orientations and beam weights for conformal radiotherapy using mixed integer programming. *Phys. Med. Biol.*, 48(24): 4065–4076, 2003.
47. Langer M, Morrill S, Brown R, Lee O, and Lane RA. Comparison of mixed integer programming and fast simulated annealing for optimizing beam weights in radiation therapy. *Med. Phys.*, 23: 957–964, 1996.
48. Bednarz G, Michalski D, Houser C, Huq MS, Xiao Y, Anne PR, and Galvin JM. The use of mixed integer programming for inverse treatment planning with pre-defined beam segments. *Phys. Med. Biol.*, 47: 2235–2245, 2002.
49. Preciado-Walters F, Rardin, R, and Langer M. A coupled column generation, mixed-integer approach to optimal planning of intensity modulated radiation therapy for cancer. *Math. Prog. B*, 101: 319–338, 2004.
50. Romeijn H, Ahuja R, Dempsey J, Kumar A, and Li, J. A novel linear programming approach to fluence map optimization for intensity modulated radiation therapy treatment planning. *Phys. Med. Biol.*, 48: 3521–3542, 2003.
51. Kallman P, Lind BK, and Brahme A. An algorithm for maximizing the probability of complication-free tumor control in radiation therapy. *Phys. Med. Biol.*, 37: 871–890, 1992.
52. Dantzig GB. Convex programming, *Linear Programming and Extensions*. NJ: Princeton University Press, 1963, p. 471.
53. Deasy JO. Multiple local minima in radiotherapy optimization problems with dose-volume constraints. *Med. Phys.*, 24: 1157–1161, 1997.
54. Mattia M, Del Giudice P, and Barbara C. IMRT optimization: Variability of solutions and its radiobiological impact. *Med. Phys.*, 31: 1052–1060, 2004.
55. Dantzig GB. Basic theorems on duality (chapter 6-4) and Lagrange multipliers (chapter 6-5), *Linear Programming and Extensions*. NJ: Princeton University Press, 1963.
56. Webb S. Optimization by simulated annealing of three-dimensional conformal fields defined by multileaf collimator. *Phys. Med. Biol.*, 36: 1201–1226, 1999.

57. Morrill SM, Lane RG, and Rosen II. Constrained simulated annealing for optimized radiation therapy. *Comp. Meth. Prog. Biomed.*, 3: 135–144, 1990.
58. Mohan R, Wang X, Jackson A, Bortfeld T, Boyer AT, Kutcher G, Leibel SA, Fuks Z, and Ling CC. The potential and limitations of the inverse radiotherapy technique. *Radiother Oncol.*, 32: 232–248, 1994.
59. de Meerleer GO, Vakaet LAML, de Gersem WRT, de Wagter C, de Naeyer B, and de Neve W. Radiotherapy of prostate cancer with or without intensity modulated beams: A planning comparison. *Int. J. Radiat. Therap. Oncol. Biol. Phys.*, 47: 639–648, 2000.
60. Wu Q and Mohan R. Algorithms and functionality of an intensity modulated radiotherapy optimization system. *Med. Phys.*, 27: 701–771, 2000.
61. Deasy O. Multiple local minima in radiotherapy optimization problems with dose-volume constraints. *Med. Phys.*, 24: 1157–1161, 1997.
62. Xia P and Verhey LJ. Multileaf collimator leaf sequencing algorithm for intensity modulated beams with multiple static segments. *Med. Phys.*, 25: 1424–1434, 1998.
63. Preciado-Walters F, Langer MP, Rardin RL, and Thai V. Column generation for IMRT cancer therapy optimization with implementable segments. *Ann. Oper. Res.*, 148(1): 65–79, 2006.
64. Romeijn HE, Ahuja RK, Dempsey JF, and Kumar A. A column generation approach to radiation treatment planning using aperture modulation. *SIAM J. Optim.*, 15: 838–621, 2005.
65. Earl MA, Shephard DM, Naqvi S, Li A, and Yu CX. Inverse planning for intensity-modulated arc therapy using direct aperture optimization. *Phys. Med. Biol.*, 48: 1075–1089, 2003.
66. Keall PJ, Joshi S, Vedam SS, Siebers JV, Kini VR, and Mohan R. Four-dimensional radiotherapy planning for DMLC-based respiratory motion tracking. *Med. Phys.*, 32: 942–951, 2005.
67. Langer M, Rardin R, and Dink D. Fraction size effects in adaptive radiation therapy. *J. Radiat. Oncol. Biol. Phys.*, 66(3suppl.) p.S 124. (abstract), 2006.
68. Wu C, Jeraj R, Olivera GO, and Mackie TR. Re-optimization in adaptive radiotherapy. *Phys. Med. Biol.*, 47: 3181–3195, 2002.
69. Papiez L, Rangaraj D, and Keall PJ. Real-time DMLC IMRT delivery for mobile and deforming targets. *Med. Phys.*, 32: 3037–3048, 2005.
70. Lee EK, Langer M, Deasy JO, Rardin RL, Deye JA, and Mahoney FJ. NSF Final Report, NCI/NSF Workshop on Operations Research Applied to Radiation Therapy. *Ann. Oper. Res.*, 119: 143–146, 2003.

BIOLOGY II

Chapter 11

Introduction to Systems Biology for Mathematical Programmers

Eivind Almaas, Allen Holder, and Kevin Livingstone

CONTENTS

11.1 Introduction ... 312
11.2 General Background ... 313
11.2.1 Basic Biological Definitions ... 314
11.2.2 Basic Network and Mathematical Definitions ... 315
11.3 Gene-Regulatory Networks ... 316
11.3.1 Network Clustering ... 317
11.3.2 Network Motifs ... 318
11.4 Protein-Interaction Networks ... 319
11.4.1 Connectivity Distribution ... 320
11.4.2 Network Assortativity ... 322
11.4.3 Community Finding ... 323
11.4.4 Biology and Topology ... 325
11.5 Metabolic Networks ... 326
11.5.1 Metabolic Network Structure ... 327
11.5.2 Weighted Metabolic Networks ... 330
11.5.3 Fluxes and Metabolic Network Structure ... 333
11.6 Systems Biology and Operations Research ... 337
Acknowledgments ... 347
References ... 347

Abstract Many recent advances in biology, medicine, and health care are due to computational efforts that rely on new mathematical results. These mathematical tools lie in discrete mathematics, statistics and probability, and optimization, and when combined with savvy computational tools and an understanding of cellular biology, they are capable of remarkable results. One of the most significant areas of growth is in the field of systems biology, where we are using detailed biological information to construct models that describe larger entities. This chapter is designed to be an introduction to systems biology for individuals in operations research (OR) and mathematical programming who already know the supporting mathematics but are unaware of current research in this field.

11.1 Introduction

The field of systems biology represents a new, exciting collaboration between biology, mathematics, and computer science. In broad collaborations such as this, it is usually not the case that a single discipline benefits to the exclusion of the others, but rather each discipline is rewarded from the inventions of the interaction. Classic examples of similar interactions involved mathematics and physics, which led to the invention of Calculus, and the interaction between agronomists and statisticians that led to advances in experimental design, analysis of small sample sizes, and the development of analysis of variance. Many have argued that current problems in cellular biology are playing a similar role in mathematics and computer science today. In particular, the nexus of high-throughput data generation in biology, and increasingly sophisticated mathematical and computational tools makes systems biology an exciting and innovative field of study.

Broadly speaking, biologists want to answer overarching questions related to how organisms work. The complexity of life and the difficulties inherent to experimental science have traditionally led biologists to adopt a reductionist approach, working for example in a single species to find and characterize single causative factors. Subsequent research then finds factors that interact with the first factors, and so on. The reductionist approach has shed light on many individual components of an organism, but for all our work, we only know a small percentage of how organisms work.

The painstaking progress of the reductionist approach is now being accelerated, however, by new high-throughput technologies. The most reductionist level of an organism is its DNA sequence, and it is almost inconceivable that the structure of deoxyribonucleic acid (DNA) was discovered approximately 50 years ago, and that less than 20 years ago researchers labored to hand-sequence genes a few hundred bases at a time. Now our goal is to produce affordable, personal sequences of the three billion bases of an individual human genome in a matter of days or weeks, rather than the years it took to complete the first human genome sequence.

Although completing the human genome represented a pinnacle of achievement, it did not provide all the information needed to holistically model life's processes. To build a functional model of an organism, we need to know which proteins are actually made, at what times, in response to what environmental cues, and how these proteins interact either physically with other proteins or in metabolic pathways to create a static trait or dynamic response. Being able to characterize these higher levels of complexity is crucial: if anything, our reductionist studies have taught us that the whole is more than the sum of the parts.

Advances in technology similar to those seen in the sequencing arena are now also expanding our understanding of these higher-order questions. The current difficulty is how best to deal with the embarrassment of riches in biological information. On the whole, most biologists have not been trained in model building, data management, and computational skills. Experts in operations research (OR), however, are trained exactly in these fields and are well positioned to accelerate this exciting area of research. What OR professionals lack is an understanding of the underlying biology and how it transforms into familiar research topics. This tutorial is intended to fill this educational gap.

In the end, our goal is to have quantitative, predictive models, which describe systems from cells to entire organisms. In pursuing this goal, it is important to remember that our interest is not solely focused on understanding *Homo sapiens*. Although it is true that much of our research on the bacterium *E. coli*, the single-celled eukaryotic yeast *S. cerevisiae*, and the millimeter-sized roundworm *C. elegans* and fly *D. melanogaster* is undertaken using these as surrogates; our interest also extends to a myriad of other organisms that provide food, fiber, fuel, pharmaceuticals, etc. It now appears that collaborations between biology, mathematics, and computer science in the field of systems biology are the way by which progress towards this goal will be made.

11.2 General Background

This chapter discusses the three levels of whole-cell modeling based on interactions between genes, proteins, and metabolites. A thorough discussion of each whole-cell model exceeds the capability of this introductory chapter; so our goal in each section is to focus on key aspects of the underlying biology and the network representation, and then provide a summary of some of the insights this representation has provided.

To operate in modern biological terms, we need to understand the basic premises that support the research. This section is divided into two subsections: one that explains the guiding principle that dictates the related biological research, called the central dogma of molecular biology, and another that defines the fundamental terms of the network analysis used by systems biologists.

11.2.1 Basic Biological Definitions

The central dogma, elaborated by Francis Crick soon after his codiscovery of the structure of DNA, states that biological information flows from DNA to messenger ribonucleic acid (mRNA) to proteins. The DNA molecule that serves as the main repository of biological information is a pair of directional polymers whose monomers are denoted A, T, G, and C. Each of these monomers has a conserved portion that forms the backbone of the polymer and the variable portion that makes it an A, T, G, or C. The DNA double helix is comprised of two polymers that are oriented in opposite directions and held together by interactions between the variable parts of the monomers, A–T pairs, and G–C pairs, see Figure 11.1a. A DNA sequence

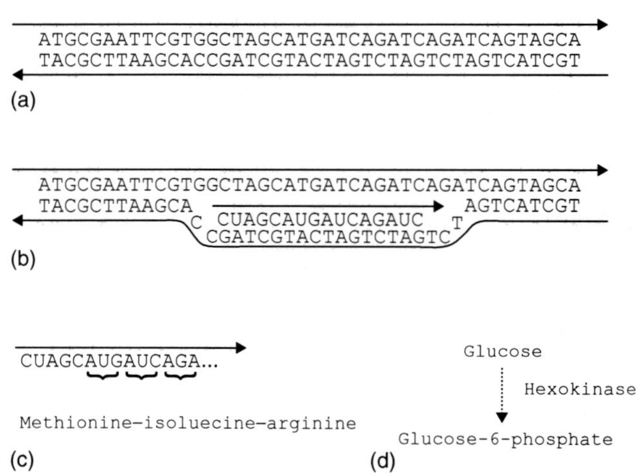

Figure 11.1 The central dogma of molecular biology. (a) DNA, (b) mRNA being made from a DNA template (transcription), (c) protein synthesis specification by mRNA (translation), (d) the enzyme hexokinase acts on the metabolite glucose in the metabolic pathway that breaks down glucose for energy production.

is usually represented by the list of letters (ATGC) read directionally along one strand, the complementary strand being implied. Each of the 46 chromosomes inside a human cell is a double helix with about 10^7–10^8 base pairs, and a gene is a known stretch of hundreds or thousands of bases of the double helix with a defined function, usually encoding a protein.

The DNA is used as a template to make an mRNA polymer by a process called transcription. The monomers of mRNA have a conserved portion that forms the backbone of the polymer, although slightly different from the corresponding DNA monomers, and four variable portions denoted A, U, G, and C. Construction of an mRNA molecule involves partial unwinding of the DNA molecule and then the exposed ATGC bases of the DNA dictate the sequence of the mRNA through interactions similar to those described above, except that where there is an A in the DNA, there will be a U in the mRNA, see Figure 11.1b. The mRNA molecule is also directional and is represented by a string of AUGC.

The mRNA intermediate of a gene is used as a template to make proteins through a process called translation. During translation, cellular machinery reads an mRNA three monomers at a time from a defined starting position, see Figure 11.1c. Each triplet determines one of the 20 amino acid monomers found in a protein or a message to stop protein synthesis. Although proteins are represented by their primary sequence using an alphabet of 20 letters, protein function is ultimately determined by the protein's three-dimensional structure, which may not be predictable based on the primary sequence alone.

Proteins are both the structural and functional workhorses of a cell that convert information stored in the DNA (the genotype) to the visible characteristics of the cell or organism (the phenotype). In this chapter, we focus on the protein's functional aspects as enzymes that take molecules and convert these to the products needed for cellular functions. These molecules are called metabolites, and Figure 11.1d gives an example of how metabolites and enzymes are organized into metabolic pathway.

Exceptions to the central dogma exist, but these subtleties and variations cannot be addressed in this presentation. Detailed descriptions of all the biological processes and exceptions can, however, be found in any current biology or genetics textbook. Despite these exceptions, the Central Dogma does appropriately model most of the information flow within a living system, so we will operate on the simplified Central Dogma throughout.

11.2.2 Basic Network and Mathematical Definitions

Networks are used in systems biology to model the relationships between cellular entities. Networks are familiar to those in OR, and this section specifies the common notation used throughout. In cases where terms vary between the two disciplines, we mention both terms but use those that are common to the field of systems biology. This convention helps those in OR understand the language of systems biology.

A network is a directed graph (V, E), where the elements of V are called nodes or vertices and the elements of $E \subseteq V \times V$ are called arcs or links. In network analysis, the direction of an arc is important, and we distinguish between (v_1, v_2) and (v_2, v_1). If we are instead referring to the graph (V, E) then direction is not important and there is no distinction between (v_1, v_2) and (v_2, v_1). In this case, the elements of E are called edges. We assume throughout that $|V| = N$ and $|E| = M$. The nodes v_1 and v_2 are adjacent if $(v_1, v_2) \in E$, and we further say that v_1 is incident to the edge (not arc) (v_1, v_2).

If $E = V \times V$ then (V, E) is complete in the sense that it contains as many arcs or edges as possible. Such graphs are defined by the size of V and are called complete and denoted K_N. We say that (V', E') is a subnetwork or subgraph of (V, E) if $V' \subseteq V$, $E' \subseteq E$, and $E' \subseteq V' \times V'$. A clique of a network or graph is a complete subnetwork or subgraph.

A network's structure is often referred to as the topology of the network, which is a bit awkward for mathematicians. A graph's topology is often described by the adjacency matrix $A = [a_{ij}]$, where $a_{ij} = 1$, if nodes i and j are adjacent and zero otherwise. For networks $a_{ij} = 1$ if $(v_i, v_j) \in E$, and $a_{ij} = -1$ if $(v_j, v_i) \in E$. In a graph, the neighborhood of a node is $N(v_i) = \{v_j : (v_i, v_j) \in E\}$ and the degree of the node is $\deg(v_i) = |N(v_i)|$. This concept naturally extends to a network where we discuss out-degree and in-degree. Much of the analysis considered by systems biologists is based on how well a graph is connected, and for this reason, the $\deg(v_i)$ is often called the connectivity of node i. Instead of $\deg(v_i)$, we denote the $\deg(v_i)$ as k_i, and for graphs we have

$$\deg(v_i) = k_i = \sum_j a_{ij}.$$

For an understood probability distribution, we let $P(x)$ be the probability of observing x. We use the typical ΛO notation and write $f(x) = O(g(x))$ if there is a λ such that $f(x) \leq \lambda g(x)$. The vector of ones is denoted by e, where length is decided by the context of its use. Other notation is introduced as needed. All terms dealing with optimization agree with those defined in the Mathematical Programming Glossary [Greenberg, 2006].

11.3 Gene-Regulatory Networks

Complex organisms exhibit dramatic differences in cellular phenotypes (characteristics). Examples of these differences are fixed differences between cell types (e.g., brain cells and liver cells) or temporarily induced differences due to environmental stimuli (e.g., increased production of melanin by skin cells after UV exposure). In general, all the cells in an organism have the same DNA, so the cause of these phenotypic differences is the variation in the amount and types of proteins present in the cell. Gene expression is the general term for this conversion of the information in the inert DNA into the functional proteins, and tight control over gene expression is what allows for different cellular phenotypes.

The majority of the control over gene expression occurs at the level of initiation of transcription (the making of the mRNA intermediate). One of the primary tools used to understand a cell, therefore, is characterization of what is called the transcriptome, the set of all genes expressed under defined conditions. Biologists can detect the levels of different mRNA molecules with precision, and new microarray technology even allows for the simultaneous measurement of the levels of all mRNAs in a cell. As will be seen later, presence of an mRNA does not always imply the presence of a functional protein, but mRNA production is a necessary first step and the correlation between mRNA and protein levels is strong enough to make mRNA quantitation a meaningful first measure for most gene expression studies.

Initiation of transcription for a gene is dependent on two factors. Production of mRNA requires a large group of proteins that unwind the DNA and facilitate the polymerization of the mRNA, and these proteins must bind to the DNA of the gene at locations called regulatory regions. Because multiple genes may have similar regulatory regions, coordinate gene expression can occur when the proteins in a cell increase the expression of all these target genes simultaneously. Coordinate repression of genes may also occur when binding of a protein to regulatory regions prevents transcription. Coordinate regulation allows groups of genes to be acted on as a unit, which is important given that many cellular actions require multiple types of proteins.

11.3.1 Network Clustering

Genetic interactions are frequently represented as networks, where the nodes correspond to genes, and a (possibly directed) link is introduced between genes A and B if the presence or absence of gene A's encoded protein enhances or suppresses the expression of gene B, or vice versa. The local properties of a gene-regulatory network are measured by how closely they resemble a clique [Watts and Strogatz, 1998]. The clustering coefficient c_i of a node, defined as

$$c_i = \frac{2}{k_i(k_i - 1)} \sum_{j,l} a_{ij} a_{il} a_{jl}, \tag{11.1}$$

measures the degree to which the neighborhood of a node resembles a complete subgraph built from triangles, and is the ratio of the actual number of triangles to possible triangles, for which node i is a member. The average clustering coefficient $\langle C \rangle = (1/N) \sum_i c_i$ provides information on the global distribution of links. A value of $\langle C \rangle$ close to unity indicates a high level of modularity, or cohesiveness of triangles, in the network, while a value close to zero indicates a lack of modularity. It is customary to test the significance of a particular $\langle C \rangle$-value by comparing it to a random-network model with the same number of nodes and edges [Albert and Barabási, 2002]. Typical random graphs have an average clustering coefficient of $\langle C_{\text{rand}} \rangle = 2M/N^2$.

Assuming that a network has a nonzero $\langle C \rangle$, we further investigate the network's large-scale modularity structure by studying the average clustering as function of degree k [Dorogovtsev et al., 2002],

$$C(k) = \frac{\sum_{\{i:k_i=k\}} c_i}{\sum_{\{i:k_i=k\}} 1}. \tag{11.2}$$

If the network shows a hierarchical modularity [Ravasz et al., 2002], the clustering $C(k) \sim 1/k$. In this case, nodes with few neighbors tend to have network-neighborhoods with high clustering, while the highly connected nodes act as bridges tying the network together.

11.3.2 Network Motifs

It has long been argued that biological systems are functionally modular [Hartwell et al., 1999], and understanding how this modularity is reflected in biological network is a primary goal. Given this modularity, additional questions arise, for example, what network modules, or partitions, carry functional information, and how does the functional modularity depend on the environmental conditions and the dynamic states of a gene-regulatory network? An interesting possibility was suggested in [Milo et al., 2004, Milo et al., 2002, Shen-Orr et al., 2002], introducing the idea of network "motifs" as the functional building blocks of a gene-regulatory network. They suggest that these networks contain particular sub-graphs, many with easily identifiable functions such as feed-forward loops, at a significantly higher frequency than should be expected by chance alone. The enrichment of biological networks with functional motifs is seen as a result of the evolutionary processes shaping the system [Milo et al., 2002]. However, the recent results in [Vazquez et al., 2004] indicate that caution is needed to determine if a motif is overexpressed. By designing random networks that matched the experimental results, they found that certain subgraphs occur at higher frequencies than in random networks without this restriction.

11.4 Protein-Interaction Networks

As stated above, the majority of the structural and functional macromolecules in a cell are proteins, and the presence of these proteins is tightly regulated by the cell mainly through initiation of transcription. Even after translation of an mRNA into a protein, however, the protein may not be functional. The activity of many proteins is influenced by modifications such as the addition of chemical groups by other proteins, binding of cofactors (which may include other proteins), or cleavage by other proteins, to name a few. These mechanisms allow for rapid cellular responses by relying on quick modifications of existing proteins rather than de novo production. Another advantage is that modification allows for coordination and amplification of a signal if a single protein can interact with many other proteins. The protein-interaction network (PIN) thus forms another level of biological organization that influences the cell.

The data needed to characterize the PIN includes determination of the set of proteins present in a cell (the proteome), the state or location of those proteins if variable, and how these proteins interact. High-throughput methods to provide these data are developing, albeit more slowly than methods used to identify mRNA levels. This disparity is due to the lack of means to artificially increase the amount of any particular protein in a sample. The technique of polymerase chain reaction allows biologists to harness the natural process of DNA replication to make millions of copies of any known DNA or mRNA molecule in a biological sample, but no comparable technology exists for proteins.

When working with biological samples, specfic proteins can be detected using either antibodies or techniques that separate proteins based on biochemical properties and then calculate a molecular weight and compare that to a database of known protein weights. These techniques can sometimes reveal whether or not a protein has a phosphate group attached, for example, or whether it is in a particular subcellular location. Interactions between proteins can be determined (1) by assays that use antibodies to pull proteins out of cellular extracts and look for proteins that are associated with the protein removed, (2) by assays that use synthetic hybrid proteins that produce a visible result if two proteins physically interact, or (3) by in vitro or in vivo experimentation with cells or organisms that have had specific genes mutated. Bioinformatics provides another method whereby protein function may be inferred by comparing the sequence of a protein to genes with known function. There are functional regions of proteins, called domains, that occur in many different proteins, which can be detected in the DNA or protein sequence. If a protein contains a sequence similar to a known functional domain, the protein is also assumed to have that functionality.

In constructing a graph to represent the PIN, the individual proteins are the nodes, and the existence of an interaction between a pair of proteins corresponds to an edge between the nodes. As seen above, there are many ways in which proteins may physically interact. Relations between proteins may also be established by examining mRNA profiles, for example. If the mRNA profiles of two proteins have a high correlation, we assume the corresponding proteins are related and include the edge even if there is not a physical interaction. Each of these techniques provides different information, and combinations thereof are thus important for a more complete characterization of the proteome and the protein–protein interactions that occur.

11.4.1 Connectivity Distribution

Analyzing systems as disparate as the World Wide Web and a PIN has revealed surprising similarities in their structural organization. One simple characterization is the average number of nearest neighbors, or average degree. In a PIN, this corresponds to an average protein's number of interaction partners.

The average degree is simply $\langle k \rangle = (1/N) \sum_{ij} a_{ij}$. However, this measure does not provide detailed insight into the structure of a network. To gain further insight into the structure of a PIN, we study the connectivity, or degree, distribution $P(k)$, which is the number of nodes of degree k. From this measure, we determine the variation in connectivities on the network. Such distributions were studied by Erdös and Rényi [Bollobás, 2001], who showed that random graphs lead to a Poisson distribution. However, for many real networks, $P(k)$ does not have a Poisson-type behavior as predicted by the Erdös–Rényi random graph model. Instead, $P(k)$ frequently adheres to a heavy-tailed distribution often modeled as a power-law $P(k) \sim k^{-\alpha}$ [Albert and Barabási, 2002]. This is the case for the PIN of the yeast *S. cerevisiae*, the nematode *C. elegans*, and the fruit fly *D. melanogaster* in Figure 11.2 (see also Table 11.1).

It is interesting to note that if the connectivity distribution had been single peaked, such as Poisson or Gaussian, the notion of a typical node, as described by the average degree $\langle k \rangle$, would have been valid. However, this is not the case for a heavy-tailed PIN. In these networks, the majority of the nodes only have a few interaction partners while they coexist with nodes that participate in hundreds of interactions. Consequently, there is no typical node. Such networks are typically called scale-free, and nodes with a large number of interactions are called hubs. Hub proteins often have biological properties that are significantly different from non-hub proteins.

One of the most popular network models to capture the heterogeneity of the connectivity distribution was proposed by Barabási and

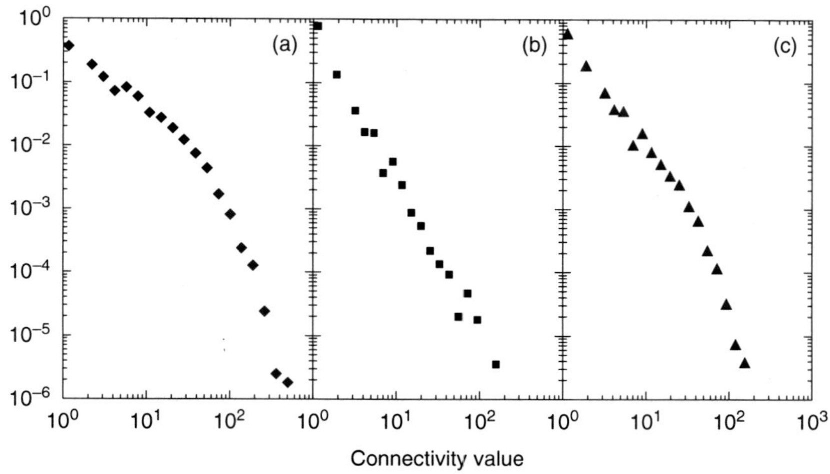

Figure 11.2 Connectivity distribution $P(k)$ for the protein-interaction networks of (a) the yeast *S. cerevisiae*, (b) the nematode *C. elegans*, and (c) the fly *D. melanogaster* (From BioGrid 2006, version 2.0.2.0, http://www.thebiogrid.org/).

Albert [Barabási and Albert, 1999]. It is similar to the network model by Price [Price, 1965] (see [Newman, 2003b] for a detailed discussion). These models are based on the notion that in a growing network new nodes are not connected with uniform probability to already existing nodes. Instead, new nodes have a higher chance of connecting to those with many neighbors than to nodes with few. This is often called the "rich get richer" effect or preferential attachment. If the chance of connecting to an already existing node i is linearly proportional to the degree, the resulting connectivity distribution is a power-law with an exponent of 3 [Albert and Barabási, 2002, Newman, 2003b].

Table 11.1 Properties of Three Whole-Organism Protein-Interaction Networks

Organism	Nodes	$\langle k \rangle$	S	$\langle C \rangle$	$\langle C_{\text{rand}} \rangle$	ρ
S. cerevisiae	5298	19.04	5294	0.154	0.0036	−0.040
C. elegans	2774	3.14	2551	0.020	0.0011	−0.159
D. melanogaster	7490	6.67	7372	0.030	0.00089	−0.039

Source: From BioGrid 2006, version 2.0.2.0, http://www.thebiogrid.org/

Note: For each network, we have indicated size, average node connectivity $\langle k \rangle$, size of the giant component S, average clustering $\langle C \rangle$, average clustering for a comparable Erdös–Rényi random network $\langle C_{\text{rand}} \rangle$, and assortativity ρ, which is defined momentarily

11.4.2 Network Assortativity

In many real networks, properties of adjacent nodes are correlated. In particular, it is often the case that the connectivities of neighboring nodes are correlated, making $P(k_i, k_j) \neq P(k_i)P(k_j)$. Several methods have been developed [Maslov and Sneppen, 2002, Newman, 2002, Newman, 2003a, Pastor-Satorras et al., 2001] to measure these connectivity correlations, and we highlight two such methods.

The first method of [Pastor-Satorras et al., 2001] measures connectivity correlations by calculating the average nearest-neighbor degree:

$$k_{nn,i} = \frac{1}{k_i} \sum_j k_j a_{ij}. \tag{11.3}$$

Consequently, $k_{nn,i}$ measures the affinity with which a node i connects to other nodes of either high or low degrees. In Figure 11.3, we have plotted $k_{nn}(k)$, which is defined by

$$k_{nn}(k) = \frac{\sum_{\{i:k_i=k\}} k_{nn,i}}{\sum_{\{i:k_i=k\}} 1}. \tag{11.4}$$

So, $k_{nn}(k)$ is the average neighborhood degree for nodes with connectivity k, If $k_{nn}(k)$ is an increasing function of k, the network shows an assortative mixing and high-degree nodes preferentially tend to be connected to other high-degree nodes. For the opposite situation, where $k_{nn}(k)$ is a decreasing function of k (as in Figure 11.3b), low-degree nodes tend to be connected to high-degree nodes, and the network is disassortative. This is typically the case for computer networks, where a limited

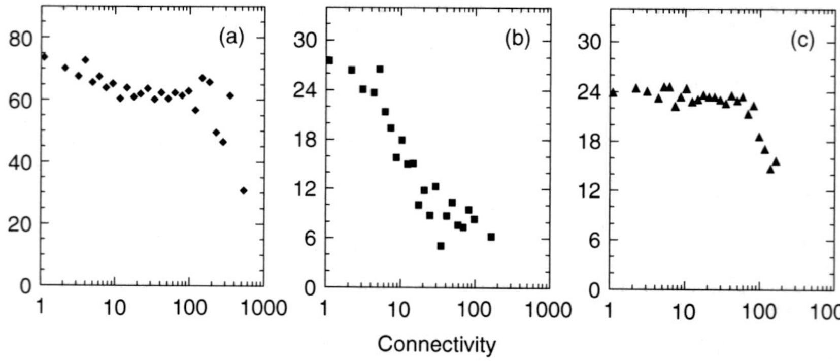

Figure 11.3 Average nearest-neighbor connectivity $k_{nn}(k)$ for the protein-interaction networks of (a) *S. cerevisiae* (b) *C. elegans*, and (c) *D. melanogaster* [BioGrid, 2006].

number of servers are connected to a large number of individual computers [Pastor-Satorras et al., 2001].

The second method of measuring degree–degree correlations collapses the distribution $P(k)$ into a single value called the assortativity of the graph [Newman, 2002]. This index is a Pearson correlation in nearest-neighbor degrees, defined as

$$\rho = \frac{M \sum_i j_i k_i - \left[\frac{1}{2} \sum_i (j_i + k_i)\right]^2}{\frac{1}{2} M \sum_i (j_i^2 + k_i^2) - \left[\frac{1}{2} \sum_i (j_i + k_i)\right]^2}, \quad (11.5)$$

where the sums are over edges, and the numbers j_i and k_i are the connectivities of the two nodes connected by edge i. The distribution $k_{nn}(k)$ and the assortativity index ρ are related as follows. If $k_{nn}(k)$ is uniform, then $\rho = 0$. However, if $k_{nn}(k)$ is increasing or decreasing then ρ is positive or negative, respectively. The magnitude of ρ indicates the strength of the correlation. It is straightforward to develop similar expressions for directed networks [Newman, 2003a].

The last column of Table 11.1 shows the assortativity ρ for three whole-organism PINs. As expected, the trends displayed in Figure 11.3 agree with the assortativity correlations calculated from Equation 11.5. In particular, panels (a) and (c) show no clear increasing or decreasing trend in $k_{nn}(k)$, which agrees with the calculated assortativity values close to zero. Taken together, these two methods offer detailed insights into the connectivity correlations of a network.

11.4.3 Community Finding

The network properties just discussed are based on characteristics of individual nodes, such as clustering, average degree, and connectivity. As stated previously, a long-standing hypothesis is that biological systems are modular, meaning that they consist of separable functional units. The idea of a community is different from the previous properties because it considers the entire network. By carefully analyzing a network, we identify modules as collections of nodes that are tightly connected when compared to the full network. These modules are often biologically significant. For instance, because proteins that exist in a cell as a complex are commonly members of the same functional class, we expect a tightly connected region to indicate a single functional class [Pereira-Leal et al., 2004, Poyatos and Hurst., 2004, Schwikowski B. and Fields, 2000, Wuchty et al., 2003].

Several methods are currently available to detect community structures. Many of these were developed by sociologists, who have long been interested in community analysis. Unfortunately, these methods typically were designed for small networks and are not tractable on networks consisting of several thousands of nodes. Many of these methods are related to a measure called betweenness-centrality (BC), which is related to shortest paths [Brandes, 2001, Freeman, 1977, Newman, 2001, Wasserman and Faust, 1994].

The study of shortest paths on networks is the source of the term small-world [Watts and Strogatz, 1998]. The length of the average shortest path, ℓ, between two nodes can be calculated using a breadth-first search, which has complexity $O(NM)$. In a random network, such as that of the Erdös–Rényi model, the average shortest distance scales with the network size as $\ell \sim \ln(N)$ [Bollobás, 2001]. The betweenness-centrality of an edge or node is the fraction of shortest paths that pass through the node or edge (see [Newman, 2001] for a detailed discussion).

A typical algorithm based on BC is to recursively remove the edge with the largest BC value, followed by recalculating the BC values for the remaining network. The complexity of such an algorithm is $O(N^3)$. Approximations where the BC values are only calculated for the initial network are much faster, but the gain in computational run-time reduces accuracy.

There are alternatives to the BC approach, and we discuss two such methods. These techniques have the advantage of rapidly identifying communities on large networks with high accuracy. The first method is due to Newman [Newman, 2004] and is described as agglomerative hierarchical clustering. Let Q be the following measure of network modularity for any node partition

$$Q = \sum_i \left[e_{ii} - \left(\sum_j e_{ij} \right)^2 \right], \tag{11.6}$$

where e_{ij} is the fraction of edges in the network connecting nodes from module i to those of module j. This measures the number of intercommunity links relative to that of a random occurrence. A value near zero suggests that there is little information in the chosen partition, whereas a value greater than 0.3 indicates significant modularity [Newman, 2004].

Newman suggests optimizing Q heuristically by starting with N communities (one for each node) and joining the two that render the highest value of Q, which may increase or decrease the current value. When all nodes have been joined into a single module, the algorithm is finished and the optimal value of Q indicates a collection of communities. This approach is $O(N^2)$ and has been successfully applied to systems with more than $50,000$ nodes. Furthermore, it is possible to generalize this community detection algorithm to incorporate varying link-strengths.

The second alternative to the BC method is called k-clique percolation [Palla et al., 2005]. Unlike the method just described, this technique does not require that each node belong to a unique community. For many networks this is favorable. For example, a protein may have multiple functions and naturally belong to many communities.

This method is based on the observation that a community often decomposes into nearly complete subgraphs that share nodes. Consequently, this method is based on the k-clique. A network module is defined as the union of all k-cliques (for a fixed k) that share $k - 1$ nodes, and thus are adjacent on the network. An alternative description is that of a rolling k-clique, only moving one node at the time.

A further benefit of k-clique percolation is that it allows a higher-level representation of a network. We may collapse the graph so that each community is a node, and two communities are connected if they have a nonempty intersection. This makes it possible to introduce a scalable map of the network that represents the communities at different levels of magnification, with the first level, highest magnification, corresponding to the actual nodes, the second level to communities, the third level to communities of communities, etc.

11.4.4 Biology and Topology

So far we have discussed topological properties of PINs without emphasizing the connection between network representations and biological information. The first indication that a PIN might carry biological information arose from questions of robustness [Albert et al., 2000], which demonstrated that networks with heavy-tailed connectivity distributions were robust against random failures yet fragile when an attack occurred at a highly connected node.

Molecular biology techniques allow for the experimental disruption of single genes, and examination of the phenotypes of these modified organisms can reveal whether the disrupted gene is essential for survival of the organism under a set of defined conditions. In fact, a large-scale experimental study in *S. cerevisiae* shows that only 18.7 percent of the total number of genes are essential on disruption or removal [Giaever et al., 2002], while a study on *E. coli* found 13.7 percent of the genes are essential [Gerdes et al., 2003]. Motivated by these experimental observations of network fragility, Barabási and coworkers investigated the possibility of correlations between a protein's connectivity and phenotypic essentiality, discovering an increased likelihood for highly connected proteins to be essential [Jeong et al., 2001]. In other words, a protein that has a large number of interaction partners is more likely to be involved in an essential cellular function, often called the centrality-lethality rule. Although recently debated, the centrality-lethality result is considered robust [Batada et al., 2006].

A recent study suggests that this increased lethality of highly connected proteins can be explained by a simple mechanism [He and Zhang, 2006]. The idea is to support the centrality-lethality rule by assuming essential nodes and links are randomly distributed on the network. The function of an essential link is carried out by the interaction of the incident proteins, and both nodes are essential. This model generates the centrality-lethality rule through the simple fact that it is more likely for a hub to be part of an essential link than a low-degree node. By choosing the essential link and node fractions appropriately, it is possible to fit the observed centrality-lethality rule within experimental error bars [He and Zhang, 2006].

Because highly connected proteins occupy a special role in the network, it is interesting to ask whether hub proteins evolve at a different pace from proteins with only a few interaction partners. The rationale for this question is that change to hub proteins might be constrained due to their interactions. Although initial results were contradictory [Coulomb et al., 2005], a recent more decisive study [Batada et al., 2006] showed these results could be explained by subtle biases in the methods used to generate the PINs. After accounting for the equal density of active domains in hub and non-hub proteins, it was shown that there are not significant differences in mean rate of protein evolution. The hub proteins of *S. cerevisiae* did, however, contain a higher number of phosphorylation sites than non-hub proteins and showed a marked trend of being encoded by mRNA's with short half-lives. Taken together, this indicates that highly connected proteins are subject to much tighter control, being part of a dynamic, short-lived protein complex [Batada et al., 2006].

We have focused on static aspects of a PIN, but proteins are constantly produced and degraded and many interactions occur in specific cellular locations, such as the cellular membrane. A more realistic depiction would address the temporal and spatial aspects of the situation. Whole-organism protein-expression arrays are currently unavailable, and the chosen substitute has been mRNA expression arrays. The recent analysis in [Han et al., 2004] indicates that highly connected nodes in the *S. cerevisiae* PIN can be either date-hubs, binding to their partners at different times or locations, or party-hubs interacting with most of their neighbors simultaneously. Including temporal aspects such as this allows us to investigate information flow because the temporal activation of protein transcription is reflective of evolved regulatory mechanisms that ensure proper cellular responses to external stimuli.

11.5 Metabolic Networks

Life depends on the ability to import molecules from the environment and convert these to the needed metabolites. These conversions are carried out by enzymes that catalyze (facilitate) specific conversions of starting molecules (reactants) into products. There may be several intermediary steps from intial reactants to the ultimate product, each carried out by a different enzyme, and the set of all these component reactants, products, reactions, and enzymes form a metabolic pathway. Metabolic pathways can be classified as either anabolic pathways that construct needed molecules or catabolic pathways that break down molecules to provide necessary reactants.

The different reactions and catalyzing enzymes vary tremendously. As seen in the previous section, the enzymes may or may not be active depending on the presence of cofactors, modification state, etc. Another difference between enzymes is in their rates of catalysis, which may vary over orders of magnitude. Variation in these reaction rates affects the overall rate of flow (flux) of metabolites in a particular pathway.

From the reactant perspective, a particular type of molecule may participate in only one reaction or be used in several different reactions. A reaction may require one or more reactants, and the ratios (stoichiometry) of those reactants may vary. Finally, while for the most part metabolic pathways can be assumed to be one-way, there are cases of reversible reactions in a cell and cyclic reaction pathways that take a reactant through a series of intermediates but end up regenerating the initial reactant.

A cell's metabolism is the sum of all the reactions it carries out. It is important to recognize that while a cell has the potential to carry out many reactions, the actual reactions that are being carried out at any one time depend heavily on the cell's environment. For example, differential gene regulation in a bacterial cell will lead to different enzymes being present under aerobic (oxygen present) vs anaerobic (oxygen absent) conditions or when glucose or lactose are present as the main carbon source.

11.5.1 Metabolic Network Structure

To represent a cell's metabolism with a network, we need to assign meaning to the nodes and links. The network abstraction is not unique, and Figure 11.4 depicts several representations of a simple metabolic network. The three reactions of the metabolism are found in Figure 11.4a. In the first reaction $A + B \rightarrow C + D$, we say that A and B are reactants and C and D are products. The most common representation of this metabolism is represented in Figure 11.4c, where metabolites are nodes that are connected with an undirected link if they participate as reactant and product in a reaction. Note that a link does not represent a single reaction, as two metabolites may appear in multiple reactions. An example is shown in Figure 11.4a, where metabolites A and D co-occur in reactions R1 and R3, and the edge or arc between A and D corresponds to both reactions. Furthermore, one reaction appears as multiple edges or arcs (see Figure 11.4).

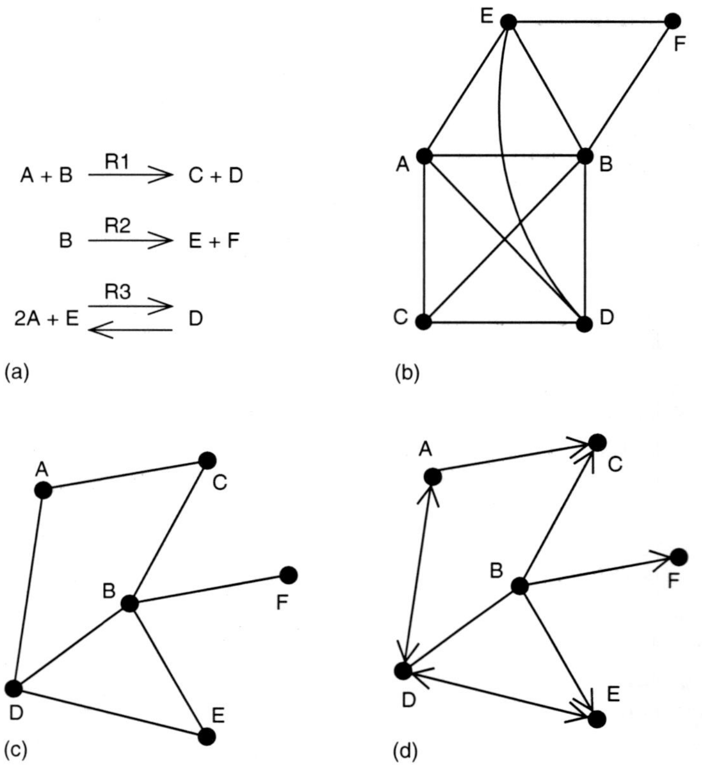

Figure 11.4 Cellular metabolism can be represented as a network. (a) A simplied metabolic reaction set. Network description of this reaction set: (b) connecting all metabolites in a single reaction with undirected links; (c) substrates are only connected to products with undirected links; (d) same as (c) with directed links.

An alternative representation that is particularly important for the discussions that follow is a bipartite network in which the nodes represent either metabolites or reactions. Allowing the set of reactions to be R and the set of metabolites to be M, we are interested in the bipartite network (R, M, E), where $(i, r) \in E$ if metabolite i is a reactant of reaction r and $(r, i) \in E$ if metabolite i is a product of reaction r. A depiction is seen in Figure 11.9.

Different network representations have different statistical properties. Using the bacterial metabolism in *E. coli* as an example, Figure 11.5 shows the differences in the connectivity distribution, $P(k)$, for the three network representations detailed in Figure 11.4. Note that $P(k)$ is heavy tailed in Figure 11.5; however, the result is not as simple when using a bipartite network representation. In this case, it is possible to distinguish

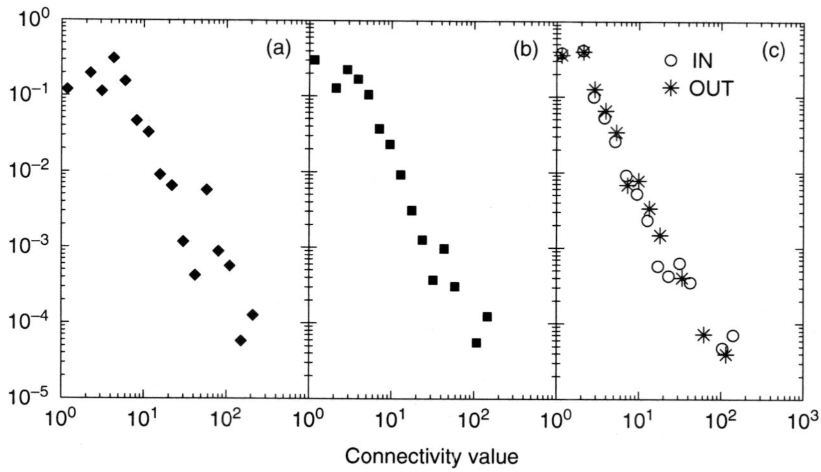

Figure 11.5 Connectivity distributions $P(k)$ of *E. coli* metabolism using the three metabolic network representations in Figure 11.4. (a) corresponds to panel (b); (b) to panel (c); (c) to panel (d).

metabolites and enzymes. For the metabolites, the connectivity distribution is still heavy tailed, although the enzyme distribution is exponential. This is not surprising, as cofactors, such as ATP or NADP may participate in hundreds of reactions while an enzyme has a limited number of active domains. To further contrast and compare biases of different network representations, Table 11.2 shows the average clustering coefficient $\langle C \rangle$ and the assortativity index ρ for three organisms using the representations in Figure 11.4b and c. The clustering and assortativity corresponding to Figure 11.4b is significantly higher than that of Figure 11.4c because it introduces a fully connected subgraph for each reaction.

Table 11.2 Average Clustering and Assortativity for Three Organismal Metabolic Networks Using the Network Representations Described in Panels Given by Figure 11.4b and c—Network Model Indicated with a Subscript

Organism	N	M_b	M_c	$\langle C \rangle_b$	$\langle C \rangle_c$	ρ_b	ρ_c
H. pylori	489	4058	1920	0.72	0.28	−0.285	−0.261
E. coli	540	3753	1867	0.66	0.20	−0.251	−0.217
S. cerevisiae	1064	6941	4031	0.67	0.23	−0.182	−0.150

11.5.2 Weighted Metabolic Networks

The majority of network studies have focused on topological properties and not on the rate of metabolic activity, which can vary significantly from reaction to reaction. This important function is not captured by topological approaches, and to develop an understanding of how the structure of a metabolic network affects metabolic activity; it is necessary to include this information in the network description. We require a meaningful understanding to consider the intensity (strength) between metabolites, the direction (when applicable), and the temporal aspects of the interactions. Although not much is known about the temporal aspects of metabolic activity, recent results [Blank et al., 2005, Cannizzaro et al., 2004, Canonaco et al., 2001, Emmerling et al., 2002, Fischer and Sauer, 2003, Fischer and Sauer, 2005, Gombert et al., 2001, Sauer et al., 1999] have provided information about the relative intensities of the interactions in single-cell metabolism, which we incorporate by considering weighted links. A natural, although not unique, measurement of interaction strength is the amount of substrate being converted to a product per unit time, the flux of the reaction.

A linear optimization approach, called flux-balance analysis (FBA), enables us to calculate the flux rate for each reaction in a whole-cell metabolic network. The FBA method assumes that the concentration of all metabolites that are not subject to transport across the cell membrane is in a steady state. Let $[A_i]$ be the concentration of metabolite i and S_{ir} be the stoichiometric coefficient of metabolite i in reaction r. For example, if reaction r is $3A_1 + 2A_2 \rightarrow 2A_3$, then $S_{1r} = -3$, $S_{2r} = -2$, and $S_{3r} = 2$. If metabolite i does not appear in reaction r, we assume that $S_{ir} = 0$. Allowing ν_r be the flux of reaction r, we have that the steady-state assumption requires

$$\frac{\mathrm{d}[A_i]}{\mathrm{d}t} = \sum_r S_{ir}\nu_r = 0. \tag{11.7}$$

Any flux values satisfying this equation correspond to a stoichiometrically allowed state of the cell. To select flux values that are biologically relevant, we optimize for cellular growth. Experiments support this hypothesis in several conditions, but there are other meaningful objectives. See [Bonarius et al., 1997, Kauffman et al., 2003] for a more detailed discussion of FBA.

The recent advances in whole-genome annotation have made it possible to generate high-fidelity whole-cell metabolic networks. Metabolic models of the bacteria *H. pylori* and *E. coli*, as well as the eukaryote *S. cerevisiae*, have been used to predict essential genes [Edwards and Palsson, 2000, Schilling et al., 2002, Duarte et al., 2004, Papp et al., 2004], genetic interactions [Segre et al., 2005], and possible minimal microbial genomes [Burgard et al., 2001, Pal et al., 2006]. The fluxes from FBA measure each

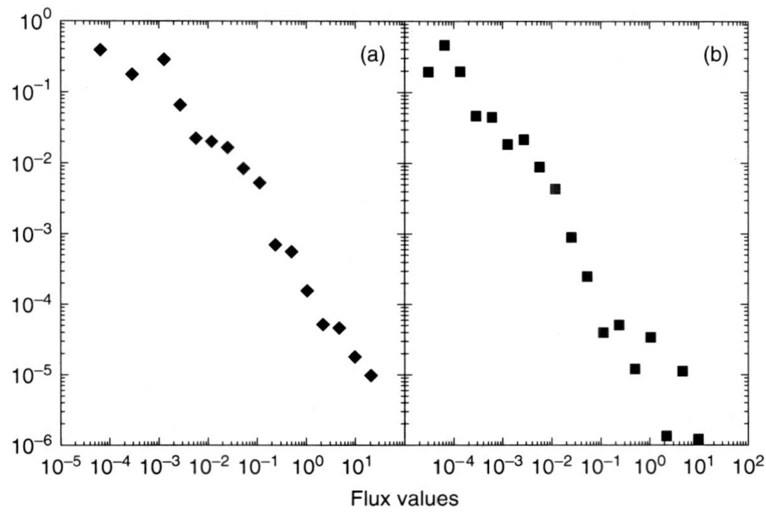

Figure 11.6 Distribution of metabolic reaction flux values (edge weights) from FBA analysis for the metabolic network of the budding yeast *S. cerevisiae* in (a) aerobic, glucose-limited and (b) aerobic, acetate-limited conditions.

reaction's relative activity. In particular, the work of [Almaas et al., 2004] demonstrates that similar to the degree distribution, the flux distribution of *E. coli* displays a strong overall inhomogeneity: reactions with fluxes spanning several orders of magnitude coexist in the same environment. The flux distribution for *S. cerevisiae* in Figure 11.6 is heavy tailed, indicating that $P(\nu) \propto \nu^{-\alpha}$. The flux exponent is predicted to be $\alpha = 1.5$ by FBA methods. In a recent experiment, the strength of the various fluxes of the central metabolism of *E. coli* was measured using nuclear magnetic resonance (NMR) methods [Emmerling et al., 2002], revealing the power-law flux dependence $P(\nu) \propto \nu^{-1}$. This power-law behavior indicates that the vast majority of reactions with small fluxes coexist with the few reactions that have large fluxes.

The FBA approach allows us to analyze a metabolism as a weighted network because each reaction is assigned a flux value. These values are node weights in the bipartite representation (R, M, E). Unfortunately, the identity of a reaction in the other network models is opaque because each reaction is a subgraph corresponding to the metabolites of the reaction. To translate the node weights ν_r of the bipartite representation to link weights of another representation, we let

$$w_{ij} = \left| \sum_r S_{ir}\nu_r + \sum_r S_{jr}\nu_r \right|,$$

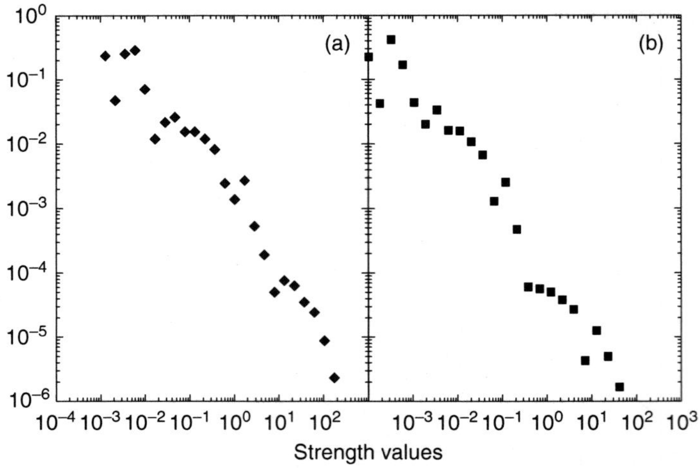

Figure 11.7 Distribution of node strength values for *S. cerevisiae* metabolism in (a) aerobic, glucose-limited, and (b) aerobic, acetate-limited conditions.

which is the aggregate rate at which metabolite i transforms into metabolite j. Generally, negative edge weights are possible and simply mean that metabolite j transforms into metabolite i.

Several measures have been introduced to study weighted networks in the context of airline transportation and the relationship between coauthors. One of these is called the strength, s_i, of a node i, defined as $s_i = \sum_j w_{ij}a_{ij}$, which is simply a weighted node degree [Barrat et al., 2004]. Figure 11.7 shows that the distribution of node strengths, $P(s)$, for the *E. coli* metabolism with glucose as the single carbon source is

$$\langle s(k) \rangle \propto k^{\beta}. \tag{11.8}$$

For networks without correlations between the node connectivity and link-weights, the weights w_{ij} are independent of i and j, and we can represent the link-weights with their average value: $w_{ij} = \langle w \rangle$, making $\beta = 1$ [Barrat et al., 2004].

We continue by generalizing the clustering coefficient to weighted networks. Because c_i indicates the local density of triangles, a similar definition with link-weights should determine if large or small weights are likely to be found in clusters. One possible definition from [Barrat et al., 2004] is

$$c_{a,i} = \frac{1}{s_i(k_i - 1)} \sum_{j,l} \frac{1}{2}(w_{ij} + w_{il})a_{ij}a_{il}a_{jl}, \tag{11.9}$$

where s_i is the node strength. The average weighted clustering is $\langle C_a \rangle = (1/N) \sum_i c_{a,i}$. If no correlations exist between weights and topology, Equation 11.9 is equal to that of the unweighted clustering (see Equation 11.1). We identify two possible scenarios. If $\langle C_a \rangle > \langle C \rangle$, large weights are predominantly distributed in local clusters, whereas if $\langle C_a \rangle < \langle C \rangle$, triangles consist of low-weight links.

Other possible definitions of a weighted clustering coefficient have been proposed [Holme et al., 2007, Onnela et al., 2005, Zhang and Horvath, 2005]. The weighted clustering coefficient expression given by Equation 11.9 only includes two weights of any triangle through node i. The following definition from [Onnela et al., 2005] extends this so that all three weights are considered

$$c_{b,i} = \frac{2}{(\max_{ij}\{w_{ij}\})k_i(k_i - 1)} \sum_{j,l} \left(w_{ij} w_{il} w_{jl}\right)^{1/3} a_{ij} a_{il} a_{jl} \qquad (11.10)$$

Notice that this is a geometric mean instead of an algebraic mean, given by Equation 11.9. The average weighted clustering is $\langle C_b \rangle = (1/N) \sum_i c_{b,i}$. Related analysis from finance has shown that Equations 11.9 and 11.10 can lead to significantly different interpretations [Onnela et al., 2005].

11.5.3 Fluxes and Metabolic Network Structure

The flux distributions of a metabolic network rely on the network topology. Some of this dependence is understood by studying the correlation between w_{ij} and k_i and k_j. The metabolic fluxes in *E. coli* scale as

$$\langle w_{ij} \rangle = \frac{\sum\limits_{\{i,j:k_i k_j = k\}} w_{ij}}{\sum\limits_{\{i:k_i k_j = k\}} 1} \propto (k_i k_j)^{\theta}, \qquad (11.11)$$

where $\theta \approx 0.5$ for metabolic fluxes in glucose-limited conditions in *S. cerevisiae* (Figure 11.8a) and *E. coli* [Macdonald et al., 2005]. However, other values for θ are possible, as demonstrated in Figure 11.8b, where we find $\theta \approx 0.7$ for acetate-limited conditions. In the case of no correlations between the connectivity k_i and k_j, we have from [Barrat et al., 2004] that the exponent θ in Equation 11.11 is related to β (Equation 11.8) in the scaling of node strength as $\beta = 1 + \theta$.

We further investigate how the flux values depend on the network topology at the single metabolite level. There are two flux structures of interest. A homogeneous local organization implies that all reactions producing (consuming) a given metabolite have comparable flux values. On the other hand, a more delocalized, or hot backbone, is expected if the

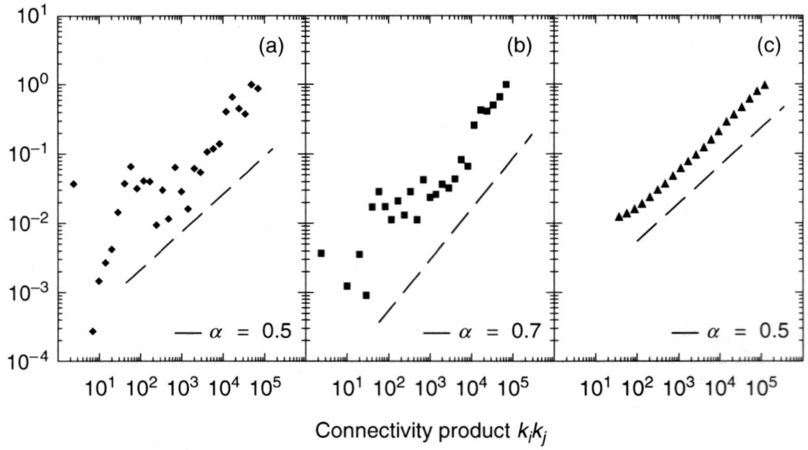

Figure 11.8 Correlation between (normalized) edge weights and local connectivity for metabolic fluxes in *S. cerevisiae* in (a) glucose-limited and (b) acetate-limited conditions, as well as (c) betweenness-centrality for the Barabasi–Albert Model [Macdonald et al., 2005]. The dashed lines serve as guides to the eye.

local flux organization is heterogeneous. To distinguish between these two scenarios, we define the following measure [Barthelemy et al., 2003, Almaas et al., 2004] for each metabolite produced or consumed by k reactions, define $Y(k, i)$ by

$$Y(k, i) = \sum_{r=1}^{k} \left(\frac{S_{ir} \nu_r}{\sum_{l=1}^{k} S_{il} \nu_l} \right)^2. \tag{11.12}$$

If all reactions producing (consuming) metabolite i have comparable values, $Y(k, i)$ scales as $1/k$. However, if a single reaction's activity dominates Equation 11.12, we expect that $Y(k, i) \sim 1$. For the two cases, where the *E. coli* metabolic performance is optimized with glucose and succinate as the only available carbon sources, we find that separately calculating $Y(k, i)$ for the in- and out-degrees follows the power-law $Y(k, i) \sim k^{-0.27}$. We interpret this as intermediate behavior between the two cases described above. However, the exponent of -0.27 indicates that the large-scale inhomogeneity observed in the overall flux distribution is increasingly valid at the level of the individual metabolites.

The local flux inhomogeneity suggests that we can identify a single reaction dominating the production or consumption of most metabolites. A simple algorithm is capable of extracting the subnetwork solely consisting of these dominate reactions, called the high-flux backbone (HFB) [Almaas et al., 2004]. The algorithm has two steps

1. For each metabolite, discard all incoming and outgoing links except the two dominating mass production.
2. From this set of reactions, keep only those reactions that appear as both maximal producer and maximal consumer.

The resulting HFB is specific to the particular choice of environmental conditions. Interestingly, the HFB mostly consists of reactions linked together, forming a giant component with a star-like topology that includes almost all metabolites produced in a specific growth environment. Only a few pathways are disconnected: although these pathways are members of the HFB, their end-products serve only as the second most important source for some other HFB metabolite. Connected reactions in the HFB largely agree with the traditional, biochemistry-based partitioning of cellular metabolism. For example, in *E. coli* all metabolites of the citric acid cycle are recovered, as well as most of the other important pathways, such as those being involved in histidine-, murein-, and purine-biosynthesis. Although the detailed nature of the HFB depends on the particular growth conditions, the HFB captures the reactions that dominate the activity of the metabolism for this condition. As such, it offers a complementary approach to the analyses in [Papin et al., 2004, Schilling et al., 2000, Schuster and Hilgetag, 1994].

Our final discussion about metabolic networks focuses on identifying the reactions that are used in varying environments, and we explore how the fluxes depend on environmental changes. Referring to Figure 11.9, we let ν_R be the collection of uptake fluxes that provide nutrients (resources, inputs, etc.) to the cell. We also let r_C be the reactions that occur within the cell (output reactions are not considered). For each ν_R, we let $r_C^*(\nu_R)$ be the point-to-set map whose image is a collection of reactions that can

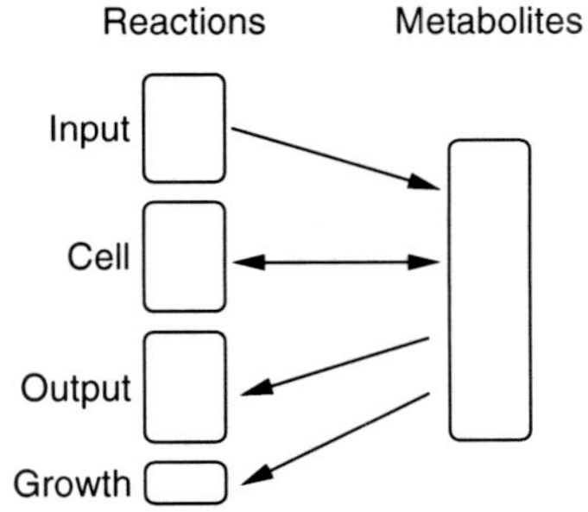

Figure 11.9 A simple bipartite representation of cellular metabolism.

have a positive flux while the cell achieves optimal growth with the input fluxes fixed at ν_R. The metabolic core is

$$\bigcap_{\nu_R \geq 0} r_C^*(\nu_R),$$

which defines the reactions that are allowed to be active in any environment when the cell achieves optimal growth.

A stochastic procedure to calculate the metabolic core is to uniformly sample the set of input fluxes and use FBA to optimize growth for each sample. If a reaction's flux is positive, we know that this flux is in $r_C^*(\nu_R)$ for the sample. Taking the intersection of these sets over the sampled inputs yields a subset of the metabolic core. The computational results in [Almaas et al., 2005] sampled $30,000$ input fluxes between 0 and 20 (20 is large enough to guarantee that a nutrient is available if needed, and hence, setting the intake fluxes to 20 assumes the cell is in an environment with unlimited resources). The metabolic core contained 138 of the 381 metabolic reactions in *H. pylori* (36.2 percent), 90 of 758 in *E. coli* (11.9 percent), and 33 of 1,172 in *S. cerevisiae* (2.8 percent).

The metabolic core is partitioned into two types of reactions. The first type consists of those that are essential for biomass formation under all environmental conditions (81 of 90 reactions in *E. coli*), whereas the second type of reaction is required only to assure optimal metabolic performance. In case of the inactivation of the second type, alternative suboptimal pathways can be used to ensure cellular survival. The metabolic core of *S. cerevisiae*, however, only contains reactions predicted by FBA to be indispensable for biomass formation under all growth conditions.

The analysis in [Almaas et al., 2005] further suggests that optimal metabolic flows adjust to environmental changes through two distinct mechanisms. The more common mechanism is flux plasticity, involving changes in the fluxes of already active reactions when the organism is shifted from one growth condition to another. For example, changing from glucose- to succinate-rich media alters the flux of 264 *E. coli* reactions by more than 20 percent. The reactions in the metabolic core always adapt to changing environmental conditions through flux plasticity. Less commonly, environmental changes induce structural plasticity, resulting in changes to the metabolism's active wiring diagram, turning on previously zero-flux reactions, and inhibiting previously active pathways. For example, when shifting *E. coli* cells from glucose- to succinate-rich media, eleven previously active reactions are turned off completely, while nine previously inactive reactions are turned on.

A similar selection of metabolic reactions was suggested by [Burgard et al., 2001]. Their minimal reaction contains the metabolic core as well as all the reactions necessary for the sustained growth on any chosen substrate. A different definition of a minimal reaction set was proposed by [Reedi and Palsson, 2004], which consists of the 201 reactions that are always active in *E. coli* for all 136 aerobic and anaerobic single carbon source minimal environments capable of sustaining optimal growth.

A reasonable speculation is that the reactions in the metabolic core play an important role in the maintenance of crucial metabolic functions because they are active under all environmental conditions. Consequently, the absence of individual core-reactions may lead to significant metabolic disruptions. This hypothesis is strengthened through cross-correlation with gene deletion data [Gerdes et al., 2003]: 74.7 percent of those *E. coli* enzymes that catalyze core metabolic reactions (i.e., core enzymes) are essential, compared with a 19.6 percent lethality fraction characterizing the noncore enzymes. A similar pattern of elevated essentiality is also observed when analyzing deletion data for *S. cerevisiae* [Giaever et al., 2002], in which essential enzymes catalyze 84 percent of the core-reactions, whereas the conditionally active enzymes have an average essentiality of only 15.6 percent [Almaas et al., 2005]. The likelihood that the cores contain such a large concentration of essential enzymes by chance is minuscule, with p-values of $3.3e^{-23}$ and $9.0e^{-13}$ for *E. coli* and yeast, respectively.

Metabolic core enzymes also stand apart from the conditionally active ones when comparing their evolutionary conservation. In comparing the set of DNA sequences of core enzymes from *E. coli* with the DNA sequences for these same enzymes in a reference set of 32 bacteria, the average amount of sequence conservation is 71.1 percent ($P < 1e^{-6}$). Similar comparisons using the set of noncore enzymes show a sequence conservation of only 47.7 percent. Despite taking into account correlations between essentiality and evolutionary conservation, one would expect the core enzymes to have a conservation level of only 63.4 percent [Almaas et al., 2005], thus showing that selection acts against excessive tinkering with these enzymes.

These results indicate that an organism depends largely on the continuous activity of the metabolic core, regardless of the environmental conditions. The conditionally active metabolic reactions represent the different ways in which a cell is capable of adjusting to utilize substrates from its environment. From a practical perspective, the core enzymes essential for biomass formation, both for optimal and suboptimal growth, may prove effective antibiotic targets given the cell's need to maintain the activity of these enzymes in all conditions.

11.6 Systems Biology and Operations Research

One of the primary research fields in OR is network optimization, including modeling, algorithms, and analysis. The variety of problems that can be modeled via a network is staggering, and numerous OR experts have spent their careers analyzing such problems. As the previous sections demonstrate, a cell's processes can be modeled with networks that highlight the interactions within a cell. This is a powerful new tool for biologists, and the experts in OR are well positioned to help advance this important science.

The goal of this section is to highlight a few of the places where systems biology and OR overlap. This is not meant to be an exhaustive exposition, which is not possible in the confines of this chapter. We encourage interested readers to look at the cited articles to begin a more thorough investigation. No matter what the particular expertise, there is likely an important and novel application in biology.

To begin, we consider the linear program that identifies the metabolic fluxes of a cell in a steady state. A simplistic but powerful depiction of the associated network is illustrated in Figure 11.9. This is a bipartite network where reactions on the left are linked to metabolites on the right. For example, if r is the reaction $A_1 + 2A_2 \rightarrow A_3 + 3A_4$, then (A_1, r), (A_2, r), (r, A_3), and (r, A_4) are arcs. The cell's inputs (resources) are modeled as reactions that transport metabolites through the cellular membrane into the cell. Similarly, the cell's outputs (products) are reactions that transport metabolites out of the cell. We let C, R, and P be matrices of the form $[S_{ir}]$, where the columns are, respectively, indexed by reactions within the cell, reactions that add resources to the cell, and reactions that terminate in products, except growth. Growth is defined as the collection of metabolites that need to pass through the cell to achieve a unit of growth, and we let G be the column vector that expresses this relationship. As an example, suppose the metabolites used to model the cell are $A_1, A_2, \ldots, A_{10}$. Then a unit of growth being defined as $2A_3 + A_7 + 3A_8$ is the same as G being $(0, 0, 2, 0, 0, 0, 1, 3, 0, 0)^T$. We point out that the matrix $[C|R|P|G]$ is similar to the biadjacency matrix, the difference being that the nonzero components are the signed stoichiometric coefficients of the associated reaction.

Although the terms used to describe this network are new to OR, the model is not. The fluxes of the reactions control the flows across the arcs, and hence the amount of metabolites in the cell. Although researchers often discuss a metabolic flow, the fluxes are not traditional flow variables because they are associated with nodes instead of edges. In particular, a positive flux can indicate that several arcs have positive flow. We let ν_C, ν_R, ν_P, and ν_G be the respective flux vectors for the reactions within the cell, the reactions that provide resources, the reactions that make products other than growth, and the amount of growth. The steady-state assumption in Equation 11.7 guarantees the conservation of metabolic flux throughout the network. This assumption essentially balances the metabolites in the cell so that they do not accrue.

Experimental results have shown that maximizing growth is a biologically relevant objective [Bonarius et al., 1997, Kauffman et al., 2003], and the linear program that achieves this is

$$\begin{aligned} \text{Max}\{z : C\nu_C + R\nu_R + P\nu_P + G\nu_G = 0, \\ 0 \le \nu_R \le u, 0 \le \nu_P, 0 \le \nu_C \le U\}, \end{aligned} \tag{11.13}$$

where u limits the cell's inputs and U bounds the flux values (each bound is the maximum rate the corresponding reaction). This linear program allows us to give a mathematical definition to a few of the terms of the previous section. Let $\mathcal{P}(u)$ be the feasible region of Equation 11.13 and make the notational convention that $\mathcal{P}(\infty)$ means the input flows are unrestricted. We also assume the $\nu = (\nu_C^T, \nu_R^T, \nu_C^T, \nu_G^T)^T$. With this notation, reaction $j \in C$ is essential (or necessary) if

$$\{\nu \in \mathcal{P}(\infty) : \nu_j = 0, \nu_G = 1\} = \emptyset. \tag{11.14}$$

So, if reaction j is turned off then it is impossible to achieve a unit of growth no matter what resources are given to the cell.

Identifying the essential reactions can be accomplished by sequentially investigating the feasibility of Equation 11.14 for each $j \in C$, which is possible by optimizing the zero function over the associated constraints. However, this tedious calculation has a more elegant solution. The question of partitioning the reactions into those that are necessary and those that are not is actually the problem of identifying the implied equalities in

$$C\nu_C + R\nu_R + P\nu_P = -G, \quad 0 \le \nu_C \le U, \ 0 \le \nu_R, \ 0 \le \nu_P. \tag{11.15}$$

The implied equalities of this system further indicate the inputs and outputs of the cell that are necessary for growth as well as those reactions that operate at their maximum rate. Identifying implied equalities has a long and important history in OR, and we point readers to refer [Greenberg, 1996] and the associated bibliography. We highlight two methods, one theoretical and one more practical.

The theoretical method relies on the concept of the optimal partition, which is a central topic in the study of interior-point algorithms. Consider the standard form linear program

$$\text{Min}\{c^T x : Ax = b, x \ge 0\}, \tag{11.16}$$

where $A \in \mathbb{R}^{m \times n}$ and we assume that there is a strictly positive feasible element (Slater's condition). Throughout the late 1980s and 1990s, interior-point algorithms were studied with regards to this problem, with the most important contribution being that these algorithms solve the problem in polynomial time. Fairly early in these investigations, it was realized that the solution produced by the most popular interior-point algorithms (called path-following interior-point algorithms) differed from the solution produced by the venerable simplex algorithm. The difference is that interior-point algorithms terminate in the strict interior of the optimal set instead of at an optimal vertex. If there is a single solution, there is no difference, but in the presence of degeneracy, the solutions are different. In particular, the solution rendered by a path following interior algorithm induces the optimal partition. Let x^* be the theoretical optimal solution produced by a path-following interior-point algorithm and define

$$B = \{i : x_i^* > 0\} \text{ and } N = \{i : x_i^* = 0\}.$$

Clearly (B, N) is a two set partition of $\{1, 2, \ldots, n\}$, but this partition uniquely defines the optimal set,

$$\mathcal{P}^* = \{x : Ax = b, x \geq 0, x_N = 0\},$$

where the set subscript indicates the subvector indexed by the elements of the set. This means that a component of x^* is zero if and only if it is required to be zero to achieve optimality. A component being positive indicates that it can be positive in an optimal solution, but some optimal solutions may have a zero at this component.

The conditions identifying the optimal partition of the linear program in Equation 11.16 are

$$Ax = b, \quad A^T y + s = c, \quad x \geq 0, \quad s \geq 0, \quad x^T s = 0, \quad x + s > 0. \qquad (11.17)$$

As any person in OR recognizes, these are the KKT (or Lagrange) conditions of optimality with the added condition that $x + s > 0$. Any (x, y, s) satisfying these conditions is called a strictly complementary solution to the linear program, and such solutions have been known to exist since 1956 [Goldman and Tucker, 1956]. Until the advent of interior-point algorithms, strictly complementary solutions held theoretical value only. If (x, y, s) satisfies all but the last condition, i.e., $x_i + s_i = 0$ for some i, then the solution is degenerate. Any pair (x_i, s_i) such that $x_i + s_i = 0$ is called a degenerate pair and the extent of degeneracy refers to the maximum number of possible degenerate pairs. Degeneracy is a topic that is mistakingly ignored in many first courses in linear programing, a pedagogical mistake that propagates misguided analysis [Jansen et al., 1997]. Understanding degeneracy provides for robust and sound analysis that appropriately explains the problem, and as we shall see momentarily, metabolic networks are highly degenerate.

A linear program that identifies the essential reactions is

$$\text{Min } \{0^T \nu_C + 0^T \nu_R + 0^T \nu_P :$$
$$C\nu_C + R\nu_R + P\nu_P = -G, \quad 0 \leq \nu_C \leq U, \ 0 \leq \nu_R, 0 \leq \nu_P\}. \quad (11.18)$$

Adapting Equation 11.17 to this problem, we see that we want to solve

$$C\nu_C + R\nu_R + P\nu_P = -G \quad (\nu_C, \nu_R, \nu_P) \geq 0, \ \nu_C \leq U,$$
$$C^T y + s^1 - \rho = 0, \quad s^1 \geq 0, \ \rho \geq 0,$$
$$S^T y + s^2 = 0, \quad s^2 \geq 0,$$
$$P^T y + s^3 = 0, \quad s^3 \geq 0,$$
$$\nu_C^T s^1 + \nu_R^T s^2 + \nu_P^T s^3 = 0,$$
$$\rho^T (U - \nu_C) = 0,$$
$$(\nu_C, \nu_R, \nu_P) + (s^1, s^2, s^3) > 0,$$
$$p + (U - \nu_C) > 0.$$

In theory, solving Equation 11.18 with a path-following interior-point algorithm should return a solution that satisfies this system. However, numerical instabilities often lead to the failure of the last two conditions, i.e., path-following interior-point algorithms regularly return a degenerate solution instead of the strictly complementary solution they theoretically should. As an example, we solved the linear program that maximizes growth for the metabolic network of yeast with two popular interior solvers, CPLEX's barrier method (with crossover turned off) and PCx. Table 11.3 indicates the difference between theory and practice that appears especially wide in this metabolic network. We point out that this problem does not address the linear program in Equation 11.18 but instead solves Equation 11.13 over $\mathcal{P}(20e)$. From a biological perspective 20 provides sufficient resources to achieve growth, so this problem is an adequate surrogate. What is important to observe from Table 11.3 is that even if variables greater 10^{-16} are declared positive, the metabolic network is at least 33 percent degenerate (the true extent of degeneracy would be the maximum number of degenerate pairs). Remember that these algorithms should theoretically provide

Table 11.3 CPLEX's Barrier Method (with Crossover Turned Off) and PCx Were Used to Maximize the Flow into the Growth Node over $\mathcal{P}(20e)$

	Tolerance for Zero	
	10^{-8}	10^{-16}
CPLEX	552/1382 (40 percent)	376/1382 (27 percent)
PCx	606/1382 (44 percent)	457/1382 (33 percent)

Note: Although the solution should be strictly complementary, both solvers terminated with highly degenerate solutions.

a solution that is void of degeneracy, which highlights the fact that this problem has interesting and difficult numeric properties.

As the numerical results show, the theoretical value of an interior-point algorithm can be undermined by numerical instabilities. So, we offer a recent alternative that was born out of the necessity for researchers to overcome the same problem when investigating the optimal design of radiotherapy treatments [Ehrgott et al., 2005]. The goal of this technique is to force variables to be positive by decreasing the largest values of a solution. When this is done iteratively, the result is called the balanced solution. To define this solution, we let sort(x) be the function that sorts the elements of x and lists them in descending order. The balanced solution is defined as the unique solution to

$$\text{lexminsort} \equiv \text{lexmin}\{\text{sort}(x) : Ax = b, x \geq 0\},$$

where lexmin is the lexicographic minimum. It is easy to show that if λe is feasible and A and b are both positive, then the solution to this problem is λe, which means that this technique correctly identified that each component of x can be positive in a feasible solution.

Adapting this idea to the metabolic network, we have

$$\text{lexminsort} \equiv \text{lexmin}\{\text{sort}(\nu_C, \nu_R, \nu_P) :$$
$$C\nu_C + R\nu_R + P\nu_P = -G, \quad U \geq \nu_C \geq 0, \quad \nu_R \geq 0, \quad \nu_P \geq 0\}.$$

This technique identifying the implied equalities is new and has not been thoroughly tested. An interesting direction for future research is to compare the speed and results of this method to those in [Almaas et al., 2005, Burgard et al., 2001, Reedi and Palsson, 2004]. We mention that there are interpretive advantages in this approach. For example, suppose that the largest value of this calculation is $\nu_i = l$. If $i \in C$, this indicates that reaction

i must have flux l to achieve a unit of growth. Similar interpretations correspond to the cases of i being in R and P.

There are only a few mathematical results regarding the calculation of lexminsort. One of these is that the solution is unique, and we let x^* be this solution. Similar to the definition of B and N, we let $\beta = \{i : x_i^* > 0\}$ and $\eta = \{i : x_i^* = 0\}$. A desirable property would be for $B = \beta$ and $N = \eta$, however, there are examples for which this is not the case. This means that x^* does not generally identify the optimal partition. Preliminary numerical studies have shown that it is often the case that x^* does induce the optimal partition, and the authors suggest that it is likely for metabolic networks. The insight comes from the fact that this method smooths out the flux values by reducing the maximum flux, which in turn should cause other fluxes to increase.

There are several questions left to be answered about the linear program in FBA. As mentioned earlier, the constraints of this problem require that the fluxes adhere to a steady-state assumption. However, a cell's state is dynamic rather than static. A major research direction is to use this technique to understand how the fluxes change as the cell's environment changes. The environment is currently modeled through the cell's inputs, and asking how the the fluxes change is a question in classical sensitivity analysis. Because the solutions are significantly degenerate, a more appropriate question is how do B and N change with regard to the upper bound vector u. This question was studied for general linear programs in [Adler and Monteiro, 1992, Holder, 2004, Holder et al., 2001, Monteiro and Mehrotra, 1996, Roos et al., 1997], but the special properties that exist in FBA are completely open. An alternative would be the modern interpretation of robust optimization, which provides complementary information to classic sensitivity analysis.

The steady-state assumption prohibits metabolite accumulation. A more realistic model would allow metabolites to accrue and then have different reactions process these metabolites. However, we do not know what objective, if any, would eliminate the extra metabolites. One simple experiment would be to replace the constraints with

$$C\nu_C + R\nu_R + P\nu_P + G\nu_G \geq 0, \quad 0 \leq \nu_C \leq U, \; 0 \leq \nu_R \leq u, \; 0 \leq \nu_P,$$

which relaxes the steady-state assumption and allows metabolites to accumulate. Maximizing growth with this set of constraints will likely show that some metabolites remain in the cell. This is not realistic, so a secondary (or tertiary, etc.) objective is likely governing the elimination of metabolites. This recasts FBA into the realm of multiple objective programming, which is likely more appropriate. This is an untapped research venue.

Another area where the degeneracy in FBA has been ignored is that of calculating the HFB. This calculation depends on an optimal solution from FBA, but the high level of degeneracy implies that the dimension of the primal or the dual solution spaces is significant. Categorizing the source of degeneracy as primal or dual for each pair would add insight to the problem. Moreover, it would be interesting to know the variability of the HFB is over the optimal set, see [Wiback et al., 2004] for related work.

Outside of FBA, we have from earlier sections that community identification is important. The algorithms used to identify communities need to be efficient due to the size of most biological networks. Mathematical programmers are trained in algorithm design and analysis, and these skills are needed. As previously mentioned, the BC measure used for many social networks is $O(N^3)$. This polynomial bound is typically considered favorable, but the cubic growth is realized in implementation, making this attack less attractive on large networks. The alternative based on Equation 11.6 is $O(N^2)$. These are both significantly better than clique finding, which is a classic NP-complete problem.

The recent suggestion of k-clique percolation [Palla et al., 2005] was published without complexity analysis, which is understandable as the first step is to locate a k-clique, and hence, the algorithm is NP-complete. However, k is typically smaller than the size of the maximum clique, and identifying a small clique is generally considered simple. This begs the question, What is the complexity of identifying a community from a known k-clique? A simple argument shows that the algorithm in Table 11.4 locates a community in $O(\Delta^k N^2)$, where $\Delta = \max\{\deg(v) : v \in V\}$. Because $\Delta \leq N$, we generally have the possibility of $O(N^{k+2})$, which is polynomial for fixed k but is worse than both of the other algorithms because $k \geq 2$. The numerical computations in [Palla et al., 2005] do not indicate that this bound is achieved in practice, and an interesting direction for future research would be to explain the difference between the theoretical complexity and the practical efficiency.

The traditional clustering techniques of k-means and k-medians can also be used to identify communities. Both of these problems are traditional facility location problems in OR and their application to biological

Table 11.4 An Algorithm to Calculate a Community from a Known k-Clique

1. Let $\mathcal{C}$ be the nodes of an initial k-clique.
2. Set $\mathcal{C}' = \emptyset$.
3. For each $v \in \mathcal{C}$:
   - a For each $v' \in N(v)\backslash\mathcal{C}$:
     - i If $|N(v') \cup N(v) \cup \mathcal{C}| \geq k - 2$, add nodes $N(v') \cup N(v) \cup \mathcal{C}$ to $\mathcal{C}'$.
4. If $\mathcal{C}' \neq \emptyset$, let $\mathcal{C} = \mathcal{C} \cup \mathcal{C}'$ and go to 2.

networks deserves attention. Although facility location is related to community finding, it is inherently different. This is because facility location is concerned with locating positions that optimize some quality of an assignment to these positions. So these problems have the two goals of grouping entities and assigning a representative to each group, which is often (but not necessarily) a member of the group. The community idea equates well to grouping, but how the representative part leads to biological information is unknown. We discuss the recent results of [Holder et al., 2006] to foreshadow some future applications of the k-median problem in systems biology.

The k-median problem is one of the four primary questions in discrete location theory (the others being the k-means problem, the uncapacitated facility location problem, and the quadratic assignment problem). Initial investigations into the problem were undertaken by Hakimi [Hakimi, 1965], and this work spawned a substantial literature [Reese, 2006]. Hakimi's original intent was to locate positions from the continium of a network or graph, i.e., facilities were allowed to be positioned on an edge or vertex. This is a graph restriction of the classic Weber problem. Assuming that positions on the graph were related by a metric, Hakimi proved two significant results: (1) there is always an optimal facility location for which the facilities are located at vertices and (2) the problem of optimally locating facilities is NP-hard in N and k. An often overlooked and misunderstood property is that the problem is polynomial for a fixed k, making it fixed-parameter tractable.

The discrete k-median problem is concerned with selecting k positions on a graph from a discrete set $\mathbb{P}$ on (V, E). The positions in $\mathbb{P}$ can be located on any edge or vertex and it is assumed that $V \subseteq \mathbb{P}$. Each pair of positions is related by a nonnegative similarity score $\mathrm{d}(p, p')$, which need not be a metric, and each node is assigned a weight $\beta(v)$. The discrete k-median problem is

$$\operatorname{Min}\left\{\sum_{p\in\mathbb{P}}\sum_{v\in V_p}\mathrm{d}(p,v)\beta(v) : \mathbb{P}\subseteq(V,E),\ |\mathbb{P}|=k\right\},$$

where

$$V_{p'} = \{v \in V : \mathrm{d}(v,p') \leq \mathrm{d}(v,p) \ \text{ for } \ p \in \mathbb{P}\}. \tag{11.19}$$

Any collection of k positions solving this problem are called medians. The nearest-neighbor condition in Equation 11.19 assigns the vertices of the graph to the medians, but unfortunately, this definition does not uniquely define $V_{p'}$ because some nodes may be equally similar to multiple medians. However, the assumption that $|\mathbb{P}| \leq |\mathbb{N}|$ allows us to list the elements of $\mathbb{P}$, and subsequently to decide ties by assigning the vertex to the position with the least index. A result in [Holder et al., 2006] similar to Hakimi's original work shows that there is always a solution of vertices.

With regards to community location, it makes sense that $\mathbb{P} = V$. However, although the similarity measure d and the node weight β are natural in many OR applications, their interpretation in a biological framework is not clear. Indeed, the communities are defined in terms of these graph characteristics, and it is likely that they can be tailored to different biological situations, yielding a flexible model. In the discussions that follow, we assume that $\mathbb{P} = V$, $\mathrm{d}(p,p') = \|p - p'\|_2$ and $\beta(v) = 1$. The use of the Euclidean norm implies the network is coordinitized in some meaningful way, which is awkward for biological networks. However, it is a place to start.

The main result of [Holder et al., 2006] shows that the discrete k-median problem is identical to a well-studied problem in data compression that optimally designs a vector quantizer. A full discussion of vector quantization is not warranted due to space limitation, and we direct interested readers to refer [Gersho and Gray, 1992]. The importance of the relationship is that it allows us to cast the graph theory problem in a way that is amenable to the efficient algorithms designed to work on the vector quantization problem. The most preeminent and significant of these techniques is the discrete Lloyd algorithm in Table 11.5. This algorithm is not an exact solution procedure because it converges to a local optimal solution. The pertinent complexity results from [Holder et al., 2006] are

- The discrete k-median problem is $O(N^{k+2})$.
- The discrete Lloyd algorithm is $O(Nk)$.

The first result shows that the worst-case complexity of the discrete k-median problem is no worse than that of the k-clique percolations. When $k \ll N$, the second result shows that the discrete Lloyd algorithm is theoretically faster than the other community-finding techniques.

Table 11.5 The Discrete Lloyd Algorithm for $\mathbb{P} = V$

1. Select an initial collection of k nodes, M.
2. Calculate the nearest neighbors V_v as in Equation 11.19 for each $v \in M$.
3. Calculate the centroid of V_v for each $v \in M$ with each node weighted with $\beta(v)$.
4. Project each centroid onto its nearest neighbor in V forming a new collection of k nodes denoted by M'.
5. If $M = M'$, stop. Otherwise, replace M with M' and go to 2.

Using the discrete k-median problem to locate communities within a biological network is promising. The questions are numerous and include

- What similarity measure and node weight are meaningful?
- Can a solution to the discrete k-median problem be found as efficiently as communities can be found with k-clique percolation in practice?
- Does the discrete Lloyd algorithm outperform other community-finding algorithms in practice?
- How do we initialize the discrete Lloyd algorithm so that it locates a global solution instead of a local solution?

We close by mentioning that although we have focused on the linear optimization problem associated with FBA and the community-finding algorithms that identify biological structures, these are but two of the many problems in systems biology that make use of standard OR techniques. The purpose of this section was to show that the problems are plentiful, important, and natural, and we encourage the involvement of the OR community. Please contact the authors if we can be of assistance.

Acknowledgments

The authors would like to thank Trinity University's Mathematics department for their generous support. We also thank Jeremy Nolan for his discussions on the complexity of k-clique percolation.

References

[Adler and Monteiro, 1992] Adler, I. and Monteiro, R. 1992. A geometric view of parametric linear programming. *Algorithmica*, 8:161–176.

[Albert and Barabási, 2002] Albert, R. and Barabási, A.-L. 2002. Statistical mechanics of complex networks. *Rev. Mod. Phys.*, 74:47.

[Albert et al., 2000] Albert, R., Jeong, H., and Barabási, A.-L. 2000. Error and attack tolerance of complex networks. *Nature*, 406:378.

[Almaas et al., 2004] Almaas, E., Kovács, B., Vicsek, T., Oltvai, Z. N., and Barabási, A.-L. 2004. Global organization of metabolic fluxes in the bacterium *Escherichia coli*. *Nature*, 427:839.

[Almaas et al., 2005] Almaas, E., Oltvai, Z. N., and Barabási, A.-L. 2005. The activity reaction core and plasticity in metabolic networks. *PLoS Comput. Biol.*, 1:e68.

[Barabási and Albert, 1999] Barabási, A.-L. and Albert, R. 1999. Emergence of scaling in random networks. *Science*, 286:509.

[Barrat et al., 2004] Barrat, A., Barthelemy, M., Pastor-Satorras, R., and Vespignani, A. 2004. The architecture of complex weighted networks. *Proc. Natl. Acad. Sci. USA*, 101:3747.

[Barthelemy et al., 2003] Barthelemy, M., Gondran, B., and Guichard, E. 2003. Spatial structure of the internet traffic. *Physica A*, 319:633.

[Batada et al., 2006] Batada, N. N., Hurst, L. D., and Tyers, M. 2006. Evolutionary and physiological importance of hub proteins. *PLoS Comp. Biol.*, 2:0748.

[BioGrid, 2006] BioGrid 2006. version 2.0.20. http://www.thebiogrid.org/.

[Blank et al., 2005] Blank, L. M., Kuepfer, L., and Sauer, U. 2005. Large-scale c-13-flux analysis reveals mechanistic principles of metabolic network robustness to null mutations in yeast. *Genome Biol.*, 6:R49.

[Bollobás, 2001] Bollobás, B. 2001. *Random Graphs*. Academic Press, New York.

[Bonarius et al., 1997] Bonarius, H. P. J., Schmid, G., and Tramper, J. 1997. Flux analysis of underdetermined metabolic networks: The quest for the missing constraints. *Trends Biotechnol.*, 15:308.

[Brandes, 2001] Brandes, U. 2001. A faster algorithm for betweenness centrality. *J. Math. Soc.*, 25:163.

[Burgard et al., 2001] Burgard, A. P., Vaidyaraman, S., and Maranas, C. D. 2001. Minimal reaction sets for *Escherichia coli* metabolism under different growth requirements and uptake environments. *Biotechnol. Progr.*, 17:791.

[Cannizzaro et al., 2004] Cannizzaro, C., Christensen, B., Nielsen, J., and von Stockar, U. 2004. Metabolic network analysis on *Phaffia rhodozyma* yeast using C-13-labeled glucose and gas chromatography–mass spectrometry. *Metab. Eng.*, 6:340.

[Canonaco et al., 2001] Canonaco, F., Hess, T. A., Heri, S., Wang, T. T., Szyperski, T., and Sauer, U. 2001. Metabolic flux response to phosphoglucose isomerase knock-out in *Escherichia coli* and impact of overexpression of the soluble transhydrogenase UdhA. *FEMS Microbiol. Lett.*, 204:247.

[Coulomb et al., 2005] Coulomb, S., Bauer, M., Bernard, D., and Marsolier-Kergoat, M. C. 2005. Gene essentiality and the topology of protein-interaction networks. *Proc. R. Soc. Lond. Ser. Biol. Sci.*, 272:1721.

[Dorogovtsev et al., 2002] Dorogovtsev, S. N., Goltsev, A. V., and Mendes, J. F. F. 2002. Pseudofractal scale-free web. *Phys. Rev. E*, 65:066122.

[Duarte et al., 2004] Duarte, N. C., Herrgard, M. J., and Palsson, B. O. 2004. Reconstruction and validation of *Saccharomyces cerevisiae* IND750, a fully compartmentalized genome-scale metabolic model. *Genome. Res.*, 14:1298.

[Edwards and Palsson, 2000] Edwards, J. S. and Palsson, B. O. 2000. The *Escherichia coli* MG1655 in silico metabolic genotype: Its definition, characteristics, and capabilities. *Proc. Natl. Acad. Sci. USA*, 97:5528.

[Ehrgott et al., 2005] Ehrgott, M., Holder, A., and Reese, J. 2005. Beam selection in radiotherapy design. Technical Report 95, Trinity University Mathematics, San Antonio, TX.

[Emmerling et al., 2002] Emmerling, M., Dauner, M., Ponti, A., Fiaux, J., Hochuli, M., Szyperski, T., Wuthrich, K., Bailey, J. E., and Sauer, U. 2002. Metabolic flux responses to pyruvate kinase knockout in *Escherichia coli*. *J. Bacteriol.*, 184:152.

[Fischer and Sauer, 2003] Fischer, E. and Sauer, U. 2003. Metabolic flux profiling of *Escherichia coli* mutants in central carbon metabolism using GC–MS. *Eur. J. Biochem.*, 270:880891.

[Fischer and Sauer, 2005] Fischer, E. and Sauer, U. 2005. Large-scale in vivo flux analysis shows rigidity and suboptimal performance of *Bacillus subtilis* metabolism. *Nat. Genet.*, 37:636.

[Freeman, 1977] Freeman, L. C. 1977. A set of measures of centrality based upon betweenness. *Sociometry*, 40:35.

[Gerdes et al., 2003] Gerdes, S., Scholle, M., Campbell, J., Balazsi, G., Ravasz, E., Daugherty, M. D., Somera, A. L., Kyrpides, N. C., Anderson, I., Gelfand, M. S., Bhattacharya, A., Kapatral, V., D'Souza, M., Baev, M. V., Grechkin, Y., Mseeh, F., Fonstein, M. Y., Overbeek, R., Barabási, A.-L., Oltvai, Z. N., and Osterman, A. L. 2003. Experimental determination and system level analysis of essential genes in *Escherichia coli* MG1655. *J. Bacteriol*, 185:5673.

[Gersho and Gray, 1992] Gersho, A. and Gray, R. M. 1992. *Vector Quantization and Signal Compression*. Kluwer International Series in Engineering and Computer Science. Kluwer Academic Publishers.

[Giaever et al., 2002] Giaever, G., Chu, A., Ni, L., Connelly, C., Riles, L., Véronneau, S., Dow, S., Lucau-Danila, A., Anderson, K., André, B., Arkin, A., Astromoff, A., El-Bakkoury, M., Bangham, R., Benito, R., Brachat, S., Campanaro, S., Curtiss, M., Davis, K.,

Deutschbauer, A., Entian, K.-D., Flaherty, P., Foury, F., Garfinkel, D. J., Gerstein, M., Gotte, D., Güldener, U., Hegemann, J. H., Hempel, S., Herman, Z., Jaramillo, D. F., Kelly, D. E., Kelly, S. L., Kötter, P., LaBonte, D., Lamb, D. C., Lan, N., Liang, H., Liao, H., Liu, L., Luo, C., Lussier, M., Mao, R., Menard, P., Ooi, S. L., Revuelta, J. L., Roberts, C. J., Rose, M., Ross-Macdonald, P., Scherens, B., Schimmack, G., Shafer, B., Shoemaker, D. D., Sookhai-Mahadeo, S., Storms, R. K., Strathern, J. N., Valle, G., Voet, M., Volckaert, G., yun Wang, C., Ward, T. R., Wilhelmy, J., Winzeler, E. A., Yang, Y., Yen, G., Youngman, E., Yu, K., Bussey, H., Boeke, J. D., Snyder, M., Philippsen, P., Davis, R. W., and Johnston, M. 2002. Functional profiling of the *Saccharomyces cerevisiae* genome. *Nature*, 418:387.

[Goldman and Tucker, 1956] Goldman, A. and Tucker, A. 1956. Theory of Linear Programming. In Kuhn, H. and Tucker, A., editors, *Linear Inequalities and Related Systems*, vol. 38, pp. 53–97. Princeton University Press, Princeton, N.J.

[Gombert et al., 2001] Gombert, A. K., dos Santos, M. M., Christensen, B., and Nielsen, J. 2001. Network identification and flux quantification in the central metabolism of *Saccharomyces cerevisiae* under different conditions of glucose repression. *J. Bacteriol.*, 183:1441.

[Greenberg, 1996] Greenberg, H. 1996. Consistency, redundancy and implied equalities in linear systems. *Annals of Mathematics and Artificial Intelligence*, 17:37–83.

[Greenberg, 2006] Greenberg, H. 1996–2006. *Mathematical Programming Glossary*. World Wide Web, http://glossary.computing.society.informs.org/. Edited by the INFORMS Computing Society.

[Hakimi, 1965] Hakimi, S. L. 1965. Optimum distribution of switching centers in a communication network and some related graph theoretic problems. *Oper. Res.*, 13(3):462–475.

[Han et al., 2004] Han, J.-D. J., Bertin, N., Hao, T., Goldberg, D. S., Berriz, G. F., Zhang, L. V., Dupuy, D., Walhout, A. J. M., Cusick, M. E., Roth, F. P., and Vidal, M. 2004. Evidence for dynamically organized modularity in the yeast protein–protein interaction network. *Nature*, 430:88.

[Hartwell et al., 1999] Hartwell, L. H., Hopfield, J. J., Leibler, S., and Murray, A. W. 1999. From molecular to modular cell biology. *Nature*, 402:C47.

[He and Zhang, 2006] He, X. and Zhang, J. 2006. Why do hubs tend to be essential in protein networks? *PLoS Genet.*, 2:0826.

[Holder, 2004] Holder, A. 2004. Simultaneous data perturbations and analytic center convergence. *SIAM J. Optim.*, 14(3):841–868.

[Holder et al., 2001] Holder, A., Sturm, J., and Zhang, S. 2001. Marginal and parametric analysis of the central optimal solution. *Inf. Syst. Oper. Res.*, 39(4):394–415.

[Holder et al., 2006] Holder, A., Lim, G., and Reese, J. 2006. The relationship between discrete vector quantization and the *p*-median problem. Technical Report 102, Trinity University Mathematics, San Antonio, TX.

[Holme et al., 2007] Holme, P., Park, S. M., Kim, B. J., and Edling, C. R. 2007. Korean university life in a network perspective: Dynamics of a large affiliation network. *Physica A*, 373:821.

[Jansen et al., 1997] Jansen, B., de Jong, J., Roos, C., and Terlaky, T. 1997. Sensitivity analysis in linear programming: Just be careful! *Eur. J. Oper. Res.*, 101:15–28.

[Jeong et al., 2001] Jeong, H., Mason, S., Barabási, A.-L., and Oltvai, Z. N. 2001. Lethality and centrality in protein networks. *Nature*, 411:41.

[Kauffman et al., 2003] Kauffman, K. J., Prakash, P., and Edwards, J. S. 2003. Advances in flux balance analysis. *Curr. Opin. Biotechnol.*, 14:491.

[Macdonald et al., 2005] Macdonald, P., Almaas, E., and Barabási, A.-L. 2005. Minimum spanning trees on weighted scale-free networks. *Europhys. Lett.*, 72:308.

[Maslov and Sneppen, 2002] Maslov, S. and Sneppen, K. 2002. Specificity and stability in topology of protein networks. *Science*, 296:910.

[Milo et al., 2002] Milo, R., Shen-Orr, S. S., Itzkovitz, S., Kashtan, N., Chklovskii, D., and Alon, U. 2002. Network motifs: Simple building blocks of complex networks. *Science*, 298:824.

[Milo et al., 2004] Milo, R., Itzkovitz, S., Kashtan, N., Levitt, R., Shen-Orr, S., Ayzenshtat, I., Sheffer, M., and Alon, U. 2004. Superfamilies of evolved and designed networks. *Science*, 303:1538.

[Monteiro and Mehrotra, 1996] Monteiro, R. and Mehrotra, S. 1996. A general parametric analysis approach and its implication to sensitivity analysis in interior point methods. *Math. Programming*, 72:65–82.

[Newman, 2001] Newman, M. E. J. 2001. Scientific collaboration networks: II. shortest paths, weighted networks, and centrality. *Phys. Rev. E*, 64:016132.

[Newman, 2002] Newman, M. E. J. 2002. Assortative mixing in networks. *Phys. Rev. Lett.*, 89:208701.

[Newman, 2003a] Newman, M. E. J. 2003a. Mixing patterns in networks. *Phys. Rev. E*, 67:026126.

[Newman, 2003b] Newman, M. E. J. 2003b. The structure and function of complex networks. *SIAM Review*, 45:167.

[Newman, 2004] Newman, M. E. J. 2004. Fast algorithm for detecting community structure in networks. *Phys. Rev. E*, 69:066133.

[Onnela et al., 2005] Onnela, J.-P., Saramäki, J., Kertész, J., and Kaski, K. 2005. Intensity and coherence of motifs in weighted complex networks. *Phys. Rev. E*, 71:065103.

[Pal et al., 2006] Pal, C., Papp, B., Lercher, M. J., Csermely, P., Oliver, S. G., and Hurst, L. D. 2006. Chance and necessity in the evolution of minimal metabolic networks. *Nature*, 440:667.

[Palla et al., 2005] Palla, G., Derényi, I., Farkas, I., and Vicsek, T. 2005. Uncovering the overlapping community structure of complex networks in nature and society. *Nature*, 435:814.

[Papin et al., 2004] Papin, J. A., Stelling, J., Price, N. D., Klamt, S., Schuster, S., and Palsson, B. O. 2004. Comparison of network-based pathway analysis methods. *Trends Biotechnol.*, 22:400.

[Papp et al., 2004] Papp, B., Pal, C., and Hurst, L. D. 2004. Metabolic network analysis of the causes and evolution of enzyme dispensability in yeast. *Nature*, 429:661.

[Pastor-Satorras et al., 2001] Pastor-Satorras, R., Vázquez, A., and Vespignani, A. 2001. Dynamical and correlation properties of the internet. *Phys. Rev. Lett.*, 87:258701.

[Pereira-Leal et al., 2004] Pereira-Leal, J., Enright, A., and Ouzounis, C. 2004. Detection of functional modules from protein interaction networks. *Proteins*, 54:49.

[Poyatos and Hurst., 2004] Poyatos, J. F. and Hurst., L. D. 2004. How biologically relevant are interaction-based modules in protein networks? *Genome Biol.*, 5:R93.

[Price, 1965] Price, D. J. d. 1965. Networks of scientific papers. *Science*, 149:510.

[Ravasz et al., 2002] Ravasz, E., Somera, A. L., Mongru, D. A., Oltvai, Z. N., and Barabási, A.-L. 2002. Hierarchical organization of modularity in metabolic networks. *Science*, 297:1551.

[Reedi and Palsson, 2004] Reedi, J. L. and Palsson, B. O. 2004. Genome-scale in silico models of *E. coli* have multiple equivalent phenotypic states: Assessment of correlated reaction subsets that comprise network states. *Genome Res.*, 14:1797.

[Reese, 2006] Reese, J. 2006. Solution methods for the *p*-median problem: An annotated bibliography. *Networks*, 48(3):125–142.

[Roos et al., 1997] Roos, C., Terlaky, T., and Vial, J.-P. 1997. *Theory and Algorithms for Linear Optimization: An Interior Point Approach*. John Wiley & Sons, New York.

[Sauer et al., 1999] Sauer, U., Lasko, D. R., Fiaux, J., Hochuli, M., Glaser, R., Szyperski, T., Wuthrich, K., and Bailey, J. E. 1999. Metabolic flux ratio analysis of genetic and environmental modulations of *Escherichia coli* central carbon metabolism. *J. Bacteriol.*, 181:6679.

[Schilling et al., 2000] Schilling, C. H., Letscher, D., and Palsson, B. O. 2000. Theory for the systemic definition of metabolic pathways and their use in interpreting metabolic function from a pathway-oriented perspective. *J. Theor. Biol.*, 203:229.

[Schilling et al., 2002] Schilling, C. H., Covert, M. W., Famili, I., Church, G. M., Edwards, J. S., and Palsson, B. O. 2002. Genome-scale metabolic model of *Helicobacter pylori* 26695. *J. Bacteriol.*, 184:4582.

[Schuster and Hilgetag, 1994] Schuster, S. and Hilgetag, C. 1994. On elementary flux modes in biochemical reaction systems at steady state. *J. Biol. Syst.*, 2:165.

[Schwikowski B. and Fields, 2000] Schwikowski, B., Uetz, P. and Fields, S. 2000. A network of protein–protein interactions in yeast. *Nat. Biotechnol.*, 18:1257.

[Segre et al., 2005] Segre, D., DeLuna, A., Church, G. M., and Kishony, R. 2005. Modular epistasis in yeast metabolism. *Nat. Genet.*, 37:77.

[Shen-Orr et al., 2002] Shen-Orr, S. S., Milo, R., Mangan, S., and Alon, U. 2002. Network motifs in the transcriptional regulation network of *Escherichia coli*. *Nat. Genet.*, 31:61.

[Vazquez et al., 2004] Vazquez, A., Dobrin, R., Sergi, D., Eckman, J.-P., Oltvai, Z. N., and Barabási, A.-L. 2004. The topological relationship between the large-scale attributes and local interaction patterns of complex networks. *Proc. Natl. Acad. Sci. USA*, 101:17940.

[Wasserman and Faust, 1994] Wasserman, S. and Faust, K. 1994. *Social Network Analysis*. Cambridge University Press, Cambridge.

[Watts and Strogatz, 1998] Watts, D. and Strogatz, S. H. 1998. Collective dynamics of 'small-world' networks. *Nature*, 393:440.

[Wiback et al., 2004] Wiback, S., Famili, I., Greenberg, H., and Palsson, B. 2004. Monte Carlo sampling can be used to determine the size and shape of the steady state flux space. *J. Theor. Biol.* 228:437–447.

[Wuchty et al., 2003] Wuchty, S., Oltvai, Z., and Barabási, A.-L. 2003. Evolutionary conservation of motif constituents in the yeast protein interaction network. *Nat. Genet.*, 35:176.

[Zhang and Horvath, 2005] Zhang, B. and Horvath, S. 2005. A general framework for weighted gene co-expression network analysis. *Stat. App. Genet. Mol. Biol.*, 4:17.

Chapter 12

Algorithms for Genomics Analysis

Eva K. Lee and Kapil Gupta

CONTENTS

12.1 Introduction 356
12.2 Phylogenetic Analysis 357
12.2.1 Methods Based on Pairwise Distance 358
12.2.1.1 Cluster Analysis: UPGMA 359
12.2.1.2 Neighbor Joining 360
12.2.2 Parsimony Methods 361
12.2.2.1 Score Computation 361
12.2.2.2 Search of Possible Tree Topologies 363
12.2.3 Maximum Likelihood Methods 364
12.2.3.1 Model of Evolution 364
12.2.3.2 Likelihood of a Tree 364
12.2.3.3 Recent Improvements 365
12.3 Multiple Sequence Alignment 366
12.3.1 Scoring Alignment 367
12.3.1.1 Independent Columns 368
12.3.1.2 Scoring Matrices 368
12.3.2 Alignment Approaches 369
12.3.3 Progressive Algorithms 369
12.3.4 Graph-Based Algorithms 370
12.3.4.1 Maximum-Weight Trace 371
12.3.4.2 Minimum-Spanning Tree and Traveling Salesman Problem 371
12.3.4.3 Eulerian Path Approach 372
12.3.5 Iterative Algorithms 372

12.3.5.1 Probabilistic Algorithms 372
12.3.5.2 Deterministic Algorithms 373
12.4 Novel Graph–Theoretical-Based Genomic Models 374
12.4.1 Definitions 376
12.4.2 Construction of a Conflict Graph from Paths of Multiple Sequences 377
12.4.3 Complexity Theory 378
12.4.4 Special Cases of MWCMS 384
12.4.5 Computational Models: Integer Programming Formulation 384
12.5 Summary 386
Acknowledgment 387
References 387

Abstract The genome of an organism not only serves as its blueprint that holds the key for diagnosing and curing diseases but also plays a pivotal role in obtaining a holistic view of its ancestry. Recent years have witnessed a large number of innovations in this field, as exemplified by the Human Genome Project. This chapter provides an overview of popular algorithms used in genome analysis and in particular explores two important and deeply interconnected problems: phylogenetic analysis and multiple sequence alignment. We also describe our novel graph–theoretical approach that encompasses a wide variety of genome sequence analysis problems within a single model.

12.1 Introduction

Genomics encompasses the study of genome in human and other organisms. The rate of innovation in this field has been breathtaking over the last decade, especially with the completion of Human Genome Project. The purpose of this chapter is to review some well-known algorithms that facilitate genome analysis. The material is presented in a way that is interesting to both the specialists working in this area and others. Thus, this review includes a brief sketch of the algorithms to facilitate a deeper understanding of the concepts involved. The list of problems related to genomics is very extensive; hence the scope of this chapter is restricted to the following two related important problems: (1) phylogenetic analysis, and (2) multiple sequence alignment. Readers interested in algorithms used in other fields of computational biology are recommended to refer to reviews by Abbas and Holmes (2004) and Blazewicz et al. (2005).

Genome refers to the complete DNA sequence contained in the cell. DNA sequence consists of the four nucleotides adenine(A), thymine(T), cytosine(C), and guanine(G). Associated with each DNA strand (sequence) is a complimentary DNA strand of the same length. The strands are complementary in that each nucleotide in one strand uniquely defines an associated nucleotide in the other: A and T are always paired, and C and G are always paired. Each pairing is referred to as a base pair, and bound complementary strands make up a DNA molecule. Typically, the number of base pairs in a DNA molecule is between thousands and billions, depending on the complexity of a given organism. For e.g., A bacterium contains about 600,000 base pairs, whereas human and mouse have some three billion base pairs. Among humans, 99.9 percent base pairs are same between any two unrelated persons. But that leaves millions of single-letter differences, which provide genetic variation between people.

Understanding the DNA sequence is extremely important. It is considered as the blueprint for an organism's structure and function. The sequence order underlies all of life's diversity, even dictating whether an organism is human or another species such as yeast or a fruit fly. It helps in understanding the evolution of mankind, identifying genetic diseases, and creating new approaches for treating and controlling those diseases. To achieve these goals, the research in genome analysis has rapidly progressed over the last decade.

The rest of this chapter is organized as follows. Section 12.2 discusses techniques used to infer the evolutionary history of species and Section 12.3 presents multiple sequence alignment problem and recent advances. In Section 12.4, we describe our research effort for advancing genomics analysis through the design of a novel graph–theoretical approach for representing a wide variety of genomic sequence analysis problems within a single model. We summarize our theoretical findings, and present computational models based on two integer programming formulations. Finally, Section 12.5 summarizes the interdependence and the pivotal role played by the above mentioned two problems in computational biology.

12.2 Phylogenetic Analysis

Phylogenetic analysis is a major aspect of genome research. It refers to the study of evolutionary relationships of a group of organisms. These hierarchical relationships among organisms arising through evolution are usually represented by a phylogenetic tree (Figure 12.1). The idea of using trees to represent evolution dates back to Darwin. Both rooted and unrooted tree representation have been used in practice [Durbin et al., 1998]. The branches of tree represent the time of divergence and the root represents the ancestral sequence (Figure 12.2).

The study of phylogenies and processes of evolution by the analysis of DNA or amino acid sequence data is called molecular phylogenetics. In this study, we will focus on methods that use DNA sequence data. There are

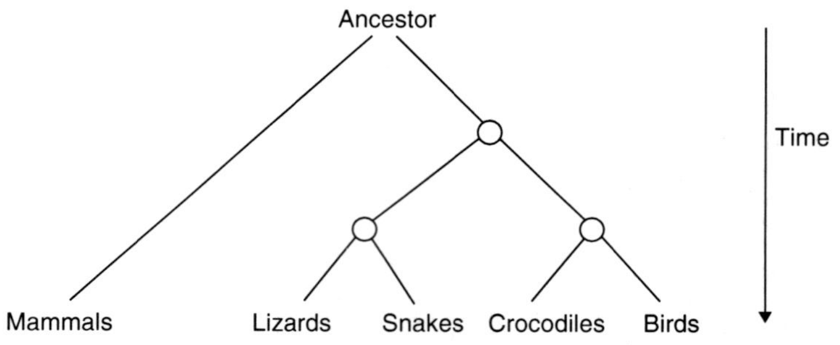

Figure 12.1 An example of evolutionary tree.

two processes involved in inferring both rooted and unrooted trees. First estimates the branching structure or topology of the tree. Second estimates the branch lengths for a given tree. Currently, there are wide varieties of methods available to conduct this analysis [Nei, 1996] [Felsenstein, 1988] [Whelan et al., 2001] [Delsuc et al., 2005]. These available approaches can be classified into three broad groups: (1) distance methods; (2) parsimony methods; and (3) maximum likelihood methods. Below, we will discuss each of them in detail.

12.2.1 Methods Based on Pairwise Distance

In distance methods, an evolutionary distance d_{ij} is computed between each pair i, j of sequences, and a phylogenetic tree is constructed

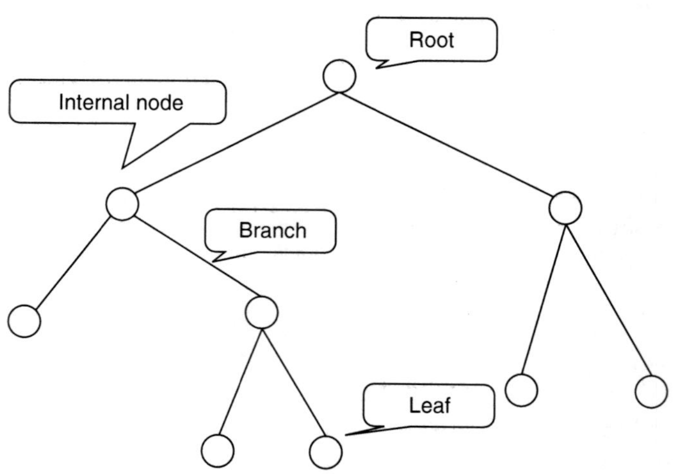

Figure 12.2 Tree terminology.

from these pairwise distances. There are many different ways of defining pairwise evolutionary distance used for this purpose. Most of the approaches estimate the number of nucleotides substitutions per site, but other measures have also been used [Tajima and Nei, 1984] [Tajima and Takezaki, 1994]. The most popular one is Jukes–Cantor distance (1969) which defines d_{ij} as $-\frac{3}{4}\log(1-\frac{4f}{3})$, where f is the fraction of sites where nucleotides differ in the pairwise alignment.

There are a large number of distance methods for constructing evolutionary trees [Waterman, 1995]. In this chapter, we discuss methods based on cluster analysis and neighbor joining.

12.2.1.1 Cluster Analysis: UPGMA

The conceptually simplest and most known distance method is UPGMA (Unweighted Pair Group Method using Arithmetic averages) developed by Sokal and Michener (1958). Given a matrix of pairwise distances between each pair of sequences, it starts with assigning each sequence to its own cluster. The distances between the clusters are defined as $d_{ij} = \frac{1}{|C_i||C_j|}\sum_{p\in C_i, q\in C_j} d(p,q)$ where C_i and C_j denote sequences in clusters i and j, respectively. At each stage in the process, the least distant pair of clusters are merged to create a new cluster. This process continues until only one cluster is left. Given n sequences, the general schema of UPGMA is shown in Algorithm 1.

The time and space complexity of UPGMA is $O(n^2)$, because there are $n-1$ iterations of complexity $O(n)$. A number of approaches have been developed which are motivated by UPGMA. Li (1981) developed a similar

Algorithm 1 UPGMA

INPUT: Distance matrix d_{ij}, $1 \leq i,j \leq n$
for $i = 1$ to n **do**
  Define singleton cluster C_i comprising of sequence i
  Place cluster C_i as a tree leaf at height zero
end for
repeat
  Determine two clusters i, j such that d_{ij} is minimal.
  Merge these two clusters to form a new cluster k having distance from other clusters defined as the weighted average of the comprised two clusters. If C_k is the union of two clusters C_i and C_j, and if C_l is any other cluster, then $d_{kl} = \frac{d_{il}|C_i| + d_{jl}|C_j|}{|C_i|+|C_j|}$
  Define a node k at height $\frac{d_{ij}}{2}$ with daughter nodes i and j
until just a single cluster remains

approach which also makes corrections for unequal rates of evolution among lineages. Klotz and Blanken (1981) presented a method where a present-day sequence serves as an ancestor to determine the tree regardless of the rates of evolution of the sequences involved.

12.2.1.2 Neighbor Joining

Neighbor joining (NJ) is another very popular algorithm based on pairwise distances [Saitou and Nei, 1987]. This approach yields an unrooted tree and overcomes the assumption of UPGMA method that the same rate of evolution applies to each branch.

Given a matrix of pairwise distances between each pair of sequences d_{ij}, it first defines modified distance matrix $\bar{d}_{ij}$. This matrix is calculated by subtracting average distances to all other sequences from the d_{ij} and thus compensating for long edges. In each stage, the two nearest nodes (minimal $\bar{d}_{ij}$) of the tree are chosen and defined as neighbors in the tree. This is done recursively until all of the nodes are paired together. Given n sequences, the general schema of NJ is shown in Algorithm 2.

Neighbor joining has an execution time of $O(n^2)$, like UPGMA. It has given extremely good results in practice and is computationally efficient [Saitou and Nei, 1987] [Takahashi and Nei, 2000]. Many practitioners have developed algorithms based on this approach. Gascuel (1997) improved the NJ approach by using a simple first-order model of the variances and covariances of evolutionary distance estimates. Bruno et al. (2000)

Algorithm 2 Neighbor Joining

INPUT: Distance matrix d_{ij}, $1 \leq i, j \leq n$
for $i = 1$ to n **do**
    Assign sequence i to the set of leaf nodes of the tree (T)
end for
Set list of active nodes $(L) = T$
repeat
    Calculate modified distance matrix $\bar{d}_{ij} = d_{ij} - (r_i + r_j)$, where $r_i = \frac{1}{|L|-2} \sum_{k \in L} d_{ik}$
    Find the pair i, j in L having minimal value of $\bar{d}_{ij}$
    Define a new node u and set $d_{uk} = \frac{1}{2}(d_{ik} + d_{jk} - d_{ij})$, for all k in L
    Add u to T joining nodes i, j with edges of length given by: $d_{iu} = \frac{1}{2}(d_{ij} + r_i - r_j)$,
    $d_{ju} = d_{ij} - d_{iu}$
    Remove i and j from L and add u
until Only two nodes remain in L
Connect remaining two nodes i and j by a branch of length d_{ij}

developed a weighted NJ using a likelihood-based approach. Goeffon et al. (2005) investigated a local search algorithm under the Maximum Parsimony criterion by introducing a new subtree swapping neighborhood with an effective array-based tree representation.

12.2.2 Parsimony Methods

In science, notion of parsimony refers to the preference of simpler hypotheses over complicated ones. In the parsimony approach for tree building, the goal is to identify the phylogeny that requires the fewest necessary changes to explain the differences among the observed sequences. Of the existing numerical approaches for reconstructing ancestral relationships directly from sequence data, this approach is the most popular one. Unlike distance-based methods which build tree, it evaluates all possible trees and gives each a score based on the number of evolutionary changes that are needed to explain the observed sequences. The most parsimonious tree is the one that requires the fewest evolutionary changes for all sequences to derive from a common ancestor [Swofford and Olsen, 1990]. As an example, consider the trees in Figures 12.3 and 12.4. The tree in Figure 12.3 requires only one evolutionary change (marked by the ⋆) compare to the tree in Figure 12.4 which requires two changes. Thus, Figure 12.3 is the more parsimonious tree.

There are two distinct components in parsimony methods: given a labeled tree, determine the score; determine global minimum score by evaluating all possible trees, as discussed below.

12.2.2.1 Score Computation

Given a set of nucleotide sequences, parsimony methods treat each site (position) independently. The algorithm evaluates the score at each

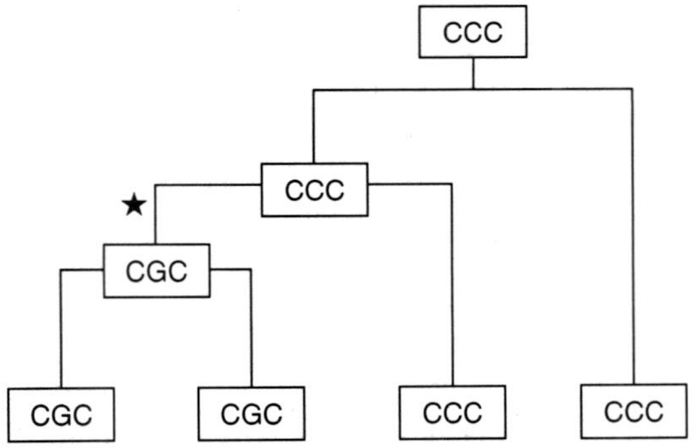

Figure 12.3 Parsimony tree 1.

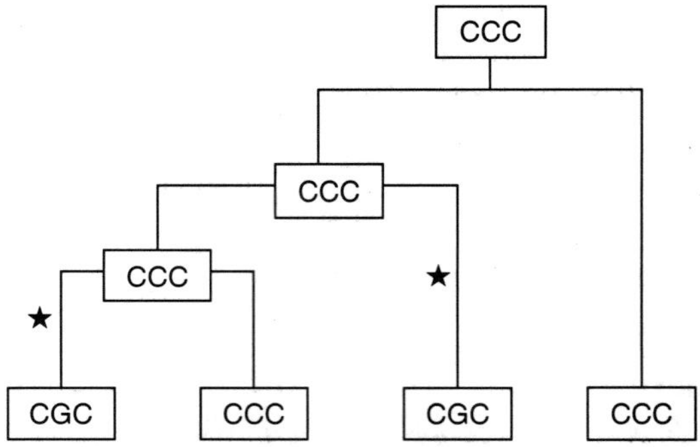

Figure 12.4 Parsimony tree 2.

position and then sums them up over all the positions. As an example, suppose we have the following three aligned nucleotide sequences.

$$CCC$$

$$GGC$$

$$CGC$$

Then, for a given tree topology, we would calculate the minimal number of changes required at each of the three sites and then sum them up. Here, we investigate a traditional parsimony algorithm developed by Fitch (1971), where number of substitutions required is taken as score. For a particular topology, this approach starts by placing nucleotides at the leaves and traverse toward the root of the tree. At each node, the nucleotides common to all of the descendant nodes are placed. If this set is empty then the union set is placed at this node. This continues until root of the tree is reached. The number of union sets equal the number of substitutions required. The general schema for every position is shown in Algorithm 3.

Figure 12.5 shows the set R_k obtained by above algorithm. The computation is done for the first site of the three sequences shown above. The minimal score given by the algorithm is 1.

There are a wide variety of approaches developed by modifying Fitch's algorithm [Swofford and Maddison, 1987]. Sankoff and Cedergren (1983) presented a generalized parsimony method which does not just count the number of substitutions, but assigns a weighted cost for each substitution.

Ronquist (1998) improved the computational time by including strategies for rapid evaluation of tree lengths and increase the exhaustiveness of branch swapping while searching topologies.

Algorithm 3 Parsimony: Score Computation

Each leaf l is labeled with set R_l having observed nucleotide at that position.
Score $S = 0$
for all Internal node k with children i and j having labels R_i and R_j **do**
    $R_k = R_i \bigcap R_j$
    if $R_k = \emptyset$ **then**
        $R_k = R_i \bigcup R_j$
        $S = S + 1$
    end if
end for
Minimal score $= S$

12.2.2.2 Search of Possible Tree Topologies

The number of possible tree topologies dramatically increases with the number of sequences. Consequently, in practice, usually only a subset of them are examined using efficient search strategies. The most commonly used is branch and bound methods to select branching patterns [Purdom et al., 2000]. For large-scale problems, heuristic methods are typically used [Swofford and Olsen, 1990]. These exact and heuristic tree search strategies are implemented in various softwares like PHYLIP (Phylogeny Inference Package) and MEGA (Molecular Evolutionary Genetic Analysis) [Felsenstein, 1989] [Kumar et al., 1994].

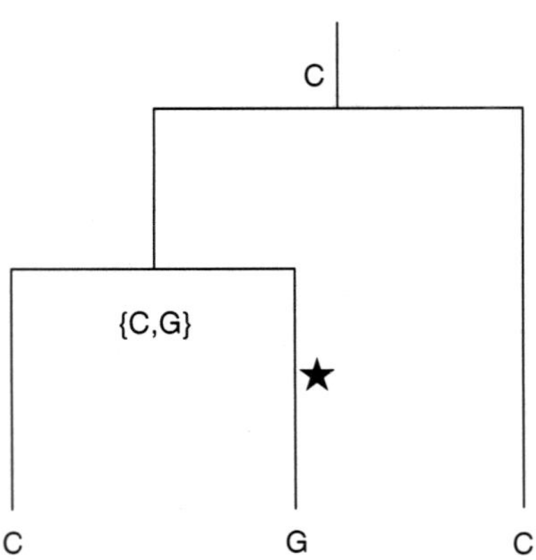

Figure 12.5 The sets R_k for the first site of given three sequences.

12.2.3 Maximum Likelihood Methods

The method of maximum likelihood (ML) is one of the most popular statistical tool used in practice. In molecular phylogenetics, maximum likelihood methods find the tree that has the highest probability of generating observed sequences, given an explicit model of evolution. The method was first introduced by Felsenstein (1981). We discuss herein both the evolution models and the calculation of tree likelihood.

12.2.3.1 Model of Evolution

A model of evolution refers to various events like mutation, that changes one sequence to the another over a period. It is required to determine the probability of a sequence S_2 arising from an ancestral sequence S_1 over a period t. Various sophisticated models of evolution have been suggested, but simple models like Jukes–Cantor are preferred in ML methods.

Jukes and Cantor (1969) model assumes that all nucleotides (A, C, T, G) undergo mutation with equal probability, and change to all of the other three possible nucleotides with same probability. If the mutation rate is 3α per unit time per site, the mutation matrix P_{ij} (probability that nucleotide i changes to j in unit time), takes the form

$$\begin{pmatrix} 1-3\alpha & \alpha & \alpha & \alpha \\ \alpha & 1-3\alpha & \alpha & \alpha \\ \alpha & \alpha & 1-3\alpha & \alpha \\ \alpha & \alpha & \alpha & 1-3\alpha \end{pmatrix}$$

Above matrix is integrated to evaluate mutation rates over time t and then used to calculate $P(nt_2|nt_1, t)$, defined as the probability of nucleotide nt_1 being substituted by nt_2 over time t.

Various other evolution models like Kimura Model have also been mentioned in literature [Kimura, 1980] [Bos and Posada, 2005].

12.2.3.2 Likelihood of a Tree

Likelihood of tree is calculated as the probability of observing a set of sequences given the tree.

$$L(\text{tree}) = \text{Probability}[\text{sequences}|\text{tree}]$$

We begin with the simple case of two sequences S^1 and S^2 of length n having a common ancestor "**a**" as shown in Figure 12.6. It is assumed that all different sites (positions) evolve independently, and thus the total likelihood is calculated as the product of likelihood of all sites [Clote and Backofen, 2000]. Here, the likelihood of each site is obtained using substitution probabilities based on evolution model.

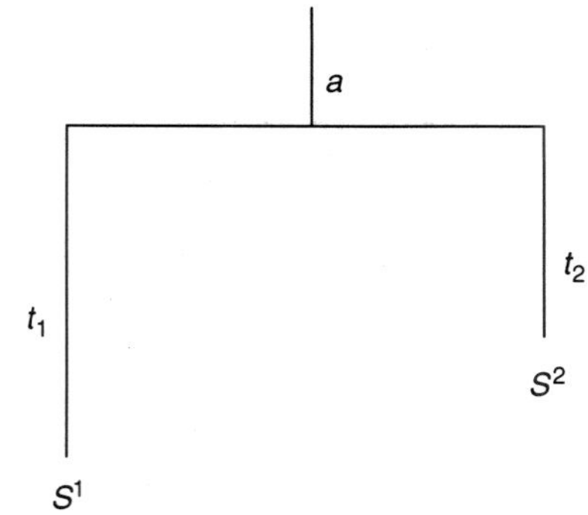

Figure 12.6 A simple tree.

Given q_a = equilibrium distribution of nucleotide a, the likelihood for simple tree in Figure 12.6 is calculated as $L(\text{tree}) = P(S^1, S^2) = \prod_{i=1}^{n} P(S_i^1, S_i^2)$ where $P(S_i^1, S_i^2) = \sum_a q_a P(S_i^1|a)P(S_i^2|a)$. To generalize this approach for m sequences, it is assumed that diverged sequences evolve independently after diverging. Hence, likelihood for every node in tree depends only on its immediate ancestral node and a recursive procedure is used to evaluate likelihood of the tree. The conditional likelihood $L_{k,a}$ is defined as the likelihood of the subtree rooted at node k, given that nucleotide at node k is a. The general schema for every site is shown in Algorithm 4. The likelihood is then maximized over all possible tree topologies and branch lengths.

12.2.3.3 Recent Improvements

ML approach has received great attention due to the existence of powerful statistical tools. It has been made more sophisticated using advance tree

Algorithm 4 Likelihood: Computation at Given Site

for all leaf l **do**
  if leaf has nucleotide a at that site **then**
    $L_{l,a} = 1$
  else
    $L_{l,a} = 0$
  end if
end for
for all Internal node k with children i and j **do**
  Define the conditional likelihood $L_{k,a} = \sum_{b,c}[P(b|a)L_{i,b}][P(c|a)L_{j,c}]$
end for
Likelihood at given site $= \sum_a q_a L_{root,a}$

search algorithms, sequence evolution models, and statistical approaches. Yang (1993) have extended it to the case where the rate of nucleotide substitutions differ over sites. Huelsenbeck (1997) incorporated the improvements in substitution models [Huelsenbeck and Crandall, 1997]. Piontkivska (2004) evaluated the use of various substitution models in ML approach and inferred that simple models are comparable in both efficiency and reliability with complex models.

Enormously large number of possible tree topologies especially, while working with large number of sequences, makes this approach computationally intensive [Takahashi and Nei, 2000]. It has been proved that reconstructing the ML tree is NP-hard even for certain approximations [Chor and Tuller, 2005]. To reduce computational time, Guindon and Gascuel (2003) developed a simple hill-climbing algorithm based on the maximum likelihood principle that adjusts tree topology and branch lengths simultaneously. Recently, parallel computation is being used to address huge computational requirement. Stamatakis et al. (2005) have used OpenMP-parallelization for symmetric multiprocessing machines and Keane et al. (2005) developed distributed platform for phylogeny reconstruction by maximum likelihood.

12.3 Multiple Sequence Alignment

Multiple sequence alignment (MSA) is arguably among the most studied and difficult problems in computational biology. It has been a vital tool because it compactly represents conserved or variable features among the family members. Alignment also allows character-based analysis compared to distance-based analysis and thus helps to elucidate evolutionary relationships better. Consequently, it plays a pivotal role in a wide range of sequence analysis problems like identifying conserved motifs among given sequences; predicting secondary and tertiary structure of protein sequences; and molecular phylogenetic analysis. It is also used for sequence comparison to find similarity of a new sequence with preexisting ones. This helps in gathering information about function and structure of new found sequences from the existing ones in databases like GenBank in United States and EMBL in Europe.

The MSA problem can be stated formally as follows. Let $\sum$ be the alphabet and let $\hat{\sum} = \sum \bigcup \{-\}$, where "−" is a symbol to represent gaps in sequences. For DNA sequences, alphabet $\hat{\sum} = \{A, T, C, G, -\}$.

An alignment for N sequences $S_1, \ldots, S_N$ is given by a set $\hat{S} = \{\hat{S_1}, \ldots, \hat{S_N}\}$ over the alphabet $\hat{\sum}$ which satisfy the following two properties: (1) the strings in $\hat{S}$ are of the same length, and (2) S_i can be obtained from $\hat{S_i}$ by

C	C	C	—		C	—	C	C
C	G	G	C		C	G	G	C
C	G	—	C		C	G	C	—

Figure 12.7 Two possible alignments for given three sequences.

removing the gaps. Thus, an alignment in which each string $\hat{S}_i$ has length K can be interpreted as an alignment matrix of N rows and K columns where row i corresponds to sequence S_i. Alphabets that are placed into the same column of alignment matrix are said to be aligned with each other.

Figure 12.7 shows two possible alignments for given three sequences: $S_1 = \text{CCC}$, $S_2 = \text{CGGC}$, and $S_3 = \text{CGC}$.

For two sequences, optimal MSA is easily obtained using dynamic programming (Needleman–Wunsch algorithm). Unfortunately, the problem becomes much harder for more than two sequences, and optimal solution can be found only for a limited number of sequences of moderate length (approximately 100) [Bonizzoni and Vedova, 2001]. Researchers have tried to solve it by generalizing dynamic programming approach to a multidimensional space. However, this approach has huge time and memory requirements and thus cannot be used in practice even for small problems of five sequences of length 100 each. This algorithm has been improved by identifying the portion of hyperspace which does not contribute to the solution and exclude it from the computation [Carrillo and Lipman, 1988]. But, even this approach of Carrillo and Lipman implemented in MSA program, can only align up to ten sequences [Lipman et al., 1989]. Although, Gupta et al. (1995) improved the space and time usage of this approach, it cannot align large data sets. To reduce the huge time and memory expenses, wide variety of heuristic approaches for multiple sequence alignment have been developed [Notredame, 2002].

There are two components of finding MSA: (1) searching over all the possible multiple alignments, and (2) scoring each of them to find the best one.

The problem becomes more complex for remotely related homologous sequences, i.e., sequences which are not derived from a common ancestor [Gotoh, 1999]. Numerous approaches have been proposed, but the quest for an approach which is accurate and fast continues. It must be remembered that even the choice of sequences and calculating the score of alignment is a nontrivial task and is an active research field in itself.

12.3.1 Scoring Alignment

There is no unanimous way of characterizing an alignment as the correct one and the strategy depends on biological context. Different alignments are possible and we never know for sure which alignment is correct. Thus, one scores every alignment according to an appropriate objective function and alignments with the higher scores are deemed to be better. A typical alignment scoring scheme consists of the following steps.

12.3.1.1 Independent Columns

The score of alignment is calculated in terms of columns of alignments. The individual columns are assumed to be independent and thus the total score of an alignment is a simple summation over column scores. Thus, score for an alignment $\text{score}(A) = \sum_j \text{score}(A_j)$, where A_j is column j of the multiple alignment A. Now, score for every column j is calculated as sum-of-pairs (SP) function using the scoring matrices described below. The SP score for column A_j is obtained as $\text{score}(A_j) = \sum_{k<l} \text{score}(A_j^k, A_j^l)$ where A_j^k and A_j^l are nucleotides in column j of alignment corresponding to sequences k and l, respectively. If gap costs are linear, score(nucleotide, −) and score(−, nucleotide) will be the insertion cost. But, this approach would not differentiate between opening a gap and its extension. So, affine gap penalties are often used where gap opening and extension penalty are treated as two different parameters. The correct value of both of these parameters is a major concern because their values can be set only empirically [Vingron and Waterman, 1994]. Also most schemes used in practice score columns as weighted sum of pairwise substitutions instead of just addition as described before. The weights are decided in accordance with the amount of independent information each sequence possesses [Altschul et al., 1989].

Both the assumption of treating every column independent and using SP score for column has limitations. The problem increases as number of sequences increases.

12.3.1.2 Scoring Matrices

Any alignment can be obtained by performing three evolution operations: insertion, deletion, and substitution. It is assumed that all the different operations occur independently and thus, the complete score is evaluated as the sum of scores from every operation. Insertion and deletion scores are calculated as either linear or affine gap penalty. Substitutions scores are stored as substitution score matrix, which contains score for every pair of nucleotides. Thus, these scores $S(A, B)$ can be treated as the score of aligning nucleotide A with B.

These substitution score matrices can be obtained in various ways. One could adopt an ad hoc approach of setting up a score matrix which produces good alignments for a given set of sequences. The second approach would be more fundamental and look into physical and chemical properties of nucleotides. If two nucleotides are similar in their properties, they would be more likely to be substituted by one another. The third and the most prominent one is a statistical approach where maximum likelihood principle is used in conjunction with probabilistic models of evolution [Altschul, 1991].

12.3.2 Alignment Approaches

The number of different approaches for MSA problem has steadily increased over the last decade and thus being exhaustive will not be possible. In this chapter, we will emphasize on the most widely used class of algorithms and the new emerging and most promising approaches.

1. Progressive alignment algorithms: most widely used type of algorithm based on using pairwise alignment information of input sequences. It assumes that input sequences are phylogenetically related, and uses these relationships to guide the alignment [Chenna et al., 2003].
2. Graph-based algorithms: a new trend where graph-based models are used to approach this problem.
3. Iterative alignment algorithms: typically, an alignment is produced and is then refined through series of iterations until no more improvement can be made.

12.3.3 Progressive Algorithms

Progressive alignment constitutes one of the most simplest and effective ways for multiple alignment. This strategy was introduced by various researchers like Waterman and Perlwitz (1984). Among all the progressive algorithms, ClustalW is the most famous one. It is a noniterative, deterministic algorithm that attempts to optimize the weighted SP with affine gap penalties [Thompson et al., 1994].

The typical progressive algorithm schema is as follows:

- Compute distance between all pairs of given sequences by aligning them. The distances represent divergence of each pair of sequences. These distances could be calculated by fast approximation methods or the slower but more precise methods like complete dynamic programming. Because for given N sequences, $\frac{N(N-1)}{2}$ pairwise scores have to be calculated and the scores are used just for construction of a guide tree and not the alignment itself, it is desirable to use approximation methods like k tuple matches.

- Find a guide tree from the distance matrix. This is typically achieved using clustering algorithms discussed in construction of an evolutionary tree. Once again, because the aim is to get the alignment and not the tree itself, approximation methods are used to construct the evolution trees.
- Align sequences progressively according to the branching order in the guide tree. The basic idea is to start from the leaves of the guide tree toward its root and to use series of pairwise alignments to align larger and larger groups of sequences. Some algorithms have only single growing alignment to which every remaining sequence is aligned whereas other approaches align subgroup of sequences and then merge the alignments.

There are three main shortcomings of the progressive algorithms.

- There does not exist an undisputable best way of ordering the given sequences.
- Once a sequence has been aligned, that alignment will not be modified even if it conflicts with sequences added later in the process. Hence, the order in which sequences are added becomes very crucial, and because there is no undisputable best way to order the sequences, this approach returns suboptimal solutions.
- For a given set of n sequences, $\binom{n}{2}$ pairwise alignments are generated; but while computing the final multiple alignment, most of these algorithms use fewer than n pairwise alignments. Thus, the resulting multiple alignment agrees with only a small amount of information available in the data.

Therefore, there is a growing need for an algorithm to align extremely divergent sequences whose pairwise alignments are likely to be incorrect. To address all these issues, some techniques have been developed; although they are innovative, it is understandable that they have their own assumptions and drawbacks.

12.3.4 Graph-Based Algorithms

Over the last few years, the field of genomics has undergone evolutionary changes with a rapid increase in new solution strategies. The use of graph-based models is easily seen as one of the most emerging and far-reaching trend. Just and Vedova (2004) use relation between facility location problem and sequence alignment to prove the NP-hardness of MSA. In this section, we review the most prominent integer programming (IP) approaches for finding multiple sequence alignment.

12.3.4.1 Maximum-Weight Trace

Kececioglu et al. (2000) use a solution of the maximum trace problem to construct alignment. The algorithm starts with calculating all pairwise alignments and using them to find a trace. To achieve this, given n sequences, an input alignment graph $G = (V, E)$ is constructed. It is an n-partite graph whose vertex set V represents the characters of given sequences and edge set E represents the pairs of characters matched in the pairwise alignments. The subset of matching in E realized by an alignment is called a trace.

Alignment graph $G = (V, E)$ is extended to a mixed graph $G' = (V, E, A)$ by adding arc set A which connects character of every sequence to the next character in the same sequence. The objective of the algorithm is to find maximum weight trace by finding cycles termed as critical mixed cycles in graph G' such that they satisfy sequence alignment properties [Reinert et al., 1997].

The IP model for this problem is formulated as

$$\text{Max} \sum_{e \in E} w_e x_e \tag{12.1}$$

$$\text{s.t.} \sum_{e \in P \cap E} x_e = |E \cap P| - 1 \forall \text{ critical mixed cycles } P \text{ in } G'$$

$$x_e \in \{0, 1\} \text{ for all } e \in E \tag{12.2}$$

An implementation of a branch-and-cut algorithm is used to solve the above problem. Various valid inequalities for the polytope are added as cuts, some of which are facet-defining. The algorithm is capable of giving an exact solution under the SP objective function with linear gap costs. Kececioglu et al. have made a significant contribution by introducing a polyhedral approach capable of obtaining exact solutions for a subclass of MSA. However, this methodology has its own drawbacks like not being able to capture the order of insertions and deletions between two matchings and affine gap costs. Recently, Althaus et al. (2006) have proposed a general model using this approach in which arbitrary gap costs are allowed.

12.3.4.2 Minimum-Spanning Tree and Traveling Salesman Problem

Shyu et al. (2004) explore the use of minimum spanning trees to determine the order of sequences [Shyu et al., 2004]. The idea of the approach is to preserve the most informative distances among the set of given sequences. The criterion used is meaningful and capable of working better than the traditional criteria like those in SP. The algorithm itself is very efficient for practical usage, and can be easily implemented. However, it fails to address the issue of using all the information in pairwise alignments, because it only uses the score and not the pairwise alignments themselves. Moreover, this approach has all the drawbacks of the progressive strategy.

A similar approach has also been developed by Korostensky and Gonnet (1999) using traveling salesman problem (TSP). In this technique, a circular sum measure is used instead of SP score. The cities in TSP correspond to the sequences and the scores of pairwise alignment are taken as the distances. The problem is to find the longest tour where each sequence is visited exactly once [Korostensky and Gonnet, 2000].

12.3.4.3 Eulerian Path Approach

Zhang and Waterman (2003) proposed a new approach motivated by the Eulerian method for fragment assembly in DNA sequencing. In their work, a consensus sequence is found and later pairwise alignments are obtained between each input sequence and consensus sequence. Finally, MSA is obtained according to these pairwise alignments. The most significant advantage of this method is linear time and memory cost for finding the consensus sequence. And, if the consensus sequence is the closest one to all given sequences, good quality alignment can be obtained in a reasonable amount of time. Once again, this approach suffers from the prominent drawback of the progressive strategy and issues in graph formation while finding the consensus sequence.

12.3.5 Iterative Algorithms

The main shortcoming of the progressive strategy is the failure to remove errors in the alignment, which are introduced early. The iterative algorithms are developed precisely to overcome this flaw. They are based on the idea of reconsidering and realigning previously aligned sequences with the goal of improving the overall alignment score. Each modification step is an iteration to improve the quality of the alignment.

These available approaches can be classified into two broad categories: probabilistic iterative algorithms and deterministic iterative algorithms. We will briefly discuss them below.

12.3.5.1 Probabilistic Algorithms

We will discuss both the traditional probabilistic optimization approaches like genetic algorithm and relatively recent approaches based on Bayesian idea.

12.3.5.1.1 Simulated Annealing and Genetic Algorithm

Simulated annealing (SA) and genetic algorithm (GA) are very popular stochastic methods for solving complex optimization problems. Although they are often viewed as separate and competing paradigms, both of them are iterative algorithms which search for new solutions near to already known good solutions. The fundamental difference between SA and GA is that SA performs a local move only on one solution to create a new solution whereas GA also creates solutions by combining information from two different solutions. Performance comparison between SA and GA varies with the problem and representation used.

The algorithms start with an initial alignment and alignment score is taken to be the objective function [Notredame and Higgins, 1996]. Various operations like mutation, insertion, and substitution constitute the local move which are used to get new solution from existing ones. Flexibility in scoring systems and ability to correct for errors introduced during early phase make these approaches desirable [Kim et al., 1994].

12.3.5.1.2 Hidden Markov Model and Gibbs Sampler

Hidden Markov Model (HMM) and Gibbs Sampler are relatively recent approaches which views MSA in a statistical context. Both of them use the central Bayesian idea of simultaneously maximizing the data and the model. Gibbs Sampler finds motifs using local alignment techniques [Lawrence et al., 1993]. It is essentially similar to HMM with no insert and delete states.

HMM is a statistical model based on Markov process, which has gained importance in various fields, related to pattern recognition. It determines the hidden parameters of the system based on the observable parameters of the model. For MSA, HMM consists of three types of states: match states, insert states, and delete states [Krogh et al., 1994]. Each state has its own emission probability of nucleotides and transition probability to other states. The standard expectation-maximization (EM) algorithm or gradient descent algorithms are used to train the model and evaluate the parameters.

Although HMM has been successfully used in other areas, it faces a lot of challenges. There need to be some minimum number of sequences (approximately 50) required to train the model and HMM can be easily trapped in local optima like other hill-climbing approaches [Hughey and Krogh, 1996].

12.3.5.2 Deterministic Algorithms

A deterministic iterative algorithm starts with an initial alignment and then attempts to improve it. This helps in overcoming the drawback of progressive alignment strategy where partial alignments are frozen [Barton and Sternberg, 1987]. Typical schema is as follows:

- Given N sequences $S_1, S_2, \ldots, S_N$, find alignment A.
- Remove sequence S_1 from alignment A and realign it to the profile of other aligned sequences $S_2, \ldots, S_N$ to get new alignment A'.
- Calculate the score of the new alignment A' and if better, replace A by A'.
- Remove sequence S_2 from A' and realign it. Continue this procedure for $S_3, \ldots, S_N$.
- Repeat the realignment steps until alignment score converges or number of iterations reaches the user specified limit.

Many iteration strategies which enable very accurate alignments have been developed [Wallace et al., 2005]. The aim is to reduce the greedy nature of the algorithm and avoid getting trapped in a local optima. One approach is to remove and realign every sequence to the rest in each iteration. Then, the alignment with the best score is taken to be the input for the next iteration. The other famous approach is to randomly split set of sequences into two sets, which are then realigned.

Some researchers have incorporated the iterative strategy in progressive alignment procedure itself. For instance, a double iteration loop has been used to make the alignment, guide tree, and sequence weights mutually consistent [Gotoh, 1996]. Recently, Chakrabarti et al. (2006) have developed an approach which provides a fast and accurate method for refining existing block-based alignments.

12.4 Novel Graph–Theoretical-Based Genomic Models

In this section, we present our research effort of a novel graph–theoretical approach for representing a wide variety of genomic sequence analysis problems within a single model [Lee et al., 2006]. The model allows incorporation of the operations—insertion, deletion, and substitution, and various parameters, such as relative distances and weights. Conceptually, we refer the problem as the minimum weight common mutated sequence (MWCMS) problem. The MWCMS model has many applications, including multiple sequence alignment problem, the phylogenetic analysis, the DNA sequencing problem, and sequence comparison problem, which encompass a core set of very difficult problems in computational biology. Thus, the model presented in this section lays out a mathematical modeling framework that allows one to investigate theoretical and computational issues, and to forge new advances for these distinct, but related problems.

DNA sequencing refers to determining the exact order of nucleotide sequences in a segment of DNA. This was the greatest technical challenge in the Human Genome Project. Achieving this goal has helped reveal the estimated 30,000 human genes that are the basic physical and functional units of heredity. The resulting DNA sequence maps are being used by scientists to explore human biology and other complex phenomena.

The structure of a DNA strand (sequence) is determined by experimentation. Typically, short sequences are determined to be in the strand, and the identified short sequences are then connected to form a long sequence. Recent advances attempting to identify DNA strand structure involve sequencing by hybridization [Bains and Smith, 1988] [Idury and Waterman, 1995]. Sequencing by hybridization is the process where every possible sequence of length n (4^n possibilities) is compared to a full DNA strand. Practical values for n are 8–12. Each short string either binds or does not bind to the full strand. Biologists can thus determine exactly which short strings are contained in the DNA strand and which are not.

However, the experiment does not identify the exact location of each short string in the full strand. Hence, an important issue involves how these short strings are connected together to form the complete strand. This problem can be viewed as a shortest common superstring problem and has been studied extensively [Maier and Storer, 1977] [Garey and Johnson, 1979] [Gallant et al., 1980]. Unfortunately, errors may arise during sequencing experiments. Three types of errors are deletions (a letter appears in an input string that should not be in the final sequence), insertions (a letter is missing from an input string), and substitutions (a letter in an input string should be substituted with another letter). The MWCMS problem can be used to model and solve this shortest common superstring problem while addressing the issue of possible errors.

Sequence comparison is one of the most crucial problems faced by researchers in the area of bioinformatics. The sequence patterns are conserved during evolution. Given a new sequence, it will be of interest to understand how much similarity it has with preexisting sequences. Significant similarity between two sequences implies similarities in their structure or function. There are many DNA databases containing DNA sequences and their function. The major ones are GenBank in the United States and the EMBL data library in Europe. If one finds a new sequence similar to existing ones in these databases, one can transfer information about the function and structure [Waterman, 1995]. Hence, an algorithm for sequence comparison which is efficient for large number of sequences will play a pivotal role in rapid sequence analysis. The MWCMS problem can be used to address this issue.

12.4.1 Definitions

Our motivation for first defining the problem arose from the desire to help quantify the concept of best representative sequence in the evolutionary distance problem. The evolutionary distance problem involves finding the DNA sequence of the most likely ancestor associated with a given set of DNA sequences from distinct but similar organisms. In other words, find the DNA strand that best represents a possible ancestor, if each of the organisms evolved from the same ancestor. Changes that contribute to differences between the given sequences and the ancestor are referred to as insertions, deletions, and substitutions. These operations account for both evolutionary mutations and experimental errors in sequencing. Mathematically, given two sequences S and B, let $ord(S, B)$ be an ordered collection of insertions, deletions, and substitutions to convert sequence S to B. (For any two sequences S and B, there are an infinite number of collections $ord(S, B)$.) Let $w(ord(S, B))$ be the weight of the conversion from S to B, where the weight is the sum of an expression involving values η, δ and $\psi \in \Re^+$ which represent the weights associated with a single insertion, deletion, and substitution, respectively. Let $ord^*(S, B)$ be such that $w(ord^*(S, B)) \leq w(ord(S, B))$ for all $ord(S, B)$. Define $d(S, B) = w(ord^*(S, B))$. Formally, MWCMS can be stated as

Problem MWCMS: Given positive weights η, δ and ψ corresponding to a single insertion, deletion, and substitution, respectively, a positive threshold κ, and finite sequences $S_1, \ldots, S_m$ from a finite alphabet, does there exist a sequence B such that $\sum_{i=1}^{m} d(S_i, B) \leq \kappa$?

We have defined the MWCMS problem—which incorporates the notions of insertion, deletion, and substitution—to help quantify the concept of best representative sequence in the evolutionary distance problem. We now make precise the operations of insertion, deletion, and substitution. Let $S = \{s_1, \ldots, s_n\}$ be a finite sequence of letters from a finite alphabet.

1. An insertion of an element x in position i of the sequence S is characterized by the addition of x between elements s_i and s_{i+1}. An insertion carries an associated penalty cost of η.
2. A deletion of an element in position i of S amounts to deleting s_i from the sequence S. The penalty for deletion is represented by δ.
3. A substitution of an element in position i of S amounts to replacing s_i with another letter from the alphabet. The penalty for substitution is represented by ψ.

We remark that a penalty cost for an operation could, more generally, depend on the position where the operation is performed or the element to be inserted/deleted/substituted.

Let $S_1 = \{s_{11}, \ldots, s_{1m}\}$ and $S_2 = \{s_{21}, \ldots, s_{2n}\}$ be two finite sequences of letters from a finite alphabet Σ. We say that the relative distance between elements s_{1i} and s_{2j} is k if $|i - j| = k$. We define a k-restrictive bipartite graph as a graph $G_k = (V_1, V_2, E_k)$ such that the nodes in V_1 and V_2 correspond respectively to each of the elements from the first and the second sequences. We assume the nodes in V_i are ordered in the same order as they appear in the sequence S_i. There is an edge between nodes $u \in V_1$ and $v \in V_2$ if u and v are identical (i.e., same letter of the alphabet Σ) and if the relative distance between these two elements is less than or equal to k. The problem of identifying the greatest similarity between these two sequences can then be approached as the problem of finding a maximum cardinality matching between the associated node sets, subject to restrictions on which matchings are allowed. In particular, one must take into consideration the ordering of nodes so as to preserve the relative occurrence of the elements in the matching. In addition, matchings that have edge crossings must be prevented. When $k = \max\{|S_1|, |S_2|\} - 1$, we denote the graph by $G = (V_1, V_2, E)$, and the problem is equivalent to the well-studied longest common subsequence (LCS) problem for two sequences, which is polynomial-time solvable [Garey and Johnson, 1979].

12.4.2 Construction of a Conflict Graph from Paths of Multiple Sequences

Let $S_i, i = 1, \ldots, m$, be a collection of finite sequences, each of length n, over a common alphabet Σ. Let $G_k = (V_1, \ldots V_m, E_1, E_2, \ldots, E_{m-1})$ be the k-restrictive multilayer graph in which each element in S_i forms a distinct node in V_i. Assume the nodes in V_i are ordered in the same order as they appear in the sequence S_i. E_i denotes the set of edges between nodes in V_i and V_{i+1}. There is an edge between nodes $u \in V_i$ and $v \in V_{i+1}$ if and only if u and v are the same letter in the alphabet Σ, and the relative distance between them is less than or equal to k. The multiple sequence comparison problem involves finding the LCS within the sequences $S_i, i = 1, \ldots, m$. We call a path $P = p_1, p_2, \ldots, p_m$ a complete path in G_k if $p_i \in V_i$ and $p_i p_{i+1} \in E_i$. Two complete paths are said to be parallel if their node sets are disjoint and the edges do not cross. Hence, a set of parallel complete paths in G_k corresponds to a feasible solution to LCS on the collection of sequences $S_i, i = 1, \ldots, m$. We say that two complete paths P_1, P_2 cross if they are not parallel. We remark that the LCS problem with the number of sequences bounded, is polynomial-time solvable using dynamic programming [Garey and Johnson, 1979]. In general, the problem remains $\mathcal{NP}$-complete.

We can incorporate insertions by generating new paths which include inserted nodes on various layers. The weight for such a new path will be affected by the total number of insertions in the path. In particular, if L is a common subsequence for S_i and $|S_i| = n$ for all $i = 1, \ldots, m$, then the total number of unmatched elements remaining will be $m(n - |L|)$. These elements can be deleted completely, or for a given unmatched element, one can increase the size of L by 1 by appropriately inserting this element into various sequences. By doing so, the number of unmatched elements decreases. Let l be the number of insertions needed to generate a new complete path. Then the number of unmatched elements will decrease by $m - l$. If we assume that at the end of the sequencing process all unmatched elements will be deleted, then the penalty for generating this new complete path will be given by $l\eta - (m - l)\delta$.

We next define the concept of conflict graph relative to complete paths in G_k.

Definition 1. *Let $\mathcal{P} = \{P_1, \ldots, P_s\}$ be a finite collection of complete paths in G_k.* The conflict graph $\mathcal{C}_{\mathcal{P}} = (V_{\mathcal{P}}, E_{\mathcal{P}})$ associated with $\mathcal{P}$ is constructed as follows:

- $V_{\mathcal{P}} = \{P_1, \ldots, P_s\}$.
- There is an edge between two nodes P_i and P_j in $V_{\mathcal{P}}$ if and only if P_i and P_j cross each other.

This definition applies to any multilayer graph in general. Note that any stable set of nodes in $\mathcal{C}_{\mathcal{P}}$ corresponds to a set of parallel complete paths for G_k, and thereby to a feasible solution to LCS on the collection of sequences S_i, $i = 1, \ldots, m$.

We remark that when $m = 2$, the resulting conflict graph is weakly triangulated, and thus is perfect. For $m > 2$, the conflict graph can contain an antihole of size 6. However, these complete paths can be viewed as continuous functions on the interval 0–1, thus by construction, $\mathcal{C}_{\mathcal{P}}$ is perfect [Golumbic et al., 1983].

12.4.3 Complexity Theory

Recall that the notation $ord(S, B)$, $w(ord(S, B))$, $ord^*(S, B)$, and the formal definition of problem MWCMS were given in Section 12.4.1. As an optimization problem, MWCMS can be stated as follows. Given a set of input sequences, problem MWCMS seeks to mutate every input sequence to the same a priori unknown sequence using the operations of insertion, deletion, and substitution; weights are assigned for each operation, and the total weight associated with all mutations is to be minimized. Levenshtein first considered a special case of this problem by changing a single input sequence to another sequence using insertions, deletions, and substitutions [Levenshtein, 1966]. Our study involves changing multiple input sequences to arrive at an a priori unknown common sequence.

Given positive weights η, δ, and ψ corresponding, respectively to insertions, deletions, and substitutions and any two sequences S and B, clearly any $ord^*(S, B)$ will never contain more than $|B|$ insertions or substitutions. Proving that MWCMS is in $\mathcal{NP}$ is not obvious. While one can transform MWCMS to special applications (as described in beginning of Section 12.4) to conclude that it is in NP, here we prove it directly for the general case. One needs to be able to evaluate $d(S, B)$ in polynomial time for any two sequences S and B. We next construct a graph that can be used to establish the existence of a polynomial-time algorithm for obtaining $d(S, B)$. The constructs and arguments used here typify those used to establish many of the results presented in this paper. It is noteworthy that the notions of both conflict graph and perfect graph come into play.

Let Σ be a finite alphabet, and define Σ-cross to be a directed bipartite graph consisting of $|\Sigma|$ vertices in each bipartition such that each vertex in the bipartition represents a distinct element in Σ. There is an arc between two vertices if the vertices correspond to the same element in Σ, and the geometric layout is rigidly constructed so that every arc crosses every other arc. This graph will be used as a supernode for insertion and substitution operations in our model. Figure 12.8 shows an example for Σ-cross when $\Sigma = \{A, C, G, T\}$.

We now construct a 3-layer supergraph, G_L, using the sequences S and B along with the Σ-cross graphs. Layers 1 and 2 consist of exactly $|B|(|S| + 1) + |S|$ Σ-crosses. The first $|B|$ Σ-crosses represent potential

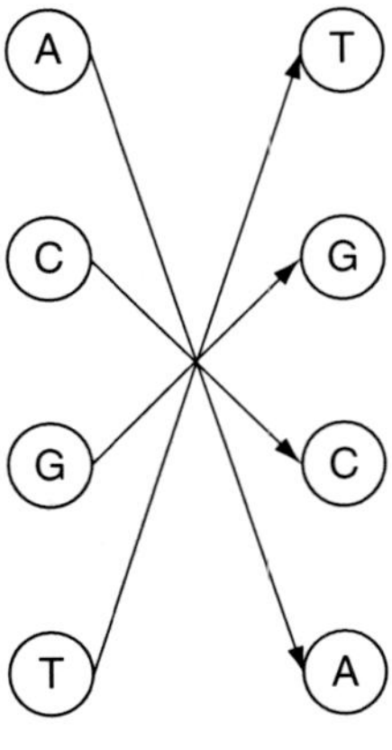

Figure 12.8 An example of $\sum$-cross when $\sum = \{A, C, G, T\}$.

insertions before the first letter in S. The next Σ-cross represents either the first letter of S or a substitution of this letter. The next $|B|$ Σ-crosses represent potential insertions between the first and second letters of S. And this is followed by a Σ-cross representing either the second letter of S or a substitution of this letter. This continues for each letter in S with the final $|B|$ Σ-crosses representing up to $|B|$ insertions after the last letter in S. Each Σ-cross is called either an insertion supernode or a substitution supernode, according to what it represents. The weight of all of the arcs in an insertion supernode is η. An arc in a substitution supernode has weight $-\delta$ if the arc represents the original letter in the sequences, or $\psi - \delta$ if the arc represents a substitution of the original letter. Layer 3 consists of the vertices represented by B. A vertex in layer 2 is connected to a vertex in layer 3 if they have the same letter. The weight of every arc between layers 2 and 3 is $M \leq -(\eta + \delta + \psi)$. A sample of a 3-layer supergraph is given in Figure 12.9. The bold arcs are used to denote the original letters in S (the weight of these arcs is $-\delta$). For simplicity, we omit the first two insertion supernodes before the first letter G. The first supernode thus represents the letter G from the original sequence which allows for substitution. The second and third supernodes correspond to insertion supernodes, and the fourth supernode corresponds to the letter C and allows substitution as well. There are two more insertion supernodes which are omitted from the graph.

The main step in proving $d(S, B)$ to be polynomial-time solvable for any sequences S and B involves the use of the conflict graph as defined in Definition 1. We state some preliminary theoretical results below. Detail proofs can be found in Lee et al. (2006).

Lemma 1. *The following statements are equivalent:*

(i) There exists a conversion from S to B using no more than a total of $|B|$ insertions or substitutions.
(ii) There exists a set of noncrossing complete paths in the associated 3-layer supergraph G_L of size $|B|$.
(iii) There exists a node packing of size $|B|$ in the associated conflict graph $\mathcal{C}$.

Lemma 2. *Calculating $d(S, B)$ for any sequences S and B can be accomplished in polynomial time.*

The 3-layer supergraph can be generalized to multilayer when multiple sequences are considered. Clearly, such multilayer supergraphs are much too large for practical purposes, yet polynomiality is preserved in the construction, and it is therefore sufficient. We can now arrive at the result that MWCMS is in $\mathcal{NP}$.

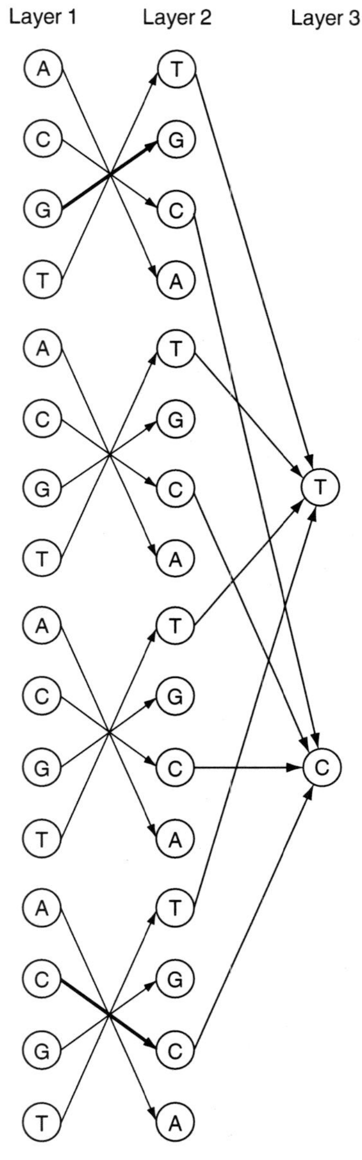

Figure 12.9 An example of the 3-layer supergraph for converting the sequence S = GC to B = TC. Bold arcs are used to denote the original letters in S (the weight of these arcs is $-\delta$). For simplicity, we omit the first two insertion supernodes before the first letter G. The first supernode in this figure thus represents the letter G from the original sequence which allows for substitution. The second and third supernodes correspond to insertions, and the fourth supernode corresponds to the letter C and allows substitution as well. There are two more insertion supernodes which are omitted from the graph.

THEOREM 1. *MWCMS is in $\mathcal{NP}$.*

To prove that MWCMS is polynomial-time solvable when the number of input sequences is bounded by a positive constant, the following lemma is crucial, though trivial.

LEMMA 3. *Given η, δ, $\psi \in \Re^+$, an optimal solution B to any MWCMS problem has the following properties. B has no substitutions from letters other than the original letters in an S_i, and B will never have an element which is inserted in every sequence (in the same location). Therefore, there are at most $\sum_{i=1}^{m}|S_i|$ insertions in any sequence.*

In addition, we also require the construction of a (directed) $2m$-layer supergraph, G_L^m, similar to the 3-layer supergraph, G_L.

Given sequences $S_1, \ldots, S_m$, generate a $2m$-layer (directed) graph $G_L^m = (V, E)$ as follows. Layers $2i - 1$ and $2i$ consist of $(\sum_{j=1}^{m}|S_j|)(|S_i| + 1) + |S_i|$ copies of Σ-crosses for $i = 1, \ldots, m$, constructed in exactly the same manner as layers 1 and 2 of the 3-layer supergraph using the input sequence S_i. The first $\sum_{j=1}^{m}|S_j|$ Σ-crosses represent the possibility that $\sum_{j=1}^{m}|S_j|$ different letters can be inserted before the first element in S_i. The next Σ-cross corresponds to either the first letter in S_i or a substitution of this letter. This is repeated $|S_i|$ times (for each letter in S_i), and the final $\sum_{j=1}^{m}|S_j|$ Σ-crosses represent insertions after the final letter in S_i. Thus, the first $\sum_{j=1}^{m}|S_j|$ Σ-crosses represent the insertion supernodes, followed by one Σ-cross representing a letter in S_i or a substitution supernode, and so forth. An arc exists from a vertex in layer $2i$ to a vertex in layer $2i+1$ if the vertices correspond to the same letter. Observe that G_L^m is an acyclic directed graph which is polynomial in the size of the input sequences. Assign every arc between layers $2i$ and $2i+1$ a weight of 0. There are three different weights for arcs between layers $2i - 1$ and $2i$ each corresponding to an insertion, deletion, or substitution. The assignment of weights on such arcs is analogous to the assignment in G_L: a weight of η is assigned to every arc contained in an insertion supernode; and an arc in a substitution supernode is assigned a weight of $-\delta$ if it corresponds to the original letter, or $\psi - \delta$, otherwise.

Figure 12.10 shows a sample graph for two sequences: $S_1 = GC$ and $S_2 = TG$. Observe that at most two insertions are needed in an optimal solution; thus we can reduce the number of Σ-crosses as insertion supernodes from $\sum_{i=1}^{2}|S_i| = 4$ to 2. For simplicity, in the graph shown in Figure 12.10, we have not included the two insertion supernodes before the first letter nor those after the last letter of each sequence. Thus, in the figure, the first Σ-cross represents the substitution supernode associated with the first letter in S_1. The second and third Σ-crosses represent two insertion supernodes. And the last Σ-cross represents the substitution supernode associated with the second letter in S_1. For simplicity, we include only arcs

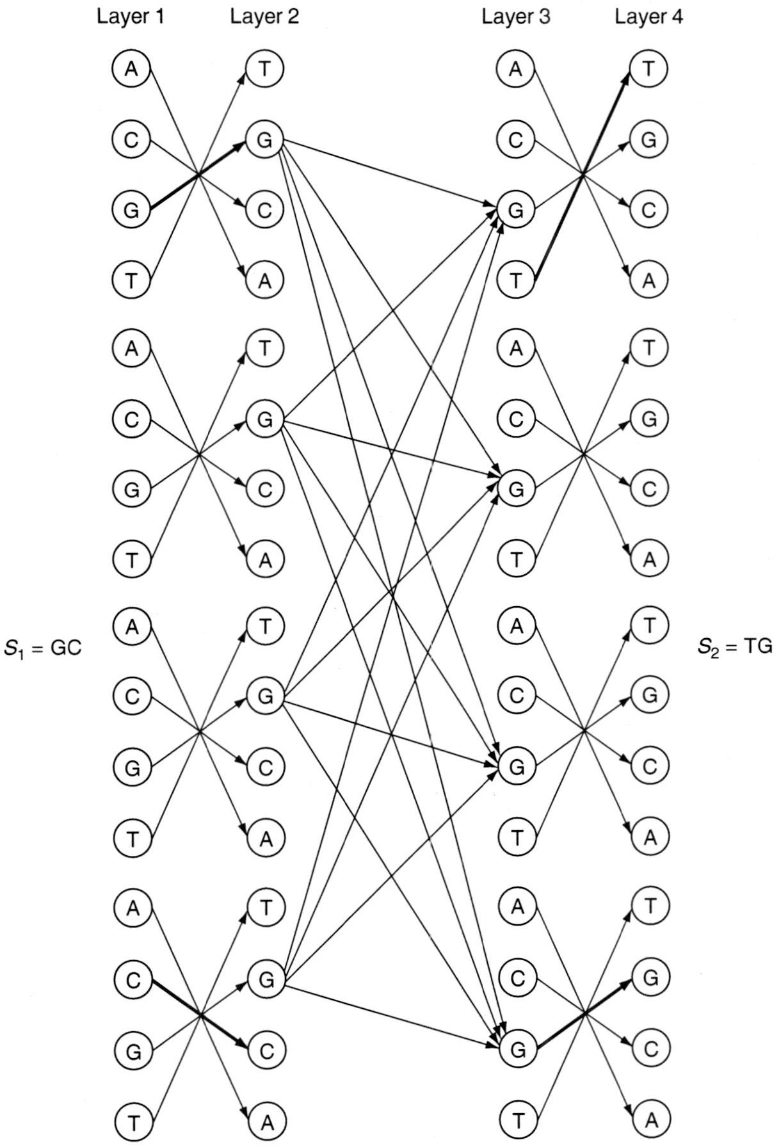

Figure 12.10 A sample graph G_L^m of MWCMS with S_1 = GC to S_2 = TG where $\sum = \{A, C, G, T\}$.

connecting vertices associated to the element G between layers 2 and 3. The arcs for other vertices follow similarly.

A conflict graph $\mathcal{C}$ associated with G_L^m can be generated by finding all complete paths (paths from layer 1 to layer $2m$) in G_L^m. These complete paths correspond to the set of vertices in $\mathcal{C}$, as in Definition 1. If we assign a weight to each vertex equal to the weight of the associated complete path, then the following result can be established.

Theorem 2. *Every node packing in $\mathcal{C}$ represents a candidate solution to MWCMS if and only if at most $\sum_{i=1}^{m} |S_i|$ letters can be inserted between any two original letters. Furthermore, the weight of the node packing is equal to the weight of the MWCMS $- \sum_{i=1}^{m} |S_i|\delta$.*

The supergraph G_L^m and its associated conflict graph are fundamental to our proof of the following theorem on polynomial-time solvability of a restricted version of problem MWCMS.

Theorem 3. *Problem MWCMS restricted to instances for which the number of sequences is bounded by a positive constant is polynomial-time solvable.*

12.4.4 Special Cases of MWCMS

MWCMS encompasses a very broad class of problems. In computational biology as discussed in this chapter, first and foremost, it represents a model for phylogenetic analysis. MWCMS as defined is the most likely ancestor problem, and the concept of 3-layer supergraph as described in Section 12.4.3 describes the evolutionary distance problem. An optimal solution to a multiple sequence alignment instance can be found using the solution of the MWCMS problem obtained on the $2m$-layer supergraph, G_L^m. The alignment is the character matrix obtained by placing together the given sequences incorporating the insertions into the solution of the MWCMS problem. Furthermore, DNA sequencing can be viewed as the shortest common superstring problem, although sequence comparison of a given sequence B to a collection of N sequences $S_1, \ldots, S_N$ is the MWCMS problem itself.

Broader than the computational biology applications, special cases of MWCMS include shortest common supersequences (SCSQ), LCS, and shortest common superstring (SCST); these problems are of interest in their own right as combinatorial optimization problems and for their role in complexity theory.

12.4.5 Computational Models: Integer Programming Formulation

The construction of the multilayer supergraphs described in our theoretical study lays the foundation and provides direction for computational models and solution strategies that we will explore in future research. Although the theoretical results obtained are polynomial time in nature, they present computational challenges. In many cases, calculating the worst-case scenario is not trivial. Furthermore, the polynomial-time result of a node-packing problem for a perfect graph by Grötschel et al. Grötschel et al. [1984] Grötschel et al. [1988] is existential in nature, and relies on the polynomial-time nature of the ellipsoid algorithm. The process itself involves solving an IP relaxation multiple times. In our case, the variables of the IP generated are the complete paths in the multilayer supergraph, $\mathrm{G}_{\mathrm{L}}^m$. Formally, the integer program corresponding to our conflict graph can be stated as follows:

Let x_p be the binary variable denoting the use or nonuse of the complete path p with weight w_p. Then the corresponding node-packing problem is

$$\begin{aligned} &\text{Min} \sum w_p x_p \\ &\text{s.t. } x_p + x_q \leq 1 \quad \text{if complete paths } p \text{ and } q \text{ cross} \\ &\qquad x_p \in \{0, 1\} \quad \text{for all complete paths } p \text{ in } \mathrm{G}_{\mathrm{L}}^m. \end{aligned} \tag{MIP1}$$

We call the inequality $x_p + x_q \leq 1$ an adjacency constraint. A natural approach to improve the solution time to (MIP1) is to decrease the size of the graph $\mathrm{G}_{\mathrm{L}}^m$ and thus the number of variables. Reductions in the size of $\mathrm{G}_{\mathrm{L}}^m$ can be accomplished for SCST, LCS, and SCSQ problems. Among these three problems, the graph $\mathrm{G}_{\mathrm{L}}^m$ is smallest for LCS. In LCS, all insertion and substitution supernodes can be eliminated.

Our theoretical results thus far rely on the creation of all complete paths. Clearly, the typical number of complete paths will be in the order of n^m, where $n = \max |S_i|$. In this case, an instance with three sequences and 300 letters in each sequence generates more than one million variables. Hence, an exact formulation with all complete paths is impractical in general. A simultaneous column and row generation approach within a parallel implementation may lead to computational advances related to this formulation.

An alternative formulation can be obtained by examining $\mathrm{G}_{\mathrm{L}}^m$ from a network perspective using arcs (instead of complete paths) in $\mathrm{G}_{\mathrm{L}}^m$ as variables. Namely, let $x_{i,j}$ denote the use or nonuse of arc (i, j) in the final sequence with $c_{i,j}$ the cost of the arc in $\mathrm{G}_{\mathrm{L}}^m$. The network formulation can be stated as

$$\text{Min} \sum_{(i,j)\in E} c_{i,j} x_{i,j}$$

$$\text{s.t. } \sum_{i:(i,j)\in E} x_{i,j} = \sum_{k:(j,k)\in E} x_{j,k} \quad \text{for all } j \in V \text{ in layers } 2, \ldots, 2m-1$$

$$x_{i,j} + x_{k,l} \leq 1 \quad \text{for all crossing arcs } (i,j) \text{ and } (k,l) \in E$$

$$x_{i,j} \in \{0,1\} \quad \text{for all } (i,j) \in E \qquad \text{(MIP2)}$$

The first set of constraints ensures flow in equals flow out in all vertices contained in sequences $2, \ldots, m-1$ (complete paths). The second set of constraints ensures that no two arcs cross each other. This model grows linearly in the number of sequences. This alternative integer programming formulation is still large, but is manageable for even fairly large instances.

Utilizing a collection of DNA sequences (each with 40,000 base pairs in length) from a bacteria, and a collection of short sequences associated with genes found in breast cancer patients, computational tests of our graph–theoretical models are underway. We seek to develop computational strategies to provide reasonable running times for evolutionary distance problem instances derived from these data. In an initial test, when three sequences each with 100 letters are used, the initial linear program requires more than 10,000 seconds to solve when tight constraints are employed (in this case, each adjacency constraint is replaced by a maximal clique constraint). Our ongoing computational effort will focus on developing and investigating solution techniques for practical problem instances, including those based on the above two IP formulations, as well as development of fast heuristic procedures.

In Lee et al. (2006), we outline a simple yet practical heuristic based on (MIP2) that we developed for solving the multiple sequence alignment problem; and we report on preliminary tests of the algorithm using different sets of sequence data. Motivation for the heuristic is derived from the desire to reduce computational time through various strategies for reducing the number of variables in (MIP2).

12.5 Summary

Multiple sequence alignment and phylogenetic analysis are deeply interconnected problems in computational biology. A good multiple alignment is crucial for reliable reconstruction of the phylogenetic tree [Phillips et al., 2000]. On the other hand, most of the multiple alignment methods require a phylogenetic tree as the guide tree for progressive iteration.

Thus, the evolutionary tree construction might be biased by the guide tree used for obtaining the alignment. To avoid this pitfall, various algorithms have been developed which simultaneously find alignment and phylogenetic relationship among given sequences. Sankoff and Cedergren (1983) developed a parsimony-based algorithm using a character-substitution model of gaps. The algorithm is guaranteed to find evolutionary tree and alignment which minimizes tree-based parsimony cost. Hein (1989) also developed a parsimony-type algorithm but uses an affine gap cost which is more realistic than the character-substitution gap model. This algorithm is also faster than Sankoff and Cedergreen's approach but makes simplifying assumptions in choosing ancestral sequences.

Like parsimony methods for finding a phylogenetic tree, both of the above approaches require search over all possible trees to find the global optimum. This makes these algorithms computationally very intensive. Hence, there has been a strong focus on developing an efficient algorithm that considers both alignment and tree. Vingron and Haeseler (1997) have developed an approach based on three-way alignment of pre-aligned groups of sequences. It also allows change in the alignment made early in the course of computation. Many softwares, like MEGA, are trying to develop an efficient integrated computing environment that allows both sequence alignment and evolutionary analysis [Kumar et al., 2004].

We address this issue of simultaneously finding alignment and phylogenetic relationships by presenting a novel graph–theoretical approach. Indeed, our model can be easily tailored to find theoretically provable optimum solutions to a wide range of crucial sequence analysis problems. These sequence analysis problems are proven to be NP-hard, and thus understandably present computational challenges. To strike a balance between time and quality-of-solution, a variety of parameters are provided. Ongoing research efforts explore development of efficient computational models and solution strategies in a massive parallel environment.

Acknowledgment

This research is partially supported by grants from the National Science Foundation.

References

[Abbas and Holmes, 2004] Abbas, A. and Holmes, S. 2004. Bioinformatics and management science: Some common tools and techniques. *Operations Research*, 52(2):165–190.

[Althaus et al., 2006] Althaus, E., Caprara, A., Lenhof, H., and Reinert, K. 2006. A branch-and-cut algorithm for multiple sequence alignment. *Mathematical Programming*, 105(2–3):387–425.

[Altschul, 1991] Altschul, S. 1991. Amino acid substitution matrices from an information theoretic perspective. *Journal of Molecular Biology*, 219(3):555–565.

[Altschul et al., 1989] Altschul, S. F., Carroll, R. J., and Lipman, D. J. 1989. Weights for data related by a tree. *Journal of Molecular Biology*, 207(4):647–653.

[Bains and Smith, 1988] Bains, W. and Smith, G. 1988. A novel nethod for DNA sequence determination. *Journal of Theoretical Biology*, 135:303–307.

[Barton and Sternberg, 1987] Barton, G. J. and Sternberg, M. J. E. 1987. A strategy for the rapid multiple alignment of protein sequences: Confidence levels from tertiary structure comparisons. *Journal of Molecular Biology*, 198:327–337.

[Blazewicz et al., 2005] Blazewicz, J., Formanowicz, P., and Kasprzak, M. 2005. Selected combinatorial problems of computational biology. *European Journal of Operational Research*, 161:585–597.

[Bonizzoni and Vedova, 2001] Bonizzoni, P. and Vedova, G. 2001. The complexity of multiple sequence alignment with SP-score that is a metric. *Theoretical Computer Science*, 259:63–79.

[Bos and Posada, 2005] Bos, D. and Posada, D. 2005. Using models of nucleotide evolution to build phylogenetic trees. *Developmental and Comparative Immunology*, 29(3):211–227.

[Bruno et al., 2000] Bruno, W. J., Socci, N. D., and Halpern, A. L. 2000. Weighted neighbor joining: A likelihood-based approach to distance-based phylogeny reconstruction. *Molecular Biology and Evolution*, 17:189–197.

[Carrillo and Lipman, 1988] Carrillo, H. and Lipman, D. 1988. The multiple sequence alignment problem in biology. *SIAM Journal on Applied Mathematics*, 48(5):1073–1082.

[Chakrabarti et al., 2006] Chakrabarti, S., Lanczycki, C. J., Panchenko, A. R., Przytycka, T. M., Thiessen, P. A., and Bryant, S. H. 2006. Refining multiple sequence alignments with conserved core regions. *Nucleic Acids Research*, 34(9):2598–2606.

[Chenna et al., 2003] Chenna, R., Sugawara, H., Koike, T., Lopez, R., Gibson, T. J., Higgins, D. G., and Thompson, J. D. 2003. Multiple sequence alignment with the clustal series of programs. *Nucleic Acids Research*, 31(13):3497–3500.

[Chor and Tuller, 2005] Chor, B. and Tuller, T. 2005. Maximum likelihood of evolutionary trees: Hardness and approximation. *Bioinformatics*, 21(Suppl. 1):I97–I106.

[Clote and Backofen, 2000] Clote, P. and Backofen, R. 2000. *Computational Molecular Biology: An Introduction*. John Wiley & Sons Ltd., New York.

[Delsuc et al., 2005] Delsuc, F., Brinkmann, H., and Philippe, H. 2005. Phylogenomics and the reconstruction of the tree of life. *Nature Reviews Genetics*, 6(5):361–375.

[Durbin et al., 1998] Durbin, R., Eddy, S., Krogh, A., and Mitchison, G. 1998. *Biological Sequence Analysis*. Cambridge University Press, Cambridge.

[Felsenstein, 1981] Felsenstein, J. 1981. Evolutionary trees from DNA sequences: A maximum likelihood approach. *Journal of Molecular Evolution*, 17(6):368–376.

[Felsenstein, 1988] Felsenstein, J. 1988. Phylogenies from molecular sequences: Inference and reliability. *Annual Review of Genetics*, 22:521–565.

[Felsenstein, 1989] Felsenstein, J. 1989. PHYLIP—phylogeny inference package (version 3.2). *Cladistics*, 5:164–166.

[Fitch, 1971] Fitch, W. M. 1971. Towards defining the course of evolution: Minimum change for a specific tree topology. *Systematic Zoology*, 20:406–416.

[Gallant et al., 1980] Gallant, J., Maider, D., and Storer, J. 1980. On finding minimal length superstrings. *Journal of Computer and System Sciences*, 20:50–58.

[Garey and Johnson, 1979] Garey, M. and Johnson, D. 1979. *Computers and Intractability: A Guide to the Theory of NP-Completeness*. W.H. Freeman, San Francisco, CA.

[Gascuel, 1997] Gascuel, O. 1997. BIONJ: An improved version of the NJ algorithm based on a simple model of sequence data. *Molecular Biology and Evolution*, 14(7):685–695.

[Goeffon et al., 2005] Goeffon, A., Richer, J., and Hao, J. 2005. *Local Search for the Maximum Parsimony Problem*. LNCS, 3612:678–683.

[Golumbic et al., 1983] Golumbic, M. C., Rotem, D., and Urrutia, J. 1983. Comparability graphs and intersection graphs. *Discrete Mathematics*, 43:37–46.

[Gotoh, 1996] Gotoh, O. 1996. Significant improvement in accuracy of multiple protein sequence alignments by iterative refinement as assessed by reference to structural alignments. *Journal of Molecular Biology*, 264(4):823–838.

[Gotoh, 1999] Gotoh, O. 1999. Multiple sequence alignment: Algorithms and applications. *Advances in Biophysics*, 36:159–206.

[Grötschel et al., 1984] Grötschel, M., Lovász, L., and Schrijver, A. 1984. Polynomial algorithms for perfect graphs. *Annals of Discrete Mathematics*, 21:325–356.

[Grötschel et al., 1988] Grötschel, M., Lovász, L., and Schrijver, A. 1988. *Geometric Algorithms and Combinatorial Optimization*. Springer-Verlag, New York.

[Guindon and Gascuel, 2003] Guindon, S. and Gascuel, O. 2003. A simple, fast, and accurate algorithm to estimate large phylogenies by maximum likelihood. *Systematic Biology*, 52(5):696–704.

[Gupta et al., 1995] Gupta, S., Kececioglu, J., and Schaeffer, A. 1995. Improving the practical space and time efficiency of the shortest-paths approach to sum-of-pairs multiple sequence alignment. *Journal of Computational Biology*, 2:459–472.

[Hein, 1989] Hein, J. 1989. A new method that simultaneously aligns and reconstructs ancestral sequences for any number of homologous sequences, when the phylogeny is given. *Molecular Biology and Evolution*, 6(6):649–668.

[Huelsenbeck and Crandall, 1997] Huelsenbeck, J. and Crandall, K. 1997. Phylogeny estimation and hypothesis testing using maximum likelihood. *Annual Review of Ecology and Systematics*, 28:437–466.

[Hughey and Krogh, 1996] Hughey, R. and Krogh, A. 1996. Hidden markov models for sequence analysis: Extension and analysis of the basic method. *Computer Applications in the Biosciences*, 12(2):95–107.

[Idury and Waterman, 1995] Idury, R. M. and Waterman, M. S. 1995. A new algorithm for DNA sequence assembly. *Journal of Computational Biology*, 2(2):291–306.

[Jukes and Cantor, 1969] Jukes, T. H. and Cantor, C. R. 1969. Evolution of protein molecules. In Munro, H. N., editor, *Mammalian Protein Metabolism*, pages 21–123. Academic Press, New York.

[Just and Vedova, 2004] Just, W. and Vedova, G. 2004. Multiple sequence alignment as a facility-location problem. *INFORMS Journal on Computing*, 16(4):430–440.

[Keane et al., 2005] Keane, T., Naughton, T., Travers, S., McInerney, J., and McCormack, G. 2005. DPRml: Distributed phylogeny reconstruction by maximum likelihood. *Bioinformatics*, 21(7):969–974.

[Kececioglu et al., 2000] Kececioglu, J., Lenhof, H., Mehlhorn, K., Mutzel, P., Reinert, K., and Vingron, M. 2000. A polyhedral approach to sequence alignment problems. *Discrete Applied Mathematics*, 104:143–186.

[Kim et al., 1994] Kim, J., Pramanik, S., and Chung, M. J. 1994. Multiple sequence alignment using simulated annealing. *Bioinformatics*, 10(4):419–426.

[Kimura, 1980] Kimura, M. 1980. A simple method for estimating evolutionary of base substitution through comparative studies of nucleotide sequences. *Journal of Molecular Evolution*, 16:111–120.

[Klotz and Blanken, 1981] Klotz, L. and Blanken, R. 1981. A practical method for calculating evolutionary trees from sequence data. *Journal of Theoretical Biology*, 91(2):261–272.

[Korostensky and Gonnet, 1999] Korostensky, C. and Gonnet, G. H. 1999. Near optimal multiple sequence alignments using a traveling salesman problem approach. *Proceedings of the String Processing and Information Retrieval Symposium*, page 105.

[Korostensky and Gonnet, 2000] Korostensky, C. and Gonnet, G. H. 2000. Using traveling salesman problem algorithms for evolutionary tree construction. *Bioinformatics*, 16(7):619–627.

[Krogh et al., 1994] Krogh, A., Brown, M., Mian, I. S., Sjolander, K., and Haussler, D. 1994. Hidden markov models in computational biology: Applications to protein modeling. *Journal of Molecular Biology*, 235:1501–1531.

[Kumar et al., 1994] Kumar, S., Tamura, K., and Nei, M. 1994. MEGA: Molecular evolutionary genetics analysis software for microcomputers. *Computer Applications in Biosciences*, 10:189–191.

[Kumar et al., 2004] Kumar, S., Tamura, K., and Nei, M. 2004. MEGA3: Integrated software for molecular evolutionary genetics analysis and sequence alignment. *Briefings in Bioinformatics*, 5(2):150–163.

[Lawrence et al., 1993] Lawrence, C., Altschul, S., Boguski, M., Liu, J., Neuwald, A., and Wootton, J. 1993. Detecting subtle sequence signals: A Gibbs sampling strategy for multiple alignment. *Science*, 262:208–214.

[Lee et al., 2006] Lee, E. K., Easton, T., and Gupta, K. 2006. Novel evolutionary models and applications to sequence alignment problems. *Annals of Operations Research*, 148(1):167–187.

[Levenshtein, 1966] Levenshtein, V. L. 1966. Binary codes capable of correcting deletions, insertions, and reversals. *Cybernetics Control Theory*, 10(9):707–710.

[Li, 1981] Li, W. 1981. Simple method for constructing phylogenetic trees from distance matrices. *Proc. Natl. Acad. Sci. U.S.A.*, 78(2):1085–1089.

[Lipman et al., 1989] Lipman, D., Altschul, S., and Kececioglu, J. 1989. A tool for multiple sequence alignment. *Proceedings of the National Academy of Sciences of the United States of America*, 86(12):4412–4415.

[Maier and Storer, 1977] Maier, D. and Storer, J. A. 1977. A note on the complexity of the superstring problem. Technical Report 233, Princeton University, NJ.

[Nei, 1996] Nei, M. 1996. Phylogenetic analysis in molecular evolutionary genetics. *Annual Review of Genetics*, 30:371–403.

[Notredame, 2002] Notredame, C. 2002. Recent progress in multiple sequence alignment: A survey. *Pharmacogenomics*, 3(1):131–144.

[Notredame and Higgins, 1996] Notredame, C. and Higgins, D. 1996. SAGA: Sequence alignment by genetic algorithm. *Nucleic Acids Research*, 24(8):1515–1524.

[Phillips et al., 2000] Phillips, A., Janies, D., and Wheeler, W. 2000. Multiple sequence alignment in phylogenetic analysis.

Molecular Phylogenetics and Evolution, 16(3): 317–330.

[Piontkivska, 2004] Piontkivska, H. 2004. Efficiencies of maximum likelihood methods of phylogenetic inferences when different substitution models are used. *Molecular Phylogenetics and Evolution*, 31(3):865–873.

[Purdom et al., 2000] Purdom, P., Bradford, P. G., Tamura, K., and Kumar, S. 2000. Single column discrepancy and dynamic max–mini optimizations for quickly finding the most parsimonious evolutionary trees. *Bioinformatics*, 16:140–151.

[Reinert et al., 1997] Reinert, K., Lenhof, H., Mutzel, P., Mehlhorn, K., and Kececioglu, J. 1997. A branch-and-cut algorithm for multiple sequence alignment. Proceedings of the First Annual International Conference on Computational Molecular Biology (RECOMB-97), pp. 241–249.

[Ronquist, 1998] Ronquist, F. 1998. Fast fitch-parsimony algorithms for large data sets. *Cladistics*, 14:387–400.

[Saitou and Nei, 1987] Saitou, N. and Nei, M. 1987. The neighbor-joining method: A new method for reconstructing phylogenetic trees. *Molecular Biology and Evolution*, 4:406–425.

[Sankoff and Cedergren, 1983] Sankoff, D. and Cedergren, R. J. 1983. Simultaneous comparison of three or more sequences related by a tree. In Sankoff, D. and Kruskal, J. B., editors, *Time Warps, String Edits, and Macromolecules: The Theory and Practice of Sequence Comparison*, Addison-Wesley, Reading, MA, pp. 253–264.

[Shyu et al., 2004] Shyu, S. J., Tsai, Y. T., and Lee, R. 2004. The minimal spanning tree preservation approaches for DNA multiple sequence alignment and evolutionary tree construction. *Journal of Combinatorial Optimization*, 8(4):453–468.

[Sokal and Michener, 1958] Sokal, R. and Michener, C. 1958. A statistical method for evaluating systematic relationships. *University of Kansas Scientific Bulletin*, 38:1409–1438.

[Stamatakis et al., 2005] Stamatakis, A., Ott, M., and Ludwig, T. (2005). RAxML-OMP: An efficient program for phylogenetic inference on SMPS. *Proceedings of 8th International Conference on Parallel Computing Technologies (PaCT2005)*, LNCS 3606, pp. 288–302.

[Swofford and Maddison, 1987] Swofford, D. L. and Maddison, W. P. 1987. Reconstructing ancestral character states under wagner parsimony. *Mathematical Biosciences*, 87:199–229.

[Swofford and Olsen, 1990] Swofford, D. L. and Olsen, G. J. 1990. Phylogeny reconstruction. In Hillis, D. M. and Moritz, G., editors, *Molecular Systematics*, Sinauer Associates Sunderland, MA. pp. 411–501.

[Tajima and Nei, 1984] Tajima, F. and Nei, M. 1984. Estimation of evolutionary distance between nucleotide sequences. *Molecular Biology and Evolution*, 1(3):269–285.

[Tajima and Takezaki, 1994] Tajima, F. and Takezaki, N. 1994. Estimation of evolutionary distance for reconstructing molecular phylogenetic trees. *Molecular Biology and Evolution*, 11:278–286.

[Takahashi and Nei, 2000] Takahashi, K. and Nei, M. 2000. Efficiencies of fast algorithms of phylogenetic inference under the criteria of maximum parsimony, minimum evolution, and maximum likelihood when a large number of sequences are used. *Molecular Biology and Evolution*, 17:1251–1258.

[Thompson et al., 1994] Thompson, J. D., Higgins, D. G., and Gibson, T. J. 1994. CLUSTAL W: Improving the sensitivity of progressive multiple sequence alignment through sequence weighting, position-specific gap penalties and weight matrix choice. *Nucleic Acids Research*, 22(22):4673–4680.

[Vingron and Haeseler, 1997] Vingron, M. and Haeseler, A. 1997. Towards integration of multiple alignment and phylogenetic tree construction. *Journal of Computational Biology*, 4(1):23–34.

[Vingron and Waterman, 1994] Vingron, M. and Waterman, M. 1994. Sequence alignment and penalty choice. Review of concepts, case studies and implications. *Journal of Molecular Biology*, 235(1):1–12.

[Wallace et al., 2005] Wallace, I. M., O'Sullivan, O., and Higgins, D. G. 2005. Evaluation of iterative alignment algorithms for multiple alignment. *Bioinformatics*, 21(8):1408–1414.

[Waterman and Perlwitz, 1984] Waterman, M. and Perlwitz, M. 1984. Line geometries for sequence comparisons. *Bulletin of Mathematical Biology*, 46(4):567–577.

[Waterman, 1995] Waterman, M. S. 1995. *Introduction to Computational Biology: Maps, Sequences and Genomes*. Chapman and Hall, UK.

[Whelan et al., 2001] Whelan, S., Lio, P., and Goldman, N. 2001. Molecular phylogenetics: State-of-the-art methods for looking into the past. *Trends in Genetics*, 17(5):262–272.

[Yang, 1993] Yang, Z. 1993. Maximum-likelihood estimation of phylogeny from DNA sequences when substitution rates differ over sites. *Molecular Biology and Evolution*, 10(6):1396–401.

[Zhang and Waterman, 2003] Zhang, Y. and Waterman, M. 2003. An eulerian path approach to global multiple alignment for DNA sequences. *Journal of Computational Biology*, 10(6):803–819.

Chapter 13

Computational Methods for Probe Design and Selection

Cláudio N. Meneses, Panos M. Pardalos, and Michelle A. Ragle

CONTENTS

13.1 Introduction 396
13.1.1 Applications 399
13.2 Probe Design and Selection 400
13.2.1 Specificity 400
13.2.2 Sensitivity 402
13.2.3 Homogeneity 403
13.2.4 Probe Length 404
13.2.5 Probe Selection Problems 405
13.2.5.1 Oligonucleotide Fingerprinting 405
13.2.5.2 Nonunique Probe Selection Problem 410
13.3 Software 411
13.4 Closing Remarks 412
References 412

Abstract There is a direct correlation between the cost and efficiency of hybridization experiments and the quality of the probes used in the experiments. Many factors must be considered in designing and selecting probes for use in experiments and these factors are generally dictated by the

This work has been partially supported by NSF, NIH, and CRDF grants.

primary goals of the experiment. In this chapter, we focus on computational methods for the design and selection of probes. We consider properties of probes such as the melting temperature, specificity, existence of secondary structures, and probe length. We also discuss various algorithms and software packages that have been developed to aid in the design and selection of probes.

13.1 Introduction

The discovery of the structure of DNA in the 1950s is attributed to James Watson, Francis Crick, Maurice Wilkins, and Rosalind Franklin. Watson, Crick, and Wilkins were awarded the 1962 Nobel Prize for Physiology or Medicine for their work. Rosalind Franklin had died in 1958. Not long after their results were made public, it was shown that the two strands of the double helix could be separated by the appropriate application of heat or solvent [24]. The reconstitution to double-stranded molecules is referred to as renaturation or hybridization and was first described by Marmur and Doty [17]. These discoveries and the property of sequence complementarity, which we will discuss shortly, form the basis of the methods used to analyze DNA and RNA [24].

DNA is a long polymer of nucleotides. Each nucleotide consists of a 5-carbon sugar, one phosphate group, and a nitrogen containing base attached to the sugar. The nucleotides are symbolized by four letters identifying their bases adenine (A) and guanine (G) (Figure 13.1), and cytosine (C) and thymine (T) (Figure 13.3).

When nucleotides are chained together by a phosphodiester bond, they form a DNA strand. The ends of the DNA strand are numbered by the carbon atom position where the next nucleotide can be attached. The 5′ end contains a phosphate group. The 3′ position of the nucleotide at the 3′ end is free for appending to the next nucleotide. DNA sequences are always read from the 5′ end to the 3′ end [3].

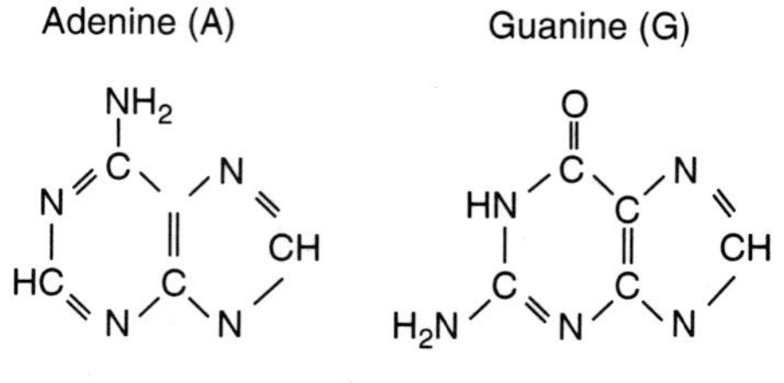

Figure 13.1 The purine bases are adenine and guanine.

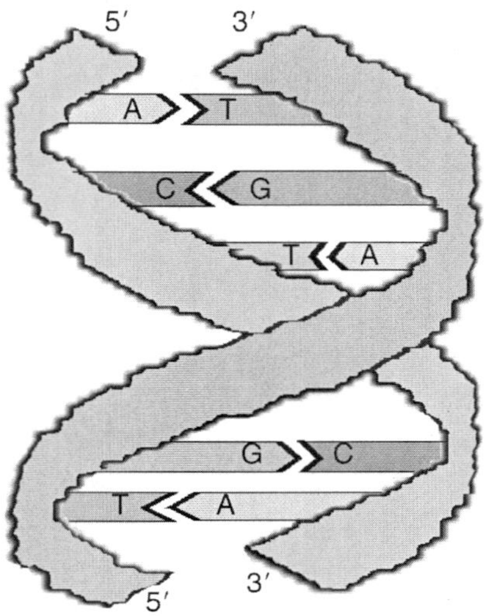

Figure 13.2 Complementary DNA strands.

The DNA molecule consists of two complementary strands, with adenine opposite thymine, and cytosine opposite guanine (Figure 13.2). One strand goes in direction from 5′ to 3′ while the other goes from 3′ to 5′ [3]. This pairing is referred to as Watson–Crick base pairing.

RNA differs from DNA in that it is almost always single stranded, and consists of the bases adenine, cytosine, guanine, and uracil. That is, the base thymine is replaced by uracil (U). Uracil is complementary to adenine [12].

Biologists can use hybridization to determine whether a specific DNA fragment is present in a DNA solution. They often use oligonucleotide probes. Oligonucleotide probes are short pieces of single-stranded DNA

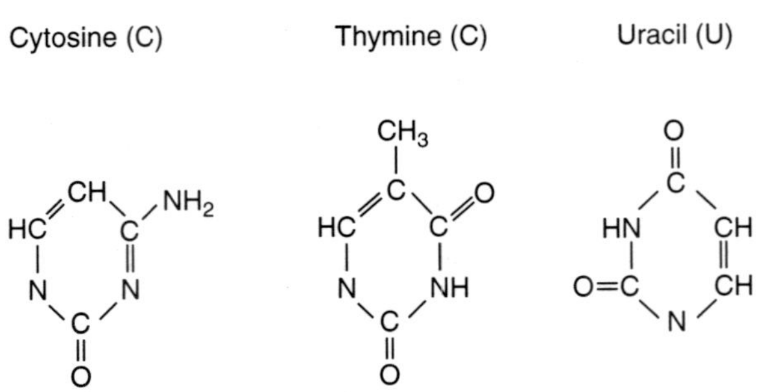

Figure 13.3 The pyrimidine bases are cytosine, thymine, and uracil.

that have a known sequence that is complementary to the DNA fragment in question. The probes have either a fluorescent or radioactive tag with fluorescent tags being more common [12]. In general, the biologist would apply the appropriate amount of heat or solvent to separate the strands of the unknown DNA, then put the DNA in contact with the probes and allow them to reconstitute. By observing whether a known probe hybridizes to the unknown DNA fragment, a biologist can determine the presence or absence of the complementary sequence. RNA can also be probed to see if a gene is on or off [12]. In 1979, synthetic oligonucleotide probes were introduced for use as hybridization probes [24]. Throughout this chapter, the term probe will be used to refer to nucleic acids of known sequence, and the term target will refer to the unknown sequence or set of sequences.

Microarrays allow for large-scale gene expression measurements. A microarray consists of a surface such as a glass slide or membrane that is spotted with DNA fragments or oligonucleotides to form an array (Figure 13.4). The value of microarrays lies in the fact that because there can be many thousand DNA molecules per array, the expression of many thousands of genes can be measured simultaneously. There are a number of microarray platforms that have been developed. Two commonly used platforms are oligonucleotide and cDNA [30].

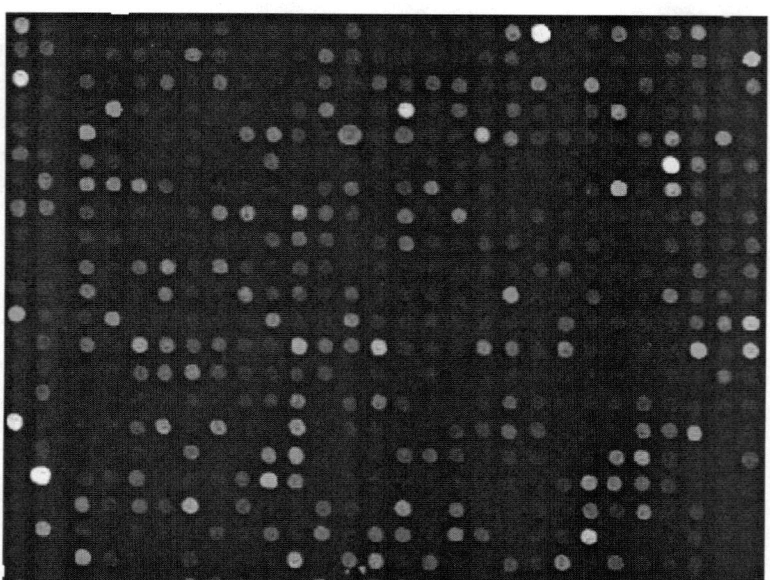

Figure 13.4 Microarray.

Once a microarray is created, a solution containing a labeled target nucleic acid sample is brought into contact with the substrate. Probes on the array hybridize to the target if there is a segment of the target sample that is Watson–Crick complementary to the probe. After some time, the substrate is washed to remove unbound and weakly bound target oligonucleotides from the sample. The microarray is scanned after the remaining target molecules have been stained with a fluorophore. The expression levels are then measured based upon detection of the fluorescent signal [10,30].

The focus of this chapter is on computational approaches for the optimal design and selection of probes.

13.1.1 Applications

There are many applications in biology that involve the use of probes. Here we discuss a few of those applications. One such application is sequencing by hybridization (SBH).

In 1987, Drmanac and Crkvenjakov applied for a patent for a method they had developed as an alternative approach for DNA sequencing. The method involved the use of hybridization data from a probe set containing overlapping probes to reconstruct complex DNA sequences. The method is referred to as SBH [24]. In this procedure, microarrays are created in which each cell contains a distinct known probe. The array is then brought into contact with a solution containing many copies of the target DNA that is to be sequenced. Each of the copies will have been tagged with a fluorescent or radioactive marker. As described in Section 13.1, probes on the array hybridize to the target if there is a segment of the target that is Watson–Crick complementary to the probe. After all of the DNA copies that have not hybridized are washed off the array, the subset of probes that hybridize to the target, called the spectrum of the sequence, is identified by observing which cells of the array are tagged. The spectrum is then used to reconstruct the DNA sequence by a combinatorial sequencing algorithm [5,8,9].

The first SBH design required chips that contained all 4^k strings of length k. The problem was reduced to finding a Eulerian path on a directed graph. Since then, several new approaches have been proposed.

In another application, diagnostic probes can be designed to detect bacterial infections. Given DNA sequences from a group of closely related pathogenic bacteria, the idea is to find a string that is a substring of each of the bacterial sequences without being a substring of the host's DNA sequence. Probes are designed to hybridize to these substring (target) sequences. The probes can then be used to detect the presence of at least one of the bacterial species [15].

Rash and Gusfield addressed the string barcoding problem in Ref. [25]. They developed a method to find a probe set of minimal cardinality for use in the detection of viral-size pathogens. We discuss the Rash and Gusfield paper in Section 13.2.5.1.

Antisense oligonucleotides are short synthetic oligonucleotides that are designed to hybridize to RNA. Once they bind to the target RNA, these oligonucleotides prevent expression of the encoded protein product. There are complicated challenges involved in the process; however, antisense oligonucleotides do show promise as therapeutic agents for the treatment of human diseases [32].

Sung and Lee cite an example in Ref. [31] in which DNA microarrays can be used to help identify the presence of alternate forms of, or an irregular expression in, a gene that results in resistance to chemotherapy.

These are just a few of the applications that involve the use of probes. It is clear that the optimal design and selection of probes serves a valuable function in many diverse areas.

13.2 Probe Design and Selection

In designing probes for use in experiments, it is necessary to consider many issues in addition to the Watson–Crick base pairing. For instance, it is important to design probes that will hybridize under the same conditions as the target sample. Probe length, as well as the composition and order of the bases in the sequence, is also an important factor to consider. The objective of the experiment in which the probes are to be used is of primary importance in determining the probe design and selection strategy. In all cases, good quality probes have the following three properties: specificity, sensitivity, and homogeneity [30,31,33].

13.2.1 Specificity

Specific probes are unique to each gene in the genome. They will return a weak signal when their Watson–Crick complement is not present in the sample. Specificity minimizes cross-hybridization of probes to other targets. If probes are not sufficiently specific, false positive results are more likely to occur due to nonspecific cross-hybridization. A false positive result occurs when probes are expected not to bind with a clone but the signal intensity is high. Oligonucleotide microarrays are not as prone to cross-hybridization of one probe to multiple targets as cDNA microarrays are [30,31].

In Ref. [31], Sung and Lee discuss the application of filters in their Find-Probe algorithm to select the best probes. To determine if a probe is specific enough, they check the Hamming distance between a probe and every subsequence in the genome. They define the Hamming distance of two strings s and t to be the number of positions where the characters at corresponding positions of the two strings differ. To be accepted as a valid candidate probe, the Hamming distance must be greater than a given constant. A naive approach would require a scan through the entire length-n genome for every length-m probe to determine the Hamming distance. Such an approach would take $O(mn^2)$ time. Sung and Lee instead employ the Pigeonhole Principle to reduce running time [31].

As mentioned, specificity minimizes cross-hybridization. Probes containing low-complexity sequences, i.e., repetitive sequences, should be avoided because they are likely to cross-hybridize to other targets. Several software programs have been developed for the design of probes. RepeatMasker is a software that detects repeat sequences of all types. It accepts as input a DNA or RNA sequence and returns the sequence with repeats replaced by N's. If designing oligonucleotide probes for use on a microarray, one would avoid selecting probes in regions masked with N's [30].

A homology search algorithm identifies sequences in a selected database that match part or all of a specific sequence referred to as the query sequence. BLAST (Basic Local Alignment Search Tool) is commonly used to search for similarities, but many other programs are also available. To use BLAST to check potential probes, you would input the probe's sequence as the query sequence and then select a database containing other target genes that you do not want to bind to the probe for comparison. The output produced by BLAST includes a hit list that tells the user if the input sequence is similar to a sequence contained in the selected database. The hit list has a bit score column that gives the statistical significance of the alignment. The higher the bit score is, the more similar are the two sequences. In most cases, the bit score is twice the length of the longest perfect alignment. If several probes are being compared for use in a microarray, preference is generally given to probes with lower bit scores. BLAST also has a low-complexity filter [30].

Wang and Seed [34] developed a program called OligoPicker for the selection of oligonucleotide probes for protein coding sequences. In their selection scheme, they make the rejection of contiguous sequence identity the primary filter, the reason being that they hypothesize that contiguous base pairing is the single most-important determinant of cross-hybridization. They use a hash table to quickly find repetitive sequence stretches of 10–20 m in length. They also screen probes based upon their BLAST scores to reduce the likelihood of cross-hybridization due to global similarity. In cases where sequences cannot be represented by a single unique probe, probes are selected from regions that cross-hybridize to the smallest number of other sequences in the sample universe [34].

13.2.2 Sensitivity

Sensitive oligonucleotide probes return a strong signal when the Watson–Crick complementary target is present in the sample. Probes that form secondary structures, i.e., probes that are self-complementary and can fold back on themselves (Figure 13.5) or dimerize with neighboring identical probes, adversely affect sensitivity [30,31]. In fact, secondary structural motifs in probes of length 20 m can reduce the hybridization signal up to 50-fold [14].

A DNA sequence that is identical to its reverse complement sequence, such as TGCA, is referred to as a palindrome. Dan Gusfield describes algorithms to identify palindromes in sequences in Ref. [8].

Secondary structure in both the probe and the target sequence can significantly impair hybridization affinity [14]. Because the target sequence is often unknown, detection of target secondary structure can be considerably more challenging. A measure that is commonly used to predict the stability of secondary structure in probes is Gibbs free energy [20].

Gibbs free energy describes the energy available to do work within a system. It satisfies the equation $G = H - TS$, where H denotes enthalpy, T denotes temperature in degrees Kelvin, and S denotes entropy. Enthalpy is defined to be the sum of the internal energy plus the product of the pressure and volume, and entropy is a measure of disorder or randomness. Both enthalpy and entropy are state functions. For any process at constant pressure and temperature, the change in free energy is given by $\Delta G = \Delta H - T\Delta S$ [3]. The nearest-neighbor model is one method that is used to calculate free energy. The method is considered to be computationally feasible while still being sufficiently accurate.

Sung and Lee [31] give preference to probes with the highest free energy in their FindProbe algorithm to ensure that the probes have minimal secondary structure. Their sensitivity filter eliminates probes determined to have secondary structure. This determination is made in a single pass of each probe where they check each length-x $3'$ end of the probe to ensure that there are no more than y consecutive complementaries with the $5'$ end of the probe [31].

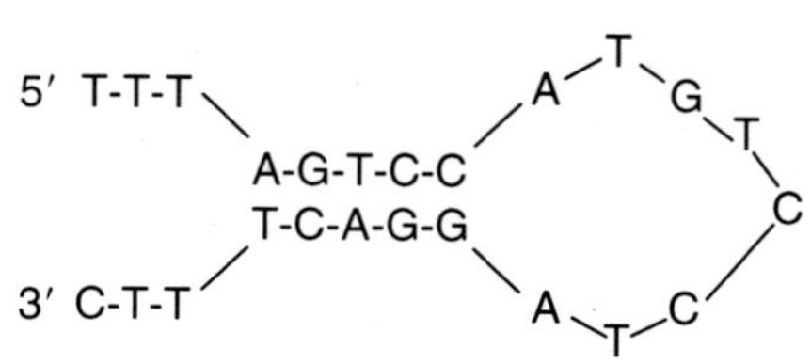

Figure 13.5 Probe TTTAGTCCATGTCCTAGGACTTTC could fold back on itself.

A program called mfold is available to predict secondary structure of DNA and RNA molecules from their sequence. It calculates thermodynamic properties using the base-stacking model. Mfold accepts as input a DNA or RNA sequence and returns, among other information, the free energy calculations and drawings of folded input molecules [30].

Li and Stormo [16] developed a program called ProbeSelect to design and select optimal DNA probes for gene expression arrays. Their approach works in two major steps. The first step consists of finding, for each gene, a set of candidate probes that will maximize the minimum number of mismatches to every other gene in the genome. The second step involves selection based upon free energy. Optimal probes are selected so that they have free energy for the intended target in an acceptable range, and a maximum difference in free energy for every other mismatched target [16].

In 2006, Pozhitkov et al. [23] published results of experiments on eukaryotic target sequences in which they compared current approaches to predicting fluorescent signal intensities to actual intensities. Their results did not support the use of thermodynamic properties to accurately predict signal intensitiy values of duplexes with rRNAs on oligonucleotide DNA microarrays. On the basis of their results, they recommended that thermodynamic criteria not be used for the design of oligonucleotide probes for species identification. They suggested instead that each probe be empirically verified to provide the best signal intensities [23].

13.2.3 Homogeneity

Isothermal or homogenous probes are selected so that they have melting temperatures that are close to the experiment temperature. This enables more uniform performance amongst all of the probes [31]. The melting temperature, denoted T_{m}, is the temperature at which half the strands are in the double-helical state and half are in the random-coil state [28]. The salt concentration of the solution and the base composition of the DNA both affect the melting temperature. DNA with many G–C pairs (high GC content) has a higher melting temperature than DNA with more A–T pairs. G–C pairs have three hydrogen bonds while A–T pairs have two hydrogen bonds [20,21].

For self-complementary oligonucleotide duplexes, T_{m} is predicted using the nearest-neighbor model as reported by SantaLucia in Refs. [28,29] by the equation

$$T_{\mathrm{m}} = \frac{\Delta H^o}{\Delta S^o + R \log C_{\mathrm{T}}}$$

where C_{T} is the total oligonucleotide strand concentration and R is the molar gas constant (1.987 cal/kmol). For non-self-complementary molecules, C_{T} is replaced by $\frac{C_{\mathrm{T}}}{4}$ [28,29].

In Ref. [31], Sung and Lee calculate the melting temperature of each probe p using a formula that is based on nearest-neighbor parameters but has a slight variation:

$$T_{\mathrm{m}}(p) = \frac{\Delta H(p)}{\Delta S(p) + R \log (C_{\mathrm{T}})} - 273.15 + 16.6 \log (\mathrm{Na}^{+})$$

where Na^{+} is the salt concentration of the solution. C_{T} is replaced by $\frac{C_{\mathrm{T}}}{4}$ when the oligonucleotides are not self-complementary. $\Delta H(p)$ and $\Delta S(p)$ represent the enthalpy and entropy for the helix formation of p, respectively. They then calculate the optimal hybridization temperature T_{h} of p by

$$T_{\mathrm{h}}(p) = T_{\mathrm{m}}(p) - 25 - 0.62(C_{\mathrm{F}})$$

where C_{F} is the formamide concentration of the solution. Finally, Sung and Lee require that the content of any single base do not exceed 50 percent [31].

Wang and Seed [34] require that the melting temperature of all probes fall within a given narrow range. They calculate the median melting temperature using the formula

$$\mathrm{Median} = 64.0 + 41 \frac{\mathrm{gcCount}}{\mathrm{oligoLength}} - \frac{600}{\mathrm{oligoLength}}$$

where gcCount is the number of Gs and Cs in the probe and the molar sodium content is 0.1 M. Probe candidates are discarded if their melting temperature is not within 5°C of the median [34].

In Ref. [30], it is noted that for determining the melting temperature and the stability of a nucleic acid duplex, the base-stacking model, which bases calculations on each base pair, supersedes the use of base composition. The base-stacking model does implicitly include base composition (GC content), but it is more complex because it also considers the order of the bases in the sequence.

13.2.4 Probe Length

To distinguish between perfectly matched duplexes and single- or two-base mismatches, short oligonucleotide probes are considered to be most suitable. If short probes are used, it is, however, common to use several different probes corresponding to a single gene to enhance the reliability of the hybridization signal [33].

Long probes tend to have more reliable hybridization properties, but the greater length increases the chance of nonspecific cross-hybridization. Probes of length 50–60 show an improved specificity and sensitivity when compared to shorter probes; however, they may not reliably distinguish single-base mismatches. In short, the intended application must be considered when determining the best probe length [33].

Chou et al. [2] have studied the effect of varying probe length and the number of probes per gene on microarray analysis. Their results suggest that probes of length approximately 150 m are optimal for accurate microarray measurement of gene expression. They also found that both short or long probes worked well if either multiple probes were used for each gene, or the probes were selected via validation by experimental hybridization [2].

13.2.5 Probe Selection Problems

In this section, we discuss several specific probe selection problems.

13.2.5.1 Oligonucleotide Fingerprinting

Oligonucleotide fingerprinting is a method for identification of cDNA or genomic DNA sequences where a vector of numeric values is assigned to the sequence according to hybridization signals obtained by hybridizing to a set of short oligonucleotide probes [11]. In this section, we discuss several papers whose focus is on the selection of probe sets for this application.

In Ref. [1], Borneman et al. discuss an application of microarrays in which a single probe is applied to a DNA microarray containing a large sample of rDNA sequences from the population being studied. This is quite different from the procedure discussed in Section 13.1.1 where a microarray contains a large number of probes. One use of this method is in the analysis of microbial communities where multiple experiments are performed using a single probe for each experiment. Clearly, the number of experiments and, as a result, the cost, are directly related to the number of probes used [1].

The probe selection problem involves the selection of a set S of probes of length l, which will be used to analyze a population $\mathcal{C}$ of m unknown clones. In the problem addressed by Borneman et al. in Ref. [1], the clones are approximately of length 1500 and the probes are of length 6–10.

A probe p is said to distinguish a pair of clones c and d if p is a substring of exactly one of c or d. The goal is to select a minimal cardinality set of probes such that each distinct pair of clones in $\mathcal{C}$ is distinguished by at least one probe in S [1].

In a hybridization experiment, the fluorescence response is linear with respect to the number of occurrences of the probe in a clone up to a certain value R. Because of this, there are different versions of the distinguishability criteria [1]. Borneman et al. consider two cases, $R = 1$ and $R = 4$, referred to as binary and nonbinary distinguishability, respectively. For example, say

$c = \text{ATTACATT}$ and $d = \text{CAATTAGGT}$. If $R = 1$, then GGT distinguishes c and d, but because ATT appears at least once in both c and d, it does not. If, however, $R = 4$, ATT does distinguish between c and d.

For a given set S of m probes, the S-fingerprint of a clone c, which is denoted $\text{fingerprint}_S(c)$, is a vector of length m which contains one entry for each probe p. Each entry denotes the minimum of the value R and the number of occurrences of the probe p in c. Note that if $R = 1$, the fingerprint will just be a binary vector with a one in position i if probe p_i is a substring of clone c, and a zero otherwise.

A set of probes S is said to distinguish two clones c and d if $\text{fingerprint}_S(c) \neq \text{fingerprint}_S(d)$. We denote by $\Delta_S \subseteq \mathcal{C}^2$ the set of pairs of clones that are distinguished by the set S, and so $|\Delta_S|$ denotes the number of distinct pairs of clones distinguished by the probe set S [1].

As noted above, the rDNA sequences of the population being studied are unknown. Therefore choosing probes to distinguish pairs of clones can be a problem. To overcome this problem, a subset $\mathcal{C}'$ of rDNA clones is randomly selected from the population. A probe set is then found for the subset $\mathcal{C}'$ and used to analyze the population [1].

In Ref. [1,18], two variations of the probe selection problem are considered. They are referred to as the Maximum Distinguishing Probe Set (MDPS) and the Minimum Cost Probe Set (MCPS) problems and are defined as follows.

Maximum Distinguishing Probe Set (MDPS)

Instance: A set $\mathcal{C} = \{c_1, c_2, \ldots, c_m\}$ of clones, a set $\mathcal{P} = \{p_1, p_2, \ldots, p_n\}$ of probes, and an integer k.

Solution: A subset $S \subseteq \mathcal{P}$, with $|S| = k$.

Measure: $|\Delta_S|$, to be maximized [1].

Minimum Cost Probe Set (MCPS)

Instance: A set $\mathcal{C} = \{c_1, c_2, \ldots, c_m\}$ of clones and set $\mathcal{P} = \{p_1, p_2, \ldots, p_n\}$ of probes.

Solution: A subset $S \subseteq \mathcal{P}$ such that $\Delta_S = \mathcal{C}^2$.

Measure: $|S|$, to be minimized [1].

Both problems MDPS and MCPS are NP-hard when the length of probes is unbounded [1]. MCPS is a special case of the Set Cover Problem.

In Ref. [1], Borneman et al. applied a simulated annealing algorithm to the MDPS problem and a Lagrangian relaxation algorithm to the MCPS problem. Both methods produced successful results.

Meneses et al. [18] addressed the MDPS and MCPS problems using a method called the Asynchronous Teams method or A-Team method. The A-Team method was proposed by Souza and Talukdar [4]. An A-Team is comprised of several different heuristic algorithms, called agents, that

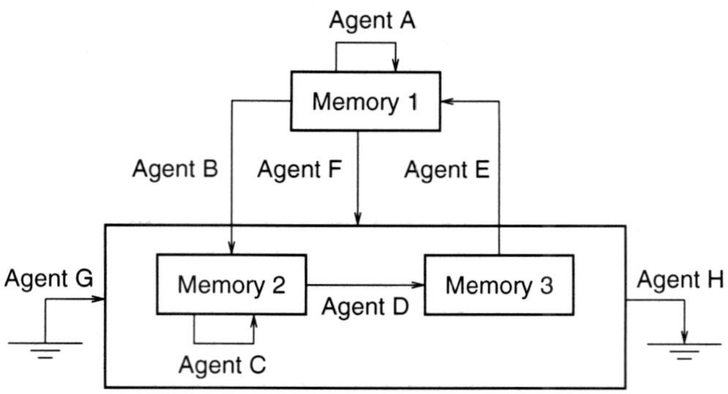

Figure 13.6 A-Team configuration.

communicate with each other by means of shared memories. The shared memories store solutions generated by agents. Each agent can make its own decisions about inputs, scheduling, and resource allocation. Figure 13.6 illustrates a generic A-Team configuration.

Meneses et al. designed agents (i.e., algorithms) for the MDPS and MCPS problems. In Ref. [1], the probe sets only contained probes of a fixed length. The algorithms designed by Meneses et al. were able to construct solutions with probes of mixed length as well as fixed lengths. They tested their approach using both real and simulated data. Their results showed that their method was able to obtain results that were comparable to those found by Borneman et al. but in a small fraction of the time.

In 2005, Fu et al. developed an improved probe selection method [6]. They found that probe sets selected with the currently used method were theoretically optimal; however, in actual biological experiments probes often did not hybridize in a consistent and predictable manner. They referred to these probes as unreliable.

Two common errors in hybridization experiments are false negatives, where probes are expected to bind with a clone but the signal intensity is low; and false positives, where probes are expected not to bind with a clone but the signal intensity is high. The occurrence of both types of errors may be related to the location of the probe target sites. The basic idea of the approach of Fu et al. [6] lies in the hypothesis that the reliability of the probe is related to its location in the clone. Their method used a probabilistic model to identify unreliable probes and eliminate them. Their results showed that application of this method significantly decreased the number of unreliable probes; in fact, 90.9 percent of unreliable probes can be eliminated [6].

Herwig et al. [11] proposed the design of probe sets for simultaneous identification of sequences by oligonucleotide fingerprinting based upon maximization of Shannon entropy. Given a set of M sequences, $s_1, \ldots, s_M$, and a single probe p, the set of sequences is partitioned into two subsets, those that match the probe sequence or its reverse complement, and those that do not. In information theory, the concept of entropy is directly related to the measure of the amount of information contained in a signal. In Ref. [11], the amount of information of a probe with respect to a set of sequences is measured by

$$-I = \sum_{i=1}^{N} p_i \log_2 p_i$$

where p_i is the proportion of sequences in subset i, and N is the number of subsets. Note that the entropy is maximized when the subsets are of equal size. Probe sets are constructed by successively selecting probes that partition sets of training sequences into subsets that are as equal in size as possible. Herwig et al. also consider the G–C content and complexity of the probes. They developed a simulation pipeline to assess the quality of the probe sets and determined that their method produced probe sets that were superior to those chosen according to maximum frequency or by a random process [11].

In Ref. [25], Rash and Gusfield considered the string barcoding problem and its application in selecting probes for hybridization experiments used to identify viral-size pathogens. The barcode they refer to is equivalent to a binary oligonucleotide fingerprint. The problem they consider is as follows. Given a set S of m strings, $S = s_1, \ldots, s_m$, find a set of substrings $P = p_1, \ldots, p_n$ of minimum cardinality such that every pair of strings in S has at least one substring in P that distinguishes the pair. Once P is found, a binary fingerprint or barcode, as described at the beginning of this section, is associated with each string in S. For a given string $s_i \in S$, the set of strings from P that are substrings of s_i is called the signature of s_i. Both the barcode and the signature should be unique for each string [25].

In their approach to the string barcoding problem, Rash and Gusfield construct and solve an integer linear program (ILP) that contains one binary variable for each substring and one inequality for each pair of strings in S. Let $i = 1, \ldots, q$ denote the pairs of the set S. The variable v_j is included in each equation where p_j distinguishes a given string pair. For example, the equation for the pair $s_1, s_2 \in S$ where $p_1, \ldots, p_k$ is the set of substrings that distinguish s_1 and s_2 and $v_1, \ldots, v_k$ is the corresponding set of variables representing the substrings is given by $v_1 + v_2 + \ldots v_k \geq 1$. Assume we have N candidate substrings and let T_i denote the set of indices of substrings in P that distinguish the pair i. We then have the following integer program.

$$\text{Min} \sum_{j=1}^{N} v_j$$

s.t.

$$\sum_{j \in T_i} v_j \geq 1 \quad \text{for all } i$$

$$v_j \in \{0, 1\} \quad \text{for all } j$$

Rash and Gusfield use suffix trees [8] to construct the initial set of candidate substrings. By using suffix trees instead of enumerating all possible distinct substrings, they begin with a much smaller set of substrings, i.e., considerably fewer variables in the ILP. They further reduce the number of variables by filtering out substrings that do not satisfy the valid length requirements [25].

Rash and Gusfield also addressed problems that may occur if the virus mutates in nature after a signature is developed and the mutated version must still be identified. To do so, they added two sets of constraints. In the first set of constraints, instead of requiring that each pair of strings in S be distinguished by at least one substring in P, they require a minimum of r substrings for each pair. That is,

$$\sum_{j \in T_i} v_j \geq r \quad \text{for all } i.$$

They found a value of $r = 5$ to be sufficient to ensure with high confidence that a signature would remain valid under reasonably high mutation rates and short generation times.

The second set of constraints enforced a minimum edit distance between pairs of strings. So, for each pair of substrings p_i and p_j that are not the minimum edit distance apart, only one of the strings is allowed to be in the solution set. To enforce this requirement, the constraint

$$p_i + p_j \leq 1$$

is added for each such pair. They determined that edit distances as small as two or four resulted in added confidence that mutations in nature not invalidate the signatures. They solved the resulting ILP using CPLEX. Viral sequences from Genbank were used to test their approach. Their approach finds provable optimum in many cases, and in even more cases it can get provably close to optimum. Their implementation was able to handle data sets containing 150–300 strings per set [25].

13.2.5.2 Nonunique Probe Selection Problem

If the target genes in a microarray experiment are closely related, it can be very difficult to find unique probes. In these cases, the use of nonunique probes that bind to more than one target may be necessary.

In Ref. [13], Klau et al. consider the selection of nonunique probes for use in experiments to detect the presence or absence of viruses or bacteria in a biological sample. They begin with a zero-one target-probe incidence matrix $H = [h_{ij}]$ where $h_{ij} = 1$ if probe j hybridizes to target i, and $h_{ij} = 0$ otherwise. The goal is to find a probe set of minimum size such that every target is covered by at least c_{min} probes and every target pair is distinguished by at least h_{min} probes. They find an initial solution using a heuristic algorithm. They then construct an ILP and use CPLEX to solve the ILP, thus further reducing the size of the probe set. In the description of an ILP that they refer to as the slave ILP, they propose a branch-and-cut approach in which constraints are added to enforce group separation in addition to pairwise separation. The results presented in the paper are, however, strictly for pairwise separation. Finally, they evaluate their results using decoding software [13].

Meneses et al. [19] developed a heuristic algorithm for the nonunique probe selection problem addressed in Ref. [13]. The algorithm begins with the target-probe incidence matrix described above and selects a minimum-sized probe set that satisfies the ILP formulation used by Klau et al.

$$\text{Min} \sum_{j=1}^{n} x_j$$

s.t.

$$\sum_{j=1}^{n} h_{ij}x_j \geq c_{min} \qquad \text{for all } i \in M \qquad \text{[Coverage]}$$

$$\sum_{j=1}^{n} |h_{ij} - h_{kj}|x_j \geq h_{min} \qquad \text{for all } (i,k) \in P \qquad \text{[Hamming distance]}$$

$$x_j \in \{0,1\} \qquad j = 1, \ldots, n,$$

where $|h_{ij} - h_{kj}|$ in the Hamming distance constraints stands for the absolute value of the difference between the real numbers h_{ij} and h_{kj}. Note that h_{ij} and h_{kj} are constants. This definition of Hamming distance is different from the Hamming distance used in the Sung and Lee paper discussed in Section 13.2.1.

The algorithm in Ref. [19] constructs feasible solutions to the ILP in two phases, a construction phase and a reduction phase. The construction phase ensures that a feasible solution is constructed, if possible, and the reduction phase reduces the size of the solution while maintaining feasibility [19].

In the construction phase, the set cover constraint is first satisfied for each target by adding probes until every target is covered by at least c_{min} probes. The algorithm next focuses on satisfying the minimum Hamming distance constraint. Probes are ordered by the number of target pairs separated by each probe and are then selected from this list until the number of target pairs failing to satisfy the minimum Hamming distance constraint reduces to a value input by the user. The algorithm then moves to a specific search for probes to separate the target pairs. Only pairwise separation of targets is considered. In the reduction phase, the probe set is reduced by selecting two probes at a time to be deleted. If the solution can remain feasible by adding zero or one probe to replace the two, the solution is updated, otherwise it remains the same.

Gasieniec et al. address probe selection in microarray design in Ref. [7]. In their approach, they first search for unique probes for an input set of target sequences. If it is not possible to find a unique probe for a given sequence, they search for a small collection of probes that together uniquely identify the sequence. They start by first using a filtering process on the whole genome to narrow down the list of candidate probes. The filtering process checks for GC content, content of single bases, length of contiguous regions of a single nucleotide, homogeneity, sensitivity, and specificity. Randomization is used to speed up the process. They developed a software implementation of their method called RandPS.

13.3 Software

There are several software packages available for the design and selection of hybridization probes. In previous sections we discussed the following programs: Sung and Lee's FindProbe; Wang and Seed's OligoPicker; Li and Stormo's ProbeSelect; and Gasieniec et al.'s RandPS. We now consider a few more probe design and selection programs.

Primer3 is software that generates both oligonucleotide probes and primer probes according to parameters entered by the user such as melting temperature, length, and GC content [27]. It is available in both an online version and a downloadable version. The online version can only work with one sequence at a time and will not do batch processing. The downloadable version does, however, allow for batch processing. Primer3 was developed at Whitehead Institute and Howard Hughes Medical Institute. The main authors of the current version are Steve Rozen and Helen Skaletsky.

Rimour et al. [26] developed a very different method for the design of microarray probes. In their approach, a probe is composed of two noncontiguous subsequences of the target sequence. The two sequences are joined by a small linker composed of randomly chosen bases. Their method is implemented in software called GoArrays.

OligoWiz 2.0 [35] is another tool developed for the design of microarray probes, but it sets itself apart from the other design tools. In addition to the usual design parameters considered such as melting temperature and GC content, their method allows for integration of sequence annotation, such as exon/intron structure and untranslated regions (UTRs).

Eric Nordberg designed YODA (Yet another Oligonucleotide Design Application) to design micorarray probes [22]. Rather than rely on the use of BLAST for sequence similarity searches, as do many software packages, YODA incorporates a custom sequence similarity search. It also allows for multiple design goals such as single-genome, multiple-genome, pathogen-host, and species or strain-identification.

13.4 Closing Remarks

We have defined and discussed various properties of probes that can be calculated by computational methods given the probe sequence. We have also considered several different algorithms and software packages that have been developed for the optimal design and selection of probes. Although there are limitations to what can be accomplished at the computational level, the value of computational methods in molecular biology cannot be overstated. It is clear that the partnership between the biological and computational sciences will pave the way for future research and discoveries.

References

1. J. Borneman, M. Chroback, G.D. Vedova, A. Figueroa, and T. Jiang. Probe selection algorithms with applications in the analysis of microbial communities. *Bioinformatics*, 17:S39–S48, 2001.
2. C. Chou, C. Chen, T. Lee, and K. Peck. Optimization of probe length and the number of probes per gene for optimal microarray analysis of gene expression. *Nucleic Acids Research*, 32(12):e99, 2004.
3. P. Clote and R. Backofen. *Computational Molecular Biology*. John Wiley and Sons Ltd, West Sussex, England, 2000.
4. P. S. de Souza and S. N. Talukdar. Asynchronous organizations for multialgorithm problems. ACM Symposium on Applied Computing, Indianapolis, 1993.
5. A.M. Frieze, F.P. Preparata, and E. Upfal. Optimal reconstruction of a sequence from its probes. *Journal of Computational Biology*, 6(3):361–368, 1999.

6. Q. Fu, J. Borneman, J. Ye, and M. Chrobak. Improved probe selection for DNA arrays using nonparametric kernel density estimation. In Proceedings of the 2005 IEEE Engineering in Medicine and Biology 27th Annual Conference, Shanghai, China, 2005.
7. L. Gasieniec, C.Y. Li, P. Sant, and P.W.H. Wong. Efficient probe selection in microarray design. 2006 IEEE Symposium on Computational Intelligence in Bioinformatics and Computational Biology, Toronto, Ontario, Canada, 2005.
8. D. Gusfield. *Algorithms on Strings, Trees, and Sequences.* Press Syndicate of the University of Cambridge, Cambridge, England, 1997.
9. E. Halperin, S. Halperin, T. Hartman, and R. Shamir. Handling long targets and errors in sequencing by hybridization. *Journal of Computational Biology*, 10(3):176–185, 2003.
10. G. Held, G. Grinstein, and Y. Tu. Relationship between gene expression and observed intensities in DNA microarrays—A modeling study. *Nucleic Acids Research*, 34(9):e70, 2006.
11. R. Herwig, A. Schmitt, M. Steinfath, J. O'Brien, H. Seidel, S. Meier-Ewert, H. Lehrach, and U. Radelof. Information theoretical probe selection for hybridisation experiments. *Bioinformatics*, 16:890–898, 2000.
12. N.C. Jones and P.A Pevzner. *An Introduction to Bioinformatics Algorithms.* MIT Press, Cambridge, MA, 2004.
13. G.W. Klau, S. Rahmann, A. Schliep, M. Vingron, and K. Reinert. Optimal robust non-unique probe selection using integer linear programming. *Bioinformatics*, 20:i186–i193, 2004.
14. R. Koehler and N. Peyret. Effects of DNA secondary structure on oligonucleotide probe binding efficiency. *Computational Biology and Chemistry*, 29:393–397, 2005.
15. J.K. Lanctot, M. Li, B. Ma, S. Wang, and L. Zhang. Distinguishing string selection problems. In *Proceedings of the Tenth Annual ACM-SIAM Symposium on Discrete Algorithms*, pp. 633–642. Society for Industrial and Applied Mathematics, Philadelphia, PA, 1999.
16. F. Li and G. Stormo. Selection of optimal DNA oligos for gene expression arrays. *Bioinformatics*, 17:1067–1076, 2001.
17. M. Marmur and P. Doty. Thermal renaturation of deoxyribonucleic acids. *Journal of Molecular Biology*, 3:585–594, 1961.
18. C.N. Meneses, P.M. Pardalos, and M.A. Ragle. Asynchronous teams for probe selection problems. In review, 2005.
19. C.N. Meneses, P.M. Pardalos, and M.A. Ragle. A new approach to the non-unique probe selection problem. *Annals of Biomedical Engineering*, 35(4):651–658, 2007.
20. O. Milenkovic and N. Kashyap. DNA codes that avoid secondary structure. In Proceedings of the 2005 IEEE International Symposium on Information Theory, Adelaide, South Australia, Australia, 2005.
21. R. Murray, D. Granner, P. Mayes, and V. Rodwell. *Harper's Illustrated Biochemistry*, 26th ed. McGraw-Hill Companies, New York, 2003.
22. E. Nordberg. YODA: selecting signature oligonucleotides. *Bioinformatics*, 21:1365–1370, 2005.
23. A. Pozhitkov, P. Noble, T. Domazet-Loso, A. Nolte, R. Sonnenberg, P. Staehler, M. Beier, and D. Tautz. Tests of rRNA hybridization to microarrays suggest that

hybridization characteristics of oligonucleotide probes for species discrimination cannot be predicted. *Nucleic Acids Research*, 34(9):e66, 2006.
24. J.B. Rampal, editor. *DNA Arrays: Methods and Protocols*. Humana Press Inc., Totowa, NJ, 2001.
25. S. Rash and D. Gusfield. String barcoding: Uncovering optimal virus signatures. In S. Istrail P. Perzner G. Meyers, S. Hannenballi and M. Waterman, editors, *Proceedings of the Sixth Annual International Conference on Computational Biology*, Washington D.C., pp. 254–261, April 2002.
26. S. Rimour, D. Hill, C. Militon, and P. Peyret. Goarrays: highly dynamic and efficient microarray probe design. *Bioinformatics*, 21(7):1094–1103, 2005.
27. S. Rozen and H. Skaletsky. Primer3 on the www for general users and for biologist programmers. In S. Krawetz and S. Misener, editors, *Bioinformatics Methods and Protocols: Methods in Molecular Biology*, pp. 365–386. Humana Press, Totowa, NJ, 2000.
28. J. SantaLucia. A unified view of polymer, dumbbell, and oligonucleotide DNA nearest-neighbor thermodynamics. In *Proceedings of the National Academy of Science*, U.S.A., volume 95, pp. 1460–1465, 1998.
29. J. SantaLucia, H. Allawi, and P. Seneviratne. Improved nearest-neighbor parameters for predicting DNA duplex stability. *Biochemistry*, 35:3555–3562, 1996.
30. D. Stekel. *Microarray Bioinformatics*. Cambridge University Press, Cambridge, England, 2003.
31. W. Sung and W. Lee. Fast and accurate probe selection algorithm for large genomes. In Proceedings of the Computational Systems Bioinformatics. IEEE Computer Society, Stanford University, California, 2003.
32. N. Templeton, editor. *Gene and Cell Therapy: Therapeutic Mechanisms and Strategies, 2nd ed.* Marcel Dekker, Inc., New York, 2004.
33. S. Tomiuk and K. Hofmann. Microarray probe selection strategies. *Briefings in Bioinformatics*, 2(4):329–340, 2001.
34. X. Wang and B. Seed. Selection of oligonucleotide probes for protein coding sequences. *Bioinformatics*, 19(7):796–802, 2003.
35. R. Wernersson and H.B. Nielsen. Oligowiz 2.0—integrating sequence feature annotation into the design of microarray probes. *Nucleic Acids Research*, 33 (web server issue):W611–W615, 2005.

Chapter 14

Implementation of Logical Analysis of Data for Oligo Probe Selection

In-Yong Jang, Kwangsoo Kim, and Hong Seo Ryoo

CONTENTS

14.1 Introduction 416
14.2 Proposed Probe Selection Method 419
14.2.1 Data Binarization 419
14.2.2 Pattern Generation 420
14.2.3 Classification Rules 426
14.3 *In Silico* Experiments 427
14.3.1 A Comparative Experiment: Classification of High and Low Risk HPV 429
14.3.2 Experiments on Genotyping Viral Pathogens 430
14.4 Concluding Remarks 436
References 436

Abstract Specializing a general framework of logical analysis of data for efficiently handling large-scale genomic data, we develop in this chapter a probe design method for selecting short oligo probes for genotyping applications. Extensively tested on genomic sequences obtained from the National Center of Biotechnology Information (NCBI) in various monospecific and polyspecific *in silico* experiments, the proposed probe design method was able to select a small number of oligo probes of length 7 or 8 nucleotides that perfectly classified all unseen testing sequences. These results demonstrate the efficacy of the proposed probe design method and illustrate the usefulness and potential a well-designed optimization-based probe selection method has in genotyping applications.

14.1 Introduction

Between November 1, 2002 and July 31, 2003, severe acute respiratory syndrome (SARS) virus infected 8096 people and proved fatal to 774 worldwide.* The avian influenza (AI) virus subtype H5N1 alone infected 152 people worldwide between 2003 and November 2006, and 154 died of the disease.† Luckily, none of the outbreaks of SARS and AI infections at the beginning of the new millennium brought about the worst-case scenario. Alarmingly, influenza experts seem to agree that another pandemic may be imminent [34] and, as of this writing, a fearful AI H5N1 virus continues to spread in part of Asia.

A microarray or a DNA chip is a small glass or silica surface bearing DNA probes. Probes are single-stranded reverse transcribed mRNAs, each located at a specific spot of the chip for hybridization with its Watson–Crick complementary sequence in a target to form the double helix [25,28]. Microarrays currently use two forms of probes, namely, oligonucleotide (shortly, oligo) and cDNA, and have prevalently been used in the analysis of gene expression levels, which measures the amount of gene expression in a cell by observing hybridization of mRNA to different probes, each targeting a specific gene. With the ability to identify a specific target in a biological sample, microarrays are also well suited for detecting biological agents for genetic and chronic disease [7,10,15,17]. Furthermore, as viral pathogens can be detected at the molecular and genomic level much before the onset of physical symptoms in a patient, the microarray technology can be used for an early detection of patients infected with viral pathogens [26,31,32,35].

The success of microarrays depends on the quality of probes that are tethered on the chip. Having an optimized set of probes is beneficial for two obvious reasons. One, the background hybridization is minimized, hence true gene expression levels can be more accurately determined [16]. The other, as the number of oligos needed per gene is minimized, the cost of each microarray is minimized or the number of genes on each

* From http://www.who.int/csr/sars/country/table2004_04_21/en/index.html, accessed on December 11 2006.

† From http://www.who.int/csr/disease/avian_influenza/country/cases_table_2006_01_25/en/index.html, accessed on December 11, 2006.

chip is increased, yielding oligo fingerprinting, a much faster and more cost-efficient technique [2,16]. Short probes consisting of 15–25 nucleotides (nt), usually, are used in genotyping applications [28]. Having short optimal probes means a high genotyping accuracy in terms of both sensitivity and specificity [16,26], hence can play a key role in genotyping applications. For example, in a pandemic, an effective method for selecting short optimal probes may be used in the mass production of a cost-efficient device for screening for the disease in suspected or susceptible hosts. Reverse genetics would be the most rapid means by which to produce an antigenically matched vaccine in a pandemic [34]. An effective probe selection methodology can identify conserved regions of a viral family, hence may prove useful in the preparation of a vaccine via reverse genetics. Furthermore, the methodology can promote the availability of affordable home testing kits for accurate and confidential diagnosis of genetic and infectious disease and allow advanced and adequate medical treatment planning for patients.

A well-studied problem in machine learning and data mining deals with the discovery of a classification rule for different types of data. The probe design, say, for genotyping applications, can be roughly stated as selecting oligo probes for detecting a specific disease-agent in genomic sequences, hence falls into the realm of classical classification. Thus far, this interesting problem at the intersection of molecular biology and optimization has received relatively little attention from the optimization community, and systematic oligo design methods proposed so far are based on a simple greedy procedure [11], the set covering (SC)–based classification methodology [2,23], support vector machines [15], a genetic algorithm [7,14], and mixed integer and linear programming (MILP) [12], briefly summarizing.

From the perspective of numerical optimization, genomic data presents an unprecedented challenge for supervised learning approaches for a number of reasons. To name a few, first, genomic data is long sequence over the nucleic acid alphabet $\Sigma = \{A,C,G,T\}$. Second, for example, the complexity of viral flora, owing to constantly evolving viral serotypes, requires a supervised learning theory to be trained on a large collection of target and nontarget samples. That is, a typical training set contains a large number of large-scale samples. Furthermore, a supervised learning framework usually requires a systematic pairing or differencing between each target and nontarget samples during the course of training a decision rule [2,3,12,23]. Owing to these and the nature of general data analysis and classification [19], a supervised learning approach to classification of genomic data without specialized features for efficiently handling large-scale data is confronted by a formidable challenge.

On the basis of a general framework of logical analysis of data (LAD) from Ref. [24], we develop in this chapter a probe design method for selecting short oligo probes of length l nt, where $l \in [6,10]$. To list some advantages of selecting oligo probes by the proposed method, first, the method selects probes via sequential solution of a small number of compact SC instances, which offers a great advantage from computational point of view. To be more specific, consider classification of two types of data and suppose that a training set is comprised of m^+ target and m^- nontarget sequences. The size of the SC training instances solved by the proposed method is minimum of m^+ and m^- orders of magnitude smaller than optimization learning models used in Refs. [2,12,23], for instance. Second, the method uses the sequence information only and selects probes via optimization based on principles of probability and statistics. That is, the probability of an l−mer (oligo of length l) appearing in a single sequence by chance is $(0.25)^l$. Unless statistically significant, an l−mer appearing in multiple samples of one type and none or only a few of the sequences of the other type by chance is extremely small. Third, the proposed method does not rely on any extra tool, such as BLASTn [1], a local sequence alignment search tool that is commonly used for probe selection [26,32,33], or the existence of preselected representative probes [26]. This makes the method truly stand-alone and free of problems that may possibly be caused by limitations associated with external factors. As mentioned earlier, the proposed probe design selects optimal probes via sequential solution of SC instances. Although SC is $\mathcal{NP}$−complete [8], its wide practical applications have invited an array of efficient (meta-)heuristic solution procedures to be developed. Therefore, last, the proposed method is readily implementable for efficient selection of oligo probes.

This chapter is organized as follows. In Section 14.2, we specialize a LAD Framework from Ref. [24] for efficiently analyzing genomic sequences and develop an effective method for selecting short oligo probes. In Section 14.3, we test the proposed probe design algorithm in various *in silico* genotyping experiments using viral genomic sequences and report superb experimental results. To summarize, in all monospecific and polyspecific genotyping experiments on classification of viral pathogens using genomic sequences obtained from the National Center of Biotechnology Information (NCBI) Web site, the proposed probe design method selected a small number of probes of length 7 or 8 nt that perfectly classified all unseen testing sequences. Classifying the noisy human papillomavirus (HPV) sequences from the Los Alamos Laboratory by high and low risk types, the proposed probe design method selected optimal probes in a few CPU seconds that classified the testing sequences with 90.6 percent accuracy. For comparison, Refs. [7,22] experimented with the same HPV dataset and reported the classification accuracy of 85.6 percent and 81.1 percent, respectively. These *in silico* results demonstrate efficacy and efficiency of the proposed oligo design method and further illustrate the usefulness and potential of a well-designed optimization-based probe design method in the forthcoming era of biotechnology. Finally, Section 14.4 concludes the chapter with a few remarks.

Before proceeding, we refer interested readers to refer [25,28,31] for background in microarray analysis and its usage in the diagnosis of infectious disease. Furthermore, as classification of more than two types of data can be accomplished by sequential classification of two types of data (see Refs. [6,29,30] and Section 14.3 below), we present the material below in the context of the classification of $+$ and $-$ types of data for convenience and without loss of generality.

14.2 Proposed Probe Selection Method

The backbone of the proposed procedure is LAD. LAD is a relatively new supervised learning methodology that is based on Boolean logic, combinatorics, and optimization. A typical implementation of LAD analyzes data on hand via four sequential stages of data binarization, support feature selection, pattern generation, and classification rule formation. As a Boolean logic-based, LAD first converts all nonbinary data into equivalent binary observations. A $+$ $(-)$ 'pattern' in LAD is defined as a conjunction of one or more binary attributes or their negations that distinguishes one or more $+$ $(-)$ type observations from all $-$ $(+)$ observations. The number of attributes used in a pattern is called the degree of the pattern. As seen from the definition, patterns hold the structural information hidden in data. After patterns are generated, they are aggregated into a partially-defined Boolean discriminant function/rule to generalize the discovered knowledge to classify new observations.

Referring readers to Refs. [3,9,24] for more background in LAD, we design a LAD-based method below for efficiently handling and analyzing large-scale genomic data and selecting optimal oligo probes for genotyping applications.

14.2.1 Data Binarization

Let there be m^+ and m^- sample observations of type $+$ (target) and $-$ (nontarget), respectively. For $\bullet \in \{+,-\}$, let us use $\bar{\bullet}$ to denote the complementary element of $\bullet$ with respect to the set $\{+, -\}$. Let $S^\bullet$ denote the index set of $m^\bullet$ sample sequences for $\bullet \in \{+, -\}$.

A DNA sequence is a sequence of nucleic acids A, C, G, and T, and the training sequences need to be converted into Boolean sequences of 0 and 1 before LAD can be applied. Toward this end, we first choose an integer value for l, usually $l \in [6, 10]$ (see Section 14.3), generate all 4^l possible $l-$mers over the four nucleic acid letters and then number them consecutively from 1 to 4^l by a mapping scheme. Next, each $l-$mer is selected in turn and every training sample is fingerprinted with the oligo for its presence or absence. That is, with oligo j, we scan each sequence p_i, $i \in S^+ \cup S^-$, from the beginning of the sequence and shifting to the right by a base and stamp

$$p_{ij} = \begin{cases} 1, & \text{if oligo } j \text{ is present in sequence } i\text{; and} \\ 0, & \text{otherwise.} \end{cases}$$

After this, the oligos that appear in all or none of the training sequences can be deleted from further consideration. We re-number the surviving $l-$mers consecutively from 1 to n and replace the original training sequences described in the nucleic acid alphabets by their Boolean representations. Let $N = \{1, \ldots, n\}$.

14.2.2 Pattern Generation

The data is now described by n attributes $a_j \in \{0, 1\}$, $j \in N$. For observation p_i, $i \in S^\bullet$, $\bullet \in \{+, -\}$, let p_{ij} denote the binary value the jth attribute takes in this observation. Let l_j denote the literal of binary attribute a_j. Then, $l_j = a_j$ ($l_j = \bar{a}_j$) instructs to take (negate) the value of a_j in all sequences. A term t is a conjunction of literals. Given a term t, let $N_t \subseteq N$ denote the index of literals included in the term. Then, we have $t = \bigwedge_{j \in N_t} l_j$. A $\bullet$ pattern is a term that satisfies $t(p_i) := \prod_{\substack{l_j = a_j, \\ j \in N_t}} p_{ij} \prod_{\substack{l_j = \bar{a}_j, \\ j \in N_t}} \bar{p}_{ij} = 1$ for at least one p_i, $i \in S^\bullet$, and $t(p_k) = 0$ for all p_k, $k \in S^{\bar{\bullet}}$. Note here that N_t of a $\bullet$ pattern identifies probes that collectively distinguish one or more $\bullet$ sequences from the sequences of the other type.

To aid in presentation, let us temporarily introduce n additional features a_{n+j}, $j \in N$, and use a_{n+j} to negate a_j. Let $N' = \{1, \ldots, 2n\}$ and let us introduce a binary decision variable x_j for $a_j, j \in N'$, to determine whether to include l_j in a pattern. Ref. [24] formulated a compact MILP Model below with respect to a reference sample p_i, $i \in S^\bullet$, $\bullet \in \{+, -\}$:

$$
(\text{MILP-2.i}^\bullet) \quad \left| \begin{array}{lll}
z_{2.i} = & \text{Min}_{\boldsymbol{x},\boldsymbol{y},d} & \displaystyle\sum_{l \in S^\bullet \setminus \{i\}} y_l \\
 & \text{s.t.} & \displaystyle\sum_{j \in J_i} x_j = d \\
 & & \displaystyle\sum_{j \in J_i} p_{lj} x_j + y_l \geq d, \quad l \in S^\bullet \setminus \{i\} \\
 & & \displaystyle\sum_{j \in J_i} p_{lj} x_j \leq d - 1, \quad l \in S^{\bar{\bullet}} \\
 & & 1 \leq d \leq n \\
 & & \boldsymbol{x} \in \{0, 1\}^n \\
 & & \boldsymbol{0} \leq \boldsymbol{y} \leq \boldsymbol{n},
\end{array} \right.
$$

where $J_i := \{j \in N' : p_{ij} = 1\}$ for p_i, $i \in S^\bullet$. Consider the following.

Lemma 1. *Let $(\boldsymbol{x}, \boldsymbol{y}, d)$ denote a feasible solution of (MILP-2.i$^\bullet$). Let $N_t = \{j \in J_i : x_j = 1\}$. Then,*

$$\mathcal{P} := \bigwedge_{j \in J_i, x_j = 1} a_j$$

forms a $\bullet$ pattern.

Proof. First, via the first constraint of (MILP-2.i$^\bullet$) and the definition of J_i, we trivially have

$$\mathcal{P}(p_i) = \prod_{j \in N_t} p_{ij} = 1$$

for the reference observation p_i, $i \in S^\bullet$. Next, the second set of hard constraints yields that at least one of $p_{lj} = 0$ for $j \in N_t$ for each p_l, $l \in S^{\bar{\bullet}}$. This gives

$$\mathcal{P}(p_l) = \prod_{j \in N_t} p_{lj} = 0$$

for all p_l, $l \in S^{\bar{\bullet}}$, and completes the proof. □

Lemma 1 shows that any feasible solution of (MILP-2.i$^\bullet$) can be used to form a $\bullet$ pattern. Now, note that if $y_l = 0$ for $l \in S^\bullet \setminus \{i\}$ in the solution, then the $\bullet$ pattern $\mathcal{P}$ formed also distinguishes p_l from the $\bar{\bullet}$ observations. Therefore, with the objective of minimizing the sum of y_l's, the MILP Model can be understood as a way to generate a $\bullet$ pattern that distinguishes (more or less) a maximum number of $\bullet$ observations from the $\bar{\bullet}$ observations.

As easily seen, the number of 1's in the (optimal) solution determines the degree of the pattern generated.

As demonstrated in Ref. [24], this model efficiently generates patterns of all degree with equal ease, provided that the number of training samples used is moderate and that n is not a big number. Genomic data is large scale in nature, however. Furthermore, owing to constantly evolving viral serotypes, the complexity of viral flora is high and this requires large numbers of target and nontarget viral samples to be used for selecting optimal genotyping probes. Adding to these, the difficulties associated with numerical solution of MILP in general, we see that (MILP-2.i$^\bullet$) presents no practical way of selecting genotyping probes.

With the need to develop a more efficient pattern generation scheme, we select a reference sequence p_i, $i \in S^\bullet$, $\bullet \in \{+, -\}$, and set

$$a_j^{(i,k)} = \begin{cases} 1, & \text{if } p_{ij} \neq p_{kj}; \text{ and} \\ 0, & \text{otherwise,} \end{cases} \tag{14.1}$$

for $k \in S^{\bar{\bullet}}$ and $j \in N$. Next, we set

$$a_j^{(i,l)} = \begin{cases} 1, & \text{if } p_{ij} = p_{lj}; \text{ and} \\ 0, & \text{otherwise,} \end{cases}$$

for $l \in S^\bullet$ and $j \in N$. Now, consider the SC Model

$$(\mathrm{SC}_i^\bullet) \quad \left| \begin{array}{lll} \mathrm{Min}_{\boldsymbol{x},\boldsymbol{y}} & \sum_{j \in N} c_j x_j + \sum_{l \in S^\bullet \setminus \{i\}} y_l & \\ \text{s.t.} & \sum_{j \in N} a_j^{(i,l)} x_j + y_l \geq 1, & l \in S^\bullet \setminus \{i\} \\ & \sum_{j \in N} a_j^{(i,k)} x_j \geq 1, & k \in S^{\bar{\bullet}} \\ & x_j \in \{0, 1\}, & j \in N \\ & y_l \in \{0, 1\}, & l \in S^\bullet \setminus \{i\}, \end{array} \right.$$

where c_j $(j \in N)$ are positive real numbers (refer to Remark 4 below.)

Theorem 1. *Let $(\boldsymbol{x}, \boldsymbol{y})$ denote a feasible solution of $(\mathrm{SC}_i^\bullet)$. Then,*

$$\mathcal{P} := \bigwedge_{\substack{x_l=1,\\ p_{il}^\bullet=1}} a_l \bigwedge_{\substack{x_l=1,\\ p_{il}^\bullet=0}} \bar{a}_l \tag{14.2}$$

forms a $\bullet$ LAD pattern.

PROOF. To show the result, we need to show that the conjunction of literals formed via Equation 14.2 distinguishes at least one $\bullet$ observation from all $\bar{\bullet}$ observations. Toward the end, recall that $p_{ik} = 1(0)$ indicates the presence (absence) and the absence (presence) of probe k in the reference sequence selected p_i, $i \in S^\bullet$, and in p_k, $k \in S^{\bar{\bullet}}$, respectively. With the cover $(\boldsymbol{x}, \boldsymbol{y})$ of $(\text{SC}_i^\bullet)$ on hand, let us subdivide the index set $N_t = \{j \in N : x_j = 1\}$ into two subsets $N_t^1 := \{j \in N_t : p_{ij} = 1\}$ and $N_t^0 := \{j \in N_t : p_{ij} = 0\}$. Observe now that

$$\mathcal{P}(p_i) = \prod_{l \in N_t^1} p_{il} \prod_{l \in N_t^0} \bar{p}_{il} = 1$$

for p_i, $i \in S^\bullet$, hence $\mathcal{P}(p_i) = 1$ for at least one $\bullet$ observation.

Note in Equation 14.1 that $a_j^{(i,k)} = 1$ if $p_{ij} \neq p_{kj}$ for $k \in S^{\bar{\bullet}}$. That is, $a_j^{(i,k)} = 1$ implies that exactly one of p_{ij} and p_{kj} equals 1 for p_i and p_k, $k \in S^{\bar{\bullet}}$. The cover $(\boldsymbol{x}, \boldsymbol{y})$ of $(\text{SC}_i^\bullet)$ by definition satisfies all constraints of $(\text{SC}_i^\bullet)$, and the hard constraints of the problem in the second set of cover inequalities require that at least one x_l in the cover is set to 1 among $l \in N$ with $a_l^{(i,k)} = 1$ for all $k \in S^{\bar{\bullet}}$. This in turn implies that at least one p_{kl} for $l \in N_t^1$ or $\bar{p}_{kl}$ for $l \in N_t^0$ equals 0 for all p_k, $k \in S^{\bar{\bullet}}$ and yields

$$\mathcal{P}(p_k) = \prod_{l \in N_t^1} p_{kl} \prod_{l \in N_t^0} \bar{p}_{kl} = 0$$

for all p_k, $k \in S^{\bar{\bullet}}$, hence $\mathcal{P}(pk)$ for all $\bar{\bullet}$ observations. □

Note that $\mathcal{P}$ generated on the solution $(\boldsymbol{x}, \boldsymbol{y})$ of $(\text{SC}_i^\bullet)$ via Equation 14.2 also satisfies $\mathcal{P}(p_l) = 1$ for all $l \in S^\bullet \setminus \{i\}$ with $y_l = 0$. The following result is immediate.

LEMMA 2. *With a feasible solution* $(\boldsymbol{x}, \boldsymbol{y})$ *of* $(SC_i^\bullet)$*, let* $N_t = \{j \in N : x_j = 1\}$. *Then,* $y_l = 0$ *for* $l \in S^\bullet \setminus \{i\}$ *if and only if* $p_{lk} = p_{ik}$ *for all* $k \in N_t$.

As (MILP-2.i$^\bullet$), $(\text{SC}_i^\bullet)$ is also formulated in reference to p_i for some $i \in S^\bullet$ and finds a cover that distinguishes most $\bullet$ observations from the $\bar{\bullet}$ observations. Therefore, although not identical, $(\text{SC}_i^\bullet)$ can be seen as an SC version of (MILP-2.i$^\bullet$). Although smaller than the MILP Model by only one constraint and one integer variable, $(\text{SC}_i^\bullet)$ has a much simpler structure and is defined only in terms of 0–1 variables. In addition, owing to having a wide range of practical applications, SC has invited the development of an array of efficient (meta-)heuristic solution procedures ([5] and references therein) and any of these can be used for solving $(\text{SC}_i^\bullet)$ (refer to Remark 1 below.)

From the computational point of view, therefore, ($SC_i^\bullet$) is much preferred over its MILP counterpart.

Note that ($SC_i^\bullet$) is defined by $m^+ + m^- - 1$ cover inequalities and $n + m^\bullet - 1$ binary variables. Also, recall that n is large for genomic sequences and the analysis of viral sequences requires large numbers of target and nontarget sequences, that is, m^+ and m^- are also large numbers. To develop a more compact SC-based probe selection model, we select a reference sequence p_i, $i \in S^\bullet$, $\bullet \in \{+, -\}$, and set the values of $a_j^{(i,k)}$ for $k \in S^{\bar{\bullet}}$ and $j \in N$ via Equation 14.1. Consider the following SC Model:

$$(\text{SC-pg}_i^\bullet) \quad \left| \begin{array}{lll} \text{Min}_{\mathbf{x}} \sum_{j \in N} c_j x_j & & \\ \text{s.t.} \sum_{j \in N} a_j^{(i,k)} x_j \geq 1, & & k = 1, \ldots, m^{\bar{\bullet}} \\ x_j \in \{0, 1\}, & & j \in N. \end{array} \right.$$

where c_j's are positive reals (again, refer to Remark 4 below.)

THEOREM 2. *Let $\boldsymbol{x}$ denote a feasible solution of (SC-pg$_i^\bullet$). Then, $\mathcal{P}$ generated on $\boldsymbol{x}$ via Equation 14.2 forms a $\bullet$ LAD pattern.*

PROOF. Same as the proof for Theorem 1. □

We immediately have the following result that can be used for efficiently identifying the $\bullet$ observations that are also distinguished from the $\bar{\bullet}$ observations by the pattern generated on the solution of (SC-pg$_i^\bullet$).

LEMMA 3. *With a feasible solution $\boldsymbol{x}$ of (SC-pg$_i^\bullet$), generate a $\bullet$ pattern $\mathcal{P}$ via Equation 14.2. Then, $\mathcal{P}$ distinguishes every $\bullet$ sequence p_l, $l \in S^\bullet$, with $p_{lk} = p_{ik}$ for all $k \in N_t$ from the $\bar{\bullet}$ observations, where $N_t = \{j \in N : x_j = 1\}$.*

Note that (SC-pg$_i^\bullet$) can be considered as a relaxation of ($SC_i^\bullet$): to see this, project ($SC_i^\bullet$) onto the space of $\boldsymbol{x}$. Generally speaking, therefore, a feasible solution of ($SC_i^\bullet$) has more x_j's set to 1 in it than in a feasible solution of ($SC_i^\bullet$) formulated on the same data, hence tends to generate a higher degree pattern that generally explains a difference between the target and nontarget sequences. As more $\bullet$ observations are distinguished from the $\bar{\bullet}$ observations at a time by a solution of ($SC_i^\bullet$), it is formulated and solved for a less number of times for generating a set of $\bullet$ patterns that collectively distinguish all $\bullet$ observations from the $\bar{\bullet}$ data in a dataset under analysis (refer to the oligo selection procedure detailed below). On the other hand, (SC-pg$_i^\bullet$) generates per solution a lower degree pattern that explains the specific difference between the reference $\bullet$ observation and the $\bar{\bullet}$ sequences and, hence, is formulated and solved for a more number of times for generating a set of $\bullet$ patterns. Overall, the two models select about the same number of probes. However, as (SC-pg$_i^\bullet$) is much smaller in size, hence is more efficiently solved, and because a high specificity is desired in genotyping applications, we prefer (SC-pg$_i^\bullet$) for selecting genotyping oligo probes.

Using (SC-pg$_i^\bullet$), we design one simple oligo probe selection procedure below, where $P^\bullet$ denotes the set of $\bullet$ patterns generated so far.

procedure `SC-pg`
begin
  for $\bullet \in \{+, -\}$ **do**
    set $P^\bullet = \emptyset$ and $S \leftarrow S^\bullet$.
    while $S \neq \emptyset$ **do**
      - randomly choose p_i, $i \in S$, and formulate (SC-pg$_i^\bullet$).
      - solve (SC-pg$_i^\bullet$).
      - generate a $\bullet$ pattern $\mathcal{P}$ via (14.2).
      - set $P^\bullet \leftarrow P^\bullet \cup \{\mathcal{P}\}$.
      - set $S \leftarrow S \backslash \{i\} \backslash \{j \in S, j \neq i : p_{jk} = p_{ik}, \forall k \in N_i\}$.
    end while
  end for
end

The following is immediate.

THEOREM 3. ***procedure*** *`SC-pg` terminates finitely.*

A few remarks are due now.

Remark 1. *Simply put, the number of 1's in the covers generated via* ***procedure*** *`SC-pg` determines the number of probes to be used for a specific genotyping purpose. In other words, the quality of an SC solution determines the cost of genotyping applications.*

SC is a well-known $\mathcal{NP}$−complete problem [8]. Owing to having a wide range of practical applications (despite its simple structure), SC has invited an array of (meta-heuristic solution procedures to be developed for its efficient heuristic solution [5] and references therein) and any of these can be used for solving (SC-pg$_i^\bullet$). In fact, the genotyping accuracy is not affected at all as long as the covers found are near-optimal and good enough (see results in the following section) and this was the rationale behind our developing SC-based probe selection models in this chapter: recall that probe selection is a large-scale combinatorial optimization problem in nature.

Furthermore, the efficiency of SC heuristic solution procedures allows one to apply ***procedure*** *`SC-pg` or the similar directly to the binarized data to generate patterns without going through the feature selection phase. This* *is another benefit the SC-based pattern generation offers over its MILP counterparts from Ref. [24] or the standard term-enumeration-based procedure for generating patterns in the LAD literature [3].*

Remark 2. *If constraint j in (SC-pg$_i^\bullet$) has all zero coefficients, the SC instance is infeasible. This case arises when the reference sequence p_i, $i \in S^\bullet$ and the sequence p_j, $j \in S^{\bar{\bullet}}$ have identical 0–1 fingerprints, which is a contradiction. Supervised learning methodologies, including LAD, presume for the existence of a classification function that each unique sequence in the training set belongs to exactly one of the two classes. When this holds, contradiction-free 0–1 clones of the original data can always be obtained by using oligos of longer length for data binarization.*

Remark 3. *If desired, the hybridization affinity of probes can be ensured in a number of ways, including the following. First, during data binarization, one can remove from further consideration each $l-$mer with the GC content less than a prescribed level or with the melting temperature calculated via, for example, the formula found in Ref. [33] that falls outside a certain prescribed range from the median melting temperature of all $l-$mers generated. Next, the proposed LAD-based method can be applied to select an optimal set of probes on the surviving $l-$mers that are compatible in terms of their hybridization behavior.*

Remark 4. *(SC-pg$_i^\bullet$) is a general-purpose model and can be specialized to select a minimal set of optimal oligo probes by any quantifiable probe selection criterion. For example, one may use the longest common factors from Ref. [23] or the OVL scores from Ref. [11] for c_j values in (SG-pg$_i^\bullet$) to select probes by the (dis-)similarity preference. One may use, for example, the Shannon entropy scores from Ref. [11] for c_j's and incorporate the complexity of oligos in probe selection.*

14.2.3 Classification Rules

Denote by $P_1^+, \ldots, P_{n_+}^+$ and $P_1^-, \ldots, P_{n_-}^-$ the positive and negative patterns, respectively, generated via **procedure** *SC-pg*. In classifying unseen + (target) and − (nontarget) sequences, we use three decision rules. First, in polyspecific genotyping applications (see, for example, Experiment 4 in Section 14.3.2), we form the standard LAD classification rule [3]

$$\Delta := \sum_{i=1}^{n_+} \frac{\omega_i^+}{|S^+|} P_i^+ - \sum_{i=1}^{n_-} \frac{\omega_i^-}{|S^-|} P_i^-, \tag{14.3}$$

where $\omega_i^\bullet$ denotes the number of $\bullet$ training sequences covered by $P_i^\bullet$ and assign class $+$ $(-)$ to new sequence p if $\Delta(p) > 0$ $(\Delta(p) < 0)$. We fail to classify sequence p if $\Delta(p) = 0$.

For the monospecific genotyping, we use a strict classification rule. Specifically, for classification of two viral (sub-)types (see, for example, Experiment 1 in Section 14.3.2), we form a decision rule by

$$\Delta^+ = \sum_{i=1}^{n_+} P_i^+ \text{ and } \Delta^- := \sum_{i=1}^{n_-} P_i^- \tag{14.4}$$

and assign p to class $\bullet$ if $\Delta^\bullet(p) > 0$ while $\Delta^{\bar{\bullet}}(p) = 0$. When $\Delta^\bullet(p) > 0$ and $\Delta^{\bar{\bullet}}(p) > 0$ or when $\Delta^\bullet(p) = 0$ and $\Delta^{\bar{\bullet}}(p) = 0$, we fail in classifying the sequence.

For the monospecific classification of more than two viral (sub-)types $k = 1, \ldots, m$ (see, for example, Experiment 7 in Section 14.3.2), we use the decision rule

$$\Delta^k = \sum_{i=1}^{n_k} P_i^k, \tag{14.5}$$

where $P_1^k, \ldots, P_{n_k}^k$ are the probe(s) selected to for virus (sub-)type k, and assign p to class k if $\Delta^k(p) > 0$ while $\Delta^i(p) = 0$ for all $i = 1, \ldots, m, i \neq k$. When $\Delta(p) > 0$ for more than two virus types or $\Delta^k = 0$ for all k, then we fail to assign a class to sequence p.

14.3 *In Silico* Experiments

In this section, we extensively test the proposed probe design for classification of viral disease-agents in *in silico* setting. To make these experiments as realistic as possible, we design each of these experiments based on information from the literature and the official Website of the World Health Organization (WHO) and use viral genomic sequences obtained from the NCBI and HPV sequences from the Los Alamos National Laboratory. To be more specific about the data used, we obtained the HPV data from the Los Alamos National Laboratory site for illustrative and comparative purposes. The data corresponds to the 72 high and low risk HPV sequences that are used in Refs. [7,22]. Although some of these manually classified virus sequences contain classification errors [7], we used the data with its classification from [22] to allow a comparison among our result and results reported in [7,22]. For the experiments on genotyping viral pathogens, we used genomic sequences of SARS virus, influenza virus classified by

Table 14.1 Viral Sequences Used in Experiments

Viral Sequence	Number	Length		
		Min.	Avg. ± 1 Std. Dev.	Max.
Human papillomavirus (HPV)				
- High risk HPV	18	449	7365 ± 1730	7989
- Low risk HPV	54	455	7198 ± 1683	8027
SARS coronavirus	105	29350	29692 ± 91	29765
Coronavirus	39	9203	29013 ± 3569	31526
Other virus				
- Human respiratory syncytial virus	10	13933	15091 ± 386	15226
- Human adenovirus	32	34125	35215 ± 618	36015
- Human parainfluenza virus	4	15646	15652 ± 3	15654
- Human rhinovirus (A, B)	8	7102	7157 ± 36	7212
- Influenza virus (A, B, C)	53	838	1701 ± 527	2368
Influenza virus hemagglutinin (H) subtype				
- H1	137	1698	1749 ± 24	1778
- H3	660	1695	1735 ± 21	1768
- H5	148	1677	1721 ± 25	1779
- H7	77	1659	1690 ± 27	1792
- H9	93	1683	1704 ± 26	1742
- H else (2, 4, 6, 8, 11, 12, 13, 16)	65	1689	1742 ± 29	1773
Influenza virus neuraminidase (N) subtype				
- N1	218	1344	1410 ± 39	1463
- N2	1050	1341	1434 ± 28	1467
- N3	44	1326	1411 ± 29	1460
- N else (4, 5, 6, 7, 8, 9)	64	1341	1434 ± 25	1467

their hemagglutinin (H) and neuraminidase (N) types (influenza viruses are typed according to their H and N surface glycoproteins), coronavirus and other viral agents of disease with SARS-like symptoms. In Table 14.1, we provide the number and the length (the minimum, average ±1 standard deviation and maximum length) of each type of the genomic data used in our experiments.

In analyzing data in an experiment, we first decided on a length of oligos to use by calculating the smallest integer value l such that 4^l became larger than or equal to the average of the lengths of target and nontarget sequences of the experiment. Then, 4^l candidate oligos were generated to fingerprint and binarize the data. If the length of oligos turned out to be not long enough during the pattern generation stage (see Remark 2), the data binarization stage was repeated with the value of l incremented by 1 and this process was repeated until the binary representations of the data became contradiction free. Next, **procedure** *SC-pg* was applied to generate patterns, hence probes. In applying **procedure** *SC-pg* in these *in silico* experiments, we did not consider any oligo picking criterion that is nontheoretical in nature (refer to Remark 4) and selected a minimal set of oligo probes with using $c_j = 1$ for all $j \in N$. For solving the unicost (SC-$\text{pg}_i^\bullet$)'s generated, we used for ease of implementation the textbook heuristic procedure [21] that selects one variable at a time by the rule

$$k \leftarrow \text{argmax}\left\{j \in N,\ x_j = 0 : \left|I_j \cap M_u\right|\right\},$$

where I_j denotes the index of rows k with $a_j^{(i,k)} = 1$ and M_u denotes the set of rows that are not yet covered by the partial cover $\boldsymbol{x}$ on hand.

In each of the experiments in this section, to fairly assess the classification capabilities of oligo probes selected by the proposed probe design procedure, we

1. Randomly selected 90 percent of the target and of the nontarget data to form a training set of sequences
2. Binarized the training data
3. Selected optimal oligo probes on the training data via **procedure** *SC-pg*
4. Formed a classification rule given by one of (Equations 14.3 through 14.5) with the selected oligo probes
5. Used the classification rule to (sub-)type each of the reserved testing sequences, comprised of the remaining 10 percent of the target and the nontarget sequences
6. Repeated steps above 20 times to obtain the average testing performance and other relevant information of the experiment

The computational platform used for experiments was an Intel 2.66 GHz Pentium Linux PC with 512 Mb of memory.

14.3.1 A Comparative Experiment: Classification of High and Low Risk HPV

Infection with HPV is the main cause of cervical cancer, the second most common cancer in women worldwide [4,20]. There are more than 80 identified types of HPV and the genital HPV types are subdivided into high and low risk types: low risk HPV types are responsible for most common sexually transmitted viral infections, whereas high risk HPV types are a crucial etiological factor for the development of cervical cancer [18].

We applied the proposed probed design method on the 72 HPV sequences downloaded from the Los Alamos National Laboratory with their classification found in Table 3 of Ref. [22]. The selected probes were used to form a decision rule by Equation 14.3 and tested for their classification capability.

Results from this polyspecific probe selection experiment are provided in Table 14.2. The target (+) and the nontarget (−) virus types of the experiments are specified in the first column of Table 14.2 and other tables of this section. Then, the tables provide two bits of information on the candidate oligos, namely, the length l and the average and the standard deviation of the number of features generated and used in the 20 runs of each experiment for data binarization and for pattern generation: recall that we skip the feature selection stage of LAD (see Remark 1). Provided next in the tables is the information on the number of probes selected in the format "the average ±1 standard deviation" and information on the LAD patterns generated. Finally, the testing performance of the probes selected is provided in the format "the average ±1 standard deviation" of the percentage of the correct classifications of the unseen sequences.

Briefly summarizing, the proposed probe design method selected probes on the HPV data in a few CPU seconds that tested 90.6 percent accurate in classifying unseen HPV samples. For comparison, the same HPV dataset was used in Refs. [7,22] for the classification of HPV by high and low risk types. In brief, the probe design methods given in Refs. [7,22] required several CPU hours of computation and selected probes that obtained 85.6 percent and 81.1 percent correct classification rates, respectively.

Before moving on, we note that the sequences belonging to the target and the nontarget groups in this experiment all have different HPV subtypes (see Table 3 in Ref. [22].) The combination of all target and nontarget sequences being different from one another and the presence of noise in the data (the classification errors) gave rise to selecting a relatively large number of polyspecific probes in this experiment.

14.3.2 Experiments on Genotyping Viral Pathogens

The proposed probe design method was extensively tested on genomic viral sequences from NCBI for selecting monospecific and polyspecific probes for screening for SARS and AI in a number of different binary and multicategory experimental setting and performed superbly on all counts. We summarize the results from some of these experiments in this section.

Before proceeding, we briefly illustrate the benefit of probe selection via (SC-pg$_i^\bullet$) from the computational point of view with Experiment 5 below. For the purpose, let us first recall that that probe selection is a combinatorial optimization problem. Therefore, for the selection of oligo

Table 14.2 Polyspecific Classification of High and Low Risk Human Papillomavirus (HPV)

Experiment	*l*–mers Used		Probes Selected		Testing
	l	Number[a]	Number[a]	Patterns	Accuracy[ab]
High risk human papillomavirus (HPV) (+) vs	8	58359.9 ± 130.4	18.7 ± 1.7	Degree 1 and 2 patterns	90.6 ± 9.8
Low risk HPV (−)			22.8 ± 1.6	Degree 1 and 2 patterns	

[a] In format average ± standard deviation.
[b] Percentage of correct classifications of testing/unseen data.

probes for differentiating lethal AI virus H5 and H9 from the other AI virus H subtypes in Experiment 5, a supervised learning method based on a complete pairwise differencing of the target and nontarget training sequences [2,3,12,23] would require solving one or more combinatorial optimization problems with between $(148 + 93) \times (137 + 660 + 77 + 65) = 226,299$ and $137 \times 660 \times 148 \times 77 \times 93 \times 65 \approx 6.23 \times 10^{12}$ rows (refer to Table 14.1 above for the numbers of the target and nontarget viral sequences) and with at least 39,056 0–1 decision variables (see Table 14.7 for the average number of l−mers generated in this experiment). For Experiment 5, we note in comparison that the largest (SC-pg$_i^\bullet$) instance generated and solved by **procedure** *SC-pg* had $\max\{148 + 93, 137 + 660 + 77 + 65\} = 939$ rows and 39,056 columns.

Experiment 1. SARS virus vs coronavirus

SARS virus is phylogenetically most closely related to group 2 coronavirus [27]. Hundred and Five SARS sequences and thirty nine coronavirus samples were used to select one monospecific probe for screening for SARS. Used in a classification rule (Equation 14.4), the SARS probe and one probe selected for coronavirus together perfectly classified all testing sequences (see Table 14.3).

Experiment 2. SARS virus vs influenza virus

This experiment simulates a SARS pandemic, where suspected patients with SARS-like symptoms are screened for the disease. We used the 105 SARS virus sequences and 108 samples of other influenza virus types (the other virus in Table 14.1) in this experiment and selected polyspecific probes. Used in a classification rule (Equation 14.3), these probes collectively gave the perfect classification of all testing sequences (see Table 14.4).

Experiment 3. Classification of pathogenic AI virus H7 and other influenza virus H subtypes

AI virus H7N7 is highly pathogenic with the capacity to pass from human-to-human, and this raised concerns for a possible viral reassortment

Table 14.3 Monospecific Classification of SARS Virus and Coronavirus, a Phylogenetically Closest Sibling of SARS

	l−mers Used		Probes Selected		
Experiment 1	l	Number[a]	Number[a]	Patterns Generated	Testing Accuracy[ab]
SARS virus (+)	8	57745.3 ± 306.1	1 ± 0	Degree 1	100 ± 0
Coronavirus (−)			1 ± 0	Degree 1	

[a] In format average ± standard deviation.

[b] Percentage of correct classifications of testing/unseen data.

Table 14.4 Classification of SARS Virus and Influenza Virus That Cause Disease with SARS-Like Symptoms

	l–mers Used		Probes Selected		
Experiment 2	*l*	Number[a]	Number[a]	Patterns Generated	Testing Accuracy[ab]
SARS virus (+)	8	64141.5 ± 36.5	1 ± 0	Degree 1	100 ± 0
Influenza virus (−)			10.1 ± 0.8	Degree 1 only	

[a] In format average ± standard deviation.

[b] Percentage of correct classifications of testing/unseen data.

with human influenza H1N1 and H3N2 strains during a large outbreak of H7N7 infection in the Netherlands in 2003 [13,34].

On the basis of information from Ref. [13], we replicated the classification of H7 and other influenza virus H subtypes in this experiment by using 77 H7 sequences and 1103 other H subtype samples. Polyspecific probes were selected and tested in a classification rule Equation 14.3 to give the perfect classification rate (see Table 14.5).

Experiment 4. Classification of pathogenic AI virus H5 and H7 and other influenza virus H subtypes

H5 and H7 have an ominous capacity to pass from human-to-human (http://www.who.int, [34]). This experiment, using 225 H5 and H7 viral samples and 955 other H subtype sequences, selected polyspecific probes for detecting the two pathogenic H subtypes of the AI virus from the other influenza virus H subtypes and vice versa. A classification rule was formed by Equation 14.3 for testing the selected probes, and we obtained the perfect testing result (see Table 14.6).

Experiment 5. Classification of lethal AI virus H5 and H9 and other influenza virus H subtypes

Table 14.5 Classification of Highly Pathogenic H7 AI Virus (with Capacity to Pass from Human-to-Human) and Other H Subtypes of Influenza Virus

	l–mers Used		Probes Selected		
Experiment 3	*l*	Number[a]	Number[a]	Patterns Generated	Testing Accuracy[ab]
H7 (+)	7	14724.2 ± 30.9	1 ± 0	Degree 1	100 ± 0
Other H strains (−)			7 ± 1	Degree 1 only	

[a] In format average ± standard deviation.

[b] Percentage of correct classifications of testing/unseen data.

Table 14.6 Classification of Highly Pathogenic H5 and H7 AI Virus and Other H Subtypes of Influenza Virus

	l–mers Used		Probes Selected		
Experiment 4	l	Number[a]	Number[a]	Patterns Generated	Testing Accuracy[ab]
H5 and H7 (+)	8	39164.4 ± 333	15.7 ± 1.5	Degree 1 only	100 ± 0
Other H strains (−)			27.6 ± 1.2	Degree 1 only	

[a] In format average ± standard deviation.
[b] Percentage of correct classifications of testing/unseen data.

AI virus H5 and H9 subtypes cause a most fatal form of the disease [13], and they were separated from the other H subtypes of influenza virus in this experiment. 241 H5 and H9 target sequences and 1010 other H subtype sequences were used to select polyspecific probes for detecting AI virus H5 and H9 subtypes from the rest. In a classification rule Equation 14.3, the selected probes collectively classified all testing sequences correctly (see Table 14.7).

Experiment 6. Monospecific Classification of SARS, human influenza H1, human influenza H3, AI virus H5, and AI virus H7

This multicategory classification experiment selects monospecific probes for distinguishing one from another a few notorious viral pathogens. We used 103 SARS virus, 137 human influenza virus H1, 660 human influenza virus H3, 148 lethal AI virus H5, and 77 pathogenic AI virus H7 sequences and selected monospecific probes for each virus type in sequential binary classification of one type against the rest. The selected probes were tested in a classification rule (Equation 14.5) to classify the testing sequences p by a strict decision rule of "assign class i to p only if one or more probes selected for virus type i is found in p while none of the probes selected for the other types are not" and gave the perfect classification result. (See Table 14.8. Note that only a small number of monospecific probes were selected, as in Experiment 1).

Table 14.7 Classification of Fatal H5 and H9 AI Virus and Other H Subtypes of Influenza Virus

	l–mers Used		Probes Selected		
Experiment 5	l	Number[a]	Number[a]	Patterns Generated	Testing Accuracy[ab]
H5 and H9 (+)	8	39056 ± 398.3	6.7 ± 0.5	Degree 1 only	100 ± 0
Other H strains (−)			21.6 ± 1.3	Degree 1 only	

[a] In format average ± standard deviation.
[b] Percentage of correct classifications of testing/unseen data.

Table 14.8 Monospecific Classification of SARS Virus and H1, H3, H5, and H7 Subtypes of Influenza Virus

	l–mers Used		Probes Selected		
Experiment 6	*l*	Number[a]	Number[a]	Patterns Generated	Testing Accuracy[ab]
SARS virus	8	45259 ± 527	1 ± 0	Degree 1	100 ± 0
H1			1 ± 0	Degree 1	
H3			2.9 ± 0.3	Degree 1 only	
H5			3 ± 0	Degree 1 only	
H7			1 ± 0	Degree 1	

[a] In format average ± standard deviation.
[b] Percentage of correct classifications of testing/unseen data.

Experiment 7. Monospecific Classification of N1, N2, and N3 influenza virus

The statement "monospecific neuraminidase (NA) subtype probes were insufficiently diverse to allow confident NA subtype assignment" from Ref. [26] motivated us to design this experiment on multicategory and monospecific classification of influenza virus by N subtypes. We used the three influenza virus N subtypes with 30 or more samples in Table 14.1 and selected monospecific probes for their classification. Tested in a classification rule (Equation 14.5), the selected probes performed perfectly in classifying all testing sequences. (See Table 14.9. Note again that only a small number of monospecific probes were selected and proved needed in this experiment, as in the other two monospecific genotyping experiments, Experiments 1 and 6).

Table 14.9 Monospecific Classification of N1, N2, and N3 Subtypes of Influenza Virus

	l–mers Used		Probes Selected		
Experiment 7	*l*	Number[a]	Number[a]	Patterns Generated	Testing Accuracy[ab]
N1	7	13151 ± 39.3	3 ± 0	Degree 1	100 ± 0
N2			3.7 ± 0.5	Degree 1 only	
N3			1 ± 0	Degree 1	

[a] In format average ± standard deviation.
[b] Percentage of correct classifications of testing/unseen data.

14.4 Concluding Remarks

The problem of probe design for hybridization-based experiments is an interesting problem lying at the intersection of molecular biology and optimization but has received relatively little attention from the OR community. In this chapter, we specialized a general LAD Framework from Ref. [24] for efficiently handling large-scale genomic data and developed a probe design method for selecting short oligo probes for genotyping applications. Extensively tested on genomic sequences obtained from the NCBI and the Los Alamos National Laboratory in various monospecific and polyspecific *in silico* experiments, the proposed probe design method was able to select a small number of oligo probes of length 7 or 8 nucleotides that performed superbly in classifying unseen testing sequences. These *in silico* results demonstrate the efficacy of the proposed oligo design method. Experimental results further illustrate a huge potential that a well-designed optimization-based probe design method has in hybridization-based genotyping applications.

Collaborative research activities are planned to realize the *in silico* performance of the proposed probe design method on microarrays and in real hybridization experiments. Also, we plan to investigate the possibility of exploiting frequently used oligo selection criteria [11,14,16,23] within the proposed probe design framework to further improve its effectiveness in terms of the number of probes needed.

References

1. S.F. Altschul, W. Gish, W. Miller, E.W. Myers, and D.J. Lipman. Basic local alignment search tool. *Journal of Molecular Biology*, 215:403–410, 1990.
2. J. Borneman, M. Chrobak, G.D. Vedova, A. Figueroa, and T. Jiang. Probe selection algorithms with applications in the analysis of microbial communities. *Bioinformatics*, 17(Suppl. 1):S39–S48, 2001.
3. E. Boros, P.L. Hammer, T. Ibaraki, A. Kogan, E. Mayoraz, and I. Muchnik. An implementation of logical analysis of data. *IEEE Transactions on Knowledge and Data Engineering*, 12:292–306, 2000.
4. F.X. Bosch, A. Lorincz, N. Muñoz, C.J.L.M. Meijer, and K.V. Shah. The causal relation between human papillomavirus and cervical cancer. *Journal of Clinical Pathology*, 55:244–265, April 2002.
5. A. Caprara, M. Fischetti, and P. Toth. A heuristic method for the set covering problem. *Operations Research*, 47(5):730–743, 1999.
6. C. Cortes and V.N. Vapnik. Support vector networks. *Machine Learning*, 20:273–297, 1995.
7. J.-H. Eom, S.-B. Park, and B.-T. Zhang. Genetic mining of DNA sequence structures for effective classification of the risk types of human papillomavirus (hpv). In N.R. Pal, N. Kasabov, R.K. Mudi, S. Pal, and S.K. Parui, editors, *Neural Information Processing*, LNCS 3316, Springer-Verlag, Berlin, 2004, pp. 1334–1343.

8. M.R. Garey and D.S. Johnson. *Computers and Intractability: A Guide to the Theory of $\mathcal{NP}-$Completeness.* Freeman, New York, 1979.
9. P.L. Hammer. Partially defined boolean functions and cause–effect relationships. Proceedings of the International Conference on Multi-Attribute Decision Making Via OR-Based Expert Systems, April 1986.
10. R.A. Heller, M. Schena, A. Chai, D. Shalon, T. Bedilion, J. Gilmore, D.E. Woolley, and R.W. Davis. Discovery and analysis of inflammatory disease-related genes using cDNA microarrays. *Proceedings of the National Academy of Sciences*, 94:2150–2155, March 1997.
11. R. Herwig, A.O. Schmitt, M. Steinfath, J. O'Brien, H. Seidel, S. Meier-Ewert, H. Lehrach, and U. Radelof. Information theoretical probe selection for hybridisation experiments. *Bioinformatics*, 16(10):890–898, 2000.
12. G.W. Klau, S. Rahmann, A. Schliep, M. Vingron, and K. Reinert. Optimal robust non-unique probe selection using integer linear programming. *Bioinformatics*, 20(Suppl. 1):i186–i193, August 2004.
13. M. Koopmans, B. Wilbrink, M. Conyn, G. Natrop, H. van der Nat, H. Vennema, A. Meijer, J. van Steenbergen, R. Fouchier, A. Osterhaus, and A. Bosman. Transmission of H7N7 avian influenza A virus to human beings during a large outbreak in commercial poultry farms in the Netherlands. *Lancet*, 363:587–593, February 21 2004. www.thelancet.com.
14. I.-H. Lee, S. Kim, and B.-T. Zhang. Multi-objective evolutionary probe design based on thermodynamic criteria for HPV detection. In C. Zhang, H.W. Guesgen, and W.K Yeap, editors, *PRICAI 2004: Trends in Artificial Intelligence*, LNCS 3157, Springer-Verlag, Berlin, 2004, pp. 742–750.
15. Y. Lee and C.-K. Lee. Classification of multiple cancer types by multicategory support vector machines using gene expression data. *Bioinformatics*, 19(9):1132–1139, 2003.
16. F. Li and G.D. Stormo. Selection of optimal DNA oligos for gene expression arrays. *Bioinformatics*, 17(11):1067–1076, 2001.
17. C.-H. Liu, W.-L. Ma, R. Shi, Y.-Q. Ou, B. Zhang, and W.-L. Zheng. Possibility of using DNA chip technology for diagnosis of human papillomavirus. *Journal of Biochemistry and Molecular Biology*, 36(4):349–353, 2003.
18. S.E. McFadden and L. Schumann. The role of human papillomavirus in screening for cervical cancer. *Journal of the American Academy of Nurse Practitioners*, 13:116–125, 2001.
19. N. Megiddo. On the complexity of polyhedral separability. *Discrete and Computational Geometry*, 3:325–337, 1988.
20. N. Muñoz, F.X. Bosch, S. de Sanjosé, R. Herrero, X. Castellsagué, K.V. Shah, P.J.F. Snijders, and C.J.L.M. Meijer for the International Agency for Research on Cancer Multicenter Cervical Cancer Study Group. Epidemiologic classification of human papillomavirus types associated with cervical cancer. *The New England Journal of Medicine*, 348(6):518–527, 2003.
21. G.L. Nemhauser and L.A. Wolsey. *Integer and Combinatorial Optimization.* Wiley-Interscience Series in Discrete Mathematics and Optimization. Wiley, New York, 1988.
22. S.-B. Park, S.-H. Hwang, and B.-T. Zhang. Classification of the risk types of human papillomavirus by decision trees. In Proceedings of the 4th International Conference on Intelligent Data Engineering and Automated Learning, pp. 540–544, 2003.

23. S. Rahmann. Fast large scale oligonucleotide selection using the longest common factor approach. *Journal of Bioinformatics and Computational Biology*, 1(2):343–361, 2003.
24. H.S. Ryoo and I.-Y. Jang. MILP approach to pattern generation in logical analysis of data. Submitted.
25. M. Schena. *DNA Microarray: A Practical Approach*. Oxford University Press, 1999.
26. S. Sengupta, K. Onodera, A. Lai, and U. Melcher. Molecular detection and identification of influenza viruses by oligonucleotide microarray hybridization. *Journal of Clinical Microbiology*, 41(10):4542–4550, 2003.
27. E.J. Snijder, P.J. Bredenbeek, J.C. Dobbe, V. Thiel, J. Ziebuhr, L.L.M. Poon, Y. Guan, M. Rozanov, W.J.M. Spaan, and A.E. Gorbalenya. Unique and conserved features of genome and proteome of SARS-coronavirus, an early split-off from the coronavirus group 2 lineage. *Journal of Molecular Biology*, 331:991–1004, 2003.
28. R.L. Stears, T. Martinsky, and M. Schena. Trends in microarray analysis. *Nature Medicine*, 9(1):140–145, 2003.
29. J. Ullman. *Pattern Recognition Techniques*. Crane, London, 1973.
30. V.N. Vapnik. *Statistical Learning Theory*. Wiley-Interscience, New York, 1998.
31. G. Vernet. DNA-chip technology and infectious diseases. *Virus Research*, 82: 65–71, 2002.
32. D. Wang, L. Coscoy, M. Zylberberg, P.C. Avila, H.A. Boushey, D. Ganem, and J.L. DeRisi. Microarray-based detection and genotyping of viral pathogens. *PNAS*, 99(24):15687–15692, 2002.
33. X. Wang and B. Seed. Selection of oligonucleotide probes for protein coding sequences. *Bioinformatics*, 19(7):796–802, 2003.
34. R.J. Webby and R.G. Webster. Are we ready for pandemic influenza? *Science*, 302:1519–1522, 2003.
35. Y.M. Zhou, R.Q. Yang, S.C. Tao, Z. Li, Q. Zhang, H.F. Gao, Z.W. Zhang, J.Y. Du, P.X. Zhu, L.L. Ren, L. Zhang, D. Wang, L. Guo, Y.B. Wang, Y. Guo, Y. Zhang, C.Z. Zhao, C. Wang, D. Jiang, Y.H. Liu, H.W. Yang, L. Rong, Y.J. Zhao, S. An, Z. Li, X.D. Fan, J.W. Wang, Y. Cheng, O. Liu, Z. Zheng, H.C. Zuo, Q.Z. Shan, L. Ruan, Z.X. Lu, T. Hung, and J. Cheng. The design and application of DNA chips for early detection of SARS-CoV from clinical samples. *Journal of Clinical Virology*, 33(2):123–131, 2005.

Chapter 15

New Dihedral Angle Measure for Protein Secondary Prediction

Moon K. Kim, Yunho Jang, and J. MacGregor Smith

CONTENTS

15.1 Introduction 440
15.1.1 Motivation 440
15.1.2 Outline 440
15.2 Problem Background 441
15.2.1 Protein Folding Problem 441
15.2.2 Secondary Structure Prediction 442
15.2.3 Literature Review 443
15.2.4 Ramachandran Plots 444
15.3 Mathematical Models 445
15.3.1 Steiner Trees 445
15.3.2 Definitions 445
15.3.3 Notation 445
15.3.4 Potential Energy Functions 448
15.3.5 Steiner Trees and Minimum Energy Configurations 449
15.3.6 Serine 450
15.4 Alanine Dipeptide Results 452
15.5 General Dipeptide Results 455
15.5.1 Virtual Plane 455
15.5.2 Ala-Gly Silk β-Structure 459
15.5.2.1 Experiments for Individual Acids 459
15.5.3 Pentapeptide Results 461

15.5.3.1 Leu-Ser-Phe-Ala-Ala ($N = 71$ Atoms) $\rho = 0.995210$ 461
15.6 Summary and Conclusions 462
References 463

Abstract A new dihedral angle measure is presented in this chapter for protein secondary structure modeling and prediction. This new angle measure is simpler than the Ramachandran plots because it has only one degree of freedom and varies between 0 and $\pi/2$. The origin of this new dihedral angle measure was precipitated by a Steiner tree analysis of the twist angles within the individual amino acid structures. The Steiner tree structure of the amino acids revealed certain regular twist angles for the planes of atoms as defined by the Steiner tree topology. This regularity carried over into the analysis of dipeptide structures when it was shown that a planar characterization of the $\{N, C_\alpha, C\}$ set of atoms unique for each residue would synthesize the angular measure of the ϕ, ψ angles of the Ramachandran plots, but with one degree of freedom less. Numerous experimental results with this new angle measure are presented to characterize the α-helix and β-sheet structures of proteins.

15.1 Introduction

Characterization of how proteins fold in space is one of the most difficult and challenging computational problems. Given the amino acid sequence, the primary structure, it is very important to know how the secondary structures evolve. The α- and β-sheet secondary structures, which emerge in the folding process, are the fundamental building blocks that propagate the tertiary structure. Finally, it is the tertiary structure which is the crucial conformation of interest in the protein folding problem.

15.1.1 Motivation

The motivation for this chapter is to generate new methodological concepts and tools for the analysis of the secondary structure in proteins so that these new methodologies can assist one in the tertiary structure identification for the protein folding problem. The approach based on the geometry and three-dimensional (3-D) topology of protein structures will be followed.

15.1.2 Outline

In Section 15.2, we present a review of the protein folding problem, the secondary structure prediction aspects, the literature, and the Ramachandran plots, which posit the fundamental dihedral angle conformation characteristics. In Section 15.3, we present the Steiner problem and how it relates to proteins and secondary structure prediction through the potential energy function. In Section 15.4, we demonstrate alanine dipeptide results of our Steiner tree analysis along with their minimum energy configurations. In Section 15.5, we describe the new virtual angle approach and dipeptide and pentapeptide experimental results. Section 15.6 rounds out the chapter.

15.2 Problem Background

The protein folding problem remains one of the most important and unsolved problems in the biological sciences. It also remains one of the most important computational challenges from a combinatorial optimization perspective. Developing a computer program to generate the 3-D structure of a protein from its amino acid sequence remains an important computational and biological objective. It is suggested that Steiner minimal trees (SMTs) may play an important part in the development of such a computer program.

15.2.1 Protein Folding Problem

Proteins have definite shape and structure and are not just random, unordered blobs in space. Given that there are up to 20 different amino acids that can occur along the peptide backbone with hundreds to thousands of amino acids along the backbone chain of the protein, there are an exponential number of ways that the protein can fold up in space. Thus, the complexity of the protein folding problem as a combinatorial optimization problem is well known. We briefly present the literature and then focus on how the protein folding problem may be modeled through Steiner trees.

The protein folding problem may be conceptualized from the way biologists and chemists normally view the fundamental structure of a protein. There are basically four classification schemes: (1) primary; (2) secondary; (3) tertiary; and (4) quarternary, see Ref. [5]. Brandon and Tooze [5] provide a good overview of the folding problem and they postulate that there are a set of rules that govern the way a protein folds up in space.

The primary structure refers to the linear sequence of amino acids in the chain. Secondary refers to the helices and sheets in which the acids form from the primary sequence of amino acids along the backbone chain. Tertiary refers to the complete 3-D structure of the protein and quarternary refers to the spatial relationships between the different polypeptides or subunits of the overall globular protein. It is theorized that the 3-D structure is largely determined by the primary composition level of the amino acids, as argued in Ref. [40]. We shall largely focus on this primary and secondary level and examine what Steiner minimal trees offer here. Some of the key 3-D rules are felt to be contained in this primary structure. A theoretical topological and geometric approach is followed rather than an empirical one approach for determining the 3-D structure, because the formalism of Steiner minimal trees allows one to do so.

15.2.2 Secondary Structure Prediction

The sequence of dihedral angles along the backbone structure of a protein generally determines the rough structure of the folded protein. The sequence of angle pairs (ϕ, ψ) set up the melody for the folding of the protein [32]. Figure 15.1 illustrates the (ϕ, ψ) dihedral angles of interest along the backbone of a folded protein, where the carbon and nitrogen atoms are as indicated along with the amino acid residues R_i also known as side chains. In addition, we have indicated the peptide planes (dotted lines) and the (ϕ, ψ) dihedral angles of central interest for the planes of this tripeptide structure.

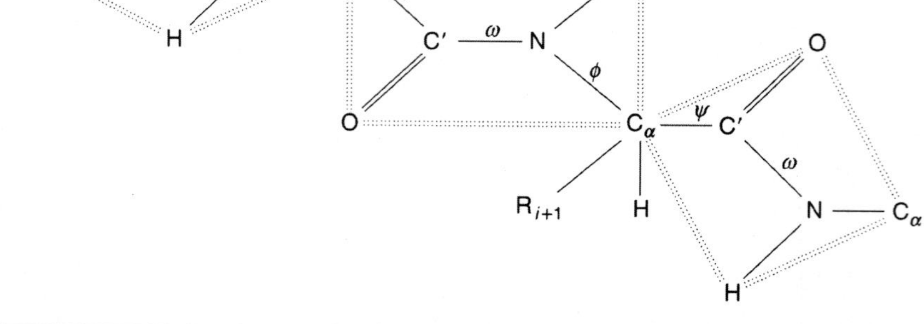

Figure 15.1 Peptide backbone dihedral angles with planes.

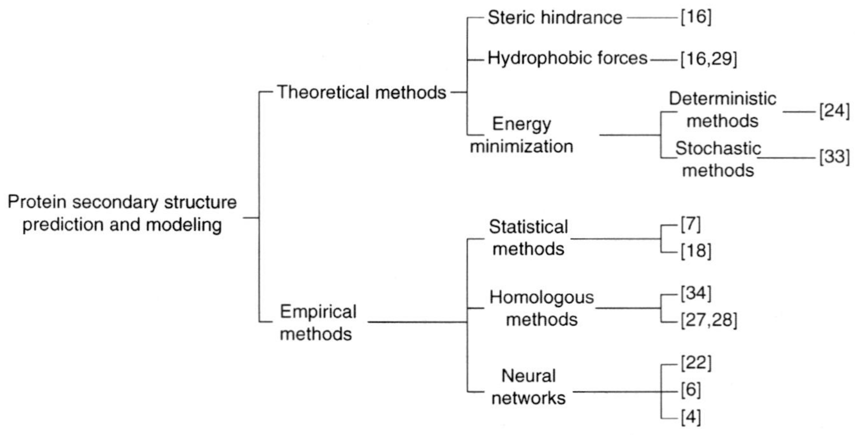

Figure 15.2 Morphological tree of structure and secondary structure prediction methods.

15.2.3 Literature Review

The literature relevant to secondary structure prediction is examined in Figure 15.2. Although prediction of secondary structure by itself will not necessarily illuminate tertiary structure, it still remains a viable area of research. There appear to be two major divisions in the research of secondary structure prediction methods: theoretical and empirical [15].

Theoretical methods make use of the physical and chemical properties of the amino acids themselves to predict structure [3] as well as energy minimization methods, while the empirical methods utilize the known sequences and properties of the structures to predict the likelihood of secondary structure.

There have been three major categories of publications on empirical methods of secondary structure prediction: (1) statistical methods, (2) sequence similarity methods, and (3) neural networks. All have had varied levels of, success, but no one single approach has clearly dominated the others.

Probably the simplest and most easily utilized methods are the statistical methods spearheaded by Chou and Fasman [7] and Garnier et al. [18]. Their approach is based on the probability of the individual amino acid residues being found in specific secondary structures (α-helix, β-sheets, or β-turns) compared to the overall probability in which they appear in protein sequence conformations. Their accuracy is reputed to be between 57 and 70 percent.

Homologous (sequence similarity) methods such as those embodied in the works of Levin et al. [27] and Reimer and Fueller [34] have an accuracy of about 65 percent.

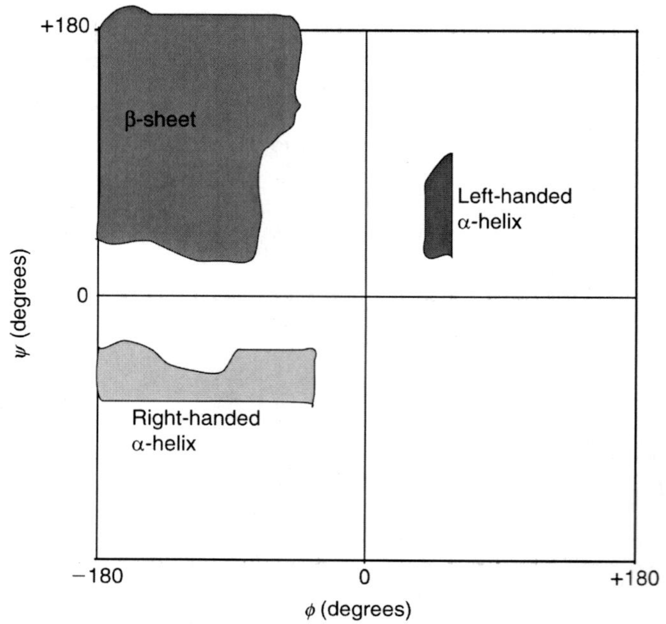

Figure 15.3 Ramachandran plots.

Neural network models such as those mentioned in Refs. [6,22] have an accuracy of around 72 percent and appear to be a bit more accurate than the former methods, yet no one method seems to dominate, and hybrid methods appear to be a wise choice. Many software tools have recently become available for secondary structure prediction (see http://us.expasy.org/tools/ and http://cmgm.stanford.edu/WWW/www-predict.html).

15.2.4 Ramachandran Plots

Given that the bond lengths, bond angles, and side chain conformations are relatively rigid, it is the dihedrahl angles of the backbone structure that represent the key elements in determining the ultimate 3-D structure of the protein [32]. Ramachandran plots (Figure 15.3) describe the basic relationship between the (ϕ, ψ) dihedral angles and the backbone planes and their possible feasible conformations. Only three possible angular conformations occur as is indicated in the figure on the left and this is due to the sterically allowed (ϕ, ψ) dihedral angles [40]. We would like to simplify the identity and range of possible dihedral angle measures so that instead of two angles, we only need one angle measure.

15.3 Mathematical Models

Combinatorial optimization mathematical models and their networks will be utilized as they provide a close analog to the geometry of folded networks of proteins and they also allow one to quantitatively assess the potential energy characteristics of proteins. In particular, Steiner tree networks will be the tool for analysis.

15.3.1 Steiner Trees

Steiner trees represent the shortest possible network in E^d to connect a given set of terminals $V = \{v_1, v_2, \ldots, v_N\}$ where possible additional points called Steiner points from a set $S = \{s_1, s_2, \ldots, s_M\}$ may be used to further reduce the Euclidean distance connecting the given points V. The problem is difficult (in fact it is $\mathcal{NP}$-Hard) because one knows neither how many Steiner points to employ nor their location in space. The focus of the chapter is on 3-D Euclidean space E^3.

15.3.2 Definitions

There are certain elemental definitions concerning Steiner trees that are important in E^3 and higher dimensions. They are the following:

- $M \leq N - 2, \ \forall E^d$ [19],
- Angles subtended at each Steiner point are equal to $2\pi/3, \ \forall E^d$ [19], and
- The ratio in the plane of dimension two $\rho_2(V) = \sqrt{3}/2$ is attained for equilateral triangles, ladders, and lattice configurations, see Ref. [13,14].

15.3.3 Notation

The following is a list of useful notation and definitions:

M	=	Number of Steiner vertices from point set S
N	=	Number of given terminal vertices from the set V
FST	=	Full Steiner tree with the maximum number of Steiner points $M = N - 2$
MST	=	Minimum spanning tree with the number of Steiner points $M = 0$
$\rho_3(V)$	=	Steiner ratio of a given terminal vertex set V, i.e., $\rho_3(V) = SMT/MST$ in E^3

To demonstrate this angle importance, let us take four terminals in space that are denoted as $\{V_i, V_j, V_k, V_\ell\}$ that are from an equilateral tetrahedron

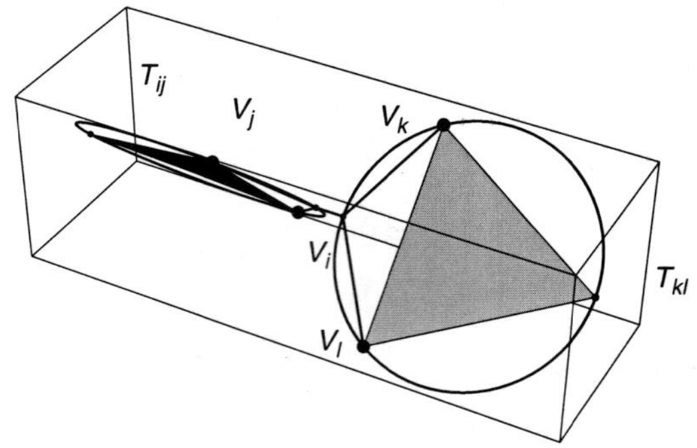

Figure 15.4 SMT solution for $N = 4$.

with cartesian coordinates (x_i, y_i, z_i), (x_j, y_j, z_j), (x_k, y_k, z_k), (x_ℓ, y_ℓ, z_ℓ), respectively:

$$\begin{array}{llll} V_i: \; 1. \;\; 1. \;\; 1.; & V_j: \; 1. \;\; -1. \;\; -1.; \\ V_k: \; -1. \;\; 1. \;\; -1.; & V_l: \; -1. \;\; -1. \;\; 1. \end{array}$$

The larger nodes in the diagrams which follow (see Figures 15.4 through 15.6) represent the terminal vertices while the smaller nodes represent either equilateral points or Steiner points.

Our objective is to find the maximum distance between the equilateral vertices of the edge pairs or, in essence, find the orientation of circles, called "Melzak circles" [31], through the equilateral reflection points that are furthest apart. The following optimization problem involving the equilateral reflection points corresponding to our given terminals $T_{ij} - T_{k\ell}$ with the Melzak circles along with the constraints ensuring the equilateral edges $e_{ij}, e_{k\ell}$ are satisfied is presented:

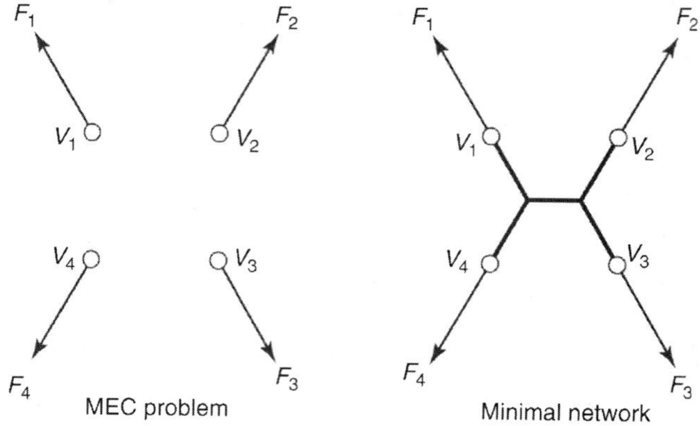

Figure 15.5 Maxwell's theorem, given V (left) SMT ≡ MEC (right).

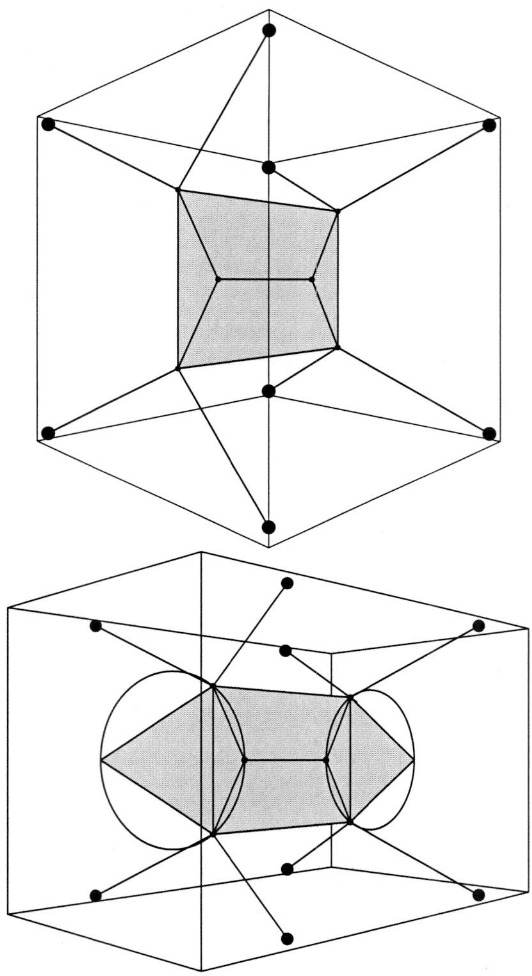

Figure 15.6 Cube with steiner plane (above) Melzak circles (below).

$$
\begin{aligned}
&\text{Max } \Phi = \|T_{ij} - T_{k\ell}\| \\
&\quad \text{s.t.} \\
&\quad\quad \left[(x_{ij} - x_i)^2 + (y_{ij} - y_i)^2 + (z_{ij} - z_i)^2\right]^{\frac{1}{2}} = e_{ij} \\
&\quad\quad \left[(x_{ij} - x_j)^2 + (y_{ij} - y_j)^2 + (z_{ij} - z_j)^2\right]^{\frac{1}{2}} = e_{ij} \\
&\quad \left[(x_{k\ell} - x_k)^2 + (y_{k\ell} - y_k)^2 + (z_{k\ell} - z_k)^2\right]^{\frac{1}{2}} = e_{k\ell} \\
&\quad \left[(x_{k\ell} - x_\ell)^2 + (y_{k\ell} - y_\ell)^2 + (z_{k\ell} - z_\ell)^2\right]^{\frac{1}{2}} = e_{k\ell}
\end{aligned}
$$

where $T_{ij} = (x_{ij}, y_{ij}, z_{ij})$ and $T_{k\ell} = (x_{k\ell}, y_{k\ell}, z_{k\ell})$.

This is a non trivial optimization problem that can be classified as $\mathcal{NP}$-Hard. The importance of the problem in relation in all that follows is captured in Figure 15.4. In Figure 15.4, the Melzak circles represent those link pairs of given terminals with equilateral vertices T_{ij} and $T_{k\ell}$. The twist angles between the planes defined by the Melzak circles are crucial to the SMT topology. For the $N = 4$ regular tetrahedron case, the twist angles between the planes are $\pi/2$. These twist angles define the Steiner tree topology for this given terminals. When one has terminals of $N = 4$ that deviate from the regular tetrahedron, then the twist angles also change and, in general, are very difficult to compute.

Therefore, as will be shown, the planes of the points (atoms) in the Steiner tree structure are an important part of the solution process for the Steiner tree topology.

15.3.4 Potential Energy Functions

It is important to know the minimum energy of a protein because it correlates strongly with the conformational structure [2]. In protein modeling, one potential energy objective function often used to measure the minimum energy configuration (MEC) is the following, where $K_{b_i}, K_{\theta_i}, A_{ij}, B_{ij}, \epsilon$ are adjustable weights; see Ref. [8]. This potential-energy objective function is based on the theoretical molecular mechanical force field model used to model most molecular structures; see Ref. [26]. It is interesting to observe that the objective function is the sum of nonlinear terms with little interaction between the terms.

$$
\begin{aligned}
E_{tot} = {} & \sum_i K_{b_i}(b_i - b_0)^2 \quad \text{bond lengths } [E_{bs}] \\
& + \sum_i K_{\theta_i}(\theta_i - \theta_0)^2 \quad \text{bond angles } [E_{ab}] \\
& + \sum_i K_{\tau_i}(\cos(3\tau_i - \gamma_0)) \quad \text{torsion angles } [E_{tor}] \\
& + \sum_i \sum_j A_{ij} d_{ij}^{-6} + B_{ij} d_{ij}^{-12} \quad \text{van der Walls } [E_{14vdW}] \\
& + \sum_i \sum_j V_i V_j / \epsilon d_{ij} \quad \text{electrostatic interactions } [E_{14e}]
\end{aligned}
$$

where

E_{bs} is the sum of energies arising from bond stretching or compression beyond the optimum bond length, and b_i, b_0 are the actual equilibrium bond lengths

E_{ab} is the sum of energies for angles that are distorted from their optimum values, and θ_i, θ_0 are the equilibrium bond angles

E_{tor} is the sum of the torsional energies that arise from rotations about each respective dihedral angle, and τ_i, γ_0 are the torsion angles
E_{14vdW} is the sum of energies due to van der Walls interactions and
E_{14e} is the electrostatic interaction

15.3.5 Steiner Trees and Minimum Energy Configurations

In this section, we address the crucial link between Steiner trees and MECs and how this linkage can be beneficial to predicting protein structure. Also, we argue that the geometric properties of Steiner trees can be used to approximate the potential energy function.

The relationship between SMTs and MECs is described by Maxwell's theorem; see Ref. [19] for proof. Let F_1, F_2, F_3, F_4 be unit forces acting at fixed terminals $\{V_1, V_2, V_3, V_4\}$, respectively. If one designs a network linking these terminals with Steiner points $\{s_1, s_2\}$ that can be moved into position, then one seeks to find the location of the Steiner points and the network where these forces will be in equilibrium. Figure 15.5 illustrates the MEC.

Theorem 1. *If one draws unit vectors from a Steiner tree in the direction of each of the lines incident to $V_1, V_2, \ldots, V_N$, and letting F_i denote the sum of the unit vectors at V_i, then in mechanical terms, F_i is the external force needed at V_i to hold the tree in equilibrium. The length of the tree SMT has the simple formula*

$$SMT = \sum_{i=1}^{N} V_i F_i$$

If the forces at the vertices are not all uniform, then the SMT acts as only a lower bound. This was discussed in some detail in our previous paper, see Ref. [39]. Notice that in Maxwell's theorem the function is separable in the force components. Thus, the FST can be decomposed into its FST components. One way to construct a FST is to identify its FST components. Thus, if we can subdivide the overall terminal point set into its FST components, then we can compute the overall SMT by constructing the SMT for its FST components.

As a simple example of what is to follow, let us examine a unit cube in 3-D space and generate its SMT (see Figure 15.6). Following that, we give a somewhat detailed examination of the Steiner structure of the amino acid serine.

Figure 15.6 illustrates the Steiner tree for $N = 8$ terminals of a unit cube. The $N-2$ Steiner points (smaller circles) lie within the cube and form a plane as indicated. The figure on the right illustrates two of the Melzak circles associated with this planar structure. Because the two Melzak circles lie in the same plane, their dihedral angle is 0°. We are generally interested in the planes generated by the Steiner trees and the dihedral angles between them.

15.3.6 Serine

Serine (Ser) is an amino acid with an uncharged polar side chain. Ser has 14 atoms. It is not essential for the human diet but is important in metabolism. Serine was first obtained from silk protein, which we will examine in more detail in later sections of the chapter. The average occurrence of Ser in proteins is ≈6.8 percent [40]. Figure 15.7 shows first the chemical structure of serine and then the optimal Steiner tree structure of Ser.

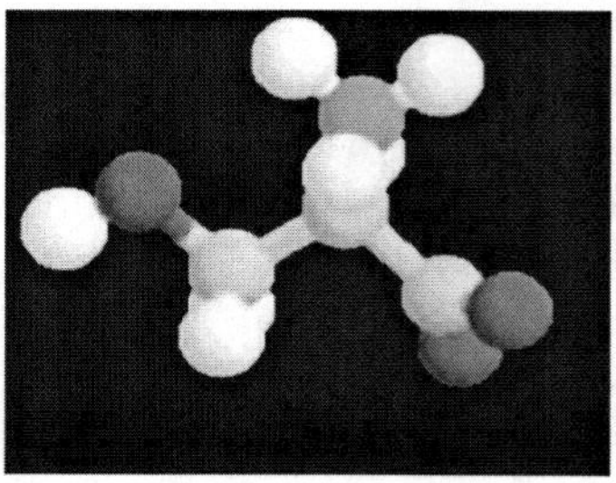

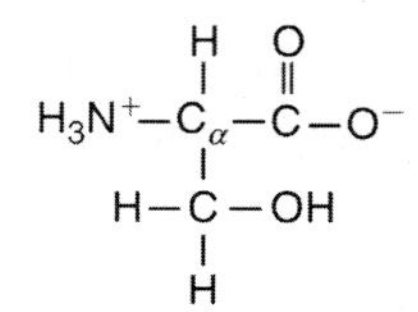

Ser chemical structure

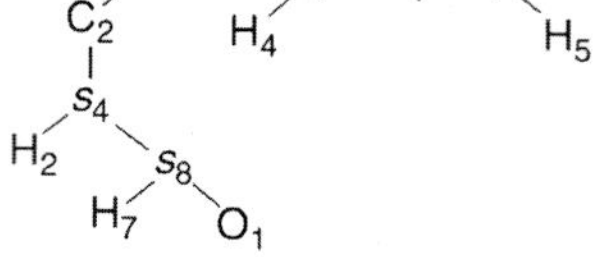

Ser Steiner topology

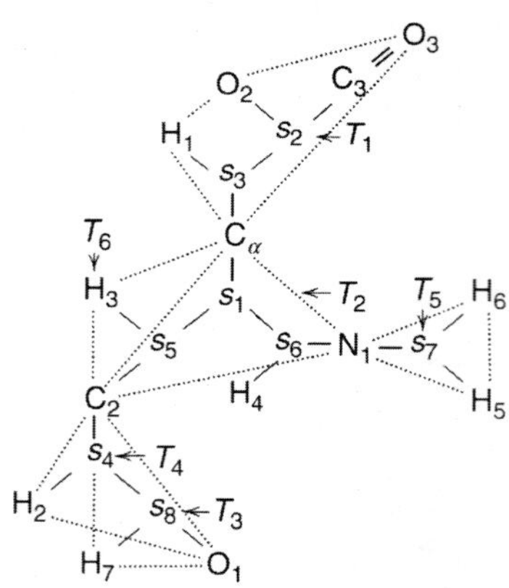

Ser Steiner planes

The bottom of Figure 15.7 illustrates the optimal Steiner structure of Ser again from two different viewpoints. The planes that can be defined for Ser from the Steiner structure are as follows:

$$\mathcal{T}_1 := \{O_3, O_2, C_\alpha, H_1\}; \quad \mathcal{P}_1 := \{O_2, C_\alpha, H_1\}$$

$$\mathcal{T}_2 := \{C_\alpha, C_2, N_1, H_3\}; \quad \mathcal{P}_2 := \{C_\alpha, C_2, N_1\}$$

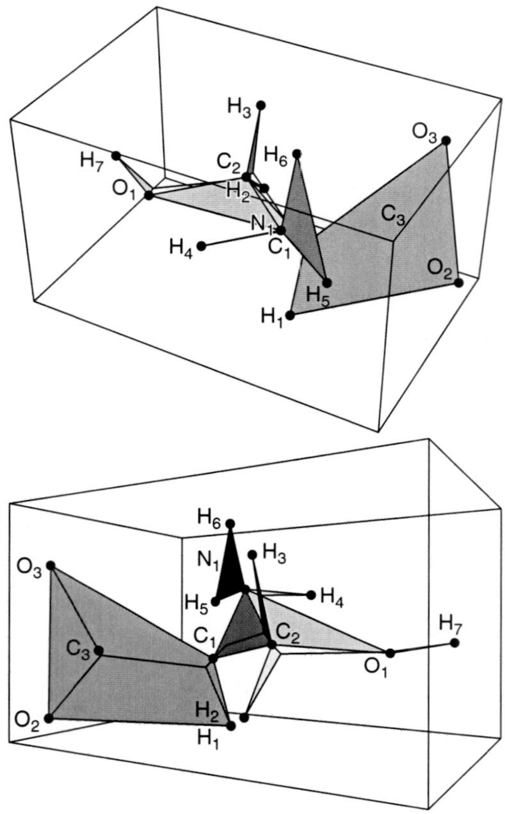

Figure 15.7 Serine Steiner topologies and optimal structure.

$$\mathcal{P}_3 := \{O_1, C_2, H_7\}; \quad \mathcal{P}_4 := \{C_2, O_1, H_2\}$$

$$\mathcal{P}_5 := \{N_1, H_5, H_6\}; \quad \mathcal{P}_6 := \{C_\alpha, C_2, H_3\}$$

Here is the MAPLE output for the twist angles along with the approximation matrix:

$\mathcal{P}$	1	2	3	4	5	6
1	[0.0]	[89.95]	[59.94]	[60.0]	[60.03]	[60.0]
2		[0.0]	[60.0]	[60.0]	[59.94]	[59.94]
3			[0.0]	[60.0]	[60.0]	[60.0]
4				[0.0]	[90.0]	[90.02]
5					[0.0]	[0.0]
6						[0.0]

$\mathcal{P}$	1	2	3	4	5	6
1	–	$\pi/2$	$\pi/3$	$\pi/3$	$\pi/3$	$\pi/3$
2		–	$\pi/3$	$\pi/3$	$\pi/3$	$\pi/3$
3			–	$\pi/3$	$\pi/3$	$\pi/3$
4				–	$\pi/2$	$\pi/2$
5						0
6						–

15.4 Alanine Dipeptide Results

Previous papers have focused on the analysis of single individual amino acids in proteins with Steiner trees such as in the analysis of serine in the previous section of this chapter. Another recent paper expounds on the torsional energy relationship between Steiner trees, planes generated by the Steiner trees, and folding and misfolding of protein structures, see Ref. [37].

We wish to look at dipeptides and pentapeptides with Steiner trees and examine the regularity properties of the dihedral angles of the planes generated by the Steiner trees.

We first examine the simplest and most well-known dipeptide, Ala-Ala, which consists of two alanine amino acids connected by a peptide bond. There are a total of 23 atoms as seen in the following schematic diagram [35].

Given the alanine dipeptide (Figure 15.8), we perturb a pair of torsion angles (ψ_1, ϕ_2), each of which can be defined by four consecutive heavy atoms along the peptide backbone such as N1a-Caa-C2a-N1b and C2a-N1b-Cab-C2b, respectively (see Figure 15.10). Table 15.1 shows the Steiner values for nine different alanine dipeptide conformations. The results of the SMT program represent the best upper bounds on the Steiner ratio that have been found in a reasonable amount of computing time. Figure 15.9 indicates the ρ values for the angular variations listed in Table 15.1. Note that the $(-180°, -60°)$ conformation has the highest Steiner ratio of 0.997588.

The SMT ρ value for the Swiss-PdbViewer energy minimization [20] configuration is optimal. To test if the inverse relationship between Steiner ratio and potential energy can be extended from single amino acid to dipeptide, we also compute energy values of those

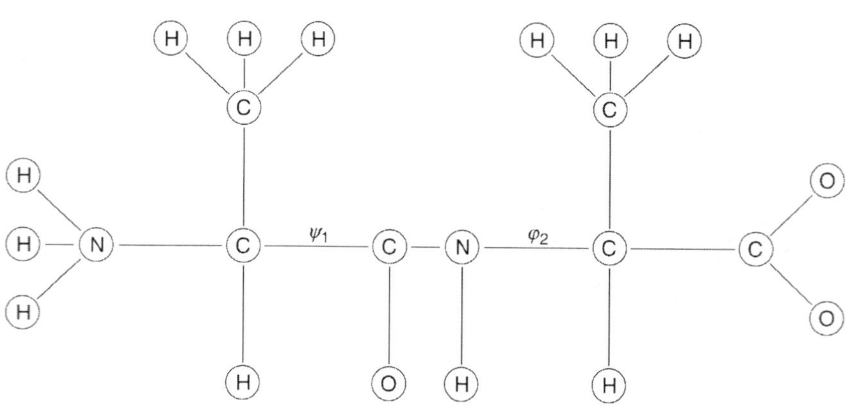

Figure 15.8 Alanine dipeptide structure.

Table 15.1 Energy Minimization and ρ Values

$(\psi - 1,$	$\phi - 2)$	Steiner ρ Value
−180	−180	.995803
−180	−60	.997588
−180	60	.995749
−60	−180	.995950
−60	−60	.995750
−60	60	.995901
60	−180	.995618
60	−60	.995778
60	60	.995811

conformations using the Swiss-PdbViewer energy calculation module [20]. The results (not displayed here) apparently show the inverse relationship. For instance, the total energy of the $(-180°, -60°)$ conformation is 480 kJ/mol $(\rho = 0.997588)$ although two other non-minimal conformations $(-60°, -180°)$ and $(60°, 60°)$ have much higher energy values of 510 kJ/mol $(\rho = 0.995950)$ and 613 kJ/mol $(\rho = 0.995811)$, respectively.

Figure 15.9 indicates the ρ values for the Ramachandran angular variations listed in Table 15.1.

Figure 15.10 presents the minimum energy conformation of alanine dipeptide with its torsion angles of $(-180°, -60°)$. From its Steiner tree topology, we have three Steiner planes such that $P_1 = \{\text{Caa}, \text{Cba}, \text{N1a}\}$; $P_2 = \{\text{Caa}, \text{O1a}, \text{Cab}, \text{N1b}\}$ and $P_3 = \{\text{Cab}, \text{C2b}, \text{Cbb}\}$ The volume of the tetrahedron P_2 is essentially zero as one can see that all four atoms are on the same plane. The twist angles among these three Steiner planes are

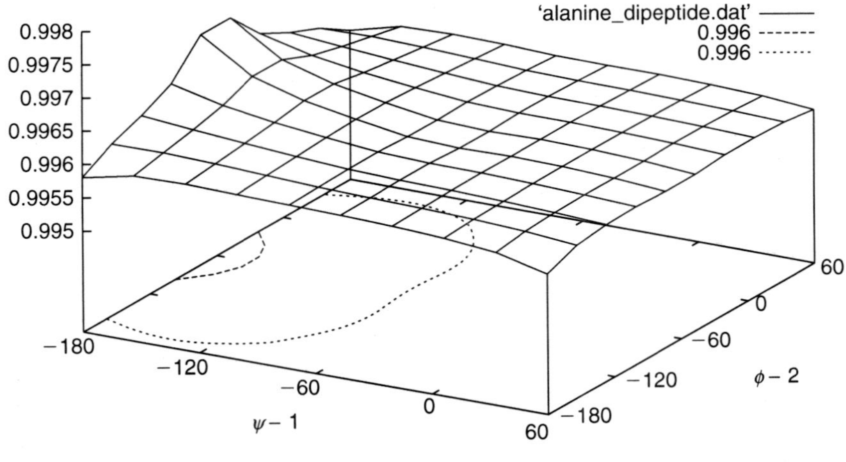

Figure 15.9 Graph of SMT ρ function.

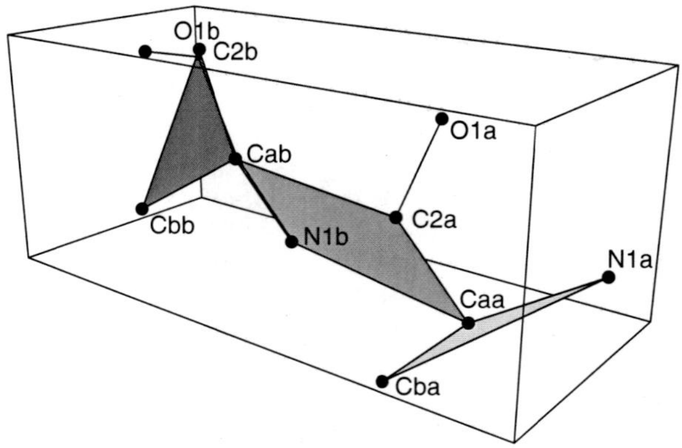

Figure 15.10 Alanine dipeptide minimum energy configuration.

given in the matrix **T** below. What is remarkable is that all the angles for the minimal conformation are essentially 60°.

$$\text{For the } (-180^\circ, -60^\circ) \text{ conformation,} \quad \mathbf{T}_1 = \begin{bmatrix} -- & \pi/3 & \pi/3 \\ & -- & \pi/3 \\ & & -- \end{bmatrix}$$

The following are samples of twist angles for other non-minimal conformations. These angles become irregular subject to the backbone torsion angle variations.

$$\text{For the } (-60^\circ, -180^\circ) \text{ conformation,} \quad \mathbf{T}_2 = \begin{bmatrix} -- & \pi/2 & \pi/3 \\ & -- & \pi/3 \\ & & -- \end{bmatrix}$$

$$\text{For the } (60^\circ, 60^\circ) \text{ conformation,} \quad \mathbf{T}_3 = \begin{bmatrix} -- & \pi/3 & \pi/3 \\ & -- & \pi/2 \\ & & -- \end{bmatrix}$$

We note that carbon and nitrogen atoms of the peptide backbone act as Steiner points as in single amino acid topology and the regularity of twist angles between Steiner planes determined by those points is strongly related to the local stability of protein chain in terms of potential energy. Aforementioned, the backbone torsion angles of dipeptide or protein in general are also determined by positions of consecutive carbon and nitrogen atoms along the backbone. Therefore, we can use Steiner planes instead of the backbone torsion angles in characterizing protein local structures. In addition, the Steiner ρ value can be a surrogate for measuring energy in the protein structure.

15.5 General Dipeptide Results

In the preceding section, the Steiner structure of the alanine dipeptide defined several planes involving the carbon and nitrogen atoms of the peptide backbone. These planes generated by the Steiner tree topology maintain certain angular relationships within the secondary structure much as in the Ramachandran plot, see Figure 15.3.

15.5.1 Virtual Plane

Motivated by this insight, we postulate that the position of Steiner points in the backbone is strongly related to the regularity of secondary structures energetically stabilized by hydrogen bonds. To clarify this hypothesis, we introduce a new virtual plane defined by the initial carbon and nitrogen atoms of each amino acid structure along the backbone. This plane is related to the conformations of amino acids of native proteins much as we have shown for the SMT protein structures in the previous section. We then measured the dihedral angle of two consecutive virtual planes (called the virtual plane angle) along the backbone. This single plane angle representation encompasses the traditional torsion angles of amino acids along the backbone (see Figure 15.11).

We chose a reference data set of 100 protein structures from the Protein Data Bank (PDB) and then statistically investigated the regularity of virtual

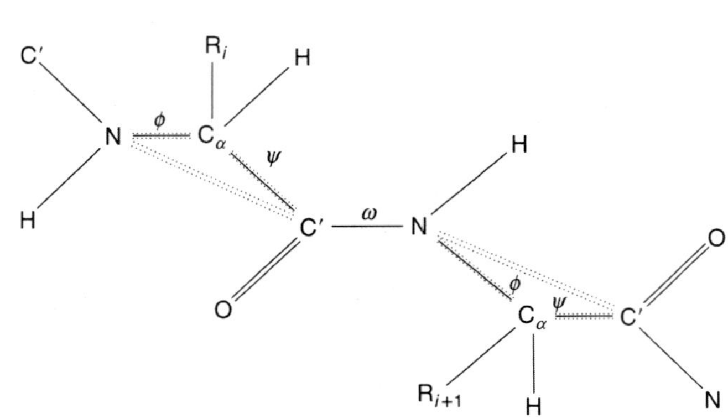

Figure 15.11 Peptide virtual dihedral angles.

plane angles depending on the type of secondary structures. We restricted the focus to α-helices (β-strands) with more than 11 (6) residue long. The first and last residues were excluded from the test to isolate terminal mobility. Consequently, we got about 3000 (500) samples of virtual plane angles for helices (strands). Figure 15.12 shows the histograms of virtual plane angles for helices and strands, respectively. Roughly speaking, most helical

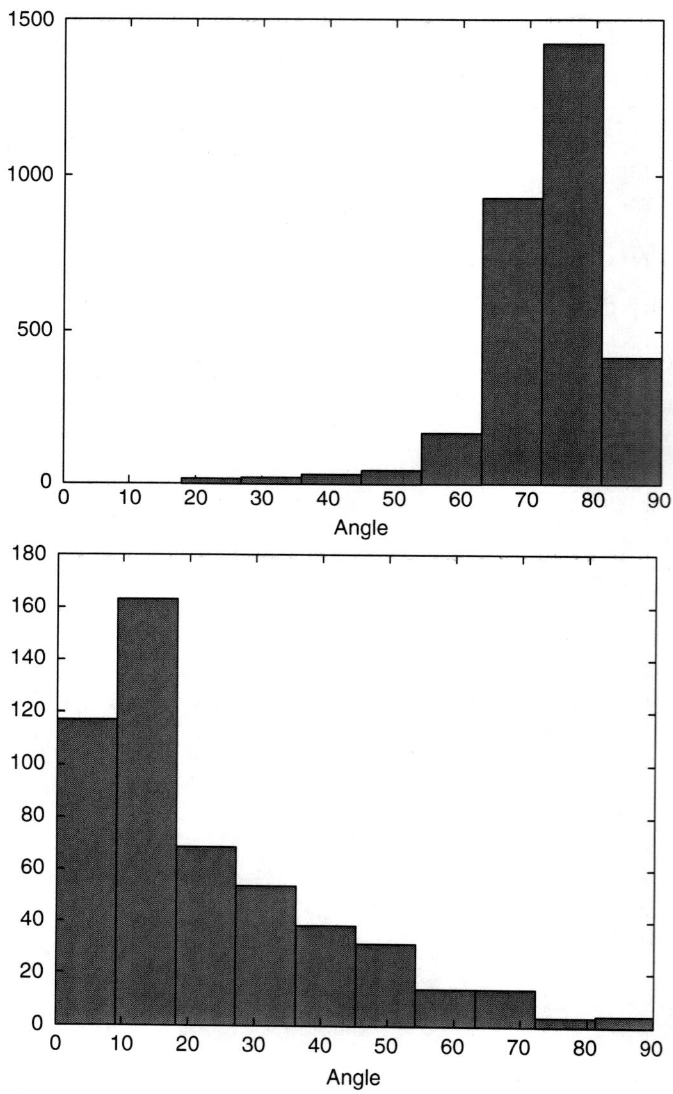

Figure 15.12 Statistical histograms of α-helix (above) and β-sheet structures (below).

conformations have large virtual plane angles (i.e., a virtual plane is almost perpendicular to the next), whereas β-strands obviously form like straight lines by taking small virtual plane angles.

Next, we modeled a 3-D standard dipeptide structure using group theory in the left of Figure 15.13 and then calculated the virtual plane angles analytically when various torsion angles were applied. We chose 60 (90) pairs

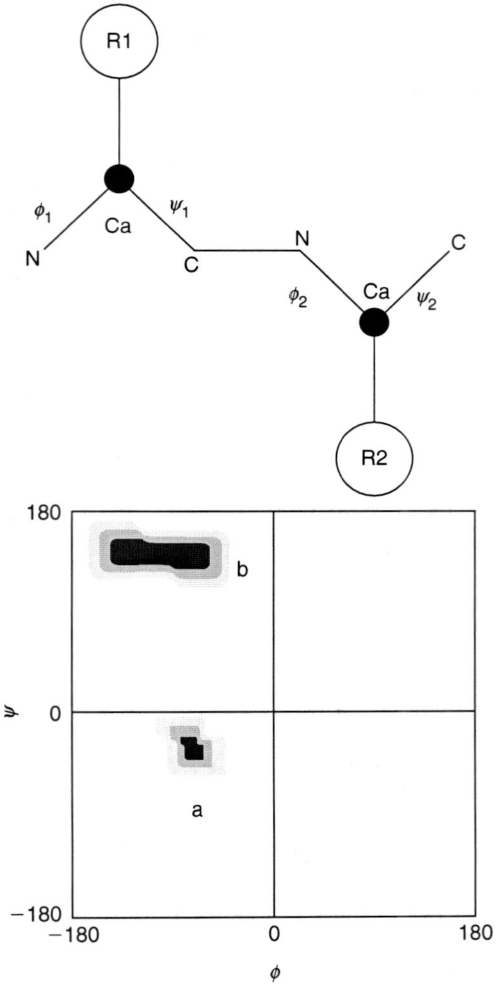

Figure 15.13 **Illustration of a 3-D dipeptide model and the Ramachandran plot (above). In the above figure, a 3-D peptide model is developed. Hydrogen and oxygen atoms are not displayed here. There are four torsion angles to be assigned. In the figure on the below, two clusters corresponding to α-helices and β-strands are chosen from the Ramachandran plot and torsion angles are sampled from there, respectively. This contour plot is adopted from [23].**

of ϕ and ψ angles from the region of Ramachandran plot corresponding to α-helices (β-strands) as shown in right of Figure 15.13. Consequently, we generated 3600 (8100) different dipeptide conformations and virtual plane angles for α-helices (β-strands).

Figure 15.14 displays the histograms of virtual plane angles from analytic data. It is observed that angular regularity of the analytic dipeptide model

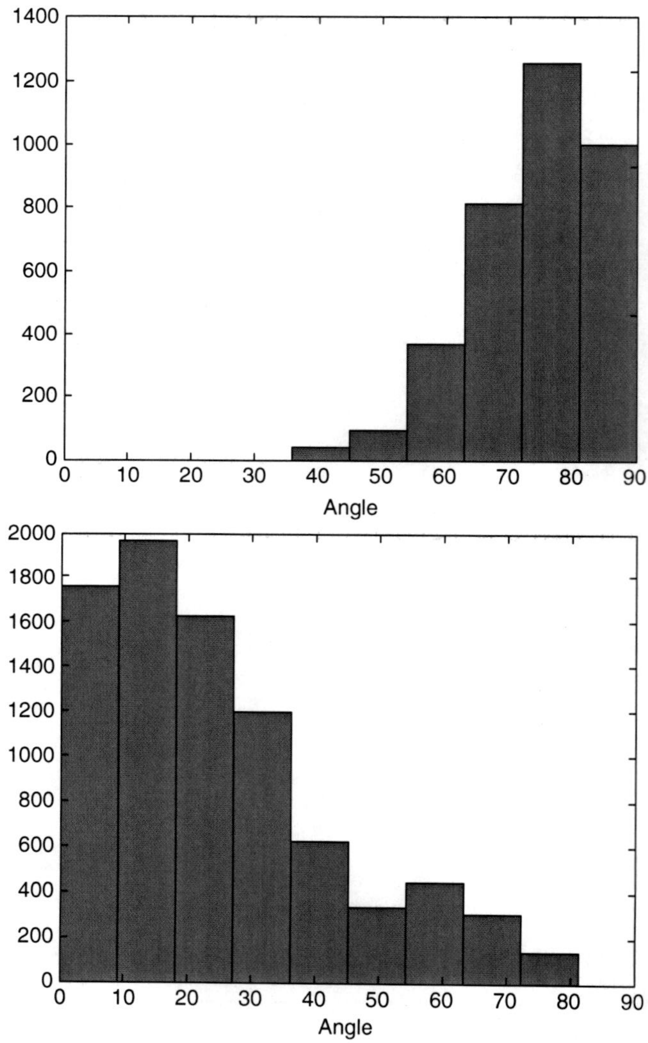

Figure 15.14 Analytical histograms of α-helix (above) and β-sheet structures (below)- histograms of the virtual plane angles are computed analytically. α-Helices (β-strands) are highly populated in the range of 70° – 90° (0° – 20°) as was observed from the statistical data in Figure 15.12.

has good agreement with statistical data even though there is small variation between statistical and analytical histograms. These errors come from the different population of torsion angles between statistical data and analytical data. The native proteins might have their unique torsion angle preferences depending on the types of secondary structures. However, the Ramachandran plot provides no detailed information about the relative population of a particular torsion angle within the same category of secondary structure, e.g., the α-helix. This limitation yields a practical assumption that the population of each sampled conformation is uniform in our 3-D model.

The concept of virtual plane angles proposed here enables the Steiner tree topology to describe the protein secondary structures, as well as side chain conformations of each amino acid. Furthermore, this single parameterization based on the protein topology encourages us to develop a computationally efficient folding prediction by searching only N dimensional conformational space for an N residue-long protein instead of using the traditional $2N$ dimensional conformational space associated with ϕ and ψ angles.

15.5.2 Ala-Gly Silk β-Structure

Silk is a protein produced in the posterior silk glands of the larva of the cultivated silk worm *Bombyx mori* for the construction of cocoons. It also occurs in the webs of a number of various spiders [40]. The silk fibers are comprised of a protein called fibroin which is a theoretical model of the protein.

The protein is constructed from layers of antiparallel β-pleated sheets, which run parallel to the fiber axis [40]. Although each chain is comprised of multiple repeats of the sequence (Gly-Ser-Gly-Ala-Gly-Ala)$_n$, the protein is often approximated by a repeating units of (Gly-Ala)$_n$. We would like to understand the alignment and the twist angles of the Steiner planes in these two acids as perhaps a key to the protein structure of fibroin.

15.5.2.1 Experiments for Individual Acids

The two protein structures of fibroin that we will examine are from a paper by Fossey et al. [17]. Although further Steiner structure details appear in Ref. [36], we are most interested in the virtual angle measurements which are reflected in Figure 15.15. In fact, the Silk2 structure has a lower energy value than the Silk1 structure.

The N_1, C_α, and C_2 atoms of each amino acid for the silk structures are identified in Table 15.2 for five of the amino acids in each silk structure. Notice that in the two tables in Table 15.2, Silk1 has more variability, i.e., a

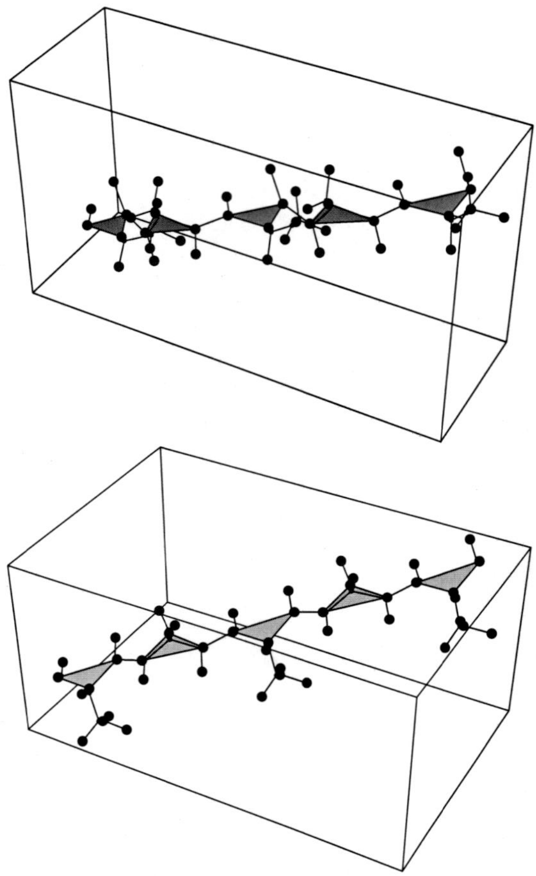

Figure 15.15 Silk1 structure (above) silk2 structure (below).

larger range in the twist angles values, than the Silk2 structure. Because they are both β-sheet structures, their twist angles are much closer to zero than in the α-helix structures. Thus, the variability of the virtual angle measures in the comparison matrix becomes a valuable measure of structural stability.

Table 15.2 Comparison of Twist Angles for Silk1 (Left) Silk2 (Right)

[0.0]	[0.0]	[6.382]	[3.969]	[10.64]	[0.0]	[4.727]	[6.382]	[4.364]	[4.364]
	[0.0]	[6.118]	[3.969]	[10.48]		[0.0]	[3.533]	[1.812]	[3.032]
		[0.0]	[3.032]	[4.364]			[0.0]	[5.316]	[1.812]
			[0.0]	[7.069]				[0.0]	[4.657]
				[0.0]					[0.0]

15.5.3 Pentapeptide Results

In a final set of experiments, a set of pentapeptide proteins will be examined. In these experiments, the properties of the planes of the acids will be utilized to get a measure useful for characterizing the overall secondary structure. As argued earlier, the N_1, C_α, and C_2 atoms of each amino acid create a virtual plane and we wish to examine the twist angles of the planes of these atoms for each acid. The reason for selecting these three atoms is that they are essentially Steiner points of the acids and they will form a force plane unique to each acid. This is a simpler way to capture the information in the Ramachandran plots (see Figure 15.13) because it involves only one degree of freedom rather than two degrees of freedom.

Why is this an equivalent representation of the backbone folding structure of the peptide? Each of the three atoms is part of the amide plane of each acid. The three atoms are the minimum number of atoms needed to define a plane in space and the three atoms together provide more information about the protein structure than simply C_α by itself. The pairs of adjacent twist angles of these planes along the peptide should reveal the underlying pattern of conformation angles for the protein.

15.5.3.1 Leu-Ser-Phe-Ala-Ala ($N = 71$ Atoms) $\rho = 0.995210$

Let us examine an α-helix structure with a different amino acids sequence from the silk structures. A native structure of Leu-Ser-Phe-Ala-Ala is chosen from the enzyme citrate synthase (PDB ID: 1CTS), residues 260 – 264, which form a part of buried helix. To test the regularity of twist angles between virtual planes, we first modeled a 3-D standard pentapeptide structure with the same sequence above and then perturbed the backbone torsion angles ψ_1 through ϕ_5 by an increment of 30° degrees. Neither ϕ_1 nor ψ_5 can be defined in this case.

Energy values of each conformation were computed to find the lowest energy conformation among the sampled structures. Torsion angles for the native structure are compared to those of the sampled lowest energy conformation in Table 15.3. If the number of sampled conformations is increased by decreasing the size of increment, it is expected to find out much closer structures to the native one. In Figure 15.16, we illustrate the

Table 15.3 Torsion Angle Comparison between the Native and the Sampled Structures

	ψ_1	ϕ_2	ψ_2	ϕ_3	ψ_3	ϕ_4	ψ_4	ϕ_5
Native	−29.5	−78.9	−35.3	−70.1	−44.5	−62.9	−28.3	−81.6
Sampled	−30.0	−90.0	−30.0	−60.0	−30.0	−60.0	−30.0	−90.0

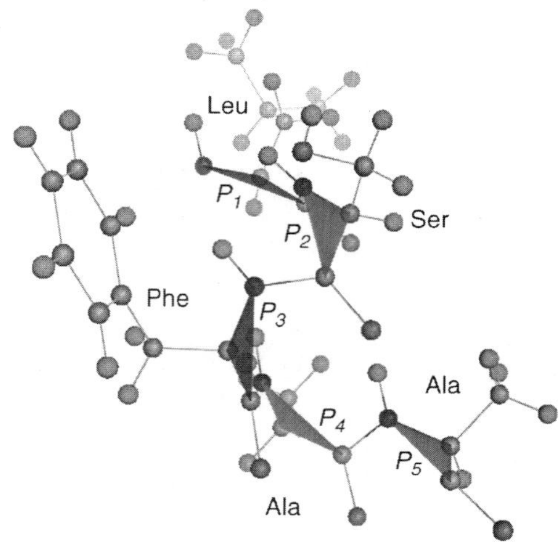

Figure 15.16 SMT of the sampled pentapeptide structure.

pentapeptide structure as well as the twist angle matrix. Notice that the consecutive planes involving the atoms of all five amino acids are very regular with twist angles of P_1, P_2, P_3, P_4, and P_5, respectively. The range of twist angles indicates that the sampled structure forms an α-helix. The Steiner ρ value is 0.99521.

Other helical segments from the same protein were also tested. The predicted conformations (not displayed here) show good agreement with the native structures and the regular twist angles indicate that they are part of a helix structure. Consequently, twist angles between consecutive virtual planes defined here enable us to characterize secondary structures in proteins faster than the conventional backbone torsion angle representation. In addition, the Steiner ratio calculation becomes a powerful tool to check the stability of local secondary structures instead of measuring their energy value. Further, these two topology-based tools can be utilized in developing a faster protein folding prediction algorithm.

15.6 Summary and Conclusions

We have presented a new dihedral angle measure relevant to the prediction of secondary structure in proteins. We have tested the angular measure for a subset of α-helix and β-structures from the PDB as well as individual dipeptide and pentapeptide structures. The new virtual angle measure is viewed as complimentary and simpler than the (ϕ, ψ) dihedral angle measure of the Ramachandran plots because it uses one degree of freedom less. Future work includes testing the methodology for identifying secondary and tertiary structure in a protein folding algorithm.

References

1. C.B. Anfinsen, E. Haber, M. Sela, and F.H. White Jr., 1961. The kinetics of formation of native ribonuclease during oxidation of the reduce polypeptide chain, *PNAS*, **47**, 1309–1314.
2. C.B. Anfinsen, 1973. Principles that govern the folding of protein chains, *Science*, **181**, 223–30.
3. V. Biou, J.F. Gibrat, J.M. Levin, B. Robson, and J. Garnier, 1988. Secondary structure prediction—Combination of three different methods, *Protein Engineering*, **2**, 185–191.
4. J. Blazewicz, P.L. Hammer, and P. Lukasiak, 2005. Predicting secondary structures of proteins, *IEEE Engineering and Biology in Medicine Magazine*, May/June, 88–94.
5. C. Brandon and J. Tooze, 1991. *Introduction to Protein Structure*, Gabriel Publishing, New York.
6. J.-M. Chandonia and M. Karplus, 1999. New methods for accurate prediction of protein secondary structure, *Proteins: Structure, Function, and Genetics*, **35**, 293–306.
7. P.Y. Chou and G.D. Fasman, 1978. Empirical predictions of protein structure, *Annual Review of Biochemistry*, **47**, 251–276.
8. F.E. Cohen, 1995. Folding the sheets: Using computational methods to predict the structure of proteins, E. Lander, Ed., *Calculating the Secrets of Life*. National Academy of Sciences, Washington D.C., pp. 236–271.
9. R.E. Dickerson and I. Geis, 1969. *The Structure and Action of Proteins*, Harper and Row, New York.
10. K.A. Dill, 1997. Additivity principles in biochemistry, *Journal of Biological Chemistry*, **272**, 701–704.
11. B.W. Dijkstra, K.H. Kalk, W.G.J. Hol, and J. Drenth, 1981. Structure of bovine pancreatic phospholipase a2 at 1.7 angstroms resolution, *Journal of Molecular Biology*, **147**, 97–123.
12. D.Z. Du, 1991. On Steiner ratio conjectures, *Annals of Operations Research*, **33**, 437–449.
13. D.Z. Du and F.K. Hwang, 1992. A proof of the Gilbert–Pollak conjecture on the Steiner ratio, *Algorithmica*, **7**, 121–135.
14. D.Z. Du, F.K. Hwang, and J.F. Weng, 1982. Steiner minimal trees on zig-zag lines, *Transaction of the American Mathematical Society*, **278**, 149–156.
15. Electronic Dissertation Library. Site visited 11/03/05. http://dis.shef.ac.uk/ruth/bib.html.
16. G.D. Fasman, 1989. The development of the prediction of protein structure, G.D. Fasman, Ed., *Prediction of Protein Structure and the Principles of Protein Conformation*, Plenum Press: London, pp. 193–316.
17. S.A. Fossey, G. Nemethy, K.D. Gibson, and H.A. Scheraga, 1991. Conformational energy studies of beta-sheets of model silk fibroin peptides. I. sheets of poly(Ala-Gly) chains, *Biopolymers*, **31**, 1529–1541.
18. J.D. Garnier, Osguthorpe, and B. Robson, 1978. Analysis of the accuracy and implications of simple methods for predicting the secondary structure of globular proteins, *Journal of Molecular Biology*, **120**, 97–120.
19. E.N. Gilbert and H.O. Pollak, 1968. Steiner minimal trees, *SIAM Journal of Applied Mathematics*, **16**, 1–29.

20. N. Guex and M.C. Peitsch, 1997. SWISS-MODEL and the Swiss-PdbViewer: An environment for comparative protein modeling, *Electrophoresis*, **18**, 2714–2723.
21. L. Holm and C. Sander, 1992. Evaluation of protein models by atomic solvation preference, *Journal of Molecular Biology* , **225**, 93–105.
22. H.W. Holley and M. Karplus, 1989. Protein secondary structure prediction with a neural network, *PNAS*, **86**, 152–156.
23. S. Hovmoller, T. Zhou, and T. Ohlson, 2002. Conformations of amino acids in proteins, *Acta Crystallographica*, **D58**, 768–776.
24. J.L. Klepsis, M.J. Pieja, and C.A. Floudas, 2003. Hybrid global optimization algorithms for protein structure prediciton: Alternating hybrids, *Biophysical Journal*, **84**, 869–882.
25. T. Lazardis and M. Karplus, 1998. Discrimination of the native and misfolded protein Models with an energy function including implicit solvation, *Journal of Molecular Biology*, **288**, 477–487.
26. A.R. Leach, 1996. *Molecular Modeling—Principles and Applications*, 2nd ed., Prentice-Hall, New York.
27. J.M. Levin, S. Pascarella, P. Argos, and J. Garnier, 1993. Quantification of secondary structure prediction improvement using multiple alignments, *Protein Engineering*, **6**, 849–854.
28. J.M. Levin, B. Robson, and J. Garnier, 1986. An algorithm for secondary structure determination in proteins based on sequence similarity, *FEBS Letters*, **205**, 303–308.
29. V.I. Lim, 1974. Algorithms for prediction of alpha-helices and beta-structural regions in globular proteins, *Journal of Molecular Biology*, **88**, 873–894.
30. A.E. Mark and W.F. van Gunsteren, 1994. Decomposition of the free energy of a system in terms of specific interactions, *Journal of Molecular Biology*, **240**, 167–176.
31. Z.A. Melzak, 1961. On the problem of Steiner, *Canadian Mathematical Bulletin*, **4**, 143–148.
32. A. Neumaier, 1997. Molecular modeling of proteins and mathemartical prediction of protein structure, *SIAM Review*, **39**, 407–460.
33. J. Pillardy, C. Czaplewski, A. Liwo, J. Lee, D.R. Ripoli, R. Kazmierkiewicz, S. Oldziej, W.J. Wedemeyer, K.D. Gibson, Y.A. Arnautova, J. Saunders, Y.-J. Ye, and H. Scheraga, 2001. Recent improvements in prediction of protein structure by global optimization of a potential energy function, *PNAS*, **98**, 2329–2333.
34. U. Reimer and G. Fuellen, 1997. Biocomputing in a Nutshell, http://www.techfak.uni-bielefeld.de/bcd/ForAll/Basics/welcome2.html.
35. R. Rodosek and D. Zupanic, 1997. Minimising the energy of the alanine dipeptide by simulated annealing, Working Paper, Imperial College, IC-Parc, London England.
36. J.M. Smith, 2006. Steiner minimal trees and twist angles in folded protein structure, *Proceedings of the BIOMAT 2005 International Symposium on Mathematical and Computational Biology*, R. Mondiani and R. Dilao Eds, World Scientitific Press: Hong Kong.
37. J.M. Smith, Y. Jang, and M.K. Kim, 2007. Steiner minimal trees, twist angles and the protein folding problem, *Proteins: Structure, Function and Bioinformatics*, **66**, 889–902.
38. D. Strickland, E. Barnes, and J. Sokol, 2005. Optimal protein structure alignment using maximum cliques, *Operation Research*, **53**, 389–402.

39. C. Stanton and J.M. Smith, 2004. Steiner trees and 3-D macromolecular conformation, *Informs Journal of Computing*, **16**, 470–485.
40. D. Voet and J.G. Voet, 1995. *Biochemistry*, 2nd ed., Wiley, New York.
41. Y. Xu, D. Xu, and E.C. Uberbacher, 1998. Computational tools for protein folding, *Journal of Computational Biology*, **5**, 597–614.

Chapter 16

Optimization of Tumor Virotherapy with Recombinant Measles Viruses

Željko Bajzer, Thomas Carr, David Dingli, and Krešimir Josić

CONTENTS

16.1 Introduction 468
16.2 Mathematical Model of Virotherapy 470
16.3 Analysis of Equilibria 474
16.4 Model Validation and Parameter Estimation 478
16.5 Simulations 480
16.6 Conclusion 488
Acknowledgments 488
References 489

Abstract Recombinant viruses based on the vaccine strain of measles virus have potent and selective activity against a wide range of tumors. Successful tumor therapy with these viruses (virotherapy) depends on efficient infection of tumor cells by the virus. Infected cells express viral proteins that allow them to fuse with neighboring cells to form syncytia. Infection halts tumor cell replication and the syncytia ultimately die. Moreover, infected cells may produce new virus particles that proceed to infect additional tumor cells. The outcome of virotherapy depends on the dynamic interactions between the uninfected tumor cells, infected cells, and the virus population. We present a model of tumor and virus interactions based on the phenomenologically established interactions between the three populations. Other similar models proposed in the literature are also discussed. The model parameters are obtained by fitting the model to experimental data. We discuss equilibrium states and explore by simulations the impact of various initial conditions and perturbations of the system in an attempt to achieve tumor eradication. We show that the total dose of virus administered and the rate at which the tumor grows play determining roles on the outcome. If tumor growth can be slowed, the minimal dose of virus needed for curative therapy can be reduced substantially. An interesting prediction of the model is that virotherapy is more effective on larger tumors when deceleration of growth occurs.

16.1 Introduction

The majority of hematopoietic neoplasms remain incurable with currently available therapies. For example, multiple myeloma (MM) has a median survival of three–four years despite the availability of high-dose therapy with stem cell transplantation and the introduction of novel agents, such as thalidomide and bortezomib [1]. Thus, there is an urgent need for novel therapeutic modalities for these disorders. Over the last few years, engineered viruses (both DNA and RNA based) have been introduced as potential cancer therapeutic agents [2,3]. Several trials have been performed with replication-selective adenoviruses in head and neck cancer [4] and metastatic colon carcinoma [5,6] while Newcastle disease virus has been given to patients with various tumor types [7]. The use of viruses for tumor therapy introduces several new concepts in the field of therapeutics because an underlying premise of tumor therapy is that the infected tumor cells become factories that generate new virus particles that infect more tumor cells in a series of waves [2]. This introduces the concept of population dynamics and the outcome of such therapy depends in a complex way on the interactions between the population of virus and tumor cells [8–12].

Our work has centered on the use of engineered viruses, derived from the Edmonston vaccine strain of measles virus (MV-Edm). The vaccine strain was chosen as a therapeutic platform because of anecdotal reports of resolution of Burkitt's lymphoma in patients who acquired wild-type measles virus infection. Moreover, the vaccine has been given to more than a billion people with an excellent safety record. MV-Edm and derivative viruses obtained by virus engineering have potent and selective oncolytic activity against a wide variety of human tumors, including non-Hodgkin lymphoma [13], MM [14], ovarian carcinoma [15], glioma [16], and breast carcinoma [17] although they leave normal tissues unharmed.

MV infection starts when the viral hemagglutinin (H) protein binds to its receptor on target cells. The H protein displayed by wild-type measles virus preferentially binds to CD150 (also known as SLAM) [18] whereas the H protein of MV-Edm preferentially interacts with CD46 [18]. Most tumor cells over-express CD46 [19,20] and this is thought to be one of the mechanisms behind the selective tropism of these viruses for tumor cells. Binding of the H protein with its cognate receptor induces conformational changes in the fusion (F) protein which in turn triggers membrane fusion between the virus particle and the target cell. The functional separation of target cell binding (via H) and fusion (via F) in MV also facilitated retargeting of the virus to specific tumors by modification of the viral H protein [21–24]. Protein engineering has also led to complete ablation of H binding to the known viral receptors (CD46 and CD150w) so that fully retargeted viruses that only infect tumor cells of interest have been generated [25–27]. In addition, MV vectors have been modified to allow noninvasive monitoring of viral gene expression by the secretion of the biologically inert soluble peptides CEA (MV-CEA) and human chorionic gonadotrophin (hCG, MV-hCG) [28]. More recently, MV-Edm has been modified to induce expression of the thyroidal sodium iodide symporter (NIS, MV-NIS) in infected cells to enable noninvasive imaging of the biodistribution and replication of the virus in vivo [29,30].

Although MV-based vectors have potent oncolytic activity, some tumors are not eliminated in vivo. Studies using MV-CEA show that although the virus efficiently infects and propagates in these tumors, the latter may persist [15,29,31]. This highlights the dynamic interplay between viral replication, tumor cell growth, and the death rate of infected tumor cells [8–12,32,33]. In this respect, MV-NIS has an advantage over the parent virus because it can be combined with beta particle emitting isotopes such as iodine (131) (131-I). The electrons emitted during isotope decay have a macroscopic path length and can destroy uninfected tumor cells with a significant bystander effect [29,34,35].

Measles virus can control tumor growth by at least two mechanisms. Infected cells express the viral H and F proteins and can interact with neighboring cells with the result that the cells fuse together. Spread of cell-to-cell fusion leads to the formation of syncytia that ultimately die, usually after a few days [14,15]. In addition, once the infected cells die, they may release free virus particles that can infect surrounding cells. Moreover, infected cells stop replicating and do not contribute to further tumor growth.

The success of tumor virotherapy depends on infection of tumor cells that serve as sites for virus amplification. Once the virus is released from infected cells, the new particles can infect additional tumor cells. Thus, the virus propagates through the tumor in a series of waves [2]. The interactions between the tumor and virus populations are complex and understanding their dynamics requires mathematical modeling. There has already been considerable work on modeling these interactions [8–11,32,33,36].

In the following, we describe our model of tumor and virus interactions that take into consideration both virus production and spread of the infection between cells. We utilize the model to evaluate various therapeutic scenarios and to test whether curative therapy is possible with virus alone or virus in combination with other agents. Patients with advanced hematological cancer are usually immunosuppressed and hence the smallest dose of virus as well as timing of therapy may be critical for an optimal response. It seems logical to try and determine the smallest possible dose of virus that can be associated with a cure, given that this will probably be associated with the lowest risk of complications. In the following, we address several critical questions pertaining to tumor therapy including

1. Is the initial tumor burden important for the outcome of therapy?
2. Is cure possible with a therapeutically achievable dose of virus?
3. Can therapy be optimized such that the dose of virus is minimized?
4. Does dose scheduling play an important role on the outcome of therapy?

We are aware of the importance of the immune response to measles virus and its potential adverse consequences on the outcome of therapy. This is a current focus of our research efforts. The model we have developed is fitted to data obtained for the growth of myeloma tumor xenografts implanted in immunocompromised mice and treated with a recombinant measles virus. Given that these mice do not have an immune response we will not consider the immune system further.

16.2 Mathematical Model of Virotherapy

Modeling the effects of any therapy on tumor growth requires a model for the growth of the untreated tumor. Usually, untreated tumor growth is reliably described by the Gompertz function [37–40], yet for some tumors the more general Bertalanffy–Richards (or generalized logistic) model is required to describe data adequately [12,41]. We use the Bertalanffy–Richards model given by

$$y' = (g/\varepsilon)y[1 - (y/K)^{\varepsilon}], \quad \varepsilon > 0, \ y(0) = y_0, \tag{16.1}$$

where

$y(t)$ is the size of the tumor cell population
$r = g/\varepsilon > 0$ is the effective growth rate constant
$K > 0$ is the carrying capacity

We note that in the limit $\varepsilon \to 0$ the Bertalanffy–Richards and Gompertz models are equivalent [42,43]. The solution of Equation 16.1 can be written in an explicit form [42–44]

$$y(t) = y_0[f^{\varepsilon} + (1 - f^{\varepsilon})e^{-gt}]^{-1/\varepsilon}, \tag{16.2}$$

where

$$f = y_0/K.$$

To model the effects of virotherapy we have to consider the dynamics of at least three interacting populations [9,12]:

- $y(t)$ – uninfected tumor cells,
- $x(t)$ – virus-infected tumor cells, and
- $v(t)$ – free infectious virus particles.

A graphical representation of the model is given in Figure 16.1.

As noted in the Introduction, the infection spreads in tumor cells either by a productive encounter of one free virus particle with one uninfected cell, or by an encounter of an infected cell (expressing the viral F and H proteins) with an uninfected cell. In the latter case, two cells fuse to form a syncytium that continues to spread acquiring new surrounding cells. Thus, the rate at which the population of uninfected cells is depleted is given by the sum $\kappa y(t)v(t) + \rho y(t)x(t)$, where $\kappa > 0$ and $\rho \geq 0$ are corresponding rate constants.

Uninfected cells are assumed to be proliferating according to the Bertalanffy–Richards model. Infected cells most probably do not proliferate

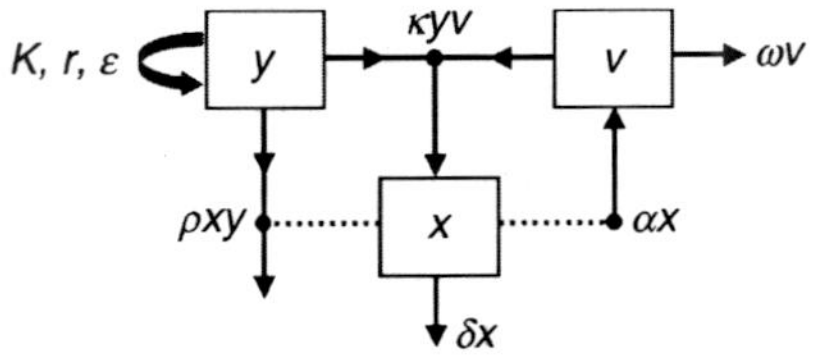

Figure 16.1 Schematic diagram of the proposed model for virotherapy. Here y denotes the populations of uninfected cells. Proliferation of these cells is described by an effective proliferation rate r, carrying capacity K, and parameter ε which characterizes the shape of the sigmoidal growth curve. Populations of infected cells and virus are denoted by x and v, respectively. Indicated rates of first and second order are explained in the text. Solid line arrows signify population influx or depletion, while dotted lines indicate that corresponding rates depend on population x.

[45], but are assumed to be dying at the effective rate $\delta x(t)$ (Figure 16.1). The rate constant δ may include the rate constant of apoptotic death ($a > 0$) and possibly a very small rate constant for proliferation (p), so that $\delta = a - p \geq 0$.

The population of free virus particles can grow when infected cells (including syncytia) release virions that have replicated within the cell. Thus the rate of free virus replication can be modeled by αx where $\alpha \geq 0$ is the corresponding rate constant. Finally, the rate of free virus elimination is modeled by the sum $\kappa y(t)v(t) + \omega v(t)$ (Figure 16.1). The first term in the sum corresponds to the rate at which virus particles enter uninfected cells. Note that we assume that one virus particle infects one cell. These particles are incapable of infecting further cells and are no longer part of the free virus population. The term $\omega v(t)$, $\omega \geq 0$, represents a rate of elimination of free virus particles by other causes including nonspecific binding and generation of defective interfering particles.

With these assumptions about rates, the virotherapy model can be represented by the following system of differential equations:

$$\begin{aligned} y' &= ry[1 - (y+x)^{\varepsilon}/K^{\varepsilon}] - (\kappa yv + \rho yx), & y(0) &= y_0, \\ x' &= \kappa yv - \delta x, & x(0) &= 0, \\ v' &= \alpha x - (\kappa yv + \omega v), & v(0) &= v_0. \end{aligned} \qquad (16.3)$$

The tumor is assumed to have grown to size y_0 when a single dose of virus v_0 is injected at time $t = 0$. The term $(y + x)^{\varepsilon}/K^{\varepsilon}$ ensures that the tumor cannot grow beyond carrying capacity K. The mathematical proof of this property of system (Equation 16.3) is analogous to the proof given in Appendix A of Ref. [12].

The rate term ρyx, which describes one possible way of infection, does not appear in the equation for x' because no new infected cell was generated in that encounter. Rather, an uninfected cell became fused with an infected cell, or with an already formed syncytium. Therefore, the population $x(t)$ is assumed to consist of single infected cells and syncytia. Experimental evidence suggests that infection is mostly due to such fusion with infected cells, rather than by free virus infection, which suggests that $\kappa \ll \rho$ (see Refs. [15,29] and Section 16.4). Although the rate constant κ in Equation 16.3 may be small compared to ρ, it still should not be completely negligible. Namely, if $\kappa = 0$, the model (Equation 16.3) breaks down because $x(t)$ may become negative. Thus we have to assume $\kappa > 0$, or reduce the model to the following:

$$\begin{aligned} y' &= ry[1 - (y+x)^{\varepsilon}/K^{\varepsilon}] - \rho yx, & y(0) &= y_0 > 0, \\ x' &= -\delta x, & x(0) &= x_0 > 0. \end{aligned} \qquad (16.4)$$

In this simplified model, it is assumed that there is an initial, fast free virus infection. This yields a number x_0 of infected cells which then infect other cells by fusion. This model did not fit the existing data (Section 16.4) and will not be considered further.

The proposed model given by Equation 16.3 is different from some previous models of virotherapy based on population dynamics [8–12,32,46]. Wodarz [9] has proposed and discussed a model in which $\rho = 0$, $\varepsilon = 1$ and the κyv term in the equation for v' is absent:

$$
\begin{aligned}
y' &= ry[1 - (y + x)/K] - dy - \kappa yv,\\
x' &= \kappa yv - \delta x,\\
v' &= \alpha x - \omega v.
\end{aligned}
$$

Here the term dy models the death rate of uninfected cells. In our model (Equation 16.3) this term is not included because it is redundant. Formally one can write

$$
ry[1 - (x + y)^{\varepsilon}/K^{\varepsilon}] - dy = \tilde{r}[1 - (x + y)^{\varepsilon}/\tilde{K}^{\varepsilon}],
$$

where $\tilde{r} = r - d$, and $\tilde{K}^{\varepsilon} = (r - d)K^{\varepsilon}/r$.

In a previous attempt to model the dynamics of infected and uninfected tumor cells, Wodarz [8,46] included the infection term βyx, but neglected the dynamics of the free virus population. He considered the following model equations:

$$
\begin{aligned}
y' &= ry[1 - (x + y)/K] - dy - \beta xy,\\
x' &= \beta xy + sx[1 - (x + y)/K] - \delta x.
\end{aligned}
\tag{16.5}
$$

Here it is assumed that both populations proliferate, although infected cells are not likely to proliferate [45]. The term βxy in Equation 16.5 implicitly models the spread of virus and it is conceptually different from our term ρxy representing the specific rate of infection via formation of syncytia.

Following Wodarz [9], in our previous model [12] we have not included the ρxy term. Also not included was the κyv term in the equation for v'. However, this term could be important, because in its absence the free virus particle count is not conserved. The models described in [10,11,32,33] are more complex (and probably more realistic) spatiotemporal models; however, they do not include the term analogous to κyv in the equation for the virus population.

Population dynamics models, such as Equation 16.3, are generally realistic when a population consists of many individuals. The goal of therapy is to reduce the total number of tumor cells, given by $u(t) = x(t)+y(t)$, to zero. However, due to the uniqueness of solutions, this goal can never be achieved in ordinary differential equations models of the type described here. We consider the tumor effectively eradicated if $u(t) < 1$ is achieved at some finite time smaller or equal to the maximal lifetime of a mouse taken to be 1000 days. Also, if the tumor burden is below a detectable amount by 1000 days, the therapy is deemed successful. A tumor consisting of less than 10^6 cells (or approximately 1 mm^3) is considered to be undetectable. The model may not be realistic when the number of tumor cells (or virus particles) is very small, so in some ways the lowest limit of one cell (or virus particle) is artificial. Yet, it can be used to estimate when the ultimate goal of virotherapy is achieved, i.e., when the tumor cell population is eliminated and there is no more free virus present.

16.3 Analysis of Equilibria

The stable states of system (Equation 16.3) which are approached as $t \to \infty$ represent the outcome of therapy if the tumor burden $x(t) + y(t)$ has not been reduced to a level below one cell at some finite time. Simulations suggest that for physiologically relevant parameters, all such states are equilibria. We therefore begin by characterizing the fixed points of the model and analyzing their stability.

In most parameter regimes, there are three equilibrium points of system (Equation 16.3). The desired outcome of therapy corresponds to the equilibrium point at the origin:

$$y_1 = 0, \quad x_1 = 0, \quad v_1 = 0. \tag{16.6}$$

This is an unstable point for biologically relevant parameters. The Jacobian of the system for this point has the eigenvalues

$$\lambda_1 = r > 0, \qquad \lambda_2 = -\delta < 0, \qquad \lambda_3 = -\omega.$$

The unstable manifold of this equilibrium is the y-axis. This instability is a consequence of the assumptions made in the model: In the absence of the virus, the number of infected cells x will remain at 0, solutions will remain on the y-axis and grow according to the Bertalanffy–Richards model.

In the absence of therapy, or if therapy fails, the tumor eventually grows to its maximal size. This is represented by the equilibrium point

$$y_2 = K, \quad x_2 = 0, \quad v_2 = 0. \tag{16.7}$$

The Jacobian of the system for this point has eigenvalues

$$\lambda_1 = -r\varepsilon/K < 0, \quad \lambda_{2,3} = -p \pm \sqrt{p^2 - q} \tag{16.8}$$

where

$$p = \delta + \kappa K + \omega_1, \quad q = (\alpha - \delta)\kappa K - \delta\omega.$$

All eigenvalues are real for biologically relevant parameters. In case $q < 0$ all eigenvalues are negative and the equilibrium is stable. When $q > 0$ it follows from Equation 16.8 that $\lambda_2 > 0$ and the equilibrium is unstable. In the unlikely case that $q = 0$ it follows $\lambda_2 = 0$ and stability is determined by higher-order terms.

Partial success of therapy is represented by the equilibrium point:

$$y_3 = \frac{\delta\omega}{(\alpha - \delta)\kappa}, \quad x_3 = \frac{1}{c}\left[1 - \left(\frac{x_3 + y_3}{K}\right)^{\varepsilon}\right], \quad v_3 = \frac{\alpha - \delta}{\omega}x_3, \tag{16.9}$$

where

$$c = \frac{\kappa(\alpha - \delta)}{r\omega} + \frac{\rho}{r}.$$

As discussed in the previous section $x(t) + y(t) \leq K$. Therefore, at this equilibrium point $y_3 < K$, since $y_3 = K$ corresponds to the equilibrium point of complete therapy failure.

For biologically relevant parameters, nonnegative values of y_3 and v_3 are obtained only for $\alpha > \delta$. In this case $c > 0$, and it is easy to show that Equation 16.9 has a unique solution for x_3. Indeed, let us denote by $\psi(\xi)$ the function

$$\psi(\xi) = [(\xi + y_3)/K]^{\varepsilon} + c\xi - 1.$$

This is a continuous function on the interval $[0, K - y_3]$ and $\psi(0) = (y_3/K)^{\varepsilon} - 1 < 0$, while $\psi(K - y_3) = c(K - y_3) > 0$. Therefore, $\psi(\xi)$ has zero within this interval, i.e., there exists at least one $x_3 \in [0, K - y_3]$ which solves the nonlinear Equation 16.9. Furthermore, this solution is unique, because if we assume two different solutions ξ_1 and ξ_2, then $\psi(\xi_1) - \psi(\xi_2) = 0$ and, therefore,

$$[(\xi_1 + y_3)/K]^{\varepsilon} - [(\xi_2 + y_3)/K]^{\varepsilon} = -c(\xi_1 - \xi_2).$$

Both $\xi_1 > \xi_2$ and $\xi_1 < \xi_2$ lead to a contradiction and, therefore, $\xi_1 = \xi_2$. Thus, the equilibrium in Equation 16.9 is uniquely defined for all biologically relevant parameters. The only exception are the cases $\omega = 0$ and $\alpha = \delta$, when Equation 16.9 does not exist, and only Equations 16.6 and 16.7

are relevant. The significance of equilibrium in Equation 16.9 is that it offers a permanent reduction of tumor burden if therapy fails to eliminate tumor cells at some finite time.

Finding the eigenvalues of the Jacobian at the equilibrium point (y_3, x_3, v_3) leads to solving a cubic equation. The Routh–Hurwitz criterion [47] for the solutions indicates that for some combinations of model parameters all the solutions can have negative real parts and, therefore, the equilibrium point can be stable. However, because equation for x_3 cannot be analytically solved, the conclusion about stability has to be reached by numerical calculations for the specific parameter values.

In the next section, it is shown that experimental data for myeloma tumor size in mice under virotherapy with MV-NIS are consistent with $\alpha = \omega = 0$. This implies that approximately no free virus is being produced in vivo and the therapeutic effect of virotherapy is only due to cell-to-cell fusion and syncytium formation. Also, this singular model implies that the free virus is not decreased due to elimination or inactivation. In this case, the equilibrium corresponding to successful therapy (Equation 16.6) and that corresponding to therapy failure (Equation 16.7) remains unchanged in location and stability.

In addition to the equilibria (Equations 16.6 and 16.7) discussed above, there is now a line of fixed points given by

$$L = \{(y, x, v) : y_3 = x_3 = 0, \quad \text{and } v = \bar{v} \text{ is arbitrary}\}. \tag{16.10}$$

These equilibria represent the situation when the tumor is eradicated, but there is a residual virus population $\bar{v}$. The size of this residual population depends on initial conditions. The Jacobian at each fixed point (Equation 16.10) has the eigenvalues

$$\lambda_1 = 0, \quad \lambda_2 = -\delta < 0, \quad \lambda_3 = r - \kappa\bar{v}. \tag{16.11}$$

The eigenvalue λ_1 corresponds to an eigenvector parallel with L, and the stability of points on L is determined by the sign of λ_2 and λ_3. Therefore, all points satisfying

$$\bar{v} < \frac{r}{\kappa}$$

are stable, and those satisfying the opposite inequality are unstable. As shown in Figure 16.2, the point $(0, 0, r/\kappa)$ lies on a separatrix between the basin of attraction of the line, and the basin of attraction of the equilibrium $(K, 0, 0)$ representing therapy failure. We note that the line given by Equation 16.10 is invariant for any parameter values, however in the case that α and ω are small and positive which still might be consistent with our

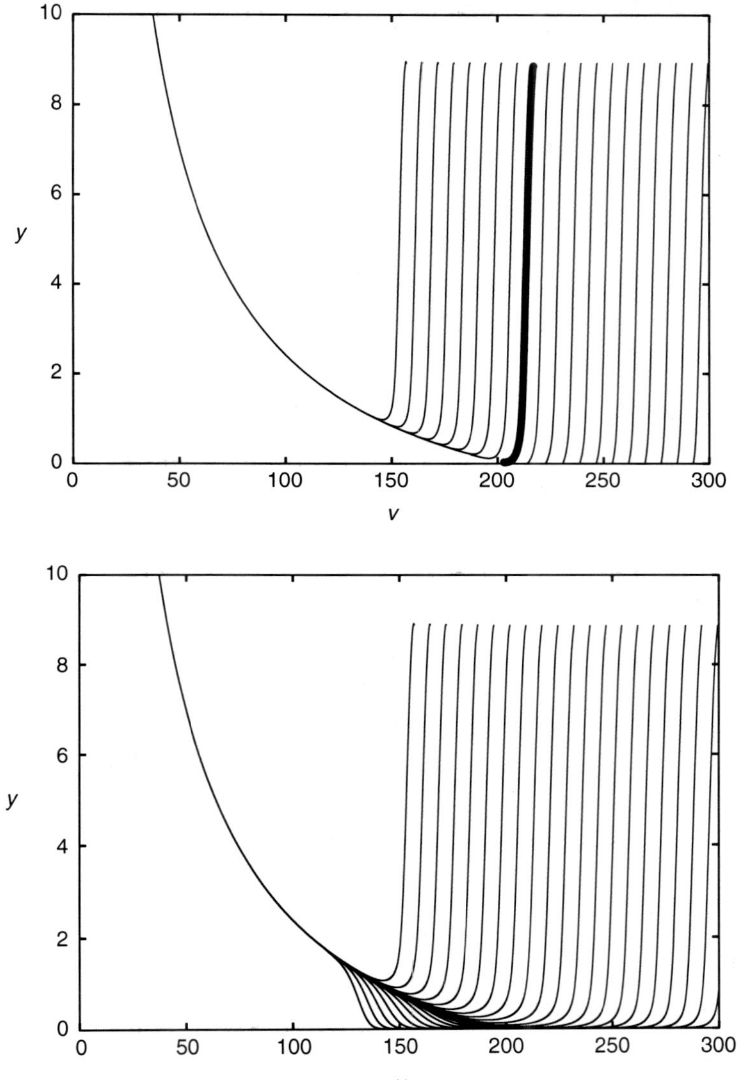

Figure 16.2 Top: The projection onto the v–y plane of a collection of orbits in the singular case $\alpha = \omega = 0$. Initial values $v(0)$ range between 155 and 305, $y(0) = 9$, and $x(0) = 0$. The orbit singled out by the heavy line lies on the separatrix between two basins of attraction. Points to the left are attracted to the equilibrium $(K, 0, 0)$ (therapeutic failure), and points to the right are attracted to the line of fixed points L (successful therapy). Other parameters are chosen as $r = 0.206$, $\rho = 0.2145$, $K = 2139$, $\varepsilon = 1.649$, $\kappa = 0.001$, $\delta = 0.5115$. Bottom: The same simulation with $\alpha = \omega = 0.001$. As discussed in Section 16.5, the line L is no longer attracting.

data (see next section), L is no longer attracting. We discuss this further in Section 16.5.

16.4 Model Validation and Parameter Estimation

The proposed model given by Equation 16.3 was validated by least square fits to available experimental data obtained for MM induced in SCID mice [29]. These data include the tumor growth curve without treatment, and the growth curve when virotherapy is introduced on day 15. Tumor size was measured as volume (in mm^3), while our model considers population of cells. In the following we will assume that 1 mm^3 corresponds to 10^6 cells and we will consider cell and virion populations y, x, v as expressed in units of 10^6.

The model was validated and parameters estimated by using the weighted nonlinear least squares method. Weighting factors were chosen as $1/\sigma_i^2$, where σ_i is the experimentally determined standard deviation for the ith data point. Technically, least squares fitting was conveniently performed in MLAB (Maryland, http://www.civilized.com, Civilized Software Inc., Bethesda, Maryland), but in cases where the minimum was difficult to find, we used our minimizer [48] in conjunction with a custom made ODE solver.

In the case of untreated tumor the fitting was relatively simple, as the analytic form of the solution is known (see Equation 16.2). A good fit was obtained with the exponent $\varepsilon \approx 1.65$ rather than $\varepsilon \approx 0$, which would imply that growth follows the Gompertz function (see Figure 16.3 for parameter estimates). By using model selection criteria we have shown that the generalized model with $\varepsilon \approx 1.65$ more adequately fits the growth data for untreated tumor then both the Gompertz and the logistic model ($\varepsilon = 1$) [12].

In the case of virotherapy, we fitted $u(t) = x(t)+y(t)$ to data. The tumor size at the beginning of therapy, y_0, was obtained from the best fit growth curve for the untreated tumor. The values of parameters r, K, ε are those obtained by fitting to the untreated tumor (Figure 16.3, also see Ref. [12]). The initial viral dose v_0 was known from the experiment. The best fit was obtained when the lower limit for the allowed values of parameters α and ω was set to zero (Figure 16.3) and the fit resulted in zero values. In terms of underlying biology, one cannot completely exclude production of free virus and its elimination. However, some in vivo experiments ([28,31]) suggest that the free virus population is not detectable, so one can infer that α is very small.

The results of our fitting suggest that the existing six data points are insufficient to determine all five model parameters; specifically α and ω appear to be the most undetermined. Thus, for example, when we chose the lower limits $\alpha = 0.9$ day^{-1}, $\omega = 0.3$ day^{-1} in the range of allowable parameter values, the minimization yielded those lower limits.

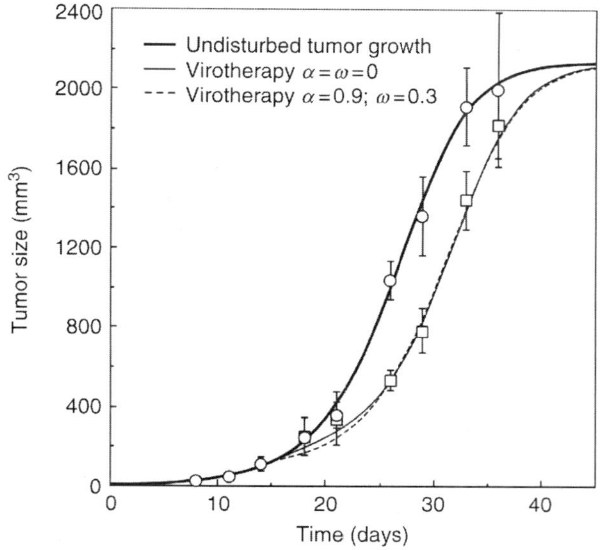

Figure 16.3 Weighted least squares fitting of the model (Equation 16.3) to growth data for multiple myeloma (MM) in SCID mice [29]. Error bars with circles denote data for untreated tumor and with squares for tumor under virotherapy. Untreated tumor data were fitted [12] using the generalized logistic model (Equation 16.2) resulting in $r = 0.2062134$ day^{-1}, $K = 2139.258$, $\varepsilon = 1.648773$. The values of parameters obtained by fitting to virotherapy data are given in Table 16.1, fits *a* and *c*. For these fits we assumed that the tumor size at the start of virotherapy on day 15 was given by the model curve for the untreated tumor, i.e., $y_0 = 126.237$. The virus dose was $v_0 = 2$.

The corresponding best fit curve passes through error bars of data points and can possibly be considered consistent with data, although χ^2 is larger (see Table 16.1). We have chosen a lower limit for ω of 0.3 day^{-1}, because some in vitro experiments suggest that approximately one-third of virus particles are inactivated per day [49]. A lower limit for α was chosen at 0.9 day^{-1} because if $\alpha > 0.9$ day^{-1} and $\omega = 0.3$ day^{-1} the best fit curve

Table 16.1 Values of Model Rate Constants (in per day) Obtained by Fitting to Virotherapy Data (see Figure 16.3). The Last Column Presents the Obtained χ^2

Fit	κ	δ	α	ω	ρ	χ^2
a	0.0009590090	0.5115017	0	0	0.2145849	1.01547
b	0.0009592272	0.5140769	0.001	0.0001	0.2153509	1.01657
c	0.0005911312	1.1189519	0.9	0.3	0.141120	1.84888

no longer passes trough the error bars, and we can consider those fits inconsistent with data.

When the values for α and ω are changed, the fits suggest that other parameters do not change dramatically (see Table 16.1). Thus, if we do the fitting with α and ω limited from below to some acceptable values, we can obtain the values of other parameters. For the sake of exploring this model by numerical simulations, we chose the values for α and ω as (a) the limiting case when they are zero, (b) small but not zero (suggested by in vivo experiments [28,31]) and low $\chi^2 v$), and (c) as large as the data allows (see discussion above).

Although based on the existing experimental data we cannot determine model parameters accurately, model given by Equation 16.3 is validated. This is not the case for the simplified model given by Equation 16.4 which yielded completely unacceptable fits.

16.5 Simulations

As discussed in Section 16.3, equilibria are important for the virotherapy outcome. However, because significant therapeutic effects have to be achieved in a finite period (1000 days for the mice considered above), it is necessary to investigate the predictions of model Equation 16.3 by numerical simulations. Therefore, in the following we discuss results of numerical simulations chosen to demonstrate significant implications for the effects of virotherapy. Throughout this section we will use the parameters given in Table 16.1, and all quantities will be measured in the units discussed in the previous section. We will start with a discussion of the singular and singularly perturbed models corresponding to fits a and b, respectively, in Table 16.1.

Figure 16.4 shows the time profile for the total tumor burden u, as well as the populations of infected cells, and the virus (x and v, respectively). As was shown in Figure 16.2, if the initial dose of virus is too low, the system rapidly approaches the equilibrium $(K, 0, 0)$ corresponding to therapeutic failure. This is illustrated in Figure 16.4a with an initial virus dose of $v(0) = 10$. There is a sharp initial increase in the number of infected tumor cells $x(t)$ followed by a decrease in the total tumor size $u(t) = y(t) + x(t)$. However, as the viral load and the number of infected cells decrease, the tumor rebounds, increasing to the level of carrying capacity.

With an initial virus dose of $v(0) = 226$, the tumor drops below the clinically detectable level $u = 1$ at $t = 6.54$, and is still undetectable at $t = 1000$ where $u(1000) = 0.88$ (see Figure 16.4b). Here $v(0) \approx 226$ is the minimum level of initial virus needed for successful therapy. In particular, if $v(0) = 225$ then $u(1000) > 1$.

Note that therapy can be successful even if the initial condition $(y(0), 0, v(0))$ is in the basin of attraction of the equilibrium $(K, 0, 0)$,

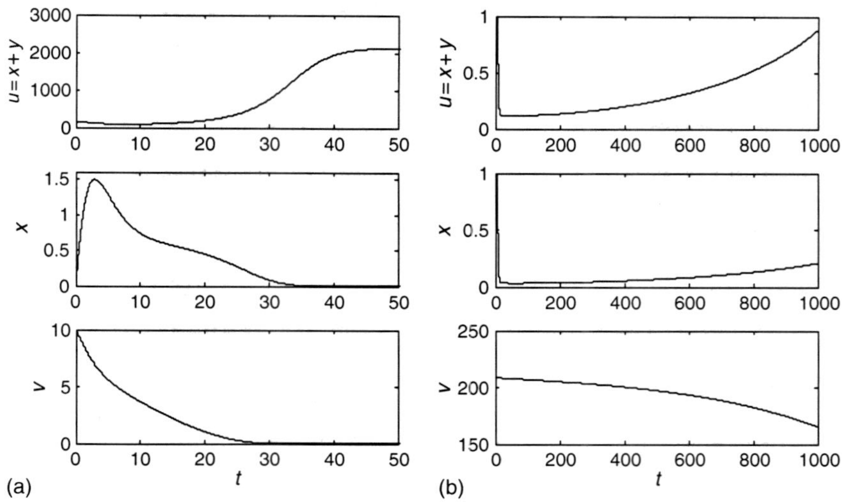

Figure 16.4 (a) Prediction of model Equation 16.3 with an initial virus dose of $v(0) = 10$. Parameter values for r, K, and ε are the same as in Figure 16.3, and $(y(0), x(0)) = (126.237, 0))$. The remaining parameters $(\kappa, \delta, \alpha, \omega, \rho)$ are from the fit a of Table 16.1. (b) A higher dose of initial virus leads to therapeutic success (see text). Note the difference in scales between panels (a) and (b).

as long as the tumor load remains undetectable up to time $t = 1000$. In fact, if continued, the orbit shown in Figure 16.4b approaches $(K, 0, 0)$ as $t \to \infty$. If the initial dose of virus is increased further to $v(0) = 236$, virotherapy is not only successful, but the initial point is in the basin of attraction of the line L. In this case, the tumor is decreasing at time $t = 1000$ and keeps decreasing if the simulation is continued (see Figure 16.5).

Figure 16.6 illustrates the effect of administering the virus in several doses, rather than at once. Figure 16.6a shows the effect of ten doses of virus, such that each dose corresponds to $v_{\text{dose}} = 10$. The doses are administered every 25 days, so that the total viral load does not increase over time. Each individual dose temporarily reduces the tumor size but the dosing schedule does not lead to long-term tumor eradication.

Similarly, Figure 16.6b shows the effect of ten doses of $v_{\text{dose}} = 29$ units of virus each, scheduled every 25 days. In this case, the tumor is undetectable when $t = 1000$ ($u(1000) < 1$). The number of doses and the period between them does not determine whether therapy will ultimately be successful or not. In general, for a fixed initial tumor size $y(0)$, the viral load must reach a certain minimal value for the therapy to be a success. This is analogous to the results in Figures 16.9 and 16.10 where we show which initial conditions lead to a successful therapy and which do not (we discuss these figures in detail below).

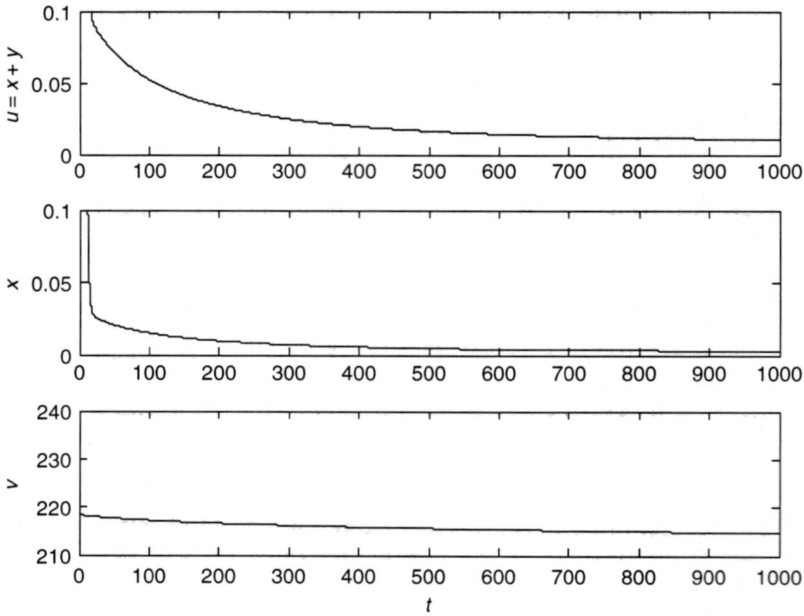

Figure 16.5 With an initial virus dose of $v(0) = 236$ the tumor size drops below the clinically detectable level $u = 1$ at $t = 6.50$, and is still decreasing at $t = 1000$, when $u(1000) = 0.01$. Conversely, if $v(0) = 235$, the tumor is undetectable but increasing at $t = 1000$. Parameter values are the same as in Figure 16.4.

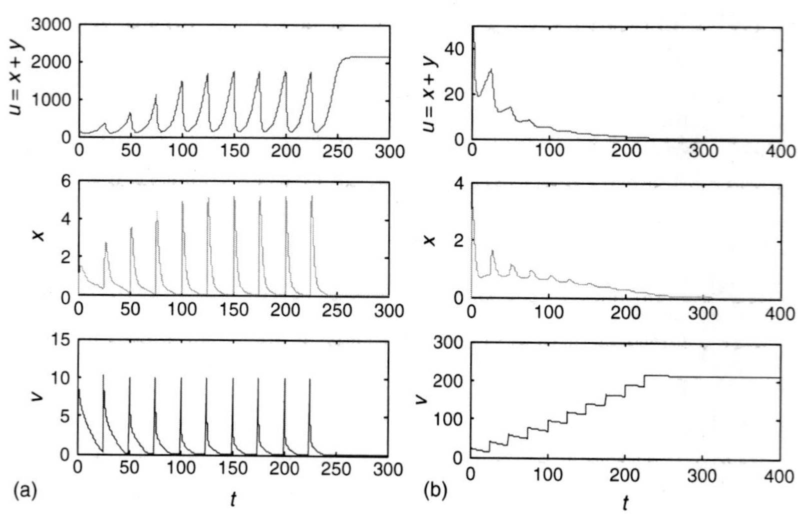

Figure 16.6 (a) The effect of administering ten doses of $v_{\text{dose}} = 10$ units of virus every 25 days. (b) The effect of administering ten doses of $v_{\text{dose}} = 29$ units of virus every 25 days. Parameter values are the same as Figure 16.4.

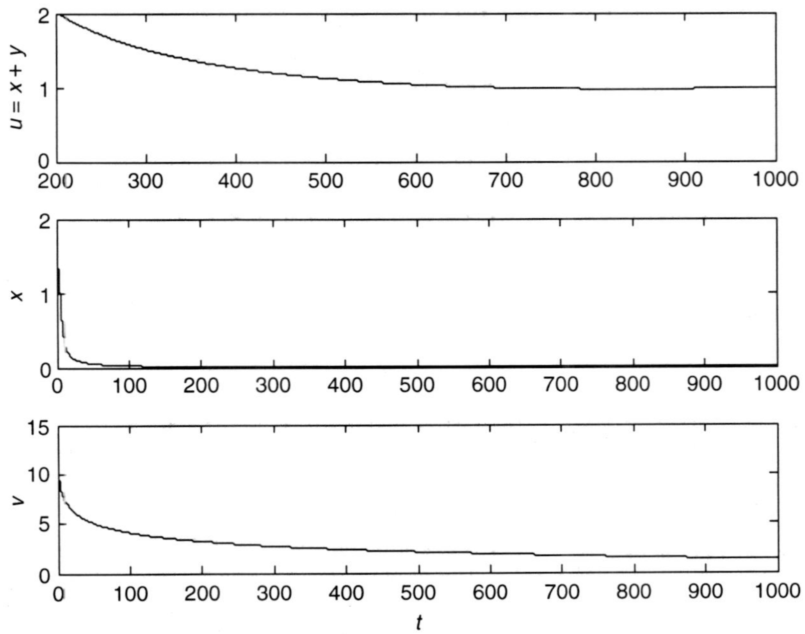

Figure 16.7 Decreasing the growth rate r has a large effect on the success of therapy. If r is decreased by two orders of magnitude compared to Figure 16.4 the dose of virus sufficient for successful therapy is reduced to $v(0) = 12$.

The growth rate r has a large effect on the outcome of therapy. Tumor growth can be slowed down by the use of inhibitors of DNA synthesis so that r can be significantly reduced. As an example, in Figure 16.7 we consider r equal to 1 percent of the fit a value and find that $v(0) = 12$ is sufficient for the tumor to be undetectable at $t = 1000$ ($u(1000) < 1$). Thus, as the rate of tumor cell replication is slowed down, the total dose of virus necessary to control the tumor decreases. However, with the smaller virus load it takes much longer for the tumor to shrink in size, specifically, $u(687.08) = 1$.

In Figure 16.8 we consider the effect of small, but nonzero virus production ($\alpha \neq 0$) and elimination ($\omega \neq 0$) using the parameter values from fit b in Table 16.1. Compared to the simulation results in Figure 16.4b, it is somewhat more difficult to achieve successful therapy requiring $v(0) = 235.0$ with $u(1000) = 0.92$ and $u = 1$ when $t = 6.48$. Other qualitative behavior regarding multiple doses and decreasing r are also the same for the parameters in fit b, as in fit a. On the other hand, increasing α relative to ω and hence increasing the viral load can lead to a successful result. This is because the relatively large rate of virus production ultimately leads to a

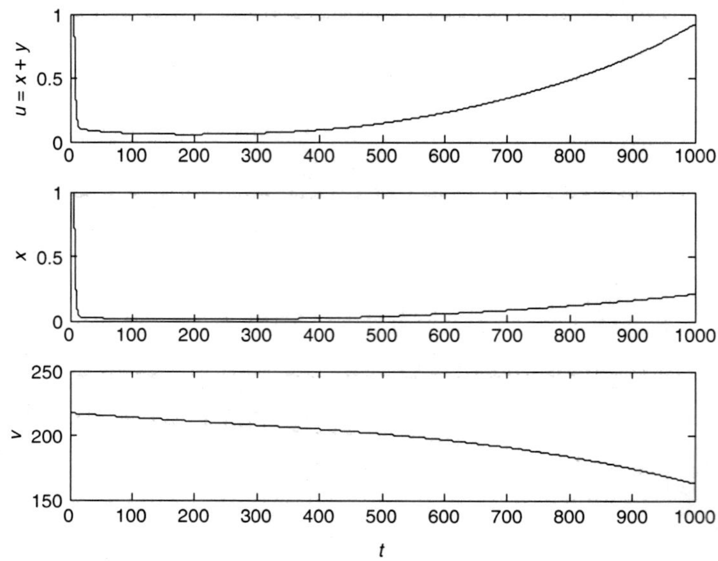

Figure 16.8 The behavior of the singularly perturbed system ($\alpha \neq 0$ and $\omega \neq 0$) is similar to that of the singular system. Compare with Figure 16.4b. Parameters are chosen according to fit *b* in Table 16.1.

large viral load; this is analogous to the large initial viral load $v(0)$ when using parameters from fit a. We can therefore conclude that the behavior of the singularly perturbed model is very similar to that of the singular model.

In Figure 16.9, we examine the effect of the initial tumor size on the final outcome of virotherapy. The shades of gray indicate the final tumor size $u(1000)$ as a function of the initial conditions $(y(0), 0, v(0))$. Of particular note is that it is sometimes preferable to allow the tumor to grow to a larger size before administering the virus. For example, for a fixed viral load of $v(0) = 210$, if the initial tumor size is $y(0) = 1600$, then therapy fails. On the other hand, if the tumor is larger with $y(0) = 1900$, then not only is the tumor undetectable at $t = 1000$, but it is essentially eliminated with $u(1000) = O(10^{-6})$. A potential explanation for this may be that a higher tumor burden at the time of virus administration increases the number of cells that are infected, leading to a higher population of tumor cells that can fuse with surrounding cells (ρ) and produce additional virus particles (α).

In Figure 16.9, we also observe an island of initial conditions where virus therapy is unsuccessful, which is surrounded by initial conditions that lead to success, $u(1000) < 1$. We do not currently have an explanation for why this isolated region exists.

In Figure 16.10, we again consider the state of the system at $t = 1000$ but with the tumor growth rate r reduced by a factor of 10. As noted

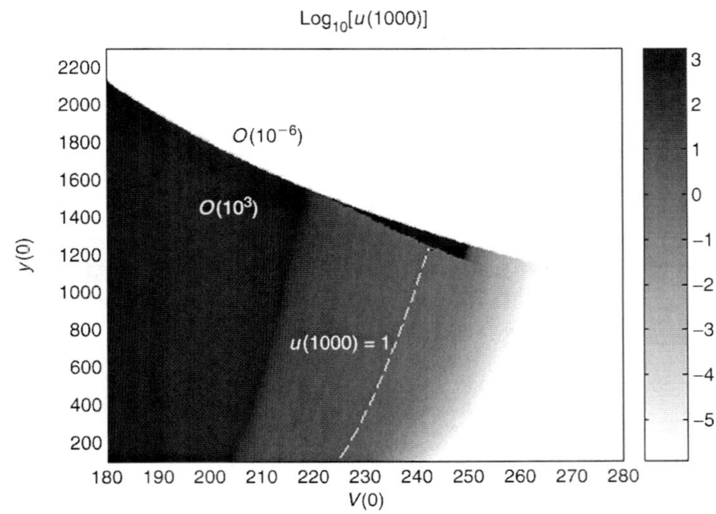

Figure 16.9 The total tumor size $u = y + x$ at $t = 1000$ as a function of the initial viral load $v(0)$ and the initial tumor size $y(0)$ using the parameters from fit *a*. The black region corresponds to initial conditions that lead to unsuccessful therapy such that $u(1000)$ has reached the level of the carrying capacity. The white region corresponds to very successful therapy such that the tumor is almost eradicated with $u(1000) = O(10^{-6})$. The dashed curve identifies when the tumor is undetectable with $u(1000) = 1$.

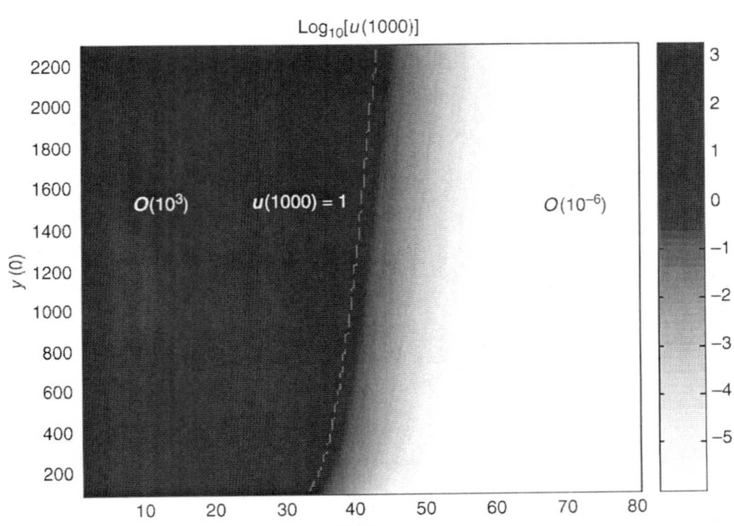

Figure 16.10 Same as Figure 16.9 but with the tumor growth rate r reduced by a factor of 10 ($r = 0.02062134$).

in Figure 16.7 when r is reduced, a lower initial dose of virus is needed to achieve successful therapy; the curve where $u(1000) = 1$ has shifted to much lower values of $v(0)$. In addition, the sharp upper boundary between $u(1000) = O(10^3)$ (black) and $O(10^{-6})$ (white) that exists in Figure 16.9 has shifted upward to higher values of u, much larger than the level of saturation, and does not appear in the figure.

On the basis of Equation 16.3, one would expect that as the ability of infected cells to fuse with uninfected tumor cells (ρ) increases, the tumor burden $u(1000)$ and $y(1000)$ should decrease. Therefore, the results in Figure 16.11 appear counterintuitive. The increase in $u(1000)$, though, is not of great significance for the therapy if $\rho > 0.2$, where 0.2 is approximately

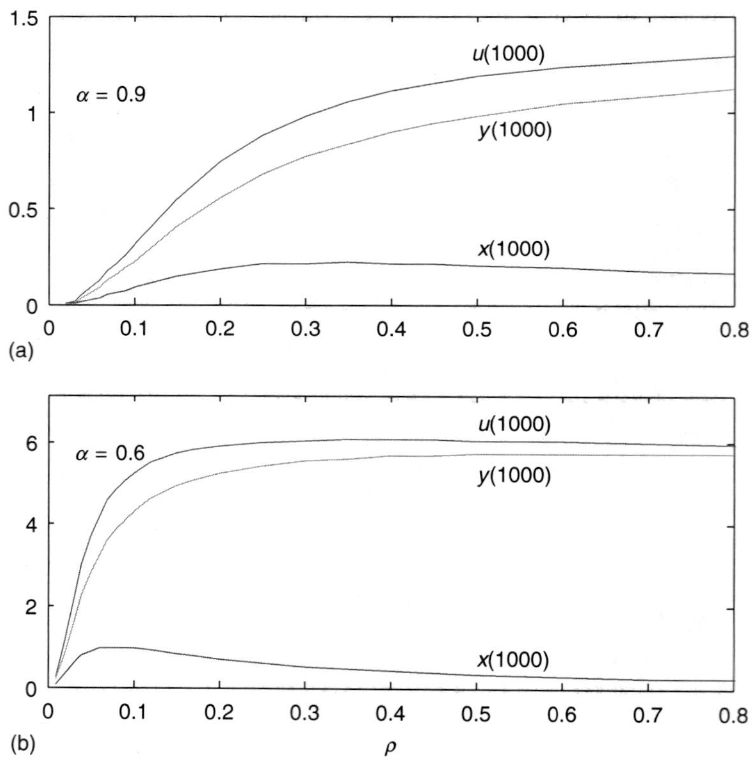

Figure 16.11 The size of the tumor at $t = 1000$ as a function of the cell fusion rate ρ. Other parameters are those of fit b with (a) $\alpha = 0.9$ and (b) $\alpha = 0.6$. In each figure there is a value of ρ such that the number of infected tumor cells $x(1000)$ is maximal. For further increases in ρ, $x(1000)$ decreases. However, in (a), the total tumor size $u(1000)$ and the uninfected tumor size $y(1000)$ are still increasing, while in (b) they are slightly decreasing.

the value for ρ obtained by fitting (see Table 16.1). We found that the equilibrium value x_3 (see Equation 16.9) slightly decreases with ρ while, of course, y_3 is constant. However at the time $t = 1000$, we are still very far from the equilibrium and the behavior is as shown in Figure 16.11. Otherwise, when α is much smaller than 0.6 and ρ sufficiently large so that the tumor burden achieves a minimum before growing to the level of carrying capacity, this minimal tumor burden indeed decreases with ρ. However, the effect is quite insignificant for ρ smaller than 1.5. It is unlikely that values of ρ higher than 1.5 are realistic.

Finally, in Figure 16.12 we show that Equation 16.3 support damped oscillatory behavior. Figure 16.12a shows strongly damped oscillations for parameter values similar to those of fit b. Only a single maximum of u is visible given the scale but the corresponding numerical data clearly exhibit very small amplitude oscillations. Figure 16.12b shows more dramatic oscillations but for parameter values away from those given in Table 16.1.

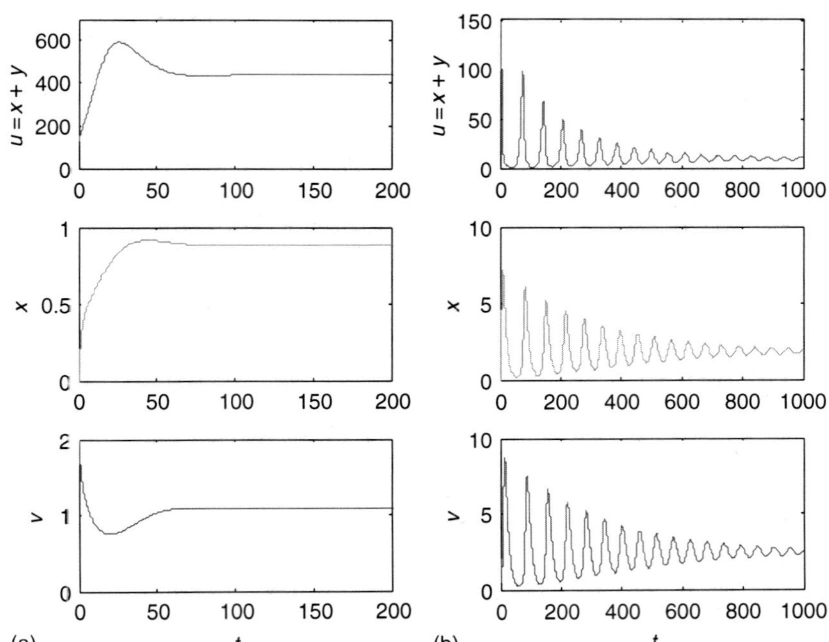

Figure 16.12 (a) Strongly damped oscillations using fit *b* parameters with $\alpha = 0.6$ and $\omega = 0.07$. (b) Weakly damped oscillations using $\kappa = 0.01$, $\rho = 0.1$, $\delta = 0.1$, $\alpha = 0.5$, and $\omega = 0.3$.

16.6 Conclusion

The availability of novel therapeutic agents such as replicating viruses for cancer therapy introduces a new paradigm in the therapy of these diseases. Therapeutic success depends on the highly specific interaction between the oncolytic virus and the tumor cell population with the dynamic considerations determining the outcome. We have presented a model to try and understand different aspects of the effect of therapy with attenuated measles viruses on tumor growth. In particular, we are able to provide partial answers to the questions posed in Section 16.1.

1. The initial tumor burden is important, but not crucial for the outcome of therapy. Contrary to intuition, a larger initial tumor burden may facilitate therapy under certain conditions, perhaps by increasing the efficiency of virus–tumor cell interactions. The result is a higher pool of infected tumor cells that proceed to generate new virus particles and fuse surrounding tumor cells stopping their replication and ultimately leading to their death.
2. It is not possible to cure the experimentally tested tumor xenografts with a therapeutically achievable dose of virus. Our analysis suggests that 226 million virus particles must be injected for the virus alone to eradicate the tumor. This is not possible in a mouse unless the virus can be concentrated significantly without loss of titer. However, if tumor growth can be slowed, the virus requirements decrease significantly and are achievable with current technology. Such an approach may offer also an additional margin of safety because the total dose of virus that will need to be injected is small and hence the risk of adverse effects will be expected to decrease.
3. If optimal virus therapy is defined as the smallest effective dose of virus that can operationally control the tumor for the lifetime of the animal, then combining virotherapy with strategies to slow tumor growth can significantly reduce the demands on the virus load needed for cure.
4. Dose scheduling does not seem to play an important role on the outcome of therapy. The main determinant of the therapeutic outcome is the total dose of virus that can be administered.

Acknowledgments

This research has been supported by NSF grants DMS-0244529 and DMS-0604229, and a Texas ARP/ATP grant to KJ. ŽB acknowledges support from Grant CA-100634 and from Mayo Clinic Cancer Center Support grant CA15083 funded by the NIH. DD is supported by the Mayo Foundation.

References

1. R. A. Kyle and S. V. Rajkumar. Multiple myeloma. *N. Engl. J. Med.*, 351(18):1860–1873, 2004.
2. D. Kirn, R. L. Martuza, and J. Zwiebel. Replication-selective virotherapy for cancer: Biological principles, risk management and future directions. *Nat. Med.*, 7(7):781–787, 2001.
3. S. J. Russell. RNA viruses as virotherapy agents. *Cancer Gene. Ther.*, 9(12):961–966, 2002.
4. J. Nemunaitis, F. Khuri, I. Ganly, J. Arseneau, M. Posner, E. Vokes, J. Kuhn, T. McCarty, S. Landers, A. Blackburn, L. Romel, B. Randlev, S. Kaye, and D. Kirn. Phase II trial of intratumoral administration of ONYX-015, a replication-selective adenovirus, in patients with refractory head and neck cancer. *J. Clin. Oncol.*, 19(2):289–298, 2001.
5. T. Reid, E. Galanis, J. Abbruzzese, D. Sze, J. Andrews, L. Romel, M. Hatfield, J. Rubin, and D. Kirn. Intra-arterial administration of a replication-selective adenovirus (dl1520) in patients with colorectal carcinoma metastatic to the liver: A phase I trial. *Gene. Ther.*, 8(21):1618–1626, 2001.
6. T. Reid, E. Galanis, J. Abbruzzese, D. Sze, L. M. Wein, J. Andrews, B. Randlev, C. Heise, M. Uprichard, M. Hatfield, L. Rome, J. Rubin, and D. Kirn. Hepatic arterial infusion of a replication-selective oncolytic adenovirus (dl1520): Phase II viral, immunologic, and clinical endpoints. *Cancer Res.*, 62(21):6070–6079, 2002.
7. A. L. Pecora, N. Rizvi, G. I. Cohen, N. J. Meropol, D. Sterman, J. L. Marshall, S. Goldberg, P. Gross, J. D. O'Neil, W. S. Groene, M. S. Roberts, H. Rabin, M. K. Bamat, and R. M. Lorence. Phase I trial of intravenous administration of pv701, an oncolytic virus, in patients with advanced solid cancers. *J. Clin. Oncol.*, 20(9):2251–2266, 2002.
8. D. Wodarz. Viruses as antitumor weapons: Defining conditions for tumor remission. *Cancer Res.*, 61(8):3501–3507, 2001.
9. D. Wodarz. Gene therapy for killing p53-negative cancer cells: Use of replicating versus nonreplicating agents. *Hum. Gene. Ther.*, 14(2):153–159, 2003.
10. J. T. Wu, H. M. Byrne, D. H. Kirn, and L. M. Wein. Modeling and analysis of a virus that replicates selectively in tumor cells. *Bull. Math. Biol.*, 63(4):731–768, 2001.
11. J. T. Wu, D. H. Kirn, and L. M. Wein. Analysis of a three-way race between tumor growth, a replication-competent virus and an immune response. *Bull. Math. Biol.*, 66(4):605–625, 2004.
12. D. Dingli, M. D. Cascino, K. Josić, S. J. Russell, and Ž. Bajzer. Mathematical modeling of cancer radiovirotherapy. *Math. Biosci.*, 199(1):55–78, 2006.
13. D. Grote, S. J. Russell, T. I. Cornu, R. Cattaneo, R. Vile, G. A. Poland, and A. K. Fielding. Live attenuated measles virus induces regression of human lymphoma xenografts in immunodeficient mice. *Blood*, 97(12):3746–3754, 2001.
14. K. W. Peng, G. J. Ahmann, L. Pham, P. R. Greipp, R. Cattaneo, and S. J. Russell. Systemic therapy of myeloma xenografts by an attenuated measles virus. *Blood*, 98(7):2002–2007, 2001.
15. K. W. Peng, C. J. TenEyck, E. Galanis, K. R. Kalli, L. C. Hartmann, and S. J. Russell. Intraperitoneal therapy of ovarian cancer using an engineered measles virus. *Cancer Res.*, 62(16):4656–4662, 2002.

16. L. K. Phuong, C. Allen, K. W. Peng, C. Giannini, S. Greiner, C. J. TenEyck, P. K. Mishra, S. I. Macura, S. J. Russell, and E. C. Galanis. Use of a vaccine strain of measles virus genetically engineered to produce carcinoembryonic antigen as a novel therapeutic agent against glioblastoma multiforme. *Cancer Res.*, 63(10):2462–2469, 2003.
17. C. J. McDonald, C. Erlichman, J. N. Ingle, G. A. Rosales, C. Allen, S. M. Greiner, M. E. Harvey, P. J. Zollman, S. J. Russell, and E. Galanis. A measles virus vaccine strain derivative as a novel oncolytic agent against breast cancer. *Breast Cancer Res. Treat.*, 99(2):177–184, 2006.
18. U. Schneider, V. von Messling, P. Devaux, and R. Cattaneo. Efficiency of measles virus entry and dissemination through different receptors. *J. Virol.*, 76(15):7460–7467, 2002.
19. B. D. Anderson, T. Nakamura, S. J. Russell, and K. W. Peng. High cd46 receptor density determines preferential killing of tumor cells by oncolytic measles virus. *Cancer Res.*, 64(14):4919–4926, 2004.
20. H. T. Ong, M. M. Timm, P. R. Greipp, T. E. Witzig, A. Dispenzieri, S. J. Russell, and K. W. Peng. Oncolytic measles virus targets high cd46 expression on multiple myeloma cells. *Exp. Hematol.*, 34(6):713–720, 2006.
21. K. W. Peng, K. A. Donovan, U. Schneider, R. Cattaneo, J. A. Lust, and S. J. Russell. Oncolytic measles viruses displaying a single-chain antibody against cd38, a myeloma cell marker. *Blood*, 101(7):2557–2562, 2003.
22. A. D. Bucheit, S. Kumar, D. M. Grote, Y. Lin, V. von Messling, R. B. Cattaneo, and A. K. Fielding. An oncolytic measles virus engineered to enter cells through the cd20 antigen. *Mol. Ther.*, 7(1):62–72, 2003.
23. U. Schneider, F. Bullough, S. Vongpunsawad, S. J. Russell, and R. Cattaneo. Recombinant measles viruses efficiently entering cells through targeted receptors. *J. Virol.*, 74(21):9928–9936, 2000.
24. A. L. Hammond, R. K. Plemper, J. Zhang, U. Schneider, S. J. Russell, and R. Cattaneo. Single-chain antibody displayed on a recombinant measles virus confers entry through the tumor-associated carcinoembryonic antigen. *J. Virol.*, 75(5):2087–2096, 2001.
25. T. Nakamura, K. W. Peng, S. Vongpunsawad, M. Harvey, H. Mizuguchi, T. Hayakawa, R. Cattaneo, and S. J. Russell. Antibody-targeted cell fusion. *Nat. Biotechnol.*, 22(3):331–336, 2004.
26. T. Nakamura, K. W. Peng, M. Harvey, S. Greiner, I. A. Lorimer, C. D. James, and S. J. Russell. Rescue and propagation of fully retargeted oncolytic measles viruses. *Nat. Biotechnol.*, 23(2):209–214, 2005.
27. E. M. Hadac, K. W. Peng, T. Nakamura, and S. J. Russell. Reengineering paramyxovirus tropism. *Virology*, 329(2):217–225, 2004.
28. K. W. Peng, S. Facteau, T. Wegman, D. O'Kane, and S. J. Russell. Non-invasive in vivo monitoring of trackable viruses expressing soluble marker peptides. *Nat. Med.*, 8(5):527–531, 2002.
29. D. Dingli, K. W. Peng, M. E. Harvey, P. R. Greipp, M. K. O'Connor, R. Cattaneo, J. C. Morris, and S. J. Russell. Image-guided radiovirotherapy for multiple myeloma using a recombinant measles virus expressing the thyroidal sodium iodide symporter. *Blood*, 103(5):1641–1646, 2004.
30. D. Dingli, B. J. Kemp, M. K. O'Connor, J. C. Morris, S. J. Russell, and V. J. Lowe. Combined I-124 positron emission tomography/computed tomography imaging

of nis gene expression in animal models of stably transfected and intravenously transfected tumor. *Mol. Imaging Biol.*, 8(1):16–23, 2006. 1536-1632.
31. K. W. Peng, E. M. Hadac, B. D. Anderson, R. Myers, M. Harvey, S. M. Greiner, D. Soeffker, M. J. Federspiel, and S. J. Russell. Pharmacokinetics of oncolytic measles virotherapy: Eventual equilibrium between virus and tumor in an ovarian cancer xenograft model. *Cancer Gene. Ther.*, 13(8):732–738, 2006.
32. Y. Tao and Q. Guo. The competitive dynamics between tumor cells, a replication-competent virus and an immune response. *J. Math. Biol.*, 51(1):37–74, 2005.
33. L. M. Wein, J. T. Wu, and D. H. Kirn. Validation and analysis of a mathematical model of a replication-competent oncolytic virus for cancer treatment: Implications for virus design and delivery. *Cancer Res.*, 63(6):1317–1324, 2003.
34. D. Dingli, S. J. Russell, and J. C. Morris. In vivo imaging and tumor therapy with the sodium iodide symporter. *J. Cell. Biochem.*, 90(6):1079–1086, 2003.
35. D. Dingli, R. M. Diaz, E. R. Bergert, M. K. O'Connor, J. C. Morris, and S. J. Russell. Genetically targeted radiotherapy for multiple myeloma. *Blood*, 102(2):489–496, 2003.
36. A. S. Novozhilov, F. S. Berezovskaya, E. V. Koonin, and G. P. Karev. Mathematical modeling of tumor therapy with oncolytic viruses: Regimes with complete tumor elimination within the framework of deterministic models. *Biol. Direct.*, 1:6, 2006.
37. M. Marušić, Ž. Bajzer, S. Vuk-Pavlović, and J. P. Freyer. Tumor growth in vivo and as multicellular spheroids compared by mathematical models. *Bull. Math. Biol.*, 56(4):617–631, 1994.
38. Ž. Bajzer, S. Vuk-Pavlović, and M. Huzak. Mathematical modeling of tumor growth kinetics. In J.A. Adam and N. Bellomo, editors, *A Survey of Models for Tumor-Immune System Dynamics*, chapter 3, Birkhäuser, Basel, 1997, pp. 89–133.
39. D. A. Cameron, A. A. Ritchie, and W. R. Miller. The relative importance of proliferation and cell death in breast cancer growth and response to tamoxifen. *Eur. J. Cancer*, 37(12):1545–1553, 2001.
40. R. Chignola, A. Schenetti, G. Andrighetto, E. Chiesa, R. Foroni, S. Sartoris, G. Tridente, and D. Liberati. Forecasting the growth of multicell tumour spheroids: Implications for the dynamic growth of solid tumours. *Cell. Prolif.*, 33(4):219–229, 2000.
41. J. A. Spratt, D. von Fournier, J. S. Spratt, and E. E. Weber. Decelerating growth and human breast cancer. *Cancer*, 71(6):2013–2019, 1993.
42. H. M. Byrne. Modelling avascular tumour growth. In L. Preziosi, editor, *Cancer Modelling and Simulation*, chapter 4, CRC Press, Boca Raton, FL, 2003 pp. 75–120.
43. Ž. Bajzer, M. Marušić, and S. Vuk-Pavlović. Conceptual frameworks for mathematical modeling of tumor growth dynamics. *Math. Comput. Modelling*, 23:31–46, 1996.
44. R. I. Fletcher. A general solution for the complete richards function. *Math. Biosci.*, 27:349–360, 1975.
45. J. Heaney, T. Barrett, and S. L. Cosby. Inhibition of in vitro leukocyte proliferation by morbilliviruses. *J. Virol.*, 76(7):3579–3584, 2002.

46. D. Wodarz and N. L. Komarova. *Computational Biology of Cancer: Lecture Notes and Mathematical Modeling*. World Scientific Publishing Company, Hackensack, NJ, 2005.
47. A. Hurwitz. On the conditions under which an equation has only roots with negative real parts. In R. Bellman and R. Kalaba, editors, *Selected Papers on Mathematical Trends in Control Theory*. Dover, New York, 1964.
48. C. Offord and Ž. Bajzer. A hybrid global optimization algorithm involving simplex and inductive search. *Lect. Notes Comput. Sci.*, 2074:680–688, 2006.
49. T. Whistler, W. J. Bellini, and P. A. Rota. Generation of defective interfering particles by two vaccine strains of measles virus. *Virology*, 220(2):480–484, 1996.

Chapter 17

Combating Microbial Resistance to Antimicrobial Agents through Dosing Regimen Optimization

Michael Nikolaou and Vincent H. Tam

CONTENTS

17.1 Antimicrobial Resistance and the Need to Optimize Dosing Regimens 494
17.2 Background 496
17.2.1 Pharmacodynamic Indices and Their Limitations 496
17.2.2 Dynamic Models of Pharmacodynamic Activity and Their Limitations for Microbial Populations of Nonuniform Susceptibility/Resistance 497
17.3 New Approach to Modeling the Effect of Antimicrobial Agents on Heterogeneous Microbial Populations 499
17.3.1 Homogeneous Microbial Population under Pharmacokinetically Realistic Antimicrobial Concentration 502
17.3.2 Heterogeneous Microbial Population under Pharmacokinetically Realistic Antimicrobial Concentration 511
17.3.3 Experimental Verification 512
17.4 Summary and Future Work 514

Acknowledgment 515
References 516

Abstract Microbial resistance to antimicrobial agents has evolved to alarming proportions. To avert potentially catastrophic consequences for public health, a concerted effort is necessary. It should include, among other elements, the development of methods that can optimize the clinical use of existing agents and accelerate the development of new ones. For both tasks, the design of effective dosing regimens that suppress the emergence and proliferation of resistant microbial populations is crucial. In this chapter, we provide a comprehensive presentation of our recent theoretical and experimental work on a mathematical modeling framework that can be used to optimize the design of such dosing regimens. Suggestions for future work are made.

17.1 Antimicrobial Resistance and the Need to Optimize Dosing Regimens

Microbial resistance to antimicrobial agents (antimicrobial resistance) has reached alarming proportions. Repeated warnings are recently heard from concerned scientists about bacterial wars, new plagues, worldwide calamities, new apocalypses, and the risk of returning to the pre-antibiotics era (Cohen, 1992; Neu, 1992; Gold and Moellering, 1996; Levy, 1998; Drlica, 2001; Landman et al., 2002; Varaldo, 2002; Levy and Marshall, 2004; Morens et al., 2004). The enormity of the problem has not escaped the attention of popular press (Di Justo, 2005; Comarow, 2006; Silberman, 2007). According to the U.S. Food and Drug Administration (FDA) "addressing the issue of antimicrobial resistance is one of the most urgent priorities in the fields of public health today" (Food and Drug Administration, 2006). To avert potentially catastrophic consequences of antimicrobial resistance, a concerted effort on many fronts is necessary. It should include, among other elements, the development of methods that can optimize the clinical use of existing agents, and accelerate the development of new agents. For both tasks, tools guiding the design of dosing regimens that suppress the emergence or proliferation of resistant microbial populations can make a significant impact. Such design tools should maximize the killing effect of agents (or combinations of agents) on heterogeneous microbial populations (composed of microbial subpopulations of varying susceptibility/resistance to the agents) while avoiding toxicity problems for host organisms. The importance of dosing regimen design for clinical use has been emphasized repeatedly (Bonhoeffer et al., 1997; Lipsitch and Levin, 1997; Lipsitch et al., 2000; Chait et al., 2007). Beyond the obvious therapeutic benefits that better design of clinicaldosing regimens would have for existing antimicrobial agents, by prolonging their efficacy through maintenance of their microbial killing effectiveness, better dosing regimen design would also make the development of new agents more attractive, by promising a longer effective period of use for a developed agent. This promise could help make antimicrobial agent development more attractive for potential developers, hopefully contributing to a welcome reversal of the dire downward trend of newly FDA approved antimicrobial agents over the last two decades (Spellberg et al., 2004). Furthermore, tools for dosing regimen design would help to directly accelerate the antimicrobial agent development process. Indeed, when developing new agents, emphasis is traditionally placed on discovering new agent candidates. As crucial as this step may be, a long (multiyear) development period ensues, until an agent is fully developed (Drusano et al., 2006). During that period of development, it is common that only a few dosing regimens are empirically tested, because of the very large number of experiments required for exhaustive testing (Sidebar 1). This practice limits our ability to realize the clinical potential of agents, either through premature abandonment of promising candidates or through inadvertent pursuit of dead ends. The critical role of selecting the right dosing regimen was dramatically exemplified in the case of daptomycin, for which selection of the right dosing regimen alone was the key differentiating factor between abandoning development in early 1980s and eventually securing FDA approval in 2003 (Sidebar 2).

Sidebar 1: Example of dosing regimen testing

To evaluate six daily doses (e.g., 0.5, 1, 2, 4, 6, and 8 g), four dosing frequencies (e.g., every 6, 8, 12, or 24 hours), four intravenous dosing administrations (e.g., intermittent infusion of 0.5, 1, and 2 hours and continuous infusion over 24 hours), and 3 durations of treatment (e.g., 5, 10, and 14 days) would require investigation of 288 ($6 \times 4 \times 4 \times 3$) regimens for a single candidate. Reduction by, for example, an order of magnitude would have obvious implications.

Sidebar 2: Importance of dosing regimens: The daptomycin case

Daptomycin (Cubicin®, Cubist Pharmaceuticals) (UCSF, 2006) was initially under development in the 1970s with an 8 hour dosing interval. Its development was abandoned in the early 1980s due to an intolerable adverse effect (muscle toxicity). However, after understanding its exposure-related killing properties and toxicity, redevelopment began in the 1990s, to finally reach FDA approval for clinical use in 2003. The key factor for FDA approval was use of a once-daily and weight-based dosing regimen. This discovery was deemed so nonobvious and important as to be awarded a patent (Oleson et al., 2005).

The preceding discussion should make clear the value of methods that can guide the design of effective dosing regimens for combating antimicrobial resistance. In this chapter, we provide a comprehensive presentation of a recent mathematical modeling framework (Nikolaou and Tam, 2006; Nikolaou et al., 2007) that can be used to optimize the design of such dosing regimens. Experimental in vitro validation on *Pseudomonas aeruginosa*, an important bacterial pathogen (Sidebar 3) is presented. However, we want to emphasize at the outset that the proposed modeling approach could be extrapolated to a variety of antimicrobial agents (e.g., antibacterials, antifungals, and antivirals) with different mechanisms of action, as well as to other pathogens (e.g., HIV, tuberculosis, anthrax, and avian influenza) with different biological characteristics (Gumbo et al., 2004; Tam et al., 2005a). In addition, the proposed mathematical framework could also be extended for use in cancer chemotherapy, by accounting for heterogeneities of cancerous cell populations (Dua et al., 2005).

In the rest of this chapter we provide a background for our work, present our findings, and conclude with suggestions for further development.

Sidebar 3: Pseudomonas aeruginosa

P. aeruginosa is associated with serious nosocomial infections such as pneumonia and sepsis. It exploits multiple mechanisms of resistance to various antimicrobial agents (such as efflux pumps, β-lactamases production, porin channel deletion, multifunctional group transferases, and target site mutation) (Livermore, 2002). Some of the mechanisms of resistance are highly specific to one agent, whereas others affect a broad spectrum of antimicrobial agents, and confer different levels of resistance. Resistance to first-line agents (such as β-lactams and fluoroquinolones) has been reported and is becoming more prevalent (Landman et al., 2002; Neuhauser et al., 2003). There are very few agents in the advanced stage of development designed to target multidrug-resistant Gram-negative bacteria, and none is expected to be available for clinical use in the next decade. Therefore, the need to develop antimicrobial agents against *P. aeruginosa* is imperative (Talbot et al., 2006).

17.2 Background

17.2.1 Pharmacodynamic Indices and Their Limitations

Because the complex pharmacodynamic interaction between an antimicrobial agent and a microbial population defies detailed first-principles modeling, surrogate pharmacodynamic indices, such as the minimum inhibitory concentration (MIC) (Figure 17.1), are used to guide empirical testing of dosing regimens (e.g., Andes and Craig, 1998; Louie et al., 1998; Nicolau

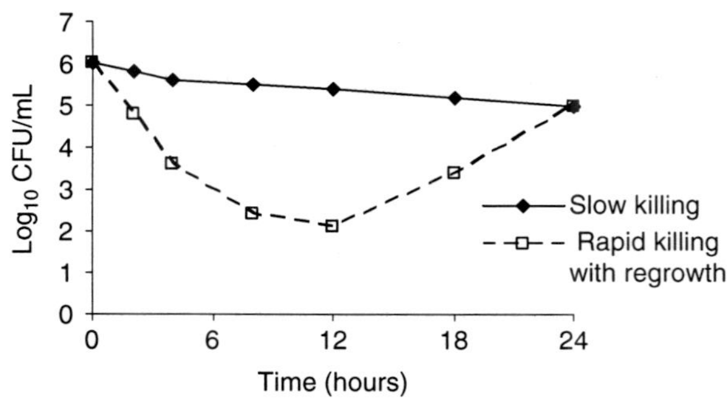

Figure 17.1 Two hypothetical killing profiles of microbial population size resulting in the same value at 24 hours after exposure to an antimicrobial agent at time-invariant concentration (Tam et al., 2005c). The corresponding microbial populations would be very different in each case (slow decline [diamonds] versus rapid decline followed by regrowth [squares]).

et al., 2000; Louie et al., 2001; Tam et al., 2002; Dandekar et al., 2003; Andes et al., 2004; Maglio et al., 2004; Miyazaki et al., 2004). This can be problematic. For example, according to the standard definition of MIC and related surrogate pharmacodynamic indices, the two populations in Figure 17.1 would correspond to the same MIC (there is no visible growth at exactly 24 hours) although they are otherwise clearly different. The second population could well grow beyond 24 hours whereas the first would probably not.

Along the same lines, a dosing regimen maintaining agent concentration above MIC would not necessarily guarantee eventual eradication of the entire population, as argued in Figure 17.2. This is because inhibition of population growth at 24 hours leaves the possibility that a small resistant subpopulation of no appreciable size during the first 24 hours may well grow later. MIC lumps all dynamic information of a time-kill experiment into a single point. Consequently, methods that make use of all available (dynamic) information from time-kill experiments would be preferable. This realization, in turn, raises the question: How is such dynamic information captured and used? We address this question in the next section.

17.2.2 Dynamic Models of Pharmacodynamic Activity and Their Limitations for Microbial Populations of Nonuniform Susceptibility/Resistance

In an effort to use the dynamic information that pharmacodynamic indices leave out, dynamic models have been formulated for homogeneous microbial populations (i.e., of uniform susceptibility or resistance) based

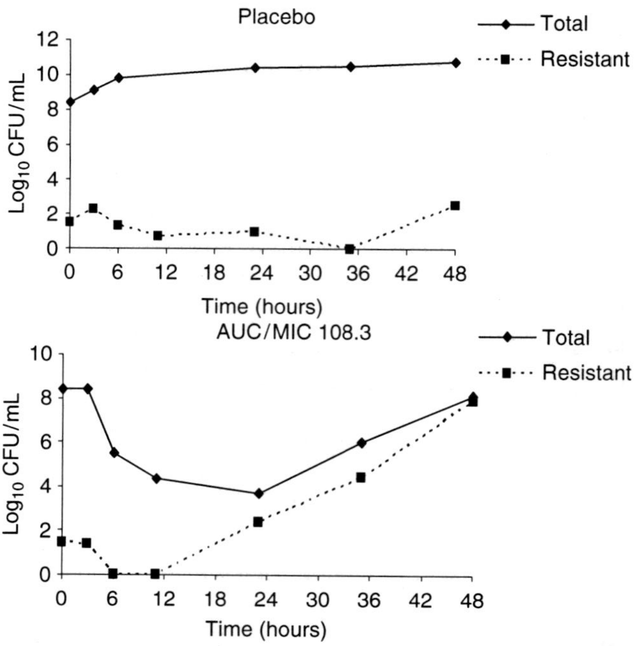

Figure 17.2 Selection of resistant *P. aeruginosa* population by garenoxacin in an in vitro hollow-fiber infection model (Figure 17.10). A population of approximately 10^8 CFU/mL bacteria was investigated (Tam et al., 2005a). In the absence of selective pressure by garenoxacin ($C(t) = 0$, top), the fraction of the resistant subpopulation remained low and relatively constant over time. In contrast, exposure to a fluctuating garenoxacin concentration $C(t)$ (bottom) as in Figure 17.4 selectively amplified the resistant subpopulation (MIC of the resistant subpopulation $\geq$ 3 x MIC of the entire population) and led to population regrowth, despite the fact that garenoxacin concentration $C(t)$ remained well above MIC during the entire period of the experiment.

on conservation principles and bacteria-agent Hill-like (Hill, 1910) kinetics (Wagner, 1968; Jusko, 1971; Giraldo et al., 2002) (Sidebar 4). When applied to heterogeneous microbial populations (i.e., of nonuniform susceptibility or resistance), such models lump subpopulations into two distinct classes: resistant and susceptible (Lipsitch and Levin, 1997; Mouton et al., 1997; Jumbe et al., 2003; Gumbo et al., 2004; Meagher et al., 2004; Campion et al., 2005; Tam et al., 2005a). Although conceptually appealing, when such models are calibrated using standard short-term data (e.g., over 24 hours [Andes and Craig, 1998; Nicolau et al., 2000; Dandekar et al., 2003; Miyazaki et al., 2004]), they may easily fail to predict the emergence of resistance—manifested as eventual population regrowth (Oliver et al.,

Sidebar 4: Basic model of effect of antimicrobial agent on homogeneous microbial population

$$\frac{dN(t)}{dt} = \underbrace{G[\overbrace{N(t)}^{\text{Population size}}]}_{\text{Physiological net growth rate}} - \underbrace{K[\overbrace{C(t)}^{\text{Agent concentration}}, N(t)]}_{\text{Kill rate due to antimicrobial agent}} \tag{17.1}$$

Example: (Hill and Michaelis–Menten dynamics)

$$G[N(t)] = K_g\left(1 - \frac{N(t)}{N_{max}}\right)N(t) \approx K_g N(t) \quad \text{if } N(t) \ll N_{max}$$

$$K[C(t), N(t)] = \underbrace{\frac{K_k C(t)^{H\}\text{ Hill factor}}}{C(t)^H + C_{50}^H}}_{\text{Kill rate constant, } r} N(t)$$

The population is eradicated if and only if $r > K_g$.
If C is time-invariant, eradication corresponds to the straight line $\ln \frac{N(t)}{N(0)} = \left(K_g - r(C)\right)t$.

2004)—as shown in Figure 17.3 (top). By lumping subpopulations of varied resistance into two distinct subpopulations (resistant and susceptible), this modeling approach essentially produces two asymptotes for the dependence of population size on time corresponding to short and long time, respectively. As Figure 17.3 (top) shows, the two asymptotes estimated by fitting 24 hour data produce overly optimistic results beyond 24 hours. For similar reasons, dynamic modeling approaches that have focused on the early time course (<60 minutes) of antimicrobial agent exposure (Regoes et al., 2004) are equally problematic.

17.3 New Approach to Modeling the Effect of Antimicrobial Agents on Heterogeneous Microbial Populations

To capture the decline–regrowth behavior of a heterogeneous microbial population, Nikolaou and Tam (2006) developed a corresponding mathematical modeling approach (excerpted in Sidebar 5) for heterogeneous microbial populations exposed to time-invariant antimicrobial agent

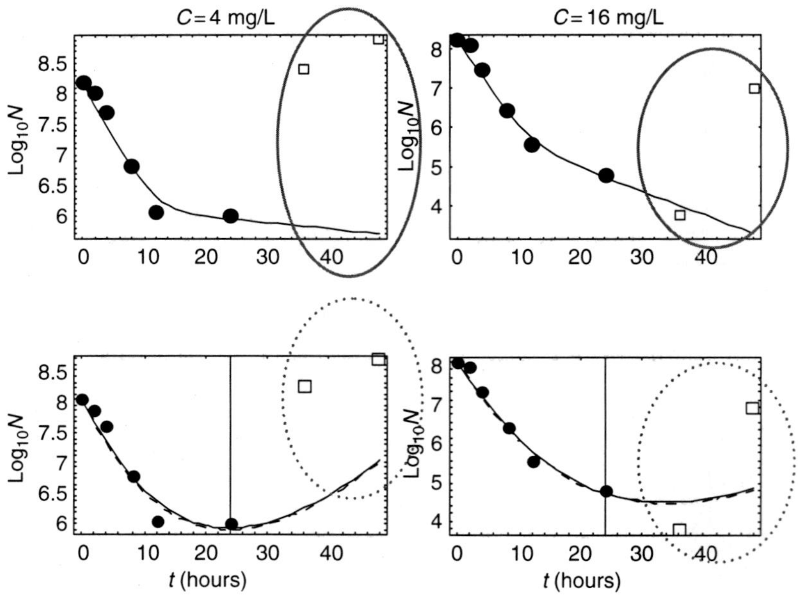

Figure 17.3 In vitro effect of the antibiotic meropenem on *Pseudomonas aeruginosa* ATCC 27853 (Tam et al., 2005b). A standard (two-subpopulation) dynamic model (top) built from 24 hour data (dots) fails to predict population regrowth beyond 24 hours (squares). By contrast, a model based on the approach developed by Nikolaou and Tam (2006) (bottom) successfully predicts regrowth using the same 24 hour data, thus suggesting that much higher antibiotic concentration is needed for eradication of the entire population.

concentrations. Dispensing with the need to rely on the asymptotic time behavior of two distinct subpopulations, this approach considers a distribution of resistance over a microbial population and employs the cumulants of that distribution. Figure 17.3 (bottom) demonstrates that this approach can successfully make use of standard 24-hour time-kill data to predict regrowth beyond 24 hours and estimate the agent concentration needed to eradicate the entire microbial population (Nikolaou and Tam, 2006).

Using the approach mentioned above, we can now address the following question, which is the main focus of this work: Given time-kill data over 24 hours at a number of time-invariant agent concentrations, what is an effective (preferably optimal) dosing regimen (daily dose and dosing interval) for time-varying agent concentration corresponding to realistic pharmacokinetics (Figure 17.4)? Optimal here refers to the smallest daily dose and corresponding dosing interval that can completely eradicate a microbial population. For lack of quantitative aids to answering the

Sidebar 5: Modeling the effect of antimicrobial agents on heterogeneous microbial populations

$$\frac{dN(t)}{dt} = [K_g - \mu(t)]N(t),$$
$$\frac{d\mu(t)}{dt} = -\sigma(t)^2, \quad \left\{\frac{d\kappa_n(t)}{dt} = -\kappa_{n+1}(t)\right\}_{n\geq 1} \quad \left(\kappa_1 \equiv \mu,\ \kappa_2 \equiv \sigma^2\right) \tag{17.2}$$

$N(t)$: microbial population size at time t

K_g: growth rate constant

$\mu(t)$, $\sigma^2(t)$: average and variance of kill rate constant over entire microbial population

$\kappa_n(t)$: n-order cumulant (Weisstein, 2005) of kill rate constant

For a distribution $f(r_i, t)$ of the kill rate constant r (Sidebar 4) cumulants are defined as

$\kappa_n(t) \hat{=} \frac{\partial^n \Psi(s,t)}{\partial s^n}$ where $\Psi(s,t) \hat{=} \ln[M(s,t)]$, $M(s,t) \hat{=} \sum_i e^{sr_i} f(r_i, t)$

Simplifying assumptions (Nikolaou and Tam, 2006) yield

$$\ln\left[\frac{N(t)}{N(0)}\right] \approx \left(K_g \underbrace{-\mu(0) + \frac{\sigma(0)^2}{A}}_{-b}\right) t + \underbrace{\frac{\sigma(0)^2}{A^2}}_{R/A}\left(e^{-At} - 1\right)$$
$$\hat{=} \left(K_g - b\right)t + \frac{R}{A}\left(e^{-At} - 1\right), \tag{17.3}$$

$$\mu(t) \approx \underbrace{\mu(0) - \frac{\sigma(0)^2}{A}}_{b} + \underbrace{\frac{\sigma(0)^2}{A}}_{R} e^{-At} \hat{=} b + Re^{-At}, \tag{17.4}$$

$$\sigma(t)^2 \approx \sigma(0)^2 e^{-At} \tag{17.5}$$

preceding question, it is common practice for antimicrobial killing action to be classified into two distinct categories: peak-concentration- or time-of-exposure-dependent (Vogelman and Craig, 1986; Craig, 1998) as shown in Figure 17.5. However, it has been widely observed that some recent antimicrobials (e.g., quinolones) do not clearly fall in either category. Therefore, a more quantitative answer to the preceding question is needed, as discussed next.

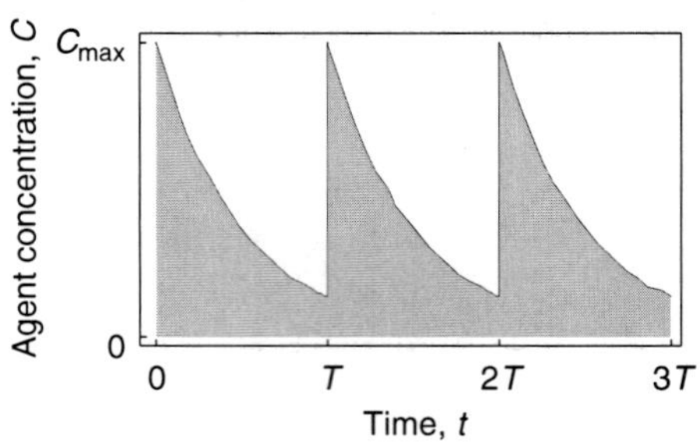

Figure 17.4 Example of a clinically relevant dosing regimen. Periodic injection of agent every T time units and its subsequent elimination result in a jump-decay periodic pattern ($C(t) = C_{max}2^{-t/t_{1/2}} = C_{max}e^{-t/\tau}$, $0 \le t < T$; $C(t+T) = C(t)$) according to typical pharmacokinetics in humans.

17.3.1 Homogeneous Microbial Population under Pharmacokinetically Realistic Antimicrobial Concentration

Assume now that the antimicrobial agent concentration does not remain time-invariant but fluctuates periodically due to periodic injection of agent every T time units and its subsequent elimination, as shown in Figure 17.4. The kill rate constant $r(C(t))$ will obviously fluctuate with the same period T. Under these conditions, it can be shown (Nikolaou et al., 2007) that the total population $N(t)$ exhibits a periodic pattern with period T, and

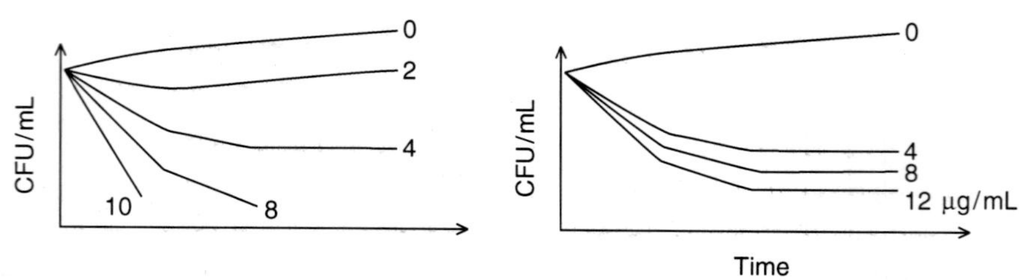

Figure 17.5 Concentration-dependent (upper) and time-dependent (lower) killing activity of antimicrobial agents. In the concentration-dependent case, killing activity depends on the concentration of the antimicrobial agent used, and suggests dosing regimens that achieve high concentrations at injection points. In the time-dependent case, killing activity quickly reaches a plateau as agent concentration increases, indicating that dosing regimens need to maintain a certain agent concentration most of the time. Increasing agent concentration will increase toxicity without appreciably increasing killing activity.

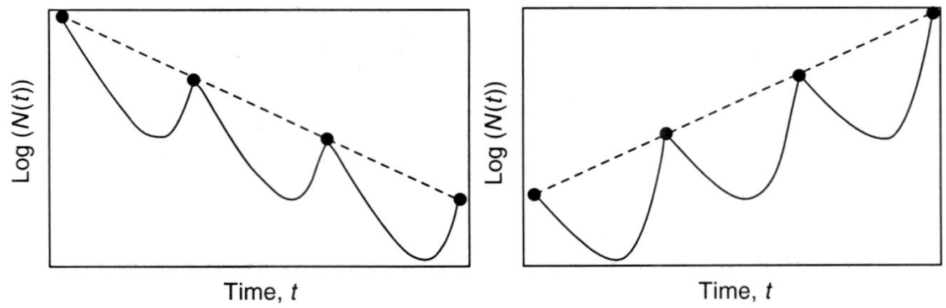

Figure 17.6 Eradication (left) or regrowth (right) of a microbial population in an environment of antimicrobial agent concentration following Sidebar 6.

the values of $\log(N(nT))$, $n = 0, 1, 2, \ldots$, lie on a straight line, akin to the case corresponding to time-invariant agent concentration (Sidebar 4) as summarized in Sidebar 6. In other words, the points $\frac{N(nT)}{N(0)}$ appear as if they were generated by a system under time-invariant agent concentration D, a fact that significantly simplifies the ensuing analysis (Figure 17.6).

Therefore, according to Sidebar 6, $\frac{D}{K_g} > 1$ implies eradication of the entire microbial population, whereas $\frac{D}{K_g} < 1$ implies eventual proliferation of the population, except for the case where eradication can occur during the first dosing interval. The latter case can occur if the minimum of $\ln N(t)$, $0 \le t \le T$, is at or below 0.

Sidebar 6: Model of effect of antimicrobial agent on heterogeneous microbial population

A homogeneous population is subjected to periodically fluctuating antimicrobial agent concentration, i.e., $C(t) = C(t + T)$. Then, it can be shown (Nikolaou et al., 2007) that

$$\ln \frac{N(t)}{N(0)} = K_g t - \left[\!\left[\frac{t}{T} \right]\!\right] DT - \int_0^{t - [\![\frac{t}{T}]\!] T} r(C(\eta))d\eta$$

where $[\![\frac{t}{T}]\!]$ is the integer part of the real number $\frac{t}{T}$, and $D \hat{=} \frac{1}{T} \int_0^T r(C(\eta))d\eta$ is the time-averaged kill rate constant.

At times $t = nT$, $n = 0, 1, 2, \ldots$, the total population satisfies the equation

$$\ln \frac{N(nT)}{N_0} = (K_g - D)nT, \; n = 0, 1, 2, \ldots$$

We can now ask: "For what dosing regimens is the condition $\frac{D}{K_g} > 1$ satisfied?" We first provide a qualitative approximate answer, followed by a quantitative answer.

Qualitatively, the value of $\frac{D}{K_g}$, to first-order approximation, can be shown to be

$$\frac{D}{K_g} \approx 1 + \frac{r'(C_{cr})}{K_g}(C_{avg} - C_{cr}) = 1 + \frac{r'(C_{cr})}{K_g} C_{cr} \left(\frac{AUC/T}{C_{cr}} - 1 \right)$$

where the area under the concentration (AUC) curve is defined as $AUC \hat{=} \int_0^T C(t)dt$. The above approximation of $\frac{D}{K_g}$ indicates that to design a dosing regimen resulting in eradication of a microbial population, the average concentration of the agent, $C_{avg} \hat{=} \frac{1}{T} \int_0^T C(\eta)d\eta$, must be above the critical concentration C_{cr}, defined as the concentration at which the kill rate constant $r(C_{cr})$ is equal to the growth rate constant K_g. It follows that the effectiveness of an agent is approximately related to the well-known pharmacokinetic/pharmacodynamic parameter AUC/MIC $\approx$ AUC/C_{cr}. However, it should be stressed that the dependence of an agent's effectiveness on AUC/MIC is only approximate. A more accurate index would have to be used to account for strong effects of higher-order derivatives in the above series expansion of $\frac{D}{K_g}$. This motivates the quantitative results presented next.

From a quantitative viewpoint, let the agent concentration follow the realistic pharmacokinetic pattern $C(t) = C_{max}e^{-kt}$, $0 \leq t < T$ (Figure 17.4) where $k = \frac{\ln 2}{t_{1/2}}$ is the agent elimination rate constant (reciprocally proportional to the half-time $t_{1/2}$) and T is the dosing interval; and let the kill rate constant follow the Michaelis–Menten/Hill kinetics in Sidebar 4. Then it can be shown (Nikolaou et al., 2007) that the value of $\frac{D}{K_g}$ can be influenced by selecting two dimensionless variables associated with the dose and dosing interval of a dosing regimen, namely the scaled average concentration $z \hat{=} \frac{C_{avg}}{C_{cr}}$ (or, equivalently, $y \hat{=} \frac{C_{avg}}{C_{50}}$) and the scaled dosing interval $x \hat{=} kT$, where C_{avg} is proportional to the administered dose (mass of agent over 24 hour period). The functional dependence of $\frac{D}{K_g}$ on x, z depends on two pharmacodynamic parameters: H and $\frac{K_k}{K_g}$.

Thus, if the values of H and $\frac{K_k}{K_g} = 1 + \left(\frac{C_{50}}{C_{cr}}\right)^H$ have been estimated from experimental time-kill data, one can visualize the agent effectiveness, i.e., value of $\frac{D}{K_g}$ in comparison to 1, as a function of the two variables that characterize a dosing regimen, namely $z \hat{=} \frac{C_{avg}}{C_{cr}}$ and $x \hat{=} kT$ (Sidebar 7). Figure 17.7 shows a small library of such patterns for different values of H and $\frac{K_k}{K_g}$, along with associated plots of the scaled kill rate constant $\frac{r(C)}{K_k}$

Sidebar 7: Killing effect as function of $z \hat{=} \frac{C_{avg}}{C_{cr}}$ *and* $x \hat{=} kT$ *for different values of* H *and* $\frac{K_k}{K_g}$

Let $C(t) = C_{max}e^{-kt}$, $0 \leq t < T$, $C(t) = C(t+T)$, and $r(C) = K_k \frac{C^H}{C^H + C_{50}^H}$. Then

$$\frac{D}{K_g} = \frac{K_k}{K_g}\frac{1}{Hx}\ln\frac{\frac{K_k}{K_g} - 1 + \left(\frac{e^x xz}{e^x - 1}\right)^H}{\frac{K_k}{K_g} - 1 + \left(\frac{xz}{e^x - 1}\right)^H}$$

$$= \left(1 + \chi_{50}^H\right)\frac{1}{Hx}\ln\frac{1 + \left(\frac{e^x xy}{e^x - 1}\right)^H}{1 + \left(\frac{xy}{e^x - 1}\right)^H}$$

where

$$\chi_{50}^H \hat{=} \left(\frac{C_{50}}{C_{cr}}\right)^H = \frac{K_k}{K_g} - 1,$$

$$z \hat{=} \frac{C_{avg}}{C_{cr}} = \frac{\text{Daily dose [mg]}}{24\ \text{[hours] clearance [L/hour]}} \Big/ C_{cr},$$

$$y \hat{=} \frac{C_{avg}}{C_{50}} = \frac{\text{Daily dose [mg]}}{24\ \text{[hours] clearance [L/hour]}} \Big/ C_{50},$$

$$x \hat{=} kT = \frac{T}{t_{1/2}}\ln 2$$

as a function of $\frac{C}{C_{cr}}$. A careful examination of these patterns for $\frac{D}{K_g}$ (lines corresponding to $\frac{D}{K_g} = 1$) reveals qualitatively different behaviors for different values of H and $\frac{K_k}{K_g}$, suggesting different designs for optimal dosing regimens. For example, for $H = 1$ and $\frac{K_k}{K_g} = 5$, it is clear that the shorter the dosing interval T (Figure 17.4), the lower the dose that can be used. Consequently, the optimal dosing regimen would be continuous infusion. This is due to the dependence of the kill rate constant r on C: A relative increase in C is associated with a lower relative increase in r. Therefore, for a periodically fluctuating profile of C around an average value C_{avg} a lot more killing power r would be lost while $C(t) < C_{avg}$ than would be gained while $C(t) > C_{avg}$. By contrast, $H = 4$ and $\frac{K_k}{K_g} = 5$ in Figure 17.7 indicates that there is an optimal value (around $kT = 5$) for the dosing interval T at the cutoff point $\frac{D}{K_g} = 1$, corresponding to the minimum dose $C_{avg}/C_{cr} \approx 1$. This is due to the presence of an inflection point in the curve corresponding to the dependence of the kill rate constant r on C: Around the inflection point, a relative increase in C is associated with a lower relative increase

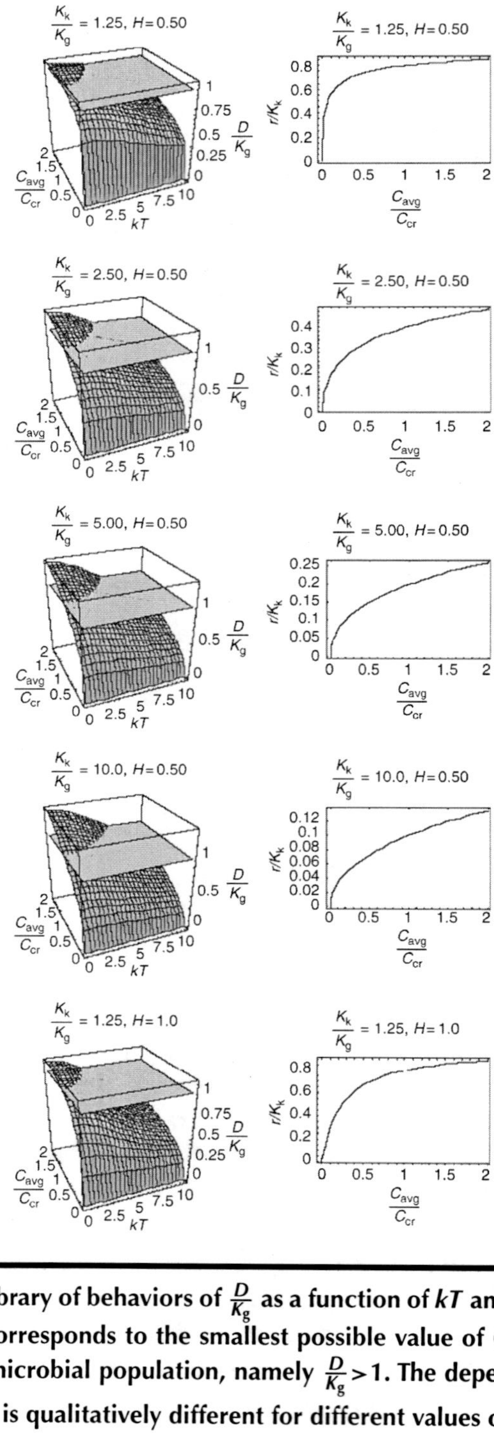

Figure 17.7 A library of behaviors of $\frac{D}{K_g}$ as a function of kT and $\frac{C_{avg}}{C_{cr}}$. The optimal dosing regimen corresponds to the smallest possible value of C_{avg} that results in eradication of a microbial population, namely $\frac{D}{K_g} > 1$. The dependence of optimal C_{avg} on $\frac{K_k}{K_g}$ and H is qualitatively different for different values of $\frac{K_k}{K_g}$ and H.

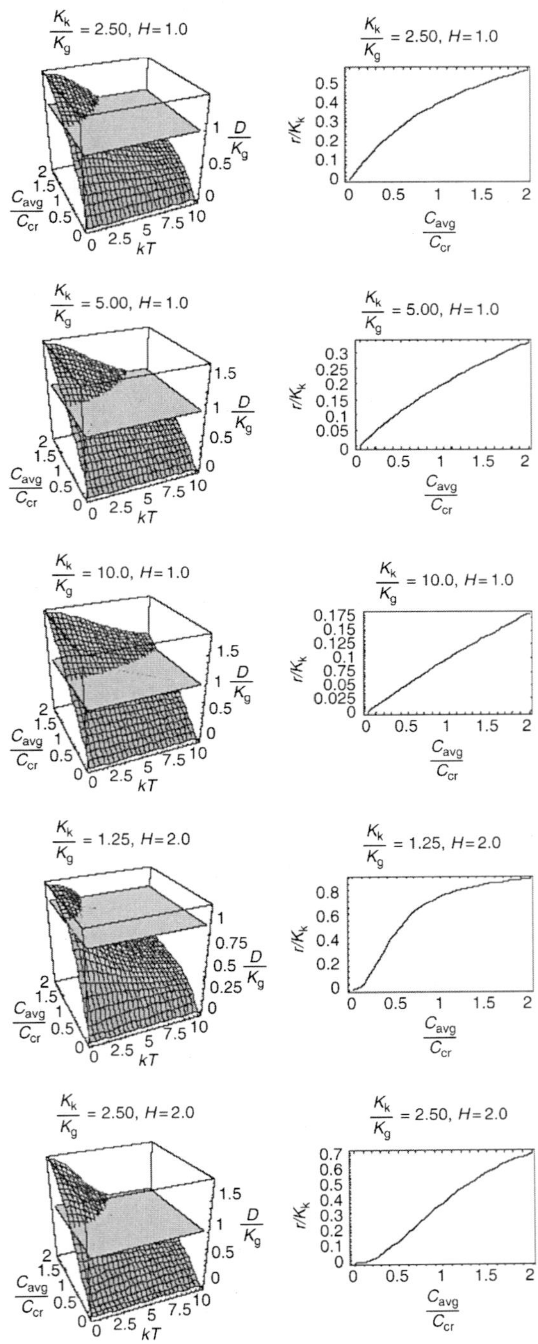

Figure 17.7 (Continued)

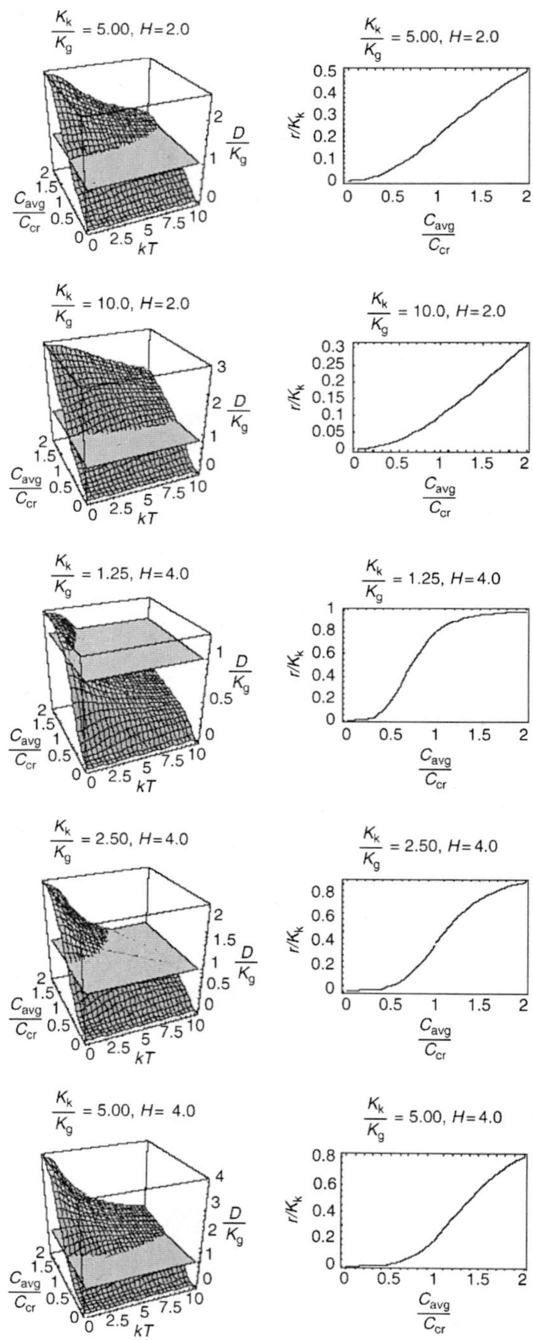

Figure 17.7 (Continued)

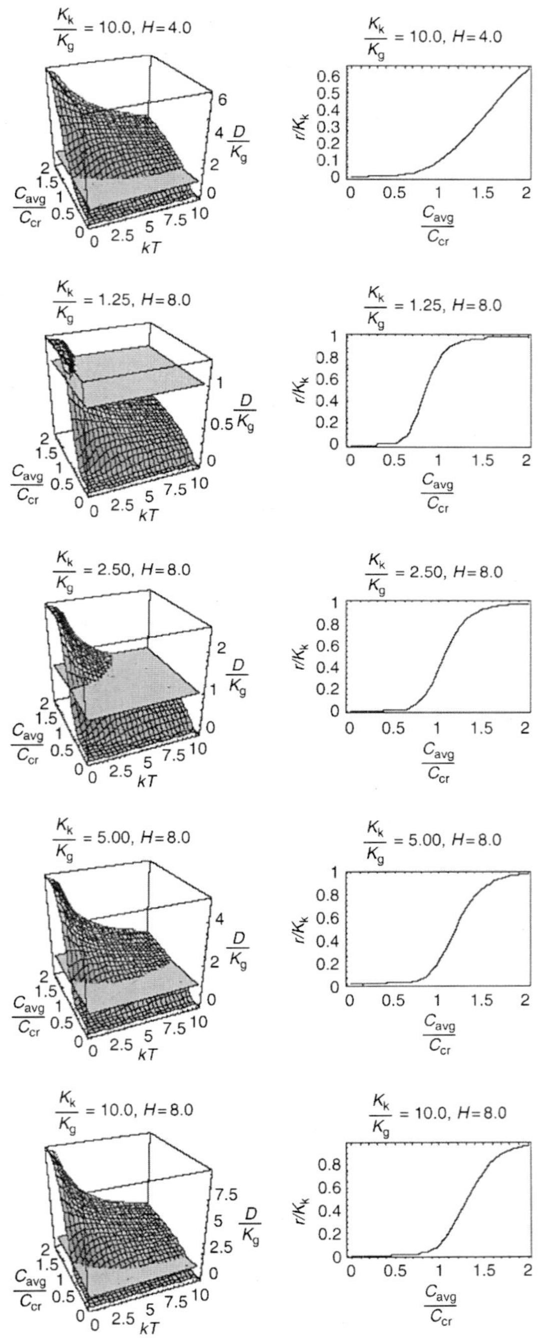

Figure 17.7 (Continued)

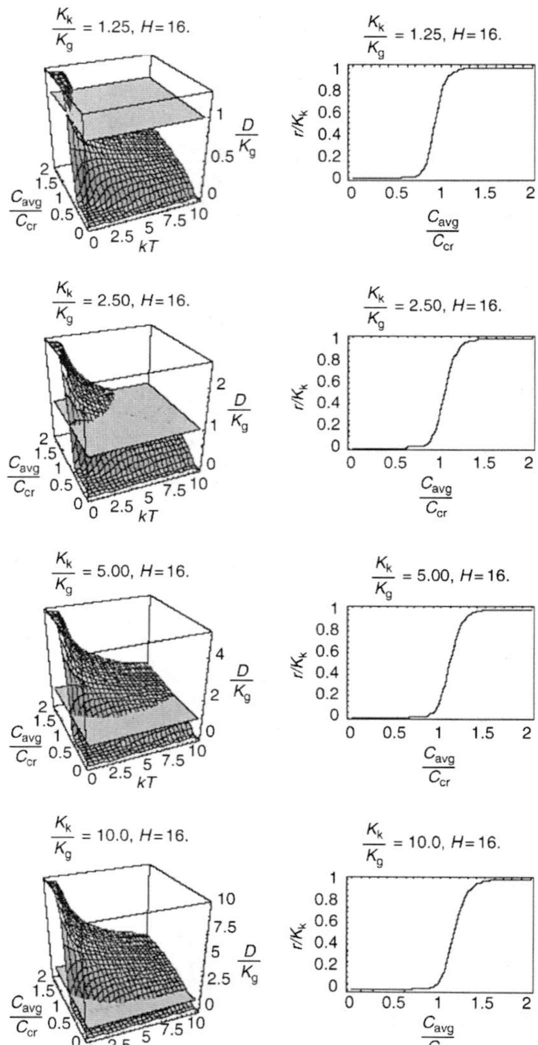

Figure 17.7 (Continued)

in r. However, a relative decline in C is also associated with a lower relative decline in r. Therefore, in balance, for a periodically fluctuating profile of C around the optimal average value C_{avg}, a lot less killing power r is lost while $C(t) < C_{\text{avg}}$ than is gained while $C(t) > C_{\text{avg}}$. Although Figure 17.7 may be sensitive to experimental errors in the estimates of H and $\frac{K_k}{K_g}$, it establishes a continuum for the model of action of antimicrobial agents, at the two ends of which are the two well-known categories, namely peak-concentration- or time-of-exposure-dependent, established by Vogelman and Craig (1986).

17.3.2 Heterogeneous Microbial Population under Pharmacokinetically Realistic Antimicrobial Concentration

The results of the preceding sections are now going to be used here to develop a method for designing optimal dosing regimens for heterogeneous microbial populations (i.e., of nonuniform susceptibility or resistance). From a theoretical viewpoint, it would be interesting to develop an equation for $\ln \frac{N(t)}{N(0)}$ analogous to that in Sidebar 6. However, the following reasoning makes this requirement unnecessary.

To design an optimal dosing regimen, it is required to find the minimum of the time-averaged agent concentration C_{avg} and corresponding dosing interval T that will eradicate a microbial population entirely. To accomplish this, it is necessary and sufficient to eradicate the most-resistant subpopulation of the microbial population, by finding the minimum of the time-averaged agent concentration C_{avg} and corresponding dosing interval T for that subpopulation. According to the analysis in Section 17.3.1, eradication of the most-resistant subpopulation means that $\frac{D}{K_g} > 1$ for that subpopulation, as suggested in Figure 17.7. Hence, the dependence of $\frac{D}{K_g}$ on dosing regimens (namely C_{avg} and T) for that subpopulation must be estimated from experimental data. Now, the analysis in Section 17.3.1 indicates that standard time-kill experiments can be used for that purpose. Indeed, for a heterogeneous population subjected to a number of time-invariant agent concentrations C, Equation 17.4 in Sidebar 5 indicates that the population-average kill rate constant will eventually reach a value b for each time-invariant agent concentration C. This C-dependent kill rate constant, b, corresponds to the most-resistant subpopulation, which will eventually dominate the entire population, and which is homogeneous, as suggested by Equation 17.5 when $t \to \infty$. Therefore, it is reasonable to assume that the functional dependence of b on C follows the kinetics discussed in Sidebar 4, namely

$$b(C) = K_b \frac{C^{H_b}}{C^{H_b} + C_{50b}^{H_b}}$$

Similarly, it can be argued (Nikolaou and Tam, 2006) that it is reasonable to postulate that

$$R(C) = K_k \frac{C^H}{C^H + C_{50}^H} - K_b \frac{C^{H_b}}{C^{H_b} + C_{50b}^{H_b}}$$

and

$$A(C) = K_A \frac{C^{H_A}}{C^{H_A} + C_{50A}^{H_A}}$$

Therefore, if experimental data is available from time-kill studies (measurements of population size at various sampling points in time, for a number of time-invariant concentrations C), then the parameters involved in the above expressions for $b(C)$, $R(C)$, and $A(C)$ can be estimated. Then, using the identified expression for $b(C)$ in place of $r(C)$ in the analysis of Sidebar 7, one can construct a surface showing the dependence of $\frac{D}{K_g}$ on C_{avg} and T, as in Figure 17.7.

17.3.3 Experimental Verification

We discuss an example of our approach (Nikolaou and Tam, 2006; Nikolaou et al., 2007) where effective dosing regimens (dose and dosing intervals)

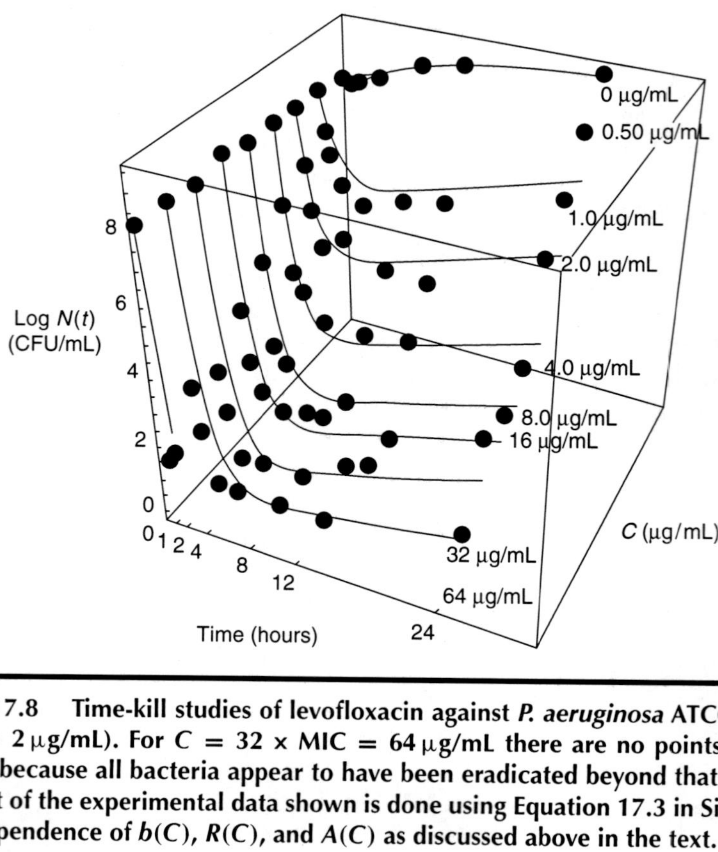

Figure 17.8 Time-kill studies of levofloxacin against *P. aeruginosa* ATCC 27853 (MIC = 2 µg/mL). For C = 32 x MIC = 64 µg/mL there are no points beyond 1 hour, because all bacteria appear to have been eradicated beyond that point in time. Fit of the experimental data shown is done using Equation 17.3 in Sidebar 5, with dependence of $b(C)$, $R(C)$, and $A(C)$ as discussed above in the text.

are characterized for levofloxacin against *P. aeruginosa*. Time-kill data is collected over 24 hours at various time-invariant concentrations of levofloxacin, and curve-fit using the approach discussed in Section 17.3.2, as shown in Figure 17.8.

Subsequent to that, the equations of Sidebar 7 are used to construct the D/K_g surface as a function of dosing regimen, namely daily dose and dosing interval for given pharmacokinetics (Figure 17.9). According to Figure 17.9, daily doses of 750 and 3000 mg daily are predicted to be ineffective and completely effective, respectively.

This was verified experimentally in a hollow-fiber in vitro infection model (Figure 17.10), as shown in Figure 17.11.

Dosing regimens predicted to be effective (corresponding to values of the index D/K_g greater than 1) or ineffective ($D/K_g < 1$) were compared to published data regarding the threshold quinolone exposure necessary to suppress resistance development of *P. aeruginosa* in a murine thigh infection model (Jumbe et al., 2003). Despite the differences between the two modeling approaches, the estimates of the levofloxacin exposure necessary for resistance suppression were consistent (approximately 2900 mg

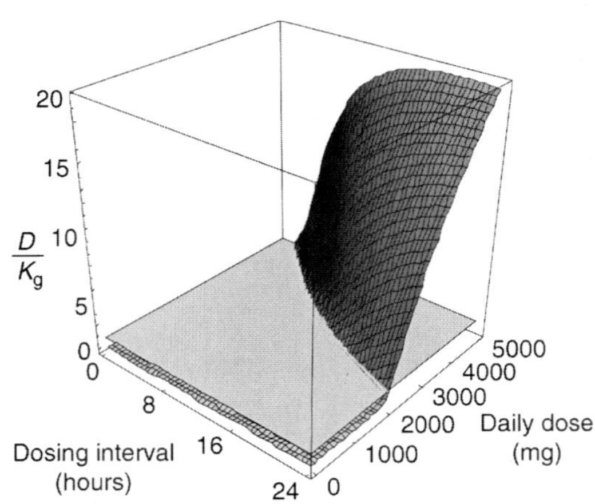

Figure 17.9 Model prediction of bactericidal effect of levofloxacin on *P. aeruginosa* for dosing regimens as in Figure 17.4 ($t_{1/2}$ = 6 hours, T = 24 hours). Dosing regimens (combinations of daily dose and dosing interval) associated with resistance suppression correspond to $D/K_g > 1$ where $D \hat{=} \frac{1}{T}\int_0^T r(C(t))dt$ is the average kill rate over T. Periodic agent injection every T = 24 hours requires above 2200 mg of levofloxacin for complete eradication of the entire bacterial population. This prediction was verified both in a hollow-fiber in vitro model (Figure 17.10) and in a murine thigh in vivo model (Jumbe et al., 2003) and is significantly higher than the standard dosing recommended for levofloxacin.

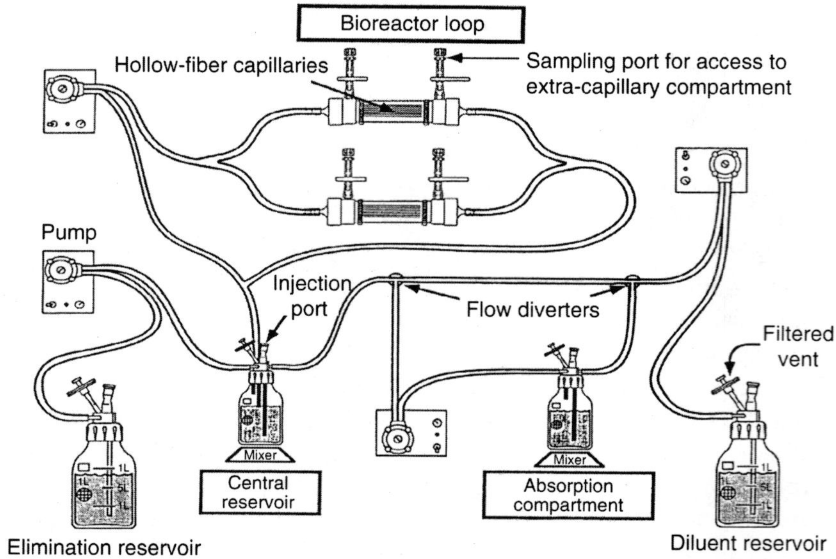

Figure 17.10 In vitro hollow-fiber infection model (Bilello et al., 1994) is a cell culture system in which a microbial population is exposed to fluctuating antimicrobial concentration, simulating human elimination and repeated dosing over a few days, corresponding to a clinical course of treatment. The system has been used by our group to investigate the preclinical potential of antimicrobial agents under development (Tam et al., 2005a; Nikolaou et al., 2007).

daily [total AUC/MIC = 157, free AUC/MIC = 110] demonstrated previously in the murine thigh infection model versus 2200 mg daily predicted by our model). The closeness of our mathematical model predictions to the murine thigh infection model data exemplifies the usefulness of the proposed approach as a tool offering guidance to optimal design of dosing regimens.

17.4 Summary and Future Work

This chapter presents a mathematical modeling framework to design optimal dosing regimens of antimicrobial agents for complete eradication of microbial populations comprising subpopulations of varying degrees of susceptibility/resistance. Preliminary experimental verification of the proposed framework was presented. Further work is needed to identify the limits of the mathematical modeling framework for various combinations of microbial populations and antimicrobial agents, identify its sensitivity

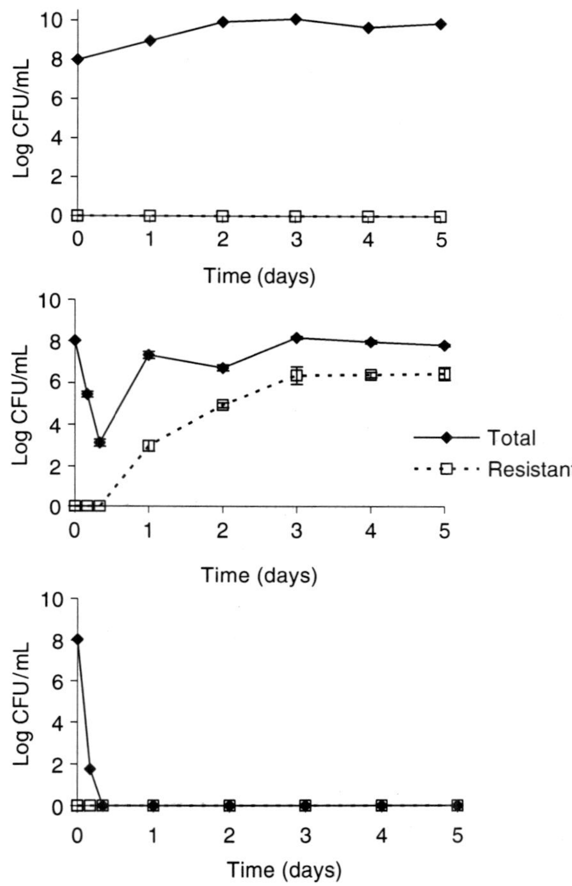

Figure 17.11 Prospective validation of the mathematical model in the hollow-fiber infection model for placebo (a), levofloxacin 750 mg (b), levofloxacin 3000 mg (c) given every 24 hours. Data presented as mean and standard deviation of duplicate samples.

to available data, develop experimental protocols for collection of better experimental data, and potentially extend the framework to other related cases such as cancer chemotherapy.

Acknowledgment

Partial support from the University of Houston through a GEAR grant and from the Johns Hopkins Center for Alternatives to Animal Testing is gratefully acknowledged.

References

Andes, D. and W. A. Craig 1998. In vivo activities of amoxicillin and amoxicillin-clavulanate against *Streptococcus pneumoniae*: application to breakpoint determinations. *Antimicrob. Agents Chemother.* **42**(9): 2375–2379.

Andes, D., K. Marchillo, R. Conklin, G. Krishna, F. Ezzet, A. Cacciapuoti, and D. Loebenberg 2004. Pharmacodynamics of a new triazole, posaconazole, in a murine model of disseminated candidiasis. *Antimicrob. Agents Chemother.* **48**(1): 137–142.

Bilello, J. A., G. Bauer, M. N. Dudley, G. A. Cole, and G. L. Drusano 1994. Effect of 2′, 3′-didehydro-3′-deoxythymidine in an in vitro hollow-fiber pharmacodynamic model system correlates with results of dose-ranging clinical studies. *Antimicrob. Agents Chemother.* **38**(6): 1386–1391.

Bonhoeffer, S., M. Lipsitch, and B. R. Levin 1997. Evaluating treatment protocols to prevent antibiotic resistance. *Proc. Natl. Acad. Sci. U S A* **94**(22): 12106–12111.

Campion, J. J., P. J. McNamara, and M. E. Evans 2005. Pharmacodynamic modeling of ciprofloxacin resistance in *Staphylococcus aureus*. *Antimicrob. Agents Chemother.* **49**(1): 209–219.

Chait, R., A. Craney, and R. Kishony 2007. Antibiotic interactions that select against resistance. *Nature* **446**(7136): 668–671.

Cohen, M. L. 1992. Epidemiology of drug-resistance—Implications for a postantimicrobial era. *Science* **257**(5073): 1050–1055.

Comarow, A. 2006. Bugs behaving badly. *U.S. News & World Report* **140**(1).

Craig, W. A. 1998. Pharmacokinetic/pharmacodynamic parameters: Rationale for antibacterial dosing of mice and men. *Clin. Infect. Dis.* **26**: 1–12.

Dandekar, P. K., P. R. Tessier, P. Williams, C. H. Nightingale, and D. P. Nicolau 2003. Pharmacodynamic profile of daptomycin against Enterococcus species and methicillin-resistant *Staphylococcus aureus* in a murine thigh infection model. *J. Antimicrob. Chemother.* **52**(3): 405–411.

Di Justo, P. 2005. The bug wars. *Wired*. October: 52–53.

Drlica, K. A. 2001. Strategy for fighting antibiotic resistance. *ASM News* **67**(1): 27–33.

Drusano, G. L., A. Louie, M. Deziel, and T. Gumbo 2006. The crisis of resistance: Identifying drug exposures to suppress amplification of resistant mutant subpopulations. *Clin. Infect. Dis.* **42**(4): 525–532.

Dua, P., V. Dua, and E. N. Pistikopoulos 2005. Optimal control of cancer by delivery of chemotherapeutic agents. AIChE Annual Meeting, Cincinnati, OH.

Food and Drug Administration 2006. Antimicrobial Resistance. 2007, from www.fda.gov/cvm/antiresistvideo.htm.

Giraldo, J., N. M. Vivas, E. Vila, and A. Badia 2002. Assessing the (a)symmetry of concentration-effect curves: Empirical versus mechanistic models. *Pharmacol. Ther.* **95**: 21–45.

Gold, H. S. and R. C. Moellering 1996. Antimicrobial-drug resistance. *N. Engl. J. Med.* **335**: 1444–1453.

Gumbo, T., A. Louie, M. R. Deziel, L. M. Parsons, M. Salfinger, and G. L. Drusano 2004. Selection of a moxifloxacin dose that suppresses drug resistance in *Mycobacterium tuberculosis*, by use of an in vitro pharmacodynamic infection model and mathematical modeling. *J. Infect. Dis.* **190**(9): 1642–1651.

Hill, A. V. 1910. The possible effects of the aggregation of the molecules of haemoglobin on its dissociation curves. *J. Physiol.* **40**: 4–7.

Jumbe, N., A. Louie, R. Leary, W. Liu, M. R. Deziel, V. H. Tam, R. Bachhawat, C. Freeman, J. B. Kahn, K. Bush, M. N. Dudley, M. H. Miller, and G. L. Drusano 2003. Application of a mathematical model to prevent in vivo amplification of antibiotic-resistant bacterial populations during therapy. *J. Clin. Invest.* **112**(2): 275–285.

Jusko, W. 1971. Pharmacodynamics of chemotherapeutic effects: dose–time-response relationships for phase-nonspecific agents. *J. Pharm. Sci.* **60**: 892–895.

Landman, D., J. M. Quale, D. Mayorga, A. Adedeji, K. Vangala, J. Ravishankar, C. Flores, and S. Brooks 2002. Citywide clonal outbreak of multiresistant *Acinetobacter baumannii* and *Pseudomonas aeruginosa* in Brooklyn, NY: The preantibiotic era has returned. *Arch. Intern. Med.* **162**: 1515–1520.

Levy, S. B. 1998. The challenge of antibiotic resistance. *Sci. Am.* **278**(3):46–53.

Levy, S. B. and B. Marshall 2004. Antibacterial resistance worldwide: causes, challenges and responses. *Nat. Med.* **10**(12): S122–S129.

Lipsitch, M. and B. R. Levin 1997. The population dynamics of antimicrobial chemotherapy. *Antimicrob. Agents Chemother.* **41**(2): 363–373.

Lipsitch, M., T. H. Bacon, J. J. Leary, R. Antia, and B. R. Levin 2000. Effects of antiviral usage on transmission dynamics of herpes simplex virus type 1 and on antiviral resistance: predictions of mathematical models. *Antimicrob. Agents Chemother.* **44**(10): 2824–2835.

Livermore, D. M. 2002. Multiple mechanisms of antimicrobial resistance in *Pseudomonas aeruginosa*: our worst nightmare? *Clin. Infect. Dis.* **34**(5): 634–640.

Louie, A., G. L. Drusano, P. Banerjee, Q. F. Liu, W. Liu, P. Kaw, M. Shayegani, H. Taber, and M. H. Miller 1998. Pharmacodynamics of fluconazole in a murine model of systemic candidiasis. *Antimicrob. Agents Chemother.* **42**(5): 1105–1109.

Louie, A., P. Kaw, W. Liu, N. Jumbe, M. H. Miller and G. L. Drusano 2001. Pharmacodynamics of daptomycin in a murine thigh model of *Staphylococcus aureus* infection. *Antimicrob. Agents Chemother.* **45**(3): 845–851.

Maglio, D., C. Ong, M. A. Banevicius, Q. Geng, C. H. Nightingale, and D.P. Nicolau 2004. Determination of the in vivo pharmacodynamic profile of cefepime against extended-spectrum-beta-lactamase-producing *Escherichia coli* at various inocula. *Antimicrob. Agents Chemother.* **48**(6): 1941–1947.

Meagher, A. K., A. Forrest, A. Dalhoff, H. Stass, and J. J. Schentag 2004. Novel pharmacokinetic–pharmacodynamic model for prediction of outcomes with an extended-release formulation of ciprofloxacin. *Antimicrob. Agents Chemother.* **48**(6): 2061–2068.

Miyazaki, S., K. Okazaki, M. Tsuji, and K. Yamaguchi 2004. Pharmacodynamics of S-3578, a novel cephem, in murine lung and systemic infection models. *Antimicrob. Agents Chemother.* **48**(2): 378–383.

Morens, D. M., G. K. Folkers, and A. S. Fauci 2004. The challenge of emerging and re-emerging infectious diseases. *Nature* **430**(08 July): 242–249.

Mouton, J. W., A. A. Vinks, and N. C. Punt 1997. Pharmacokinetic–pharmacodynamic modeling of activity of ceftazidime during continuous and intermittent infusion. *Antimicrob. Agents Chemother.* **41**(4): 733–738.

Neu, H. C. 1992. The crisis in antibiotic resistance. *Science* **257**: 1064–1073.

Neuhauser, M. M., R. A. Weinstein, R. Rydman, L. H. Danziger, G. Karam, and J. P. Quinn 2003. Antibiotic resistance among gram-negative bacilli in US intensive care units: implications for fluoroquinolone use. *Jama* **289**(7): 885–888.

Nicolau, D. P., C. O. Onyeji, M. Zhong, P. R. Tessier, M. A. Banevicius, and C. H. Nightingale 2000. Pharmacodynamic assessment of cefprozil against

Streptococcus pneumoniae: implications for breakpoint determinations. *Antimicrob. Agents Chemother.* **44**(5): 1291–1295.

Nikolaou, M., A. N. Schilling, G. Vo, K.-t. Chang, and V. H. Tam 2007. Modeling of microbial population responses to time-periodic concentrations of antimicrobial agents. *Ann. Biomed. Eng.* **DOI: 10.1007/s10439-007-9306-x**. (In press).

Nikolaou, M. and V. H. Tam 2006. A new modeling approach to the effect of antimicrobial agents on heterogeneous microbial populations. *J. Math. Biol.* **52**(2): 154–182.

Oleson, J., B. T. Frederick, and P. Francis 2005. Methods for administration of antibiotics. U. S. P. 6852689, Cubist Pharmaceuticals, Inc.

Oliver, A., B. R. Levin, C. Juan, F. Baquero, and J. Blazquez 2004. Hypermutation and the preexistence of antibiotic-resistant *Pseudomonas aeruginosa* mutants: implications for susceptibility testing and treatment of chronic infections. *Antimicrob. Agents Chemother.* **48**(11): 4226–4233.

Regoes, R. R., C. Wiuff, R. M. Zappala, K. N. Garner, F. Baquero, and B. R. Levin 2004. Pharmacodynamic functions: a multiparameter approach to the design of antibiotic treatment regimens. *Antimicrob. Agents Chemother.* **48**(10): 3670–3676.

Silberman, S. 2007. Requiem for the magic bullets. *Wired* (January 22).

Spellberg, B., J. H. Powers, E. P. Brass, L. G. Miller, and J. E. Edwards, Jr. 2004. Trends in antimicrobial drug development: implications for the future (Major Article). *Clin. Infect. Dis.* **38**(9): 1279–1286.

Talbot, G. H., J. Bradley, J. E. Edwards, Jr., D. Gilbert, M. Scheld, and J. G. Bartlett 2006. Bad bugs need drugs: an update on the development pipeline from the Antimicrobial Availability Task Force of the Infectious Diseases Society of America. *Clin. Infect. Dis.* **42**(5): 657–668.

Tam, V. H., A. Louie, M. R. Deziel, W. Liu, R. Bachhawat, D. Gajjar, D. Grasela, and G. L. Drusano 2002. Pharmacodynamics of garenoxacin in a murine thigh infection model of *Staphylococcus aureus*. Annual Meeting of the Infectious Diseases Society of America, Chicago, Illinois.

Tam, V. H., A. Louie, M. R. Deziel, W. Liu, R. Leary, and G. L. Drusano 2005a. Bacterial-population responses to drug-selective pressure: examination of garenoxacin's effect on *Pseudomonas aeruginosa*. *J. Infect. Dis.* **192**(3): 420–428.

Tam, V., A. Schilling, D. Melnick, and E. Coyle 2005b. Comparison of beta-lactams in counter-selecting resistance of *Pseudomonas aeruginosa*. *Diagn. Microbiol. Infect. Dis.* **52**(2): 145–151.

Tam, V. H., A. N. Schilling, and M. Nikolaou 2005c. Modelling time-kill studies to discern the pharmacodynamics of meropenem. *J. Antimicrob. Chemother.* **55**(5): 699–706.

UCSF 2006. Daptomycin. From http://www.ucsf.edu/idmp/whatsnew/dapto_monograph.htm.

Varaldo, P. E. 2002. Antimicrobial resistance and susceptibility testing: an evergreen topic. *J. Antimicrob. Chemother.* **50**: 1–4.

Vogelman, B. and W. Craig 1986. Kinetics of antimicrobial activity. *J. Pediatr.* **108**(2): 835–840.

Wagner, J. 1968. Kinetics of pharmacologic response I. Proposed relationships between response and drug concentration in the intact animal and man. *J. Theoret. Biol.* **20**: 173–201.

Weisstein, E. W. 2005. Cumulant. *MathWorld–A Wolfram Web Resource*, 2005, from http://mathworld.wolfram.com/Cumulant.html.

Appendix A

Tutorial Guide to Mixed-Integer Programming Models and Solution Techniques

J. Cole Smith and Z. Caner Taşkın

CONTENTS

A.1 Introduction 522
A.2 Modeling Principles 524
A.2.1 General Form 524
A.2.2 Modeling Mixed-Integer Programming Problems 525
A.3 MIP Solution Techniques 535
A.4 Example Radiation Therapy Application 541
A.4.1 Leaf Sequencing Problem 542
A.5 Conclusion 545
References 546

Abstract Mixed-integer programming theory provides a mechanism for optimizing decisions that take place in complex systems, including those encountered in biology and medicine. This chapter is intended for researchers and practitioners wanting an introduction to the field of mixed-integer programming. We begin by discussing basic mixed-integer programming formulation principles and tricks, especially with regards to the use of binary variables to form logical statements. We then discuss two core techniques, branch-and-bound and cutting plane algorithms, used to solve mixed-integer programs. We illustrate the use of mixed-integer programming in the context of several medical applications, and close with a featured study on intensity modulated radiation therapy planning.

A.1 Introduction

This chapter describes the use of mixed-integer programming in optimizing complex systems, such as those arising in biology, medicine, transportation, telecommunications, sports, and national security. Consider, for instance, an emergency that results in 100 injuries. A triage center is established to administer first aid and assign victims to one of three nearby hospitals, each of which is capable of handling a limited number of patients. Each hospital may have varying equipment and staff levels, and each will be located at a different distance from the emergency. The optimization problem that arises is to assign patients to hospitals in a way that maximizes the effectiveness of care that can be given to the victims, while obeying physical capacity restrictions imposed by the hospitals. Experts often attempt to solve these problems based on intuition and experience, but the resulting solution is almost invariably suboptimal due to the inherent complexities of such problems. In applications of critical importance, there is sufficient motivation to turn to mathematical techniques that can provably obtain a best solution.

Mixed-integer programming is a subset of the broader field of mathematical programming. Mathematical programming formulations include a set of variables, which represent actions that can be taken in the system being modeled. One then attempts to optimize (either in the minimization or maximization sense) a function of these variables, which maps each possible set of decisions into a single score that assesses the quality of the solution. These scores are often in units of currency representing total cost incurred or revenue gained etc. The limitations of the system are included as a set of constraints, which are usually stated by restricting functions of the decision variables to be equal to, not more than, or not less than, a certain numerical value. Another type of constraint can simply restrict the set of values to which a variable might be assigned.

Several applications involve decisions that are *discrete* (e.g., to which hospital an emergency patient should be assigned), while some other decisions are *continuous* in nature (e.g., determining the dosage of fluids to be administered to a patient). On the surface, the ability to enumerate all possible values that a discrete decision can take seems appealing; however, in most applications, the discrete variables are interrelated, requiring an enumeration of all combinations of values that the entire set of discrete variables can take.

What are the implications of complete enumeration techniques onprocessing time? Suppose that there exist n variables, each of which can take on a value of zero or one. Furthermore, suppose that each configuration of these variables can be evaluated (tested for feasibility to the problem constraints and scored) using n computer operations. Because there are two choices for each variable, there are 2^n configurations. Even if we are using a computer capable of processing 10 trillion operations per second (or 10 teraflops, and at the time of this writing, only 58 of the world's top 500 supercomputers are capable of such a feat), if $n = 50$, the computer will take 1.5 hours to finish enumerating all possibilities. One might be tempted to simply let the computer run all night if need be for important problems, and although this is indeed valid for the case in which $n = 50$, the computational growth rate for these problems is astounding: for $n = 60$, the computer will require 80 days to terminate, and for $n = 70$, the computer will require 262 years. Another question regards the evolution of computing speed, noting that faster computers are constantly emerging. If the program must be finished within two hours, the current 10 teraflop machine will permit the solution of problems with $n = 50$. If a quantum leap is discovered that results in the invention of 10,000 teraflop machine, this fictional computer would only be able to handle problems with $n = 60$ within two hours. Computer speedups, however impressive, are simply no match for exponential enumeration problems.

Therefore, a more efficient technique is required to solve problems containing discrete variables. Mixed-integer programming techniques do not *explicitly* examine every possible combination of discrete solutions, but instead examine a subset of possible solutions, and use optimization theory to prove that no other solution can be better than the best one found. This type of technique is referred to as *implicit enumeration*.

This chapter is not a thorough review of integer programming literature, but is intended for technical researchers who may or may not have any familiarity with linear programming, but who are looking for an entry-level introduction to modeling and solution via integer and mixed-integer programming. The text by Wolsey [18] provides an accessible account of fundamental integer programming methods and theory, while the updated classical work of Nemhauser and Wolsey [11] discusses integer programming and combinatorial theory in detail.

We discuss the general form of mixed-integer programming problems in Section A.2, and provide general tips for formulating problems as mixed-integer programs (MIPs). A brief discussion of the branch-and-bound implicit enumeration technique for solving mixed-integer programs, as is relevant to practitioners, is given in Section A.3. Next, Section A.4 provides an example of MIPs in a real radiation therapy application, illustrating the material presented in the earlier two sections. Finally, we conclude this chapter in Section A.5.

A.2 Modeling Principles

We begin this section by discussing the general form of linear and mixed-integer programming problems in Section A.2.1. We then give common steps and principles behind modeling problems of this form in Section A.2.2, and suggest a few common ways that mixed-integer variables can be used to model complex conditions arising in real-world scenarios.

A.2.1 General Form

First, we present the general form of a linear programming problem. Linear programming problems (usually called linear programs, and abbreviated as LPs) contain a set of *decision variables*, which are the unknown quantities or decisions that are to be optimized. In the context of linear and mixed-integer programming problems, the function that assesses the quality of the solution, called the objective function, should be a linear function of the decision variables. An LP will either minimize or maximize the value of the objective function. Finally, the decisions that must be made are subject to certain requirements and restrictions of a system. We enforce these restrictions by including a set of *constraints* in the model. Each constraint requires that a linear function of the decision variables is either equal to, not less than, or not more than, a scalar value. A common condition simply states that each decision variable must be nonnegative. In fact, all linear programming problems can be transformed into an equivalent minimization problem with nonnegative variables and equality constraints [1].

Thus, suppose we denote $x_1, \ldots, x_n$ to be our set of decision variables. Linear programming problems take on the form:

$$\text{Min or Max} \quad c_1x_1 + c_2x_2 + \cdots + c_nx_n \tag{A.1a}$$

$$\text{s.t.} \quad a_{11}x_1 + a_{12}x_2 + \cdots + a_{1n}x_n \; (\leq, \; =, \; \text{or} \; \geq) \; b_1 \tag{A.1b}$$

$$a_{21}x_1 + a_{22}x_2 + \cdots + a_{2n}x_n \; (\leq, \; =, \; \text{or} \; \geq) \; b_2 \tag{A.1c}$$

$$\ldots$$

$$a_{m1}x_1 + a_{m2}x_2 + \cdots + a_{mn}x_n \; (\leq, \; =, \; \text{or} \; \geq) \; b_m \tag{A.1d}$$

$$x_j \geq 0 \quad \forall j = 1, \ldots, n. \tag{A.1e}$$

Values c_j, $\forall j = 1, \ldots, n$, are referred to as objective coefficients, and are often associated with the costs associated with their corresponding decisions in minimization problems, or the revenue generated from the corresponding decisions in maximization problems. The values $b_1, \ldots, b_m$ are the right-hand-side values of the constraints, and often represent amounts of available resources (especially for $\leq$ constraints) or requirements (especially for $\geq$ constraints). The a_{ij}-values thus typically denote how much of resource/requirement i is consumed/satisfied by decision j.

Note that nonlinear terms are not allowed in the model, prohibiting for instance the multiplication of two decision variables, the maximum of several variables, or the absolute value of a variable. (These conditions are often desired, but must be achieved by different techniques.) Also, any inequalities present in the model are never strict.

Problems of the form of Equation A.1 are called linear programs because the objective function and LPs constraint functions are all linear. Integer programs (IPs) are stated in an identical fashion, except that all decision variables are constrained to take on integer values. (Hence, integer programs are sometimes called integer linear programs.) An MIP is an LP with the added restriction that some, but not necessarily all, of the variables must be integer-valued. Several studies also replace the term integer with 0–1 or binary when variables are restricted to take on either 0 or 1 values. For the purposes of this chapter, we focus on MIPs, with IPs being modeled and solved as a special case of MIPs.

A solution that satisfies all constraints is called a *feasible solution*. Feasible solutions that achieve the best objective function value (according to whether one is minimizing or maximizing) are called *optimal solutions*. Sometimes no solution exists to an MIP, and the MIP itself is called *infeasible*. On the other hand, some feasible MIPs have no optimal solution, because it is possible to achieve infinitely good objective function values with feasible solutions. Such problems are called *unbounded*.

A.2.2 Modeling Mixed-Integer Programming Problems

The modeling of complex systems using MIPs is often more of an art than a science. Typically, a three-step looped process is used to model MIPs. The first step involves defining a set of decision variables that represent choices that must be optimized in the system. The second step usually involves the statement of constraints in the model, with the third step requiring the statement of an objective function (although the last two steps can be done in either order).

It is very common, though, to recognize during model construction that the initial set of decision variables defined for the model is inadequate. Often, decision variables that seem to be the implied consequences of other actions must also be defined. The addition of new variables after an unsuccessful attempt at formulating constraints and objectives is the "loop" in the process.

The correct definition of decision variables can be especially complicated in modeling with integer variables. If one is allowed to use binary variables in a formulation, it is possible to represent yes-or-no decisions, enforce if–then statements, and even permit some sorts of nonlinearity in the model (which can be transformed to an equivalent mixed-integer LP).

To illustrate the modeling process, we consider the following three example systems.

EXAMPLE 1. An outbreak of an infectious disease has been observed in a set N of locations. There exist a set M of teams capable of investigating these outbreaks. Team $i \in M$ can conclude its investigation of the outbreak in location $j \in N$ in t_{ij} hours. Each team can either investigate zero, one, or two outbreaks. If a team investigates two outbreaks, they must travel from one location to the next. The travel time from location $j_1 \in N$ to location $j_2 \in N$ is $d_{j_1 j_2}$. Once all outbreaks have been investigated, a disease control center can take action to combat the outbreak. The goal is to minimize the amount of time necessary to complete the investigation of all locations. □

EXAMPLE 2. A medical practice is attempting to acquire a certain drug from a set M of suppliers. The practice wishes to have a stock of d_t units of this drug in month t, for $t = 1, \ldots, \bar{t}$. Purchasing one unit of the drug from supplier $i \in M$ during period $t \in \{1, \ldots, \bar{t}\}$ costs c_{it} dollars. However, to purchase drugs from supplier $i \in M$, at any period, the practice must purchase a minimum of ℓ_i units of the drug during that period. Fortunately, the practice has room for h units of inventory, and so at most h units of the drug can be stored from one period to the next. If the practice finds itself with too many units of the drug, it can simply throw away the extra supply. The goal is to minimize the cost required to purchase the drugs for time periods $1, \ldots, \bar{t}$. □

EXAMPLE 3. A nurse is assigned a set N of patients. For each patient $i \in N$, the nurse must spend p_i minutes examining the patient. The nurse must then wait somewhere between ℓ_i and u_i minutes before checking up on the patient, which requires q_i minutes. The nurse, of course, cannot be in two places at the same time, although we will assume for simplicity that the travel time to walk from one patient's room to another is zero. The objective is to minimize the total amount of time required to tend to all patients. For instance, suppose that N contains three patients:

- Patient 1's first visit lasts $p_1 = 5$ minutes. The patient will then be checked on between $\ell_1 = 30$ and $u_1 = 45$ minutes later, after which the nurse will spend $q_1 = 5$ more minutes of care on the second visit.

- Patient 2 needs $p_2 = 10$ minutes of initial care, has an inter-care gap of $\ell_2 = 25$ and $u_2 = 35$ minutes, and requires $q_2 = 5$ minutes of further care.
- Patient 3 needs $p_3 = 10$ minutes of initial care, has an inter-care gap of $\ell_3 = 30$ and $u_3 = 35$ minutes, and requires $q_3 = 10$ minutes of further care.

One solution would start treatment on patients 1, 2, and 3 in this order. The nurse's schedule would then be optimized according to the first column of Table A.1 and the total treatment time would be 65 minutes. However, if the patients are treated in the opposite order, the total treatment time becomes only 60 minutes, as shown in the second column of Table A.1. Indeed, the timing involved in this problem is quite difficult to optimize by hand. □

Attempting to formulate any of these problems as an LP is problematic. With continuous variables, the first example will likely end up assigning part of a team to one location, and part of a team to another location. The second example will not be able to represent the minimum purchase aspect of the problem. The example will not necessarily be able to keep the nurse from splitting attention simultaneously among multiple patients. Before attempting to formulate these problems, let us discuss a few common tips and tricks in modeling with integer variables.

1. *Integrality of quantities.* In staffing and purchasing decisions, it is often impossible to take fractional actions. One cannot hire, for instance, 6.5 new staff members, or purchase 1.3 hospital beds. The most obvious use of integer variables thus arises in requesting integer amounts of quantities that can only be ordered in integer amounts. In general, the optimal solution of an integer program need not be a rounded-off version of an optimal solution to an LP.

Table A.1 Two Schedules for the Nurse Scheduling Problem

Patients Ordered 1, 2, 3	Patients Ordered 3, 2, 1
Time 0–5: Patient 1	Time 0–10: Patient 3
Time 5–15: Patient 2	Time 10–20: Patient 2
Time 15–25: Patient 3	Time 20–25: Patient 1
Time 25–35: Idle	Time 25–40: Idle
Time 35–40: Patient 1	Time 40–50: Patient 3
Time 40–45: Patient 2	Time 50–55: Patient 2
Time 45–55: Idle	Time 55–60: Patient 1
Time 55–65: Patient 3	

2. *If–then statements.* Consider two continuous (i.e., possibly fractional) variables, x and y, defined so that $0 \leq x \leq 10$ and $0 \leq y \leq 10$. Suppose we wish to make a statement that if $x > 4$, then $y \leq 6$. (If $x \leq 4$, then we do not wish to further constrain y.) On the surface, because no integer quantities are requested, it does not appear that integer variables will be necessary. However, the general form of LPs as given in Equations A.1a through A.1e does not permit if–then statements like the one above. Instead, if–then statements can be enforced with the aid of a binary variable, z. We wish to make $z = 1$ if $x > 4$ (note that we make no claims on z if $x \leq 4$). This can be accomplished by adding the constraint

$$x \leq 4 + 6z, \tag{A.2}$$

because the event that $x > 4$ implies that $z = 1$. (Even if $z = 1$, the largest value for x is 10, which now makes a constraint of the form $x \leq 10$ unnecessary.) If $z = 1$, then we must also require that $y \leq 6$. This is achieved by reducing the upper bound of 10 on y to 6 if z is equal to 1 as follows:

$$y \leq 10 - 4z, \tag{A.3}$$

where once again, the bound constraint $y \leq 10$ may now be omitted. In general, suppose we wish to make the following statement: if $q_1x_1 + \cdots + q_nx_n > Q$, then $r_1x_1 + \cdots + r_nx_n \leq R$. We would include the following conditions in our model:

$$q_1x_1 + \cdots + q_nx_n \leq Q + M'z, \tag{A.4}$$

$$r_1x_1 + \cdots + r_nx_n \leq M'' - (M'' - R)z, \tag{A.5}$$

$$z \text{ binary}, \tag{A.6}$$

where M' and M'' are sufficiently large constants. These values should be just large enough to not add unintentional restrictions to the model. For instance, we are not attempting to place any hard restriction on the quantity $q_1x_1 + \cdots + q_nx_n$ (written conveniently as $\mathbf{q}^\top\mathbf{x}$ in vector form). If $z = 1$, the upper bound on $\mathbf{q}^\top\mathbf{x}$ is $Q + M'$, and hence M' must be large enough so that even if constraint given by Equation A.4 is removed from the model, $\mathbf{q}^\top\mathbf{x}$ would still never be more than $Q + M'$. Likewise, if $z = 0$, we must choose a value M'' large enough in Equation A.5 such that $\mathbf{r}^\top\mathbf{x}$ could never be more than M'' even without the restriction (Equation A.5). It is worth noting that assigning arbitrarily large values for M' and M'' is not recommended, for reasons that will become more apparent in Section A.3.

3. *Enforce at least k-out-of-p restrictions.* This situation is similar to if–then constraints in the way we model such restrictions. For a simple example, suppose we have nonnegative variables $x_1, \ldots, x_n$, and wish to require that at least three of these variables take on values of 5 or more. Then we can define variables $z_1, \ldots, z_n$, such that if $z_j = 1$, then $x_j \geq 5$, $\forall j = 1, \ldots, n$. This simple if–then constraint can easily be modeled by employing the following constraints:

$$x_j \geq 5z_j \quad \forall j = 1, \ldots, n. \tag{A.7}$$

Clearly, if $z_j = 1$, then $x_j \geq 5$. If $z_j = 0$, it is still possible for $x_j \geq 5$, but no such restrictions are enforced. We need to guarantee that three variables take on values of 5 or more, and so we add the following k-out-of-p constraint:

$$z_1 + \cdots + z_n = 3. \tag{A.8}$$

Again, this constraint does not state that exactly three variables will be at least 5, but rather that at least three variables are guaranteed to be at least 5. This same trick can be used to enforce the condition that at least k-out-of-p sets of constraints are satisfied, and so on, often by using M-values as introduced in the section on if–then constraints.

4. *Nonlinear product terms.* In some circumstances, nonlinear terms can be transformed into linear terms by the use of linear constraints. First, note that if x_j is a binary variable, then $x_j = x_j^q$ for any positive constant q. After that substitution is made, suppose that we have a nonlinear term of the form $x_1 \cdot x_2 \cdots x_k \cdot y$, where $x_1, \ldots, x_k$ are binary variables and $0 \leq y \leq u$ is another variable, either continuous or integer. That is, all but perhaps one of the terms is a binary variable. First, replace the nonlinear term with a single continuous variable, w. Using the if–then concept expressed above, note that if x_j equals zero for any $j \in \{1, \ldots, k\}$, then w equals zero as well. Also, note that w can never be more than the upper bound, u, on the y-variable. Hence, we obtain the constraints

$$w \leq ux_j \quad \forall j = 1, \ldots, k. \tag{A.9}$$

Of course, to guarantee that w equals zero in case any x_j-variable equals to zero, we must also state a nonnegativity constraint

$$w \geq 0. \tag{A.10}$$

Now, suppose that all $x_1 = \cdots = x_k = 1$. In this case, we need to add constraints that enforce the condition that $w = y$. Regardless of the x-variable values, w cannot be more than y, and so we state the constraint

$$w \leq y. \tag{A.11}$$

However, to get the constraint $w \geq y$ if $x_1 = \cdots = x_k = 1$, we include the constraint

$$w \geq u(x_1 + \cdots + x_k - k) + y. \tag{A.12}$$

If each x-variable equals to 1, then Equation A.12 states that $w \geq y$, which along with Equation A.11 guarantees that $w = y$. On the other hand, if at least one $x_j = 0$, $j \in \{1, \ldots, k\}$, then the term $u(x_1 + \cdots + x_k - k)$ is not more than $-u$, and the right-hand side of Equation A.12 is not positive; hence, Equation A.12 allows w to take on the correct value of zero (as would be enforced by Equations A.9 and A.10). As a final note, observe that even if y is an integer variable, we need not insist that w is an integer variable as well, because Equations A.9 through A.12 guarantee that $w = x_1 \cdots x_k \cdot y$, which must be an integer given integer x- and y-values.

Let us now briefly continue the examples introduced earlier in this section in the light of the modeling tricks introduced above.

Example 1, continued. Define binary decision variables x_{ij}, which equal one if team $i \in M$ investigates an outbreak at location $j \in N$, and zero otherwise. Proceeding with this initial set of variables, we would then attempt to formulate the constraints of the problem. In fact, the only restrictions encountered thus far is the fact that each location must be investigated by exactly one team, the fact that each team can investigate no more than two locations, and the fact that each x-variable must be binary. These constraints are, respectively, given by

$$\sum_{i \in M} x_{ij} = 1 \quad \forall j \in N, \tag{A.13}$$

$$\sum_{j \in N} x_{ij} \leq 2 \quad \forall i \in M, \tag{A.14}$$

$$x_{ij} \text{ binary} \quad \forall i \in M,\ j \in N. \tag{A.15}$$

Finally, we would attempt to formulate the objective function. Note that the amount of time required for team $i \in M$ to complete their investigations is the amount of time required to perform the investigations, $\sum_{j \in N} t_{ij}x_{ij}$, plus the travel time between location sites. This travel time can be represented by $\sum_{\text{all } j_1 \neq j_2 \in N} d_{j_1 j_2} x_{ij_1} x_{ij_2}$, noting that inter-location travel time is required only if a team is required to visit both sites. The objective function has the following form:

$$\text{Min maximum}_{i \in M} \left\{ \sum_{j \in N} t_{ij}x_{ij} + \sum_{\text{all } j_1 \neq j_2 \in N} d_{j_1 j_2} x_{ij_1} x_{ij_2} \right\}.$$

This function is nonlinear for two reasons. One, the investigation completion time is a nonlinear function due to the multiplication of x-variables. Two, the quantity being minimized is the maximum of the teams' investigation completion times.

At this point, we must add additional variables to the problem. Replace each nonlinear term $x_{ij_1}x_{ij_2}$ with a new variable $w_{ij_1 j_2}$, and add linearization constraints

$$w_{ij_1 j_2} \leq x_1 \text{ similar to Equation A.9,} \tag{A.16a}$$

$$w_{ij_1 j_2} \geq 0 \text{ similar to Equation A.10,} \tag{A.16b}$$

$$w_{ij_1 j_2} \leq x_2 \text{ similar to Equation A.11, treating } x_2 \text{ as ``}y\text{'',} \tag{A.16c}$$

$$w_{ij_1 j_2} \geq x_1 + x_2 - 1 \text{ similar to Equation A.12.} \tag{A.16d}$$

The objective function now becomes

$$\text{Min maximum}_{i \in M} \left\{ \sum_{j \in N} t_{ij}x_{ij} + \sum_{\text{all } j_1 \neq j_2 \in N} d_{j_1 j_2} w_{ij_1 j_2} \right\}.$$

To remove the "minimax" structure of this objective function, we rely on a common trick for linear programming. First, we add a variable θ that represents the maximum completion time. We would then minimize θ, subject to the conditions that θ must be at least as large as the completion time for team 1, θ must be at least as large as the completion time for team 2, and so on, for each team. These constraints are

$$\theta \geq \sum_{j \in N} t_{ij}x_{ij} + \sum_{\text{all } j_1 \neq j_2 \in N} d_{j_1 j_2} w_{ij_1 j_2} \quad \forall i \in M. \tag{A.17}$$

Of course, the question arises as to why θ will be *equal* to the maximum of the teams' completion times, because Equation A.17 merely enforces the condition that θ is at least as large as the maximum of these times. The answer is that the objective function will minimize θ and so θ will take on the smallest possible value permitted by Equation A.17, which will indeed be the maximum of completion times.

The overall model is then given by

$$\text{Min} \quad \theta$$

$$\text{s.t.} \quad \text{Constraints given by Equations A.13 through A.17.}$$

EXAMPLE 2, continued. An initial attempt at modeling this problem would define decision variables x_{it}, $\forall i \in M$, $t = 1, \ldots, \bar{t}$, equal to the amount of drugs purchased from supplier i in period t. It is also necessary in general to define variables g_i, $\forall i \in M$, which denote how many units of drugs are thrown away after period i (because we purchased too many from a supplier due to minimum purchase limits, and perhaps also due to inventory limits). However, we must ensure that the number of drugs purchased from supplier $i \in M$ at any period is either zero or at least ℓ_i. Hence, let us define binary decision variables z_{it}, which equal one if we purchase any supplies from supplier $i \in M$ at period $t \in \{1, \ldots, \bar{t}\}$, and zero otherwise.

However, stating the demand and inventory constraints is awkward (though possible) without another set of variables. Note that we must require $\sum_{i \in M} x_{i1} - g_1 \geq d_1$ to satisfy period 1 demand. In period 2, we have that the inventory plus the amount of drugs ordered in period 2, minus whatever is thrown away after period 2, must be at least d_2. This condition is stated as $\left(\left(\sum_{i \in M} x_{i1} - g_1\right) - d_1\right) + \sum_{i \in M} x_{i2} - g_2 \geq d_2$. For period 3, this constraint becomes $\left(\left(\left(\sum_{i \in M} x_{i1} - g_1\right) - d_1\right) + \left(\sum_{i \in M} x_{i2} - g_2\right) - d_2\right) + \sum_{i \in M} x_{i3} - g_3 \geq d_3$. and so on. It is evident that the expression quickly becomes quite large as the period t increases.

Instead, we can define inventory variables y_t, for $t = 1, \ldots, \bar{t}$, which represent the amount of drug inventory remaining after period t. From this point, it is easier to establish flow balance constraints, which state that the amount of drugs coming into the practice at each period (inventory from the previous period plus the amount purchased from suppliers during the current period) equals the drugs that leave the practice at the current period (those demanded by patients, plus those put into inventory after the period, plus those thrown away). These balance constraints are given as

$$y_{t-1} + \sum_{i \in M} x_{it} = d_t + y_t + g_t \quad \forall t = 1, \ldots, \bar{t}, \tag{A.18}$$

where y_0 is defined to be zero. The minimum purchase quantity constraints are given as

$$x_{it} \geq \ell_i z_{it} \quad \forall i \in M, \ t = 1, \ldots, \bar{t}, \tag{A.19}$$

$$x_{it} \leq M_{it} z_{it} \quad \forall i \in M, \ t = 1, \ldots, \bar{t}, \tag{A.20}$$

where M_{it} is a sufficiently large constant, $\forall i, \ t$. If $z_{it} = 1$, then $\ell_i \leq x_{it} \leq M_{it}$, while if $z_{it} = 0$, then $x_{it} = 0$. A possible value for M_{it} would be the maximum of ℓ_i and the remaining demands $\sum_{u=t}^{\bar{t}} d_u$. Finally, we require constraints that state bounds on the x-, g-, and z-variables.

$$x_{it} \geq 0 \text{ and } z_{it} \text{ binary} \quad \forall i \in M, \ t = 1, \ldots, \bar{t}, \tag{A.21}$$

$$0 \leq y_t \leq h \text{ and } g_t \geq 0 \quad \forall t = 1, \ldots, \bar{t}. \tag{A.22}$$

The overall formulation can now be stated as

$$\text{Min} \quad \sum_{i \in M} \sum_{t=1}^{\bar{t}} c_{it} x_{it} \tag{A.23}$$

s.t. Constraints given by Equations A.18 through A.22.

EXAMPLE 3, continued. There exist several methods of modeling and solving this problem. One technique defines a continuous variable f_i to be the first time that the nurse sees patient $i \in N$, and s_i to be the second time that the nurse sees patient $i \in N$. These variable definitions allow us to state the minimum and maximum gaps between the first and second visits by the nurse:

$$s_i - (f_i + p_i) \geq \ell_i \quad \forall i \in N, \tag{A.24}$$

$$s_i - (f_i + p_i) \leq u_i \quad \forall i \in N. \tag{A.25}$$

However, we must now enforce the restriction that the nurse does not tend to two or more patients at the same time. We must ensure that for each pair of nurse visits (first visits to different patients, first and second visits to different patients, and second visits to different patients), one visit starts before the other begins, or vice versa. (The two visits to the same patient are disjoint due to Equation A.24.)

Define binary variables z_{ij}^{11} to equal one if the first visit to patient $i \in M$ occurs before the first visit to patient $j \in M$, and zero otherwise. Also, define z_{ij}^{12} if the first visit to patient $i \in M$ occurs before the second visit to patient $j \in M$, and zero otherwise. Similarly, z_{ij}^{21} relates the second visit of patient $i \in M$ to the first visit of patient $j \in M$, and z_{ij}^{22} relates the second visit of patient $i \in M$ to the second visit of patient $j \in M$. The following constraints relate the z-variables to the f- and s-variables.

$$f_j - (f_i + p_i) \geq -M_{ij}^{11}(1 - z_{ij}^{11}) \quad \forall i,j \in N,\ i \neq j, \tag{A.26}$$

$$s_j - (f_i + p_i) \geq -M_{ij}^{12}(1 - z_{ij}^{12}) \quad \forall i,j \in N,\ i \neq j, \tag{A.27}$$

$$f_j - (s_i + q_i) \geq -M_{ij}^{21}(1 - z_{ij}^{21}) \quad \forall i,j \in N,\ i \neq j, \tag{A.28}$$

$$s_j - (s_i + q_i) \geq -M_{ij}^{22}(1 - z_{ij}^{22}) \quad \forall i,j \in N,\ i \neq j, \tag{A.29}$$

where the M-values once again are sufficiently large constants. For instance, Equation A.26 states that if $z_{ij}^{11} = 1$, then $f_j \geq f_i + p_i$, meaning that the nurse finishes the visit to patient i before the visit to patient j occurs. We now need to state constraints ensuring that one visit finishes before the next one starts, or vice versa.

$$z_{ij}^{11} + z_{ji}^{11} = 1 \quad \forall i,j \in N,\ i < j, \tag{A.30}$$

$$z_{ij}^{22} + z_{ji}^{22} = 1 \quad \forall i,j \in N,\ i < j, \tag{A.31}$$

$$z_{ij}^{12} + z_{ji}^{21} = 1 \quad \forall i,j \in N. \tag{A.32}$$

Constraints given by Equations A.30 and A.31, respectively, ensure that no pair of first visits and no pair of second visits overlap. Constraint given by Equation A.32 ensures that no pair of first/second visits overlaps. (In fact, the number of z-variables in the model can now be halved by substituting out z-variables according to Equations A.30 through A.32. However, we retain these variables here for ease of exposition.)

Finally, we again have a minimax objective in which the time of the final patient visit must be minimized. Again, we define θ to be the time of the final patient visit, which must equal $s_i + q_i$, for some $i \in N$. The final model is then given as

$$\text{Min} \quad \theta \tag{A.33}$$

s.t. Constraints given by Equations A.24 through A.32

$$\theta \geq s_i + q_i \quad \forall i \in N, \tag{A.34}$$

$$s_i,\ f_i \geq 0 \quad \forall i \in N, \tag{A.35}$$

$$z_{ij}^{11},\ z_{ij}^{12},\ z_{ij}^{21},\ z_{ij}^{22} \text{ binary} \quad \forall i,j \in N, i \neq j. \tag{A.36}$$

This problem is in fact adapted from a study on radar pulse interleaving, which contains similar challenges to this nurse scheduling problem. See Refs. [4,6,16] for a more thorough examination of interleaving applications and integer programming techniques.

A.3 MIP Solution Techniques

Often, there are alternative ways of modeling optimization problems as MIPs. There sometimes exist trade-offs in these different modeling approaches. Some models may be smaller (in terms of the number of constraints and variables required), but may be more difficult to solve than larger models. In fact, the difference can be significant and can make a difference in whether or not MIPs can be solved quickly enough to be practically useful. (For instance, the situation described in Example 1 mentioned in the previous section must be solved before the outbreak spreads.)

Improving the efficiency of solving MIP Models requires an understanding of how MIP solvers work. Premium MIP solvers, such as CPLEX (ILOG, Inc.), Xpress-MP (Dash Optimization), SYMPHONY and CBC (COIN-OR project), and Solver (Frontline Systems, Inc.), employ a combination of *branch-and-bound* and *cutting-plane* techniques. Although a review of how to use these software packages is well beyond the scope of this tutorial, it is important to understand the basics of MIP solution algorithms to understand the key principles in MIP modeling.

For the rest of this section, assume that we are solving a minimization MIP. (Maximization MIPs are solved in much the same fashion.) To help illustrate the branch-and-bound process, we consider the following example MIP (actually a pure IP because all variables are integers).

$$\text{Min} \quad 4x_1 + 6x_2 \tag{A.37a}$$

$$\text{s.t.} \quad 2x_1 + 2x_2 \geq 5 \tag{A.37b}$$

$$x_1 - x_2 \leq 1 \tag{A.37c}$$

$$x_1, x_2 \geq 0 \text{ and integer.} \tag{A.37d}$$

The first concept that we discuss in solving MIPs is that of *relaxations*. A relaxation of an MIP is a problem such that any solution to the MIP corresponds to a feasible solution to the relaxed problem, and each solution to the MIP has an objective function value greater than or equal to that of the corresponding solution to the relaxed problem. The most commonly used relaxation for MIPs is its LP *relaxation*, which is identical to the MIP with the exception that variable integrality restrictions are eliminated. Clearly, any integer-feasible solution to the MIP is also a solution to its LP relaxation, with matching objective function values.

Envision a bag containing several orange and blue marbles. Each marble represents a solution to the LP relaxation, but only the orange marbles also represent solutions to the MIP. Each marble has a weight, corresponding to its objective function value. The task in linear programming is to find the lightest marble. The task in solving an MIP is to find the lightest orange marble.

When describing the branch-and-bound algorithm for MIPs, it is helpful to know how LPs are solved. See Refs. [1,8,15,17] for an explanation of linear programming theory and methodology. For the purposes of this chapter, we simply note that LPs can be solved quickly (in time bounded by a polynomial of the problem's input size). Graphically, Figure A.1 illustrates the feasible region (set of all feasible solutions) to the LP relaxation of

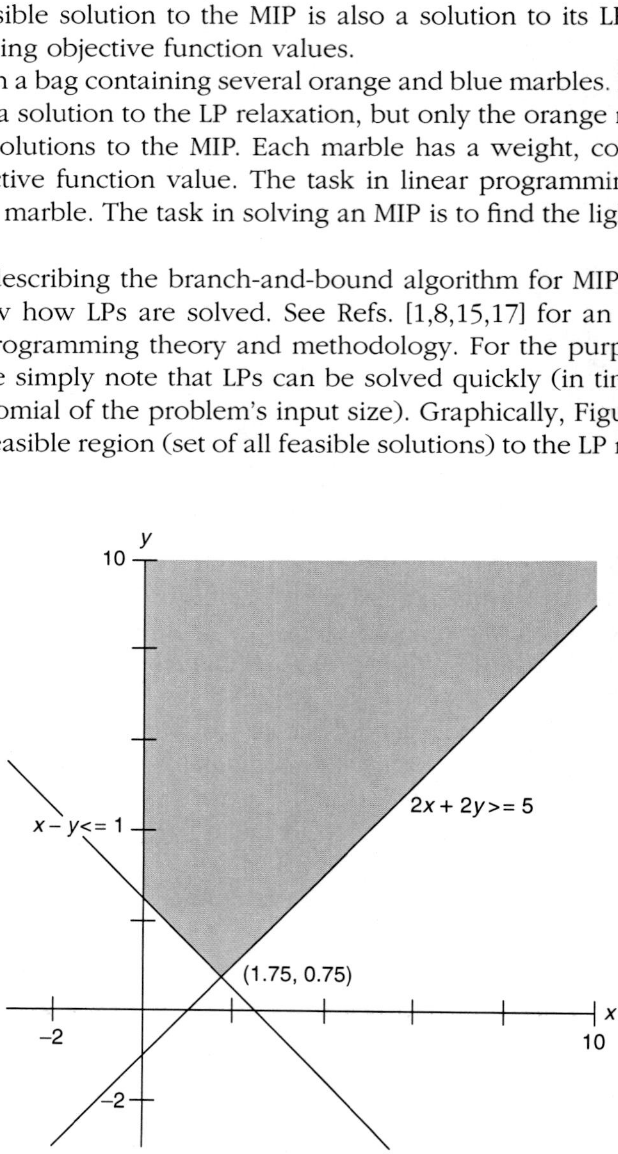

Figure A.1 Feasible region of the LP relaxation.

formulation Equation A.37. Note that the gradient of the objective function is (4,6) (taken from the coefficients of Equation A.37a). This means that the objective function is increasing in this direction, and hence we wish to follow the direction $(-4, -6)$ as far as possible in the feasible region. Put another way, think of $(-4, -6)$ as the direction of gravity, and place a pebble in the feasible region. The point to which the pebble falls, (1.75, 0.75), is the optimal solution to the LP relaxation and has an objective function value of 11.5.

Returning to the bag of marbles analogy, solving the LP relaxation has yielded a blue marble (fractional, not integer, solution) with a weight of 11.5. This implies that all orange marbles have a weight of 11.5 or more, because weight of the lightest marble in the bag was 11.5. (In general, we cannot claim that the optimal solution to the LP is unique, and so we allow for the possibility that MIP solutions exist with an identical objective function to the optimal LP solution.) The important result is that a *lower bound* on the optimal MIP solution is obtained from the LP relaxation. No solution to the MIP (Equation A.37) can be found with an objective function value of less than 11.5.

Of course, the solution (1.75, 0.75) is not a feasible solution to Equation A.37. In fact, all feasible solutions have the trait that *either* $x_1 \leq 1$ or $x_1 \geq 2$. In fact, we can split the problem (Equation A.37) into two subproblems: one in which $x_1 \leq 1$ (called region 1) and one in which $x_1 \geq 2$ (called region 2). All solutions to the original MIP are contained in exactly one of these two new subproblems. This process is *called* branching, and we could have also branched on x_2 instead, by requiring that either $x_2 \leq 0$ or $x_2 \geq 1$. Conceptually, branching is equivalent to taking our bag of marbles and splitting it into two bags. All blue marbles corresponding to solutions in which $1 < x_1 < 2$ are thrown away (importantly, this includes the lightest marble that we previously found). No orange marbles are thrown away; they either belong to the $x_1 \leq 1$ bag or the $x_1 \geq 2$ bag. Now, we will search for the lightest orange marble in each bag, compare them, and take the lightest of the two.

The feasible regions of the two new subproblems are depicted in Figure A.2. When $x_1 \leq 1$, the optimal solution is (1,1.5) with objective function value of 13. When $x_1 \geq 2$, the optimal solution is (2,1) with objective function value of 14. In the $x_1 \leq 1$ region, the lower bound is 13. In the $x_1 \leq 2$ region, though, the best solution happens to be an integer solution. That is, the lightest marble in this bag out of all marbles happens to be orange, and so it is obviously also the lightest orange marble. Therefore, the best integer solution in the $x_1 \geq 2$ region has an objective function value of 14; there is no need to further search that region. This region is said to be fathomed by integrality. We store the solution (2,1) and call it our *incumbent solution*. If no better solution is found, it will become our optimal solution.

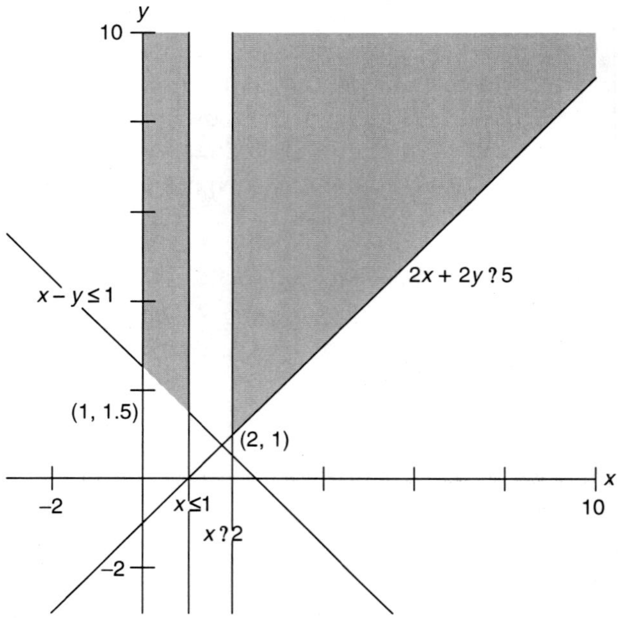

Figure A.2 Feasible regions of the subproblems.

At this point, there is one *active* region (or active node in the context of branch-and-bound trees, which we will describe shortly), which is region 1. An active region is one that has not been branched on, and that must still be explored, because there is a possibility that it contains a solution better than the incumbent solution. The initial region is not active, because we have branched on it. Region 2 is not active because we have found the best integer solution in that region. Region 1, however, is still active and must be explored. The lower bound over this region is 13. The optimal solution must have an objective function value somewhere between 13 and 14. We recursively divide region 1, in which $x_1 \leq 1$. Because the optimal solution in this region was (1,1.5), we branch by creating two new subproblems: one in which both $x_1 \leq 1$ and $x_2 \leq 1$ (called region 3), and one in which both $x_1 \leq 1$ and $x_2 \geq 2$ (called region 4). Once again, all integer solutions in region 1 are contained in either region 3 or region 4.

However, note that region 3 is empty, because the stipulation that both x_1 and x_2 are no more than 1 makes it impossible to satisfy Equation A.37b. There are therefore no integer solutions in this region either, and so we stop searching region 3. This region is said to be fathomed by infeasibility. The optimal solution to region 4's linear relaxation is (0.5,2), with objective function value of 14. We still have not found the best integer solution over region 4, but we know that the best integer solution (if one even exists)

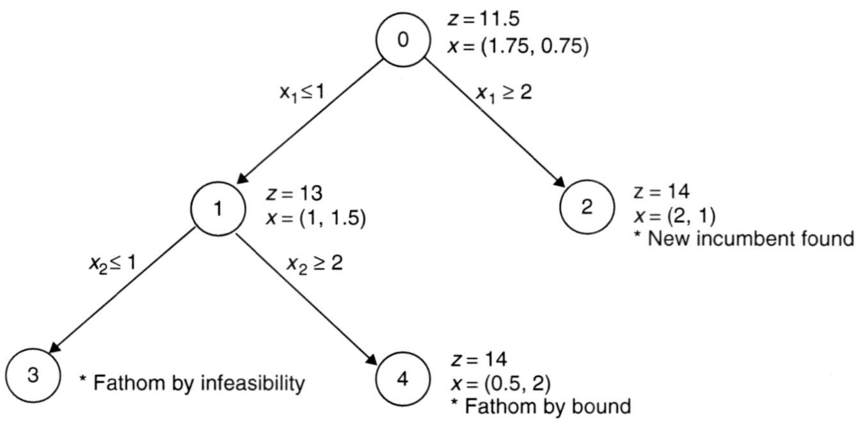

Figure A.3 Branch-and-Bound tree.

has an objective function value of 14 or more. However, our incumbent solution has an objective function value of 14. We have not found the best integer solution in region 4, but we do know that the best solution in region 4 will not improve the incumbent solution we have found. Thus, we are not interested in any integer-feasible solution in region 4, and we stop searching that region. (An alternative optimal integer solution can exist in that region, but we are not seeking to find all optimal solutions, just one.) Region 4 is said to be fathomed by bound.

Figure A.3 depicts a tree representation of this search process, which is called the branch-and-bound tree. Each node of the tree represents a feasible region. Now, there are no more regions to be examined (no more active nodes) and the algorithm terminates with the incumbent solution, (2,1), as an optimal solution.

A formal description of the branch-and-bound algorithm for minimization problems is given as follows:

Step 0. Set the incumbent objective $v = \infty$ (assuming that no initial feasible integer solution is available). Set the active node count $k = 1$ and denote the original problem as an active node. Go to Step 1.

Step 1. If $k = 0$, then stop: the incumbent solution is an optimal solution. (If there is no incumbent, i.e., $v = \infty$, then the original problem has no integer solution.) Else, if $k \geq 1$, go to Step 2.

Step 2. Choose any active node, and call it the current node. Solve the LP relaxation of the current node, and make it inactive. If there is no feasible solution, then go to Step 3. If the solution to the current node has objective value $z^* \geq v$, then go to Step 4. Else, if the solution is all integer (and $z^* < v$), then go to Step 5. Otherwise, go to Step 6.

Step 3. Fathom by infeasibility. Decrease k by 1 and return to Step 1.

Step 4. Fathom by bound. Decrease k by 1 and return to Step 1.

Step 5. Fathom by integrality. Replace the incumbent solution with the solution to the current node. Set $v = z^*$, decrease k by 1, and return to Step 1.

Step 6. Branch on the current node. Select any variable that is fractional in the LP solution to the current node. Denote this variable as x_s and denote its value in the optimal solution as f. Create two new active nodes: one by adding the constraint $x_s \leq \lfloor f \rfloor$ to the current node, and the other by adding $x_s \geq \lceil f \rceil$ to the current node. Add 1 to k (two new active nodes, minus one due to branching on the current node) and return to Step 1.

Note that in Step 0, we could run a *heuristic procedure* to quickly obtain a good-quality solution to the MIP with no guarantees on its optimality. This solution would then become our initial incumbent solution, and could possibly help conserve branch-and-bound memory requirements by increasing the rate at which active nodes are fathomed in Step 4. In Step 2, we may have several choices of active nodes on which to branch and in Step 6, we may have several choices on which variable to perform the branching operation. There has been much empirical research designed to establish good general rules to make these choices, and these rules are implemented in commercial solvers. However, for specific types of formulations, one can often improve the efficiency of the branch-and-bound algorithm by experimenting with node selection and variable branching rules.

The best-case scenario in solving a problem by branch-and-bound is that the original node yields an optimal LP solution that happens to be integer, and the algorithm terminates immediately. Indeed, in Equation A.37, if we simply add the constraint $x_1 + x_2 \geq 3$ and solve the LP relaxation, we would obtain the optimal solution (2, 1) immediately. This also underscores the importance of making the M-values introduced in the previous section as small as possible. The smaller these M-values are, the fewer fractional solutions exist in the linear programming relaxation. Paying close attention to these M-values often results in significant improvements in the computational efficiency of MIP algorithms.

Thus, a classical way to reduce the presence of fractional solutions is to find *valid inequalities*, which do not cut off any integer solutions, but do cut off some fractional solutions. In terms of marbles, these inequalities remove some blue marbles from the bag, but never orange marbles, and do so without branching into multiple bags. A *cutting plane* is a valid inequality that removes the optimal LP relaxation solution from the feasible region. In theory, MIPs can be solved without branching either by including enough valid inequalities before solving the LP relaxation, so that the LP relaxation provides an integer solution, or by looping between solving the LP relaxation, adding a cutting plane, and re-solving the LP relaxation, until the LP relaxation yields an integer solution. By themselves, these approaches are typically intractable and may suffer from numerical instability problems. However, the most effective implementations often use a combination of valid inequalities added a priori to the model, after which branch-and-bound is executed, with cutting planes periodically added to the nodes of the branch-and-bound tree. This approach is called branch-and-cut.

Valid inequality and cutting plane approaches can either be automatic or problem specific. Classical automatic approaches are summarized by Nemhauser and Wolsey [11], who provide a technical explanation to these approaches. More relevant to the material in this chapter are problem-specific valid inequality generation techniques. For instance, in Example 2 in the previous section, suppose that the demand of drugs in periods 1 and 2 is 100 units. Suppose that the minimum order quantity from each supplier is 150 units. If drugs become less expensive as time goes on, then the LP relaxation may try to place only two-thirds of a minimum order in each period, so that only 100 drugs (instead of the full complement of 150) are purchased in each period. Anticipating this class of fractional solutions, we note that at least one order must be placed in period 1 (because no orders result in unsatisfied demand). This valid inequality is stated as

$$\sum_{i \in M} z_{i1} \geq 1. \tag{A.38}$$

The addition of such inequalities to the MIP formulation often aid its performance, although occasionally they make little difference, or even worsen the performance of the branch-and-bound solver. Negative impacts usually occur when the inclusion of extra valid inequalities makes the formulation larger without sufficiently reducing its feasible region. Determining the valid inequalities that are computationally beneficial is usually a matter of trial-and-error.

A.4 Example Radiation Therapy Application

In this section, we describe an application of mixed-integer programming in Intensity Modulated Radiation Therapy (IMRT) planning. The underlying mechanism of radiotherapy is to radiate tumor tissues with high energy beams, which kill cancer cells. However, high energy radiation also kills healthy tissues through which it passes, possibly resulting in serious degradation in the patient's quality of life. IMRT is a technique designed to help solve this dilemma. IMRT delivers small amounts of doses from multiple beam angles, which intersect at tumor tissues. As a result, the tumor tissue receives enough radiation to kill the cancer cells, but the healthy tissues are spared. The IMRT planning problem is usually solved in three interdependent phases.

- Beam angle optimization (BAO): Selection of the beam angles to use
- Fluence map optimization (FMO): Determination of intensity profile to deliver from each beam angle
- Leaf sequencing: Realization of the intensity profiles under the capabilities of available machinery

Planning problems in IMRT have been investigated by several researchers [2,3,12]. The BAO problem can be formulated as an MIP Model [9,10]. The FMO problem can be formulated as a large-scale LP Model [14] or a Nonlinear Programming Model [13]. The leaf sequencing problem can be formulated as an MIP Model as we describe in this section. However, because the problem size of real-world IMRT instances is very large and these problems are inherently complex, heuristic procedures are typically used to solve real instances in a reasonable amount of time [5]. The reader is referred to Ref. [7] for a recent book chapter about mixed-integer programming applications in IMRT. In this section, we focus on the leaf sequencing problem and derive an MIP formulation for solving it optimally.

A.4.1 Leaf Sequencing Problem

We are given an $m \times n$ matrix B that consists of integers called beamlets. Matrix B represents an intensity profile that needs to be delivered from a given beam angle. We need to find an optimal way of decomposing this matrix into a set of uniform-intensity shapes that the available machinery can deliver.

Figure A.4 represents a small example fluence map. One way to decompose this map into deliverable shapes is given in Figure A.5. Define

9	4
6	1

Figure A.4 Fluence map example.

$$\begin{bmatrix} 9 & 4 \\ 6 & 1 \end{bmatrix} = 9 * \begin{bmatrix} 1 & 0 \\ 0 & 0 \end{bmatrix} + 4 * \begin{bmatrix} 0 & 1 \\ 0 & 0 \end{bmatrix} + 6 * \begin{bmatrix} 0 & 0 \\ 1 & 0 \end{bmatrix} + 1 * \begin{bmatrix} 0 & 0 \\ 0 & 1 \end{bmatrix}$$

Figure A.5 Leaf sequence 1.

beam-on time as the amount of time required to deliver the doses prescribed by a set of deliverable shapes. In this problem, we measure the beam-on time as the sum of doses. The solution represented in Figure A.5 results in four shapes and a total beam-on time of $9+4+6+1=20$ time units. Assuming that the machine requires 15 time units to switch from one shape to another, where time units are relative to one unit of beam-on time, the total time required by the configuration represented in Figure A.5 is $15 \times 4 + 20 = 80$ time units.

An alternative decomposition is given in Figure A.6, resulting in only three shapes and a total beam-on time of only 9 time units. The total time for this decomposition is only $15 \times 3 + 9 = 54$ time units. This simple example illustrates that the total time required to deliver an intensity profile for a given beam varies significantly depending on the leaf sequence employed. Because multiple beam angles are used in IMRT, the total time a patient has to spend receiving treatment, and thus the total exposure to dangerous unintentional radiation during treatment, can be reduced significantly by finding an optimal leaf sequence for each beam.

Next, we develop an MIP Model to solve a special version of the leaf sequencing problem in which all shapes used must be rectangular. Let K be the set of all rectangular shapes that can be used. Define real variables x_k that denote the amount of time the rectangle $k \in K$ is in use. Let the parameter $I_{ijk} = 1$ if the rectangle $k \in K$ covers the bixel located at coordinates (i, j) and 0 otherwise. Because the required dose to each bixel must be delivered exactly, we have the following constraint:

$$\sum_{k \in K} I_{ijk} x_k = b_{ij} \quad \forall i = 1, \ldots, m, \; j = 1, \ldots, n. \tag{A.39}$$

$$\begin{bmatrix} 9 & 4 \\ 6 & 1 \end{bmatrix} = 1 * \begin{bmatrix} 1 & 1 \\ 1 & 1 \end{bmatrix} + 5 * \begin{bmatrix} 1 & 0 \\ 1 & 0 \end{bmatrix} + 3 * \begin{bmatrix} 1 & 1 \\ 0 & 0 \end{bmatrix}$$

Figure A.6 Leaf sequence 2.

Because a component of the total time is proportional to the number of rectangles used, we need to define binary variables y_k that denote whether the rectangle $k \in K$ is in use. By definition, if $x_k > 0$ then $y_k = 1$. Using the modeling tricks introduced in Section A.2.2, we enforce the following constraint:

$$x_k \leq M_k y_k \quad \forall k \in K, \tag{A.40}$$

where M_k is a sufficiently large number. Because we want M_k to be as small as possible without imposing an artificial restriction on x_k, a good choice for M_k is the minimum value covered by the rectangle $k \in K$. In mathematical terms, this can be written as $M_k = \min_{ij}\{b_{ij} | I_{ijk} = 1\}$.

Now we can write the objective function as minimization of total treatment time. Assuming switching from one rectangle to another requires a multiple of A times a unit of beam-on time, the total treatment time is the sum of setup time for the rectangles and beam-on time. The overall model is then given by

$$\text{Min} \quad A \sum_{k \in K} y_k + \sum_{k \in K} x_k \tag{A.41a}$$

$$\text{s.t.} \quad \text{Constraints given by Equations A.39 and A.40}$$

$$x_k \geq 0 \text{ and } y_k \text{ binary} \quad \forall k \in K. \tag{A.41b}$$

The model given above can be improved by adding valid inequalities. An idea in deriving valid inequalities is to examine each bixel $(i^\star, j^\star)$, $1 \leq i^\star \leq m$, $1 \leq j^\star \leq n$, and determine whether or not a rectangle whose upper-left-hand corner is located at $(i^\star, j^\star)$ must exist in any optimal solution. We say that such a rectangle starts at that bixel. The main observation is that if the required intensity of a bixel is strictly greater than that of its neighbor to the left (i.e., if $b_{i^\star j^\star} > b_{i^\star,(j^\star-1)}$), then any rectangle that starts to the left of $(i^\star, j^\star)$ cannot deliver enough dose to $(i^\star, j^\star)$ itself. (We treat $b_{i0} = 0$ for all i.) The reason is that any rectangle that starts to the left of $(i^\star, j^\star)$ and covers it also covers the left-neighbor bixel. Because the left-neighbor bixel cannot be overdosed and because its required intensity is strictly less than that of $(i^\star, j^\star)$, the intensity of any such rectangle cannot be large enough to cover $(i^\star, j^\star)$ by itself. Therefore a rectangle that starts in the same column as $(i^\star, j^\star)$ and covers this bixel must exist in the solution to Equation A.41.

A slight extension of the same idea is based on comparing the intensity requirement of $(i^\star, j^\star)$ with the intensity requirements of the adjacent bixels on the left and above it. If the intensity requirement of $(i^\star, j^\star)$ is strictly

8	5	7	7
3	10	8	8
1	8	12	9
1	8	2	0

Figure A.7 Start index example.

greater than the sum of the intensity requirements of these two neighbors, then a rectangle must start at $(i^\star, j^\star)$. More formally

$$b_{i^\star j^\star} > b_{(i^\star-1)j^\star} + b_{i^\star(j^\star-1)} \Rightarrow \text{ a rectangle must start at } (i^\star, j^\star), \quad \text{(A.42)}$$

where $b_{0j} = 0,\ 1 \le j \le n$.

Figure A.7 illustrates this idea. Because $b_{2,2} > b_{1,2} + b_{2,1}$ we can conclude that a rectangle must start at (2,2). Note that a special case that is always satisfied by Equation A.42 occurs at the upper-left bixel (1,1), assuming $b_{11} > 0$.

We can preprocess the data before formulating the MIP Model and determine all coordinates satisfying Equation A.42. Let $\mathcal{S}$ be the set of all such coordinates, and define ST_{ij} as the set of rectangles that start at (i, j), for all $(i, j) \in \mathcal{S}$. Then we can add the following valid inequalities for each $(i, j) \in \mathcal{S}$ to model given by Equation A.41:

$$\sum_{k \in ST_{ij}} y_k \ge 1 \quad \forall (i, j) \in \mathcal{S}. \quad \text{(A.43)}$$

A.5 Conclusion

In this chapter, we discussed basic formulation and modeling principles of mixed-integer programming, fundamental MIP solution algorithms, and the use of MIPs for solving several types of problems. The ability to model different types of conditions such as integrality requirements, logical expressions (e.g., if–then and either–or relationships), and certain nonlinear expressions (e.g., minimum, maximum, absolute value, and some product terms), makes mixed-integer programming a very flexible technique for solving optimization problems. Several commercial and open source software systems for MIP are readily available. Most MIP solvers can be used as a black box, allowing the user to focus on modeling instead of solution algorithm development. However, experienced users can also interact with the solver using general-purpose programming languages such as C, C++, Java, C#, and Visual Basic. This flexibility allows users to guide the solution algorithm to exploit special structures of the problem on hand, resulting in more efficient solver performance.

Even though MIP Models are designed to find a provably optimal solution, it is possible to stop execution once a good enough solution is found. In other words, it is possible to use MIP-based algorithms as heuristics in computationally difficult problems. However, there is an important distinction between problem-specific heuristics and MIP-based algorithms: unlike the former, MIP-based algorithms are capable of measuring the quality of the solution found with respect to the (unknown) optimal solution. These features make mixed-integer programming a suitable technique for solving difficult optimization problems, including the problems in healthcare applications.

References

1. M. S. Bazaraa, J. J. Jarvis, and H. D. Sherali. *Linear Programming and Network Flows*, second edition. John Wiley & Sons, New York, 1990.
2. G. Bednarz, D. Michalski, C. Houser, M. S. Huq, Y. Xiao, P. R. Anne, and J. M. Galvin. The use of mixed-integer programming for inverse treatment planning with pre-defined field segments. *Physics in Medicine and Biology*, 47(13):2235–2245, 2002.
3. J. Deasy, E. K. Lee, T. Bortfeld, M. Langer, K. Zakarian, J. Alaly, Y. Zhang, H. Liu, R. Mohan, R. Ahuja, A. Pollack, J. Purdy, and R. Rardin. A collaboratory for radiation therapy treatment planning optimization research. *Annals of Operations Research*, 148(1):55–63, 2006.
4. M. Elshafei, H. D. Sherali, and J. C. Smith. Radar pulse interleaving for multi-target tracking. *Naval Research Logistics*, 51(4):72–94, 2004.
5. K. Engel. A new algorithm for optimal multileaf collimator field segmentation. *Discrete Applied Mathematics*, 152:35–51, 2005.
6. A. Farina and P. Neri. Multitarget interleaved tracking for phased-array radar. *IEEE Proceedings, Part F: Communications, Radar and Signal Processing*, 127(4):312–318, 1980.
7. M. Ferris, R. Meyer, and W. D'Souza. Radiation treatment planning: Mixed integer programming formulations and approaches. In G. Appa, L. Pitsoulis, and H. P. Williams, editors, *Handbook on Modelling for Discrete Optimization*, volume 88, pp. 317–340. Springer, New York, 2006.
8. F. S. Hillier and G. J. Lieberman. *Introduction to Operations Research*, eighth edition. McGraw-Hill, New York, 2005.
9. E. K. Lee, T. Fox, and I. Crocker. Integer programming applied to intensity-modulated radiation therapy treatment planning. *Annals of Operations Research*, 119:165–181, 2003.
10. G. Lim, J. Choi, and R. Mohan. Iterative solution methods for beam angle and fluence map optimization in intensity modulatedradiation therapy planning. Accepted for publication, *OR Spectrum*, 2007.
11. G. L. Nemhauser and L. A. Wolsey. *Integer and Combinatorial Optimization*. John Wiley & Sons, New York, 1999.

12. F. Preciado-Walters, R. Rardin, M. Langer, and V. Thai. A coupled column generation, mixed integer approach to optimal planning of intensity modulated radiation therapy for cancer. *Mathematical Programming*, 101(2):319–338, 2004.
13. H. E. Romeijn, R. K. Ahuja, J. F. Dempsey, and A. Kumar. A column generation approach to radiation therapy treatment planning using aperture modulation. *SIAM Journal on Optimization*, 15(3):838–862, 2005.
14. H. E. Romeijn, R. K. Ahuja, J. F. Dempsey, A. Kumar, and J. G. Li. A novel linear programming approach to fluence map optimization for intensity modulated radiation therapy treatment planning. *Physics in Medicine and Biology*, 48(21):3521–3542, 2003.
15. A. Schrijver. *Theory of Linear and Integer Programming*. John Wiley & Sons, New York, 1986.
16. H. D. Sherali and J. C. Smith. Interleaving two-phased jobs on a single machine with application to radar pulse interleaving. *Discrete Optimization*, 2(4):348–361, 2005.
17. W. L. Winston and M. Venkataramanan. *Introduction to Mathematical Programming: Applications and Algorithms*, volume 1, fourth edition. Duxbury Press, Belmont, CA, 2002.
18. L. A. Wolsey. *Integer Programming*. John Wiley & Sons, New York, 1998.

Index

A

Active path analysis and model scope, in ID, 73–78
Active trail, definition of, 71
Acute coronary syndrome (ACS), 87, 96, 99–101, 103, 107–108, 110
Acute ischemic cardiovascular pathology, 101
Acute myocardial infarction (AMI), 70, 87, 103
Add and drop algorithm, in BOO model, 245–246
Advisory Committee on Immunization Practices (ACIP), 118, 125
Aging, diagnosis of, 5
AIDS prevention and treatment, *see* HIV prevention and treatment
Alanine dipeptide, 441
  minimum energy configuration, 454
  Ramachandran angular variations, 453
  SMT ρ function, graph, 453
  SMT ρ value, 452
  Steiner ratio, 452
  Steiner values for conformations, 452
  structure, 452
  Swiss-PdbViewer energy calculation module, 453
  torsion angles, 452
Algorithm for DT-optimal leaf sequence
  with ICC, 267, 276
  with TGC and ICC, 278
  unconstrained case, 261
  without ICC, 268
Algorithms
  classes of, 369
  expectation-maximization (EM) algorithm, 373
  Fitch, 362
  genetic algorithms, 373
  graph-based algorithms, 369
  interior-point, 339
  iterative alignment, 369
  Needleman–Wunsch algorithm, 367
  neighbor joining (NJ), 360
  progressive alignment, 369
α- and β-sheets, secondary structures, 440
American Academy of Family Physicians (AAFP), 118, 125
Amino acid sequence, 357
Anderson–May model, 152
Angiogenic microvascular networks, fingerprinting, 5, 47–48, 53
Antimicrobial agent, 495, 499, 510
  effect on heterogeneous microbial populations, 499–502
Antimicrobial resistance, 494, 496
Antisense oligonucleotides, 400
Anxiety, 73
Aortic dissection (AD), 70–71, 73–77, 79, 87, 102
  influence diagram (ID)
    for network of, 74–75
    with observed clinical criteria, 75
    with possibility of esophageal perforation detection by CT, 77
    with symptoms outside of clinical decision rule, 76
Aortography, 73
Artificial intelligence, 5–6
Artificial neural networks (ANN), 31
Asian Epidemic model, 153, 166
Atherosclerois, discriminant analysis of biomarkers for prediction of, 5, 46–47, 53
Avahan Project, India, *see* India AIDS Initiative

AVERT model, 152
Avian influenza (AI), 416

B

Bacterial cell, gene regulation in, 327
Basic local alignment search tool, 401
Bayesian inference, and classification, 7–8
Bayesian networks, 63, 68
Bayes optimal rule, 31
Beam angle optimization (BAO), 542
Beam Orientation Optimization (BOO) model, 225–239
  beam orientations in, 225–226
  BOO + FMO integer program, 239–241
  feasible beam space, 226–229
    beam data generation, 228–229
    time and space considerations, 229
  greedy algorithms, 246–247
  local search algorithms, 245–246
  metaheuristics, 241–245
    evolutionary and genetic algorithms, 243
    generic evolutionary and genetic algorithms, 243–244
    particle swarm algorithm, 245
    response surface based approach, 245
    simulated annealing algorithm, 242–243
  objective function, 229–239
    beam's-eye-view (BEV) approach, 231–235
    entropy, 237
    fluence map optimization, 229–231
    Fourier transforms, 237–238
    geometric considerations, 235–236
    mean organ-at-risk data (MOD), 236–237
    path of least resistance, 238
    pseudo beam's-eye-view (pBEV), 232–233
    single- and multi-beam cost functions, 238–239
    target-eye-view (TEV), 233–235
  optimization methods, 239–247
Beam Orientation Optimization (BOO) problem, 224–225, 229, 236, 241, 243–245, 247; *see also* Beam Orientation Optimization (BOO) model
Beam's-eye-view (BEV) approach, in BOO model, 231–235
Betweenness-centrality (BC), 324
Biased classifier, argument for, 109–110
Biliary colic, 73
Bill and Melinda Gates Foundation, 149, 158, 172
Biological and medical applications
  MIP-based multigroup classification models, 40–50
    erythemato-squamous diseases diagnosis, 41–42
    heart disease diagnosis, 42
    identification of tumor shape and volume in treatment of sarcoma, 45–46
    native and angiogenic microvascular networks fingerprinting, 47–48
    pattern recognition in satellite images for determining types of soil, 49–50
    predicting aberrant CpG island methylation in human cancer, 42–44
    protein localization sites prediction, 48–49
    ultrasonic-assisted cell disruption for drug delivery, 45
Biological networks, discrete *k*-median problem to locate communities within, 347
Biologic function formulations, treatment and complication probabilities, 301–302
Biomarkers, discriminant analysis for prediction of atherosclerosis, 5, 46–47
BLAST, *see* Basic local alignment search tool
BMIP, *see* MIP model for binary attributes (BMIP)
BOO + FMO MIP model, 231, 239–247
  branch-and-bound techniques, 240–241
  commercial software systems, 241
Boundary effect attenuation, 293
Brachytherapy, 198–199, 289
Branch-and-bound techniques, 50, 240–241, 535–536, 539–540
Branch-and-cut algorithm, advantages in implementation of, 371
Breast carcinoma, 468
Burkitt's lymphoma, 468
Bystander effect, 469

C

Canadian Council for Donation and Transplantation, 189

Cancer patients
radiation therapy applications and methods, 200–218
gamma knife radiosurgery, 202–205
intensity modulated radiation therapy, 211–215
optimization planning, 197–218
optimization techniques use, 201
proton therapy, 217–218
radiation treatment planning procedure, 200–201
three-dimensional conformal radiation therapy, 205–211
tomotherapy, 216–217
treatment options for, 198
Cancer treatments, 253–254
Cardiac ischemia, 67, 73
Case-based reasoning, 70–71
Causal graphs and reasoning, 70
CD46, 469
CD150, 469
C. elegans, 320
Cell, biological organization influence, 319
Cell-to-cell fusion, 469
Centers for Disease Control and Prevention (CDC), 118
Chemotherapy, 198
Chest pain, 63, 67, 70–71, 96, 101
Chlamydia, 170
Classification; *see also* Mathematical programming-based classification models; MIP-based multigroup classification models
Bayes optimal rule, 31
Bayesian inference and, 7–8
concept and application, 5
discriminant analysis in, 5–6
discriminant functions, 8–10
matrix, 7
pattern recognition in, 5–6
rules, 6–7
statistical pattern classification, 5–6, 10
supervised learning, training and cross-validation, 6–7
terminologies related to, 5–10
Classification and regression trees (CART), 52
Classify any number of distinct groups, in predictive model, 4
Clinical cost-effectiveness analysis
for diagnosing pulmonary embolism, 82–89
influence diagrams in, 63–90
active path analysis and model scope, 73–78
computational burden and model scope, 69–73
computational complexity, 68–69
problem scope determination, 78–79
structure characteristics, 67–68
of venous thromboembolic disease, 79–82
Clinical decisions, non-Bayesian classification for, 96–110
k-clique percolation, 325
Clonogenic cell, 297
CMH model, 153
Cobalt-60 machines, 198
Cognitive science, 5
Combination vaccines, 120–122, 138–139
Combinatorial optimization, 441
Combinatorial optimization mathematical models, proteins, 445
cube with steiner plane, 447
E3 and higher dimensions, 445
Maxwell's theorem, 446, 449
Melzak circles, 446–449
$\mathcal{NP}$-Hard, 445
potential-energy objective function, 448
Serine (Ser), 450–451
SMTs and MECs, relationship, 449
SMT solution, 446
Steiner trees, 445
twist angles, 448
Commercial sex worker (CSW), 154–155, 157–168
Community from known *k*-clique, algorithm to calculate, 344
Complexity theory, 378
Compression ultrasound (CUS) testing, 81, 85–86
Computational models, for integer programming formulation, 384–386
Computed tomographic angiogram (CTA), 81, 85–87
Computed tomographic (CT) scan, 73, 200, 202
Computer-based treatment planning, 199
Conditional independence, definition of, 71
Conditional probability table (CPT), 68–69, 71, 74
Conditional Value at Risk (CVaR), 298
Conjugate gradient nonlinear approaches, 300–301
Constraints to limit rate of misclassification, in predictive model, 4

Cost-effectiveness analysis (CEA); *see also* Clinical cost-effectiveness analysis
epidemic model to support, 149–156
health and economic outcomes, 155
in India, 156–166
model scope, 151–153
multiple interventions evaluation, 156
parameters in, 154–155
population groups, 154
risk factors, 154
sensitivity analysis, 155–156
of HIV prevention and treatment, 149–156
Cost-effectiveness strategies
for pulmonary embolism diagnosis, 82–89
extended D-dimer model results, 86–89
model and results, 82–86
CpG island aberrant methylation, prediction in human cancer, 5, 32, 42–44
Cross-validation, 6–7
Cutset conditioning, 70
Cutting-plane techniques, 535–536, 541
CVaR approximation, 298

D

DAG, *see* Directed acyclic graph
DAMIP model, 50
Data binarization, 419–420
D-dimer assay
conditional probabilities for, 82–83
and venous thromboembolic disease, 79–82
D-dimer model results, for pulmonary embolism diagnosis, 86–89
Decomposition cardinality (DC), 256, 257, 279–284
computational complexity of, 256, 257, 279, 281
heuristics for, 280–281
min–max characterization of, 283
NP-hardness of, 279, 280
unconstrained case, 281
Decomposition cardinality (DC) with ICC
heuristics for, greedy step in, 284
Decomposition time (DT), 256, 257, 259–279
Deoxyribonucleic acid (DNA)
microarrays, 405
molecule, 357
repository of biological information, 314
sequencing
Eulerian method for fragment assembly in, 372
maps, 375
structure of, 312, 396
Deterministic iterative algorithm, 373–374
Diabetes, diagnosis of, 5
Digraph
path in, 283–284
vertices of, 269, 272
Diphtheria, 119–121, 138
Directed acyclic graph, 31
Direct inverse methods, for dose distribution, 301
Disability-adjusted life years (DALYs), 150, 166
Discrete support vector machine predictive models, 32–40
mixed integer programming formulations, 34–36
modeling of reserved-judgment region for general groups, 32–33
validation of model and computational efffort, 40
Discriminant analysis
in classification, 5–6
of functional perfusion data, 47–48
Diseases, diagnosing and curing using genomes, 356
D. melanogaster, 320
Dose–volume constraints, modeling of, 298–299
Dose–volume histogram (DVH) constraints, 231
Drug delivery, ultrasonic-assisted cell disruption for, 45
DTaP-HBV-IPV combination vaccine, 121
DTaP vaccine, 120
DT-optimal decomposition, 261, 262, 276, 278, 279, 281
DT-optimal leaf sequence, unconstrained case, 261, 264
DT-problem with ICC, linear program, 263, 264
Duality approach, ICC, 273
DVT (deep venous thrombosis), 79–81, 83–85, 88
Dynamic programming algorithm, 137–142
for VFSREP and VFSLBP model, 132–134

E

E. coli metabolism, connectivity distributions *P(k)* of, 329
Edmonston vaccine strain of measles virus (MV-Edm), 468
Electrocardiogram (ECG), 64, 84
ELISA D-dimer test, in venous thromboembolic disease, 81–83
Engineered viruses, as potential cancer therapeutic agents, 468
Entropy, in BOO model, 237
Equivalent uniform dose (EUD), 231
Erdös–Rényi random graph model, 320, 324
Erythemato-squamous diseases diagnosis, 5, 41–42
Eulerian path approach
  for fragment assembly in DNA sequencing, 372
Evolutionary and genetic algorithms, in BOO model, 243
Expectation-maximization (EM) algorithm, 373
External beam therapy, 289
Extraimmunization, 121–122, 125, 127–128, 131, 135–136

F

Feasible beam space, in BOO model, 226–229
  beam data generation, 228–229
  time and space considerations, 229
FindProbe algorithm, 402
Fingerprinting, angiogenic microvascular networks, 5, 47–48, 53
Fisher's linear discriminant function, 9, 14–16, 19–20, 24–25, 40
Fitch's algorithm, 362
Fluence distribution, 254, 255
Fluence map optimization (FMO), 542
  in BOO model, 229–231
  problem, 224, 229–231, 241, 247
Flux-balance analysis (FBA), 330
Formalized decision making, and quality metrics, 62–63
Fourier transforms, in BOO model, 237–238
Fusion (F) protein, 469

G

Gallstone-related pain, 73
Gamma knife radiosurgery
  for cancer patients, 198, 201–205
  MINLP models, 203
  MIP models, 203
  optimization model formulation, 203–204
  solution techniques, 205
Gastro-esophageal reflux, 73
Gene expression arrays, 403
Generalized support vector machine (GSVM), 31
General single function classification (GSFC) model, 40
Gene-regulatory networks, 316
  functional building blocks of, 318
  properties of, 317
Generic evolutionary and genetic algorithms, in BOO model, 243–244
Genetic algorithms, stochastic methods for solving complex optimization problems, 373
Genome research, phylogenetic analysis for, 357
Genome, sequence analysis, 356, 374
Genotyping viral pathogens, experiments on, 430
Gibbs free energy, 402
Gibbs Sampler, 373
Glioma, 468
Global Fund to Combat Malaria, Tuberculosis and HIV, 149
GMFC (general multiple function classification—minimizing the number of misclassifications), 24–25
Gonorrhea, 170
Gradient descent algorithms, 373
Graph-based algorithms, 369
  Eulerian path approach, 372
  minimum-spanning tree and traveling salesman problem, 371–372
  for solving maximum-weight trace problem, 371
Greedy algorithms, in BOO model, 246–247
Greedy rounding heuristic, 137–142
  for VFSLBP, 136
  for VFSREP model, 136–137

GSFC (general single function classification—minimizing the number of misclassifications), 23–25

H

Haemophilus influenzae type b, 119, 121, 138
Heartburn, 73
Heart disease
  classes of, 6
  diagnosis, 5, 42–43
Hemagglutinin (H) protein, 469
Hepatitis A, 119, 138
Hepatitis B, 119, 121, 138
Heterogeneous types of attributes as input, in predictive model, 4
Heteroscedastic model, 9
Heuristic procedure, in MIP solution techniques, 540
Heuristic tree search strategies, 363
Hidden Markov Model (HMM), 373
High-dimensional data transformation, in predictive model, 4
HIV epidemic
  in India, 156–158
  models, 152–153
HIV prevention and treatment, 148–168
  cost-effectiveness analysis, 149–150
    epidemic model to support, 151–156
  funding for, 149
  in India, 156–166
  model-based framework, 149–151
  models for CEA
    Avahan project in India, 156–166
    development of need based model, 166–168
    disease progression and treatment, 168
    epidemic model, 151–156
    homogeneous sexual behavior in, 167
    illustrative model for, 159–167, 174–175
    injection drug use, 168
    nonconstant population group size, 168
    random mixing process in, 167
    realism versus tractability, 168
    spread of other diseases, 169
    transmission risk factors in, 167
  resources, 148–149
HLA-A antigens, 183
HLA-B antigens, 183
HLA-DR antigens, 183
Homoscedastic model, 9
HPV sequences, 427
Human cancer, CpG island aberrant methylation, 5, 32, 42–44
Human chorionic gonadotrophin (hCG, MV-hCG), 469
Human immunodeficiency virus (HIV) prevention and treatment, *see* HIV prevention and treatment
Human lung carcinoma, cell motility and morphology data discriminant analysis in, 5, 44–45
Human papillomavirus (HPV), 418
Hybrid model, 13–15, 51

I

ICC, *see* Interleaf collision constraint
Immunotherapy, 198
Incremental cost-effectiveness ratio (ICER), 64–65
Incremental ray-tracing algorithms, 229
Incumbent solution, 537
India AIDS initiative
  cost-effectiveness analysis, 158–166, 168–170
    condom promotion program, 164, 166
    data for, 161–162
    illustrative model for, 159–167, 174–175
    interventions effects, 161, 163–166
    program to reduce CSW sex act, 161, 163–166
    STD prevention and treatment program, 163–166
  evaluation of, 156–166
Infectious disease, diagnosis of, 419
Influence diagrams (IDs)
  for aortic dissection (AD) network, 74–76
  Bayesian networks and, 68, 96
  in clinical cost-effectiveness analysis, 63–65
  computational complexity in, 68–69
  determining problem scope in medical decision problems, 78–79
  directed acyclic graph of, 84
  implications for clinical decision problems, 65–79
  legend and icons for, 66

model scope and
active path analysis, 73–78
computational burden, 69–73
structural characteristics of, 67–68
Influenza, 119, 138
Influenza virus, monospecific classification of, 435
Injection drug user (IDU), 154, 157
Integer linear program (ILP), 408
Integer nonlinear program (INLP) models, 215
Integer programming formulations, 357
Integer programming (IP) models, 215
for VFSLBP, 126–127
for VFSREP model, 127–128
Intensity map delivery, 302–303
Intensity map generation
mathematical optimization in, 294
for radiation therapy, 292
Intensity modulated proton therapy (IMPT), 218
Intensity modulated radiation therapy (IMRT), 254
BOO model for, 224–248
for cancer patients, 198, 201, 211–216
optimization model formulation, 212–215
beam angle and fluence map optimization, 213–214
beam segmentation optimization, 214–215
dose deposition models, 212–213
solution methods, 214–215
planning, 541–545
beam angle optimization (BAO), 542
fluence map optimization (FMO), 542
leaf sequencing problem, 542–545
Intensity-modulated radiotherapy, 290
method of, 292
Interior-point algorithms, 339
Interleaf collision constraint (ICC), 261–64, 267–269, 273, 276–280, 284
Interstitial brachytherapy, 199
Intracavitary brachytherapy, 199
Intraluminal radiation therapy, 199
Inverse treatment planning, 199
Ischemic heart disease
appication of SVM, 102–111
asymmetric cost regularization, 103
data labeling, 103
decision function accuracy assessment, 106
descriptive statistics of data set, 102
feature selection impact, 107
features identified after preprocessing of data, 104
kernel selection, 103–105
optimal penalties, 107
parameter values, 105–106
performance, 107
preprocessing and incomplete data, 103
principal components analysis (PCA), 106–107
scaling of data, 103
Isodose lines, visual inspection of, 293
Iterative alignment algorithms, 369
for improving quality of alignment, 372
probabilistic optimization approaches, 372
simulated annealing and genetic algorithm, 373
iwgAIDS model, 152

J

Jukes and Cantor model, 364
Jukes–Cantor distance, 359

K

Kidney paired donation
bipartite graph, 187
blood group incompatibilities, 183, 192
current status of, 189
designing objective, 154–156
domino paired donation, 182
finding edges for, 182–184
graph model, 179, 181
heuristic and greedy algorithm, 188
impact of, 190–192
list paired donation, 182
optimization over graphs for, 177–192
probability assumptions and data sources, 190
solution methods, 186–189
voluntary compatible pair participation, 182

L

Landsat Multi-Spectral Scanner (MSS) image, 49
Leaf sequencing problem, 542–545
Likelihood of tree, method for calculating, 364–365
LINDO optimization software, 14
Linear accelerators, 198

Linear discriminant function (LDF), 9, 14–16, 19–20, 24–25, 40
Linear programming approach, 262
Linear programming classification models, 11–20, 51
  multigroup classification, 17–20, 51
  two-group classification, 11–17, 51
Linear programming formulations, 300
Linear program with equilibrium constraints (LPEC), 26
Lloyd algorithm, 346
Local search algorithms, in BOO model, 245–246
Logical analysis of data (LAD)
  classification rule for, 426–427
  general framework of, 418
  implementation of, 419
LP models, *see* Linear programming classification models
LP relaxation, 300

M

Machine learning, 5–6
Macular degeneracy, diagnosis of, 5
Magnetic resonance imaging (MRI), 200, 202
Mathematical model
  linear programming approach, 262
  minimal DT for a matrix, 259–260
  optimal treatment, 255–258
Mathematical programming-based classification models; *see also* MIP-based multigroup classification models
  linear programming classification models, 11–20, 51
    multigroup classification, 17–20, 51
    two-group classification, 11–17, 51
  mixed integer programming classification models, 20–25, 51
    multigroup classification, 23–25, 51
    two-group classification, 20–23, 51
  nonlinear programming classification models, 25–28, 51
  for statistical pattern classification, 10–31
  support vector machine, 29–31, 51
Maximum Distinguishing Probe Set (MDPS), 406
Maximum likelihood methods, 364
MAX rounding heuristic
  for VFSLBP, 134–136
  for VFSREP model, 134–136
McGeoch's method for signal sequence recognition, 49
Mean organ-at-risk data (MOD), in BOO model, 236–237
Measles, 119
Medical decisions; *see also* Clinical cost-effectiveness analysis
  cost-effectiveness analysis for, 62–89
Medical treatment planning, 417
Metabolic networks, 326–327, 340
Metabolic network structure, 327–329
Metaheuristics, for treatment planning algorithms, 301
Metaheuristics in BOO model, 241–245
  evolutionary and genetic algorithms, 243
  generic evolutionary and genetic algorithms, 243–244
  particle swarm algorithm, 245
  response surface based approach, 245
  simulated annealing algorithm, 242–243
Microbial genomes, 330
Microvascular networks, fingerprinting of, 53
Minimal DT for a matrix, 259–261
Minimizing maximum deviation (MMD), 11–12, 14, 25, 51
Minimizing sum of deviations (MSD), 11–16, 22–23, 25, 51
Minimizing sum of interior distances (MSID), 11–12, 14, 51
Minimizing the number of misclassifications (MM) model, 19–21
Minimum Cost Probe Set (MCPS), 406
Minimum energy configurations (MEC), 449
Minimum inhibitory concentration (MIC), 496, 497
Minimum spanning trees, uses in determining order of sequences, 371–372
Minimum weight common mutated sequence (MWCMS) problem and special cases of, 374, 384
MIP-based multigroup classification models, 31–32; *see also* Mathematical programming-based classification models
  advances in, 50
  applications to medicine and biology, 31–50
  discrete support vector machine predictive models, 32–40

mixed integer programming formulations, 34–36
modeling of reserved-judgment region for general groups, 32–33
model variations, 36–40
validation of model and computational efffort, 40
model variations, 36–40
progress and challenges, 50–51
results on real-world biological and medical applications, 40–50
erythemato-squamous diseases diagnosis, 41–42
heart disease diagnosis, 42
identification of tumor shape and volume in treatment of sarcoma, 45–46
native and angiogenic microvascular networks fingerprinting, 47–48
pattern recognition in satellite images for determining types of soil, 49–50
predicting aberrant CpG island methylation in human cancer, 42–44
protein localization sites prediction, 48–49
ultrasonic-assisted cell disruption for drug delivery, 45
MIP3G Model, 25
MIP model for binary attributes (BMIP), 17, 21
MIP models, *see* Mixed-integer programming (MIP)
Mixed integer and linear programming (MILP), 300, 417
Mixed Integer Nonlinear Programming (MINLP), 203, 205
Mixed-integer programming (MIP)
classification models, 15–16, 20–25, 34–36, 51, 203, 214
multigroup classification, 23–25, 51
two-group classification, 20–23, 51
models and solution technique
general form of, 524–525
tutorial guide, 522–546
problem modeling process, 525–535
radiation therapy application, 541–545
solution techniques, 535–541
branch-and-bound techniques, 535–536, 539–540
cutting-plane techniques, 535–536, 541
heuristic procedure, 540
incumbent solution, 537
relaxations, 536
valid inequalities, 540
MM–LPEC (minimizing the number of misclassifications – linear program with equilibrium constraints), 26
Model of evolution, 364
Molecular evolutionary genetic analysis (MEGA), 363
Molecular phylogenetics, 357, 364; *see also* Amino acid sequence
Monte Carlo simulation, 121, 142–143
Morphological tree of structure and secondary structure prediction, 443
Motility and morphology, in human lung carcinoma, 5
mRNA
monomers of, 315
proteins for producing, 317
protein synthesis specification by, 314
techniques for measuring levels of, 317
translation process for making protein, 315
Multigroup classification
in LP classification models, 17–20, 51
in MIP classification models, 23–25, 51
Multileaf collimator (MLC), 302
feature of, 257
interleaf collision constraint (ICC), 261
leaf pairs of, 254, 255
principle of, 255
tongue-and-groove design, 257, 258
Multiple myeloma (MM), 468
Multiple sequence alignment (MSA), 366–367
Multisurface method tree algorithm (MSMT), 15, 18
Mumps, 119
Muscle strains, 73
MV vectors, 469
MWCMS model, 374
Myocardial infarction (MI), 70–71
Myocardial ischemia, 101, 110

N

National Cancer Institute, United States, 198
National Center of Biotechnology Information (NCBI), 415
National Immunization Program (NIP), 118
National Immunization Survey, 118

National Organ Transplantation Act (NOTA), 189
Native microvascular networks
fingerprinting, 5, 47–48
Nearest-neighbor models, 16, 52
Needleman–Wunsch algorithm, 367
Neighbor joining (NJ) algorithms, 360–361
Network flow approach, 268–273
Network structure
graph topology, 316
node connectivities in, 322
Neural network models, 444
Neuraminidase (NA), 435
Neutron beam machines, 198
Newcastle disease virus, 468
Non-Bayesian classification; *see also* Support vector machines (SVMs)
for clinical decisions, 96–111
Non-Hodgkin lymphoma, 468
Nonlinear programming (NLP) classification models, 25–28, 51, 205
Nonunique probe selection problem, 410–411
Normal tissue complication probability (NTCP), 231, 293
Novel graph–theoretical-based genomic models, 374–375
NP-hard misclassification minimization problem, 28
NP-hardness of MSA, 370
Nucleotide sequences, 361

O

Oligo fingerprinting, 417
method for identification, of cDNA or genomic DNA sequences, 405
Oligonucleotide microarrays, 400
OpenMP-parallelization method, for symmetric multiprocessing machines, 366
Optimization, algorithm for, 259
Optimization techniques, for cancer patients, 201
Optimize dosing regimens, 494; *see also* Antimicrobial resistance
daptomycin, 495
Organs-at-risk (OARs), 198, 200–201, 210, 224, 233
Orthovoltage x-ray machines, 198
Ovarian carcinoma, 468

P

Panel-reactive antibody (PRA) test, 183–184
Parametric misclassification minimization (PMM) procedure, 27–28
Parsimony-based algorithm, 386
Parsimony methods
components in, 361
for phylogenetic analysis, 358
Particle swarm algorithm, in BOO model, 245
Path of least resistance, in BOO model, 238
Pattern recognition
analysis of functional perfusion data, 47–48
in classification, 5–6
Pearson correlation, 323
Pediarix®, 138
Pediatric vaccine formularies
models for optimization, 122–128
motivation, 118–122
VFSLBP model, 123–128
computational comparison of algorithms and heuristics, 137–142
dynamic programming algorithm for, 132–134
greedy rounding heuristic for, 136–137
MAX rounding heuristic for, 134–136
VFSREP model, 123–128
computational comparison of algorithms and heuristics, 137–142
dynamic programming algorithm for, 132–134
greedy rounding heuristic for, 136–137
MAX rounding heuristic for, 134–136
Penalty function objectives, 297
Peptide, backbone dihedral angles with planes, 442
Peptides CEA (MV-CEA), 469
Peptide, Steiner trees study
dipeptide
Ala-Gly silk β-structure, 459–460
twist angles, 460
3-D dipeptide model and Ramachandran plot, 457
N dimensional conformational space, 459
virtual dihedral angles, 455
virtual plane, 455–459

angles for helices and strands, 456
pentapeptide, 461–462
SMT of sampled structure, 462
torsion angle, 461
Pertussis, 119–121, 138
Pharmacokinetically realistic antimicrobial concentration; *see also* Antimicrobial agent
experimental verification, 512–514
heterogeneous microbial population, 511–512
homogeneous microbial population, 502–510
Photon beam radiation therapy, 217
Phylogenetic analysis, 356
Phylogenetic analysis, for genome research, 357
Phylogeny Inference Package (PHYLIP), 363
Pleurisy, 73
Pneumococcus, 119, 138
Pneumonia, 73
Poisson-type behavior, of PIN, 320
Polio, 119, 121, 138
Polymerase chain reaction, 319
Polynomial-time algorithm, 379
Posterior distribution, 7–8
Progressive alignment algorithms, 369
shortcomings of, 370
Protein Data Bank (PDB), 455
Protein engineering, 469
Protein folding problems, 441
Ramachandran plots, 444
secondary structure
prediction, empirical methods, 443
proteins, tools for analysis, 440
secondary structure prediction, 442
Sequence of angle pairs (ϕ, ψ), 442
Sequence of dihedral angles, 442
Steiner problem, 441
Steiner tree analysis, 441
tertiary structure, in protein folding problem, 440
theoretical methods, 443
Protein-interaction networks (PIN), 319–321
Protein localization sites, prediction of, 5, 48–49
Proteins
coding sequences, 401
functional aspects of, 315
regulation of, 319
sequence conformations, 443
Proton beam machines, 198
Proton therapy, for cancer patients, 198, 201, 217–218
Pseudo beam's-eye-view (pBEV), objective function, 232–233
Pseudomonas aeruginosa, 496
Pulmonary embolisms (PEs) diagnosis, 73, 79–81, 83–84, 86–89, 96, 100, 102
classification system, 81
cost-effectiveness strategies, 82–89
extended D-dimer model results, 86–89
model and results, 82–86
Pulmonary mass, 73

Q

Qualitative and quantitative measurements, 6–7
Quality-adjusted life years (QALYs), 62–65, 150, 159, 161, 166
Quality metrics, and formalized decision making, 62–63

R

Radiation, 253, 254
Radiation therapy
applications and methods, 200–218
beam orientations, 299
characteristic and dose optimization in, 288, 290
degrees of optimization, 291
dose
coefficients, 294–295
distribution and treatment of, 289
distribution at each point in body, 291
homogeneity constraints, 295–296
limits over sample of points, 294
specifications, 291
volume distribution of, 295
gamma knife radiosurgery, 202–205
intensity map generation for, 292
intensity modulated radiation therapy, 211–215, 224–248
mixed-integer programming (MIP) application to, 541–545
optimization problem in, 294
optimization techniques use, 201
planning optimization for cancer patients, 197–218
proton therapy, 217–218

radiation treatment planning procedure, 200–201
three-dimensional conformal radiation therapy, 205–211
tomotherapy, 216–217
treatment of cancer therapy, 288
treatment optimization modeling for, 292
Radiation Therapy Oncology Group (RTOG), 203
Radiation treatment planning procedure, for cancer patients, 200–201
Radiotherapy treatments
energy source in, 290
optimal design of, 342
planning decision for, 291
RAGNU software package, 14
Random-network model, 318
Relaxations, in MIP solution techniques, 536
Replication-selective adenoviruses, 468
Reserved-judgment region for general groups modeling, 32–33
Restriction Landmark Genome Scanning (RLGS), 43–44
Reverse genetics, 417
Robust linear programming (RLP) model, 15, 28, 51
Rubella, 119

S

Sarcoma, identification of tumor shape and volume in treatment of, 5, 45–46
SARS virus, 432
S. cerevisiae, 320
Sequencing by hybridization (SBH), 399
Serine
chemical structure, 450
Steiner topologies and optimal structure, 451
Set covering (SC)–based classification methodology, 417
Severe acute respiratory syndrome (SARS), 416
monospecific classification of, 434
Sexually transmitted diseases (STDs), treatment program for, 150–151, 154–155, 158–169
Shannon entropy maximization, 408
Shape beam superposition, 255, 256
Shape matrix
decomposition, 254, 259, 264, 266, 267, 270, 271, 276, 277, 282, 283
minimal DT of, 274
digraph for, 273–274
flow on, 268–270
linear combination of, 256
necessary information to determine, 275–276
parameter, 271
set of, 256, 257, 279, 280, 281, 283
Shortest common supersequences (SCSQ), 384
Shortest common superstring (SCST), 384
Siddon's ray-tracing algorithms, 228–229
SimulAIDS model, 153
Simulated annealing algorithms, 301
in BOO model, 242–243
Simulated annealing (SA), 210
stochastic methods for solving complex optimization problems, 373
for treatment planning algorithms, 301
Single-and multi-beam cost functions, objective function, 238–239
Single-cell metabolism, interactions in, 330
Smith's quadratic discriminant function (QDF), 14–16, 20, 24–25
Software packages, for designing and selection of hybridization probes, 411–412
Soil types, satellite images in classification of, 5, 49–50
Statistical pattern classification, mathematical programming methods for, 5–6, 10–31
STDSIM model, 153
Steiner minimal trees (SMTs), 441, 442
Steiner ratio, 462
Steiner tree, 449
dihedral angles of planes study, 452
important in *E*3 and higher dimensions, 445
and minimum energy configurations, 449
network in *Ed*, 449
planes generated by, 455
protein folding problem, 441
structure of Ser, 450
torsional energy relationship, 452
Step-and-shoot mode, 255
Successive linearization algorithm (SLA), 28
Successive multistage classification capability, in predictive model, 4
Supervised learning, 6–7
Support vector machines (SVMs), 29–31, 34, 39–41, 51, 53, 96–99

application to ischemic heart disease, 102–111
asymmetric cost regularization, 103
data labeling, 103
decision function accuracy assessment, 106
descriptive statistics of data set, 102
feature selection impact, 107
features identified after preprocessing of data, 104
kernel selection, 103–105
optimal penalties, 107
parameter values, 105–106
performance, 107
preprocessing and incomplete data, 103
principal components analysis (PCA), 106–107
scaling of data, 103
asymmetric cost for high risk clinical decisions, 99–100
clinical background, 101–103
Synthetic hybrid proteins, 319
Syphilis, 170
Systems biology, use of networks in, 315

T

Target dose homogeneity, 231
Target-eye-view (TEV), objective function, 233–235
Teletherapy, 198
Tetanus, 119–121, 138
TGC, *see* Tongue-and-Groove Constraint
Three-dimensional conformal radiation therapy (3DCRT)
beam angle selection and wedge orientation optimization, 209–210
beam weight optimization, 208–209
for cancer patients, 198, 201, 205–211
equivalent uniform dose (EUD), 209
optimization model formulation, 208–211
solution techniques, 210–211
wedge filters, 206–207
Tomotherapy
for cancer patients, 198, 201, 216–217
optimization methods, 216–217
Tongue-and-groove constraint, 276, 277
design, 257, 258, 276
effect, 273
underdosage, 258, 259
Translation process, for protein, 315
Traveling salesman problem (TSP), 371–372
Treatment planning optimization methods, 303
Tree topologies, search of possible, 363
Trichomoniasis, 169
Tuberculosis, 154, 157, 159, 168–169
Tumor control probability (TCP), 231
biologic functions of, 293
Tumor metastasis, diagnosis of, 5
Tumor shape and volume, in treatment of sarcoma, 5, 45–46
Tumor virotherapy, 469, 470
mathematical model of, 470–474
Bertalanffy–Richards model, 470–471
equations for, 472–473
free virus particle count, 473
Gompertz models, 471
rate of free virus elimination, 472
tumor cell population, 474
model validation and parameter estimation, 478–480
best fit curve, 479–480
Gompertz and logistic model, 478
Gompertz function, 478
weighted nonlinear least squares method, 478
simulations, to investigate the predictions, 480–488
conditions lead to successful therapy, 481
damped oscillatory behavior, 487
initial conditions, virus therapy unsuccessful, 484
initial tumor size, 484
initial virus dose, 480
lower initial dose of virus, 486
rate of virus production, 483
time profile for total tumor burden, 480
total tumor size, 480
tumor burden, 486
tumor load, 481
use of inhibitors of DNA synthesis, 483
values of model rate constants, 479
virus in several doses, 481
stable states of system, 474–478
eigenvalues of Jacobian, 475, 476
equilibrium points of system, 474–475
partial success of therapy, 475
residual virus population, 476
Routh–Hurwitz criterion, 476

Two-group classification
in LP classification models, 11–17, 51
in MIP classification models, 20–23, 51

U

Ultrasonic-assisted cell disruption for drug delivery, 45
UNAIDS model, 152
United Kingdom Department of International Development, 149
United Network for Organ Sharing (UNOS), 177–178, 184, 189–190
United States Agency for International Development, 149
United States Recommended Childhood Immunization Schedule, 117–121, 124, 132, 137–140
Unstable angina, 102
Untranslated regions (UTR), 412
Unweighted pair group method using arithmetic averages (UPGMA), 359

V

Vaccine formulary selection with limited budget problem model, *see* VFSLBP model
Vaccine formulary selection with restricted extraimmunization problem model, *see* VFSREP model
Validation of model and computational efffort, in discrete support vector machine predictive models, 40
Valid inequalities, in MIP solution techniques, 540
Varicella, 119, 138
Vector quantization, 346
Venous thromboembolic (VTE) disease
clinical background and epidemiology, 79–80
and D-dimer assay, 79–82
diagnosis and treatment of, 80–81
ELISA D-dimer test, 81–83
Ventilation and perfusion (V/Q) scan, 81
VFSLBP model, 123–128
computational comparison of algorithms and heuristics, 137–142
dynamic programming algorithm for, 129, 132–134
greedy rounding heuristic for, 136
integer programming model for, 126–127
MAX rounding heuristic for, 134–136
VFSREP model, 123–128
computational comparison of algorithms and heuristics, 137–142
dynamic programming algorithm for, 129, 132–134
greedy rounding heuristic for, 136–137
integer programming model for, 127–128
MAX rounding heuristic for, 134–136
VIDAS ELISA D-dimer assay and VQ scans, 81–83
Viral genomic sequences, 427
Virotherapy, dynamic interactions of cells, 468
Virus
AI H5N1, 416
avian influenza, 416
coronavirus, 432
human papillomavirus, 418
severe acute respiratory syndrome, 416
von Heijne's method for signal sequence recognition, 49
von Heijne's Signal Peptidase II consensus sequence score, 49
Voxel ray-tracing algorithms, 228–229

W

Watson–Crick base pairing, 400
Weighted metabolic networks, 330–333
Wells' scoring rule, 82, 86
Whole-cell modeling, levels of, 313
Whole-organism protein-expression arrays, 326
WHO model, 152
Wolfe dual problem, 29–30
World Bank, 149